FIDEL

Also by Tad Szulc

FIDEL
A CRITICAL PORTRAIT

TAD SZULC

Perennial
An Imprint of HarperCollins*Publishers*

Excerpt from *Diary of the Cuban Revolution* by Carlos Franqui used by permission of
Viking Penguin Inc.

A hardcover edition of this book was published in 1986 by William Morrow.

HarperCollins books may be purchased for educational, business, or sales promotional use.
For information please write: Special Markets Department, HarperCollins Publishers Inc., 10
East 53rd Street, New York, NY 10022.

First Avon Books edition published 1987.

Reprinted in Perennial 2002.

Designed by Richard Oriolo

Library of Congress Cataloging-in-Publication Data

Szulc, Tad.
Fidel: a critical portrait.
    Includes bibliographies and index.
    1. Castro, Fidel, 1927–   . 2. Cuba—Politics and government—1959–
3. Heads of state—Cuba—Biography. I. Title.
F1788.22.C3S98   1986   972.91'064'0924[B]   86-16460

ISBN 0-380-80888-9 (pbk.)

    04 05 06 FOLIO/RRD 10 9 8 7 6 5 4

This book is for Marianne—again

CONTENTS

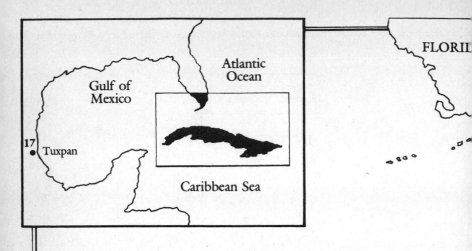

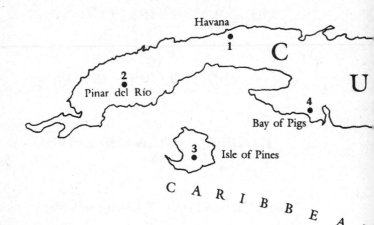

1. Havana
2. Pinar del Río—large percentage of Moncada recruits came from here
3. Isle of Pines (now Isle of Youth), Castro and company imprisoned here, 1953–1955
4. Bay of Pigs (or Playa Girón), April 17, 1961, exiles' invasion
5. Santa Clara, capital of Las Villas province—conquest climaxes Che's campaign
6. Escambray Mountains—non-Castro guerrillas in 1958/site of anti-Castro guerrillas, 1960–1965
7. Bayamo—simultaneous rebel attack on barracks, July 26, 1953
8. *Granma* landing—Los Cayuelos, December 2, 1956
9. Alegría de Pío battlefield—first Castro defeat
10. La Plata—Fidel's headquarters atop Sierra Maestra

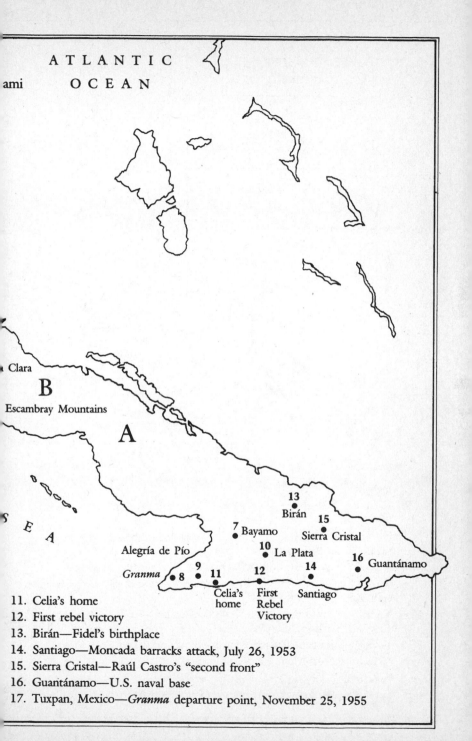

ATLANTIC

OCEAN

ami

Clara

B

Escambray Mountains

A

S

E A

13
Birán

7 Bayamo

15
Sierra Cristal

Alegría de Pío

10 La Plata

16 Guantánamo

Granma

9

8

11

12

14
Santiago

Celia's
home

First
Rebel
Victory

11. Celia's home
12. First rebel victory
13. Birán—Fidel's birthplace
14. Santiago—Moncada barracks attack, July 26, 1953
15. Sierra Cristal—Raúl Castro's "second front"
16. Guantánamo—U.S. naval base
17. Tuxpan, Mexico—*Granma* departure point, November 25, 1955

INTRODUCTION

President Fidel Castro Ruz of Cuba asked me the following question as we stood in his office winding up a long conversation shortly after midnight on February 11, 1985:

"Will your political and ideological viewpoint allow you to tell objectively my story and the revolution's story when the Cuban government and I make the necessary material available to you?" He added: "We would be taking a great risk with you."

This was at the end of five lengthy, consecutive meetings I had with President Castro at the Palace of the Revolution in Havana as I prepared to write this "portrait," and we had touched on an immense variety of themes concerning him and his life story. My reply to President Castro's question was that I didn't think total objectivity existed, but that I would commit myself to approaching this project with the greatest possible honesty. I remarked that since we both were honorable men, his ideology and mine, differing in the most absolute fashion as they did, should not interfere with the writing of an honest book. President Castro said, "You may paint me as a devil so long as you remain objective and you let my voice be heard," and we warmly shook hands.

I had my first conversations with Fidel Castro in 1959, shortly after his revolution triumphed, when I was in Havana as a correspondent for *The New York Times*. In 1961, I accompanied him on a tour of the Bay of Pigs battlefield. I returned to Cuba in January 1984 to interview President Castro for *Parade* magazine, and the idea of this book was born during a very long weekend we spent together in Havana and the countryside in endless discussions. I had reminded him that there was no serious biography of him or comprehensive study of the revolution, and that he owed it to history to remedy this lack.

We went on exchanging messages through Cuban diplomats in Washington during the balance of 1984, and we immediately agreed that this should *not* be an official or authorized biography or portrait. Instead, it would be an independent project with collaborative support by President Castro and his associates as well as access to written materials of the revolution. I spent a month in Havana early in 1985, holding a series of meetings with President Castro, then my wife and I set up shop in a house we rented in Havana for six months between March and August (where we were visited by President Castro). Our understanding did not require that the manuscript be seen by President Castro prior to publication, and therefore it was not. I am certain that when he does read it, he will disagree with many of my opinions and conclusions but that he will find the pledge of honesty to have been met. He knows, of course, that others may see him differently than he sees himself, and for me to be critical is not a violation of his trust.

Clearly, this is not a definitive biography, principally because President Castro is alive and has not completed his labors. Perhaps only the next generation of historians can attempt a full-fledged biography of this extraordinary personage. This "Critical Portrait," therefore, seeks to capture his personality and the story of his life as it is possible to reconstruct at this stage. It is not meant to be a history of the Cuban revolution, or of Cuban-American relations, and this is why I have avoided discussing in depth the achievements and the problems of the revolution. Nevertheless, Fidel Castro and his revolution are inseparable, and this portrait was sketched against the broader background of contemporary Cuban history.

To write it, I interviewed scores of Fidel Castro's friends, associates, and comrades-in-arms, in addition to my conversations with him. I have listened to a great many Cubans who have insight into the very complex personality of Fidel Castro and into the process of the revolution. I was able to see President Castro in action on occasions ranging from receptions at the Palace of the Revolution to a tour of the prison on the Isle of Youth (formerly Isle of Pines), where he had spent nearly two years as a prisoner of the Batista regime. I revisited the Bay of Pigs, and my wife and I climbed the Sierra Maestra to Fidel Castro's wartime command post to gain a sense of the environment in which he fought; we inspected the landing spot of the *Granma* that brought him and his rebels from Mexico, and the nearby battlefield where Fidel Castro's revolution almost ended three days after it began.

Among Cuban personalities I have interviewed and who have made this book possible because of the time they sacrificed were Vice-President of Cuba and Education Minister José Ramón Fernández Álvarez; Pedro Miret Prieto, a member of the Political Bureau of the Communist party and one of President Castro's oldest associates; Vice-President Carlos Rafael

Rodríguez; Culture Minister and Political Bureau member Armando Hart Dávalos; former Interior Minister and close comrade-in-arms Ramiro Valdés Menédez; Faustino Pérez and Universo Sánchez, who were with Fidel Castro at the moment of near disaster; Alfredo Guevara, whose friendship with Castro goes back to their university days and their first revolutionary experiences; former Political Bureau member and Transport Minister Guillermo García, who was the first Sierra Maestra peasant to join the Rebel Army; Blas Roca, former secretary general of the Communist party, and Fabio Grobart, one of its founders in 1925; Melba Hérnandez, who fought with Fidel in the Moncada attack and was among the first members of the revolutionary movement; Vilma Espín, president of the Cuban Women's Federation and member of the Political Bureau (and wife of Raúl Castro); and Conchita Fernández, who was Fidel Castro's personal secretary during the first years of the revolution.

It is impossible to list here all the Cuban officials, friends, and acquaintances in political and cultural fields who were of immense assistance in my research. Foreign dioplomats served as important guides, and among them I wish to mention Clara Nieto Ponce de Léon, former ambassador of Colombia in Cuba and, during our stay, director of the UNESCO office. In the Sierra Maestra, peasants who knew Castro during the war provided remarkable accounts of those days. Finally, our research and interviews in Havana were coordinated by Alfredo Ramirez Otero and Walfredo Garciga of the Ministry of External Affairs.

In the United States, conversations with Jorge Dominguez of Harvard University; Nelson Valdés of the University of New Mexico; Wayne S. Smith, who served as head of the U.S. Interests Section in Havana; George Volsky, who is a journalist in Miami and a leading expert on Cuba; and Max Lesnick, a university friend of Castro's and now a publisher in Miami were immensely useful. Numerous Cubans who knew Castro in boarding school and at the university, and are now exiled in the United States shared their recollections. My special gratitude is to The Hon. Ambler H. Moss, dean of the Graduate School of International Studies at the University of Miami, and to Dr. Jaime Suchlicki, director of the Institute of Inter-American Studies at the University of Miami, for superb research and intellectual support. Gabriela Rodríguez was a most intelligent and resourceful researcher. My wife, Marianne, lived through all of it: meetings with President Castro, entertaining Cuban friends in Havana, climbing Cuban mountains, organizing masses of material we brought back from Cuba, researching in Washington, and reading, improving and editing the manuscript.

At William Morrow and Company, my publishers, Lisa Drew was an editor with whom it was a joy to work. Morton L. Janklow and

Anne Sibbald, my literary agents, were marvelously imaginative and encouraging.

Thank you all.

—T.S.

Washington, D.C.

July 1986

FIDEL

I

THE MAN

CHAPTER

1

Advancing on his elbows and knees so slowly that his great bulk hardly seemed to move at all, the sweaty man in a torn olive-green uniform, horn-rimmed glasses on his unshaven face, slid carefully into the low canefield until he was entirely covered by a thick layer of leaves. In his right hand, he clutched a telescopic-sight rifle, a Belgian-made .30–'06-caliber weapon, his only and most beloved possession.

The tall rifleman was a thirty-year-old lawyer named Fidel Alejandro Castro Ruz, Cuba's fiercest apostle ever of a shattering social and political revolution, and now—at high noon on Thursday, December 6, 1956—he faced not only the imminent death of his dreams but his own as well.

Cubans had known Castro for years as a loud and ineffectual plotter, a loser. To the outside world, and notably to the United States next door, he was, at most, just another Caribbean troublemaker of whose existence the Eisenhower administration was not even aware.

This American ignorance reflected the traditional attitude toward Cuba, the nearest thing the United States had to a protectorate in the Western Hemisphere: Washington need not worry about Cuban politics and politicians because its proconsuls in Havana always kept them in line. The idea that within a few years Castro would establish the first Communist state in the Americas would have been dismissed as ridiculous had anyone suggested it in December of 1956.

At that moment, in fact, Fidel Castro and his absurdly small rebel group—which had landed four days earlier on the southern coast of his native Cuban province of Oriente after an almost fatal voyage from Mexico—were completely surrounded by government troops. The exhausted and famished expeditionaries had been totally routed and dispersed the previous afternoon in their first battle ashore.

The notion of surrendering to the soldiers of the dictatorship of President Fulgencio Batista Zaldívar that he and the eighty-one rebels had arrived to overthrow never occurred to Castro, the son of a tough Spaniard. On the contrary, he had the inner certainty of triumph that only visionaries feel when the odds are impossibly and virtually mathematically arrayed against them.

The last time I was in Havana to see Fidel Castro, he was nearing his sixtieth birthday, and I found him philosophizing a bit about life. Among other notions, he believed firmly that it was his natural destiny that well over a quarter-century ago, he had scaled the heights and reached the apex of power.

The subject was part of a broad conversation about history and the human condition one late evening in his office at the Palace of the Revolution, and Castro was perfectly matter-of-fact in acknowledging that some leaders are destined to play crucial roles in the affairs of men, and that, yes, he was a case in point.

He then turned to his favorite historical theme, that such leaders may affect "subjectively" the objective conditions in a country. To Fidel this is an absolutely vital point in the "correct" interpretation of the Cuban revolution inasmuch as he had succeeded in proving wrong the classical theories of the so-called "old" Cuban Communists. These Communists had insisted that a Castro-preached mass revolution in Cuba was impossible because the necessary "objective conditions," as defined by Karl Marx, did not prevail; accordingly, they turned their backs on the *Fidelista* insurrection until the closing months. Unprecedentedly, the Communists in Cuba were therefore co-opted and captured by Fidel Castro (who did not belong to the party) rather than the other way around. They had placed themselves in a situation where they had no option.

Actually, in the early days the orthodox Communists could take even less Castro's ideological heresy (or, in their view, towering arrogance) of postulating that "a man's personality can become an *objective* factor" in a changing political situation. Naturally, Fidel always had himself in mind in this context. The traditional Cuban Marxist-Leninists, with their thirty years experience as a Moscow-directed party, with activities confined to the organization of protest labor strikes or "popular front" alliances with "bourgeois" politicians (including Batista in the 1940s), could not bring themselves to believe that a single man's personality could, in effect, trigger a national revolution. Only Castro and the most faithful *Fidelistas* could believe such a thing.

It must be assumed that in 1956, the Cuban Communist party—known formally as the Popular Socialist Party and declared illegal by Batista after the coup on March 10, 1952—took its orders (and opinions) from the Kremlin. The Soviets, however, had evidently learned nothing from the Chinese civil war when Mao Zedong demonstrated that, contrary to Stalinist theory, communism could prevail only if it enjoyed full backing among the peasantry, the control of the cities was not enough.

Castro wasn't proposing a peasant revolution in Cuba, but, as the centerpiece of his strategy, he did envisage guerrilla warfare expanding with peasant support from a mountain nucleus to engulf in time the whole island—a concept the ideology-minded Communists could not absorb. Consequently, the "old" party secretly sent an emissary to Mexico in November 1956 to dissuade him from his publicly announced plans to land in Cuba that year "to be free or martyrs." Communist attitudes toward Castro at that stage and afterward describe an immensely fascinating and complicated relationship, one constituting the political backbone of the Cuban revolution that has never before been fully disclosed.

In a way that neither "old" Cuban Communists nor the United States was able to comprehend at the time—and Moscow and Washington may still not fully understand it even now—Fidel Castro built his revolution primarily on the sentiments of Cuban history. He tapped the deep roots of the mid-nineteenth-century insurrections against Spanish colonialism and its themes of nationalism, radicalism, and social-justice populism. Whatever the timing of his private allegiance to Marxism, Castro waited more than two years after victory to identify himself publicly with socialism; it may have been tactical, but it also represented a recognition of the feelings of the Cubans toward the revolution of the Sierra Maestra.

The two most worshiped political deities in socialist Cuba are José Martí, the great hero of the independence wars and one of the most brilliant thinkers in Latin America, and Karl Marx. Their portraits appear together everywhere (sometimes along with Lenin's), and it is beyond question that Martí—the man who always warned against United States ambitions in Cuba and the Caribbean—was from the outset Castro's personal role model. And, too, it is Martí's, and not Marx's, bust that stands guard at every Cuban public school, notably at the tiny schools the revolutionary regime built in the most remote mountain areas. In his speeches Castro reminds his audiences that the Cuban sense of history and nationalism was as crucial as Marxism in giving birth to the great revolution. In 1978, twenty years after his victory, he reminded his fellow Cubans and the world that "we are not only Marxist-Leninists; we are also nationalists and patriots."

With Castro ruling as the first secretary of the Cuban Communist party since 1965 (it took nearly seven years after the advent of *Fidelismo* to fashion Cuba into a full-fledged Communist state), "objective" and "subjective" concepts have nowadays acquired a clear meaning for the island's Marxist-Leninists. Fidel himself considers that *his* approach to the revolutionary strategy provided a major practical contribution to scientific Marxism; notwithstanding his exceptional intellect, he has added little of note to Marxist thought or theory. For, above all, Castro is a man of action.

At sixty, his beard and hair turning gray, Castro is searching for a new dimension of action. In the tradition of José Martí, he is taking on the mantle of the great continental leader, the elder statesman of Latin America. Presumably, Castro is looking toward new objectives because he is satisfied with his conceptual and institutional achievements as president of Cuba. If this is his judgment of his own record, history may find that it leaves much to be desired. The revolution bestowed on Cuba extraordinary gifts of social justice and equality, advances in public health and education, and an equitable distribution of the national wealth, and Fidel Castro deserves total credit for it. However, his compulsion to press ahead with new visions has left him with no patience with the day-to-day follow-up requirements of constructing a new society. His desire for absolute authority has withheld decision-making powers from his subordinates, and careful responsible management of the country and its economy remains Cuba's desperate need—to the point where the long-term success of the revolution is at issue.

In the mid-1980s, Castro set out to devote an astonishing amount of his time, private and public, to the new visions, spending endless hours at special meetings dealing with the problems of the hemisphere and holding forth on these subjects in an avalanche of speeches and interviews. Official propaganda filled the Cubans' consciousness with the myth and memory of Simón Bolívar, the nineteenth-century "Liberator" of much of South America who failed to unify the newly independent nations, demonstrating how irresistibly Castro is attracted to Bolívarian vistas. In a 1985 Havana speech, he intoned Bolívar's cry "Unity, unity . . . or anarchy will devour you," and there were again echoes of Castro's profound conviction that some men of greatness have it in them to affect the course of history.

This conviction first surfaced in Fidel Castro that December day long ago in the canefield in Oriente province under air bombardment and a barrage of automatic gunfire. Many years later, Castro would admit to an American visitor that the days and nights he spent in hiding with two companions in that canefield in a desolate area mysteriously called Alegría de Pío (Pío's Joy) was incomparably the "worst moment" in his intensely violent life. He would remark that the trap the army had sprung on him was an even "more bitter" blow than his imprisonment in 1953 for assaulting with an ill-armed rebel band a Batista military headquarters; the canefield ambush was something "I wouldn't like to expand on." Castro is not a man who cares to expand on losing.

In fact, he has never publicly discussed the Alegría de Pío incident in much detail beyond admitting "we were surprised and routed" by forces "vastly superior in numbers." Once, while we were having drinks before dinner in a hunting lodge in western Cuba, Castro drew a sketch of the Oriente coast where the rebel expedition had landed—he was explaining the awesome navigational problems they encountered, after describing the stunning inadequacy of their weapons training in Mexico—but he would not reminisce about the first days of the guerrillas ashore.

Alegría de Pío, of course, was the real turning point in Castro's life and his revolution. Cuban and world history would have evolved differently had this single individual been less determined and, most important, less lucky. Fidel's luck is a recurrent theme of his existence.

Yet, efforts to gain a sense of Fidel Castro as a human being are not easily rewarded. Castro's bearded face may be one of the best-known physiognomies in the contemporary world; his public views on every subject under the sun (he has opinions about everything from medicine to *haute cuisine*) have cascaded in the billions of words of thousands of his speeches over these long years. With the secrecy habit of a lifelong conspirator, he is a master at self-concealment, and he has discouraged systematic in-depth studies of his past and the history of the revolution—or, at least, its publication.

A deeply moody introvert despite his external persona—and a man of surprising shyness in initial personal contacts—Castro clearly desired to swathe his past, especially his early youth, in a cocoon of oblivion. He is exceedingly selective about disclosing facts that, in his judgment, may negatively affect what he thinks should be his public image. He wishes to control the outline and the paintbrush strokes in any portrait of Fidel Castro—just as he personally controls everything else in Cuba. Castro, the student of the past and the virtuoso of politics, understands the strategic importance of controlling history.

Recently Castro, Jesuit-educated, self-proclaimed Marxist-Leninist, has been spending much time debating and dictating for publication an intellectual rationalization of the values shared, in his view, by Christianity and Marxism in implementing social justice. Since John Paul II's election to the papacy, Castro has praised him often for his concern with the poor in the Third World (Cuba and the Vatican never interrupted diplomatic relations, and Castro has attended private dinners with the apostolic pro-nuncio in Havana). A meeting between the two has been discussed, and the Cuban National Ecclesiastic Encounter was held in February 1986, with his attendance, the most important Roman Catholic church conference since the revolution.

The idea of a synthesis between his brand of revolutionary socialism and Christian religion is very much on Castro's mind: intellectually when he speaks in praise of the "theology of liberation" (the powerful social justice movement in the modern Roman Catholic Church in Latin America), and mystically when he contemplates the nature of martyrdom. In lengthy conversations in 1985 with a Brazilian dominican friar, Fidel observed that "I am certain that upon the same pillars on which today reposes the sacrifice of a revolutionary, yesterday reposed the sacrifice of a martyr for his religious faith." He added that without "altruism," neither "a religious hero nor a political hero" can exist. Castro also said that "if there ever was a name more hated by reactionaries than that of a Communist, it was, in another time, the name of the Christian." Given the explosive political tensions in Latin America, and the new road taken by the young church there, Castro is engaging in more than a theoretical exercise. It fits into his strategic Bolívarian concepts, but it also brings out mysticism and a Jesuit-learned logic.

Castro's sentiments about his son, Fidelito (who was five years old when his father divorced his mother), seem to have been unusually strong when one considers that from the moment the boy was born, Castro Senior was successively a plotter (while practicing law without charging his impoverished clients), a political prisoner, an exile a *guerrillero,* and a chief of state. Castro attempted to see Fidelito under the most difficult circumstances, including having him brought illegally to Mexico while the rebel force was being trained. A Soviet-educated physicist, Fidelito, now married and with two children (though it is difficult to think of Castro as grandfather), is increasingly in the public eye, shown on national television presiding over scientific meetings and attending ceremonial receptions at the Palace of the Revolution in his capacity of secretary-general of the Cuban Nuclear Commission. In his mid-thirties, he looks exactly the way his father did at his age. It is no secret in Havana that Castro has at least one other adult child (and a grandchild) from a romantic liaison.

But Fidel Castro is a very private person, and Cubans normally do not discuss his personal life, mainly out of respect. He never remarried, and since Celia Sánchez, his absolutely devoted friend and guerrilla companion, died of cancer in 1980, nobody has replaced her in Fidel's trust and affection. Unquestionably Celia was the most important woman—and very likely the most important human being—in his life. No other woman has had her name linked publicly to Castro since Celia's death, and probably none ever will. Living among palace courtiers, Fidel seems a very lonely man.

Fidel was the fifth of nine children of Ángel Castro y Argiz, a landowner from Spain; the first two children were born of his father's first wife. Fidel Castro and his younger brother, Raúl, his designated heir as head of state and government and first secretary of the Communist party, evidently have a unique and special personal and political relationship—they shared all the revolutionary experiences—but they choose to spend little of their free time together (Raúl has a wife and children, and he is very much a family man). They are said to have had violent disagreements in the past, now seemingly forgotten. Fidel's rapport with his older brother, Ramón, and his half-sister, Lidia, is pleasant though more limited. A half-brother and two sisters in Havana see him very seldom. One sister has long lived in Mexico, and another sister—Juana—is in self-exile in Miami, denouncing Fidel at every opportunity as a Communist dictator, though during the Sierra war she was his devoted supporter.

During the guerrilla war, Castro had his family sugarcane lands set afire as an example of economic warfare against the rich of Cuba. Afterward the Castro lands were nationalized under the revolution's agrarian reform—his mother, brothers, and sisters turning their inherited land titles over to the government, although Señora de Castro was allowed to keep the large *finca* house until her death in 1963. (Fidel's father had died in 1956, and the children received inheritances.) Under the 1959 land-reform law, which established 960 acres as the maximum private holding, one half of the Castro family's wholly owned 1,920 acres were seized by the government along with the twenty-four thousand acres the Castros had been renting on a permanent basis from adjoining American-owned sugar estates. But soon thereafter they voluntarily gave up their remaining 960 acres.

The fundamental question concerning Fidel Castro, the 1959 revolution, and Cuba's transformation into a Communist state is naturally whether this whole experience was logically dictated by Cuban history or represents an extraordinary political aberration primarily instigated by his overwhelming personality. Castro, who once called for "many Robespierres" in Cuba, insists on the historical inevitability of the events along with the subjective role played by him.

Though the ultimate answer is too complex to be reduced to one-dimensional explanations, Castro's personal role in launching and guiding the revolution is certainly impressive. Even if he had fought only to obtain personal power, which most emphatically is *not* the case, the fact remains that no modern revolutionary leader or chief of state has undertaken such astounding personal risks and has been so directly engaged in the rigors of conspiracy, rebellion, and open warfare. The expeditionaries' landing off the Oriente coast, the tangled swamp and mangrove they had to cross for over a mile to reach the actual shore, the Alegría de Pío battlefield, and the tortuous, high jungle paths of the Sierra Maestra provided the most nightmarish experiences a military leader could encounter. Castro so desperately wanted to win his incredible guerrilla war that he took it all in stride, creating an army as he battled ahead.

No political leader in full possession of his mental faculties would have sailed to Cuba from Mexico the way Castro and his fanatically devoted followers did; their yacht, with eighty-two men aboard, was built for no more than a dozen persons, and not for transporting a military arsenal. No political leader without real military experience would have endured two years, often short of food, weapons, and ammunition, in the heart of the Sierra Maestra, constantly on the march with his growing band, up and down mud-packed and boulder-strewn paths through forest after forest. But less than six weeks after the Alegría de Pío disaster, Castro had the audacity to attack a Batista army detachment. This was the first victory.

Lenin sat out the start of the Russian Revolution in Zürich, risking no combat. Stalin showed his mettle mainly in holding up banks and trains, and serving time in Siberian prisons. Mao Zedong controlled vast swaths of territory and led large armies during much of the Kuomintang and Japanese wars; neither in the initial period of building his Communist base nor during the 1934–1935 Long March did Mao undergo the constant personal hardships Castro would face in Cuba. Yugoslavia's Marshal Tito operated out of well-protected headquarters, later enjoying American and British military missions' support. Vietnam's Ho Chi Minh had been in prison, but when the anti-French uprising was launched, he was not called upon to lead troops in the battlefield.

Fidel Castro's conditioning as a *guerrillero* is therefore an extremely significant element in understanding and defining his personality today. He cherishes the title of Commander in Chief and the laurel leaves of rank on his formal military uniform. The total militarization of the Cuban society is a concept stemming from his formative experiences. He learned the very hard way—in urban insurrection, Sierra war, creating from those siege conditions a siege mentality for this hostility-surrounded island where at least one half of the population is trained and organized today for defensive combat. He also learned that to survive, he must be absolutely and undeviatingly uncompromising.

CHAPTER

2

Fidel Castro's overwhelming concern that chilly Thursday in December 1956 was how to break out of the encirclement and reassemble the survivors of his battered expedition for victorious combat. This required abiding faith that he could succeed in what was by any reasonable standards a monumentally demented enterprise. The Batista regime had at its disposal a fifty-thousand-man army with cannon and armor, an air force and a navy, and a murderously efficient uniformed and secret police.

Besides, President Batista also enjoyed full United States support, including access to American arms. Tanks and artillery were shipped regularly to Cuba from U.S. ports, Batista aircraft could refuel and load explosives and napalm bombs at the U.S. naval station at Guantánamo on the coast of Oriente, and in Havana a U.S. military mission trained the Cuban armed forces. All of this was consistent with the fact that Cuba, in effect, had been an American fiefdom since Spain lost the island in the 1898 war, and that Washington discouraged any changes in the status quo.

There was no reason the United States should have understood the Castro phenomenon in 1956, any more than it understood the final Cuban war of independence that erupted in 1895. As Louis A. Pérez, Jr., an anti-Communist Cuban historian, described that insurrection: "It was a guerrilla war of national liberation aspiring to the transformation of society . . . [it] contained elements of anti-imperialism, political radicalism, agrarian reform, racial equality, and social justice." Thus the Eisenhower administration was squarely behind Batista, Vice-President Richard Nixon having visited him in Havana just a short time before Castro's December landing in Oriente. But neither Batista's firepower nor American backing impressed Castro; to him, defeats really were victories in disguise, forging and honing men's spirits and courage.

In the canefield, once he was safely concealed by the *paja* carpet of dry sugarcane leaves, only his face protruding, Fidel Castro whistled very softly. Another olive-green-clad figure slithered silently into the cane from a thicket across a narrow pathway; this was Universo Sánchez Álvarez, a big, tough peasant from the northern Matanzas province who had served as Castro's bodyguard during the clandestine invasion preparations in Mexico. Universo carried with him a telescopic-sight rifle (as Castro did), a distinct advantage in their present predicament, but he was barefoot, having lost his boots in the retreat from Alegría de Pío, and this was an awesome problem for a fighting man. Then the third fighter reached the haven of the thick cane-leaf blanket. He was Faustino Pérez Hernández, a Havana physician of slight build and one of the two Chiefs of Staff of the rebel expedition. Faustino wore combat boots, but he no longer had his weapon—a frightful minus as far as Castro was concerned. Both men were thirty-six, six years older than Fidel, but the average age in the rebel group was twenty-seven; most of them were reasonably mature individuals. As Castro remarked many years later, "There was a moment when I was Commander in Chief of myself and two others."

The three men, burying themselves under the *paja*, lay on their backs next to each other so that they could communicate in whispers. Castro cleared his throat gently and murmured triumphantly: "We are winning . . . Victory will be ours!" Universo and Faustino said nothing. Fidel Castro, Universo Sánchez, and Faustino Pérez constituted at that juncture the *entire* Rebel Army, the fighting arm of Castro's anti-Batista 26th of July Movement. The Movement had been growing in the underground of Cuban cities since 1953, the time of the disastrous and bloody attack on the Moncada barracks in Santiago, the capital of Oriente, conceived and led by Castro. He then spent twenty-one months in Batista prisons preparing himself for the next strike at the dictator. Now Castro had returned from the Mexican self-exile that followed his imprisonment to launch a guerrilla war against the regime in the rugged mountain chain of Sierra Maestra.

Actually, this Rebel Army would remain limited to the three men and two rifles for thirteen days, counting from the afternoon of the Alegría de Pío catastrophe—including five nights and days under the canefields' *paja*—until at a Sierra peasant's house they encountered Raúl Castro, the chief of one of the expedition's three platoons, who had been marching elsewhere in the mountains with four companions.

Raúl's group had five rifles and ammunition—and Fidel Castro became so excited at the growth of his "army" to eight men and seven weapons that he proclaimed in his most dramatic style that "now we have won the war . . . the days of tyranny are counted!" Four days later, seven more expeditionaries, including Ernesto "Che" Guevara de la Serna, the Argentine-born hero of the Cuban revolution who was wounded at Alegría de Pío and would die in 1967 in a guerrilla war in Bolivia, rejoined the Castro force. Fidel was ecstatic.

But on that first Thursday in Cuba, Fidel, Faustino, and Universo had concluded in a soft-whisper consultation that their safest course was to remain as long as necessary under the leafy canefield blanket. They had to elude the troopers of Squadron 12 and Squadron 13 of the Rural Guard, a heavily armed gendarmerie corps, including artillery, that had surprised and smashed the expeditionary force the previous day and was now hunting for survivors.

Although the Batista government had already announced that Fidel and Raúl Castro had been killed along with forty other rebels in combat on December 5 (a story instantly disseminated around the world by United Press International, a piece of journalism Castro never forgave the news agency), a confidential report from the Rural Guard squadron commanders to their Havana superiors that night admitted that "Dr. Fidel Castro" may have escaped. Batista obviously realized that Castro had to be located and killed no matter what: The dictatorship's political credibility—and its reputation for military effectiveness—were at stake.

Consequently, aviation was summoned to assist the Guard in the search. By midmorning of Thursday, low-flying aircraft spotted three separated groups of rebels hiding under a thick tree growth on a hill just above the Alegría de Pío battlefield. One of these groups was Fidel Castro's, and they were bombed and strafed by twin-engine B-26 light bombers at the spot where they had spent the night.

Universo Sánchez was the first to run into Fidel after the rout, and Faustino Pérez stumbled into them in the falling darkness, an hour or so later. They had heard him approach, and Fidel ordered Universo to shoot the figure in the shadows if it looked like an enemy soldier. Faustino identified himself just in the nick of time; otherwise the Rebel Army would have remained at two men. But Pérez brought the disheartening news that Che Guevara was probably dead, having been seriously wounded (actually Che's wound was minor).

The planes had missed them on the first pass, and the three rebels saw that the nearby canefields were their only salvation: They could not be out in the open. Led by Fidel, they raced toward the nearest field while the aircraft regrouped in the sky. The fugitives repeated this maneuver several times until reaching a thicket across the dividing path from the canefield Fidel thought looked the safest. They slithered to it on their bellies, and installed themselves under the large dry leaves that lay on the ground.

Castro's theory was that the Rural Guard and the aircraft would end their search of the area by evening, moving on to the next grid. This, he thought, would allow him, Faustino, and Universo to leave the canefield that same night to begin marching toward the foothills of the Sierra Maestra to the east where he knew that friendly peasants and militants of the clandestine 26th of July Movement would protect and aid them. Faustino and Universo dissuaded Fidel, however, from making a run for it right away; when he insisted on going, Universo said, "Damn it, Fidel, democratically it's two against one, so we stay."

In the meantime, from barely a few hundred yards away they could hear shouted military commands, the metallic sound of Thompson submachine guns being shifted from shoulder to hand by the soldiers, and occasional bursts of automatic gunfire. A canefield immediately to the south of the three rebels was set ablaze to flush out whoever might be concealed there, and thick, bluish smoke spread quickly over the flatland. They desperately held back their coughing, fearful that the troopers might hear it. A B-26 roared low over the burning field to strafe it.

This was about as bad as things could get for Fidel and his companions. Apart from being trapped inside a canefield, they were parched and starving. They had lost most of their supplies and equipment, including food, when *Granma,* their yacht, was shipwrecked at dawn of December 2, several hundred yards from the coast. As the eighty-two expeditionaries advanced slowly inland, they contacted the area's few peasants and charcoalmen, who received them well, sharing simple food with the rebels. Castro and his men had their last hot meal on the evening of December 4—they were also allowed to buy sausage and biscuits from the locals before moving on toward Alegría de Pío.

What gave them away to the Rural Guard was their hunger the following day. Having consumed their sausage-and-biscuit rations, the revolutionaries began breaking off sugarcane stalks as they marched, sucking them dry, then dropping the cane on the ground. Fidel did it, too, cane juice being a high-energy nutrient. But Rural Guard troopers, searching for the expeditionaries since they first landed (the Batista navy and aviation had immediately located the half-sunk *Granma* off Los Cayuelos), spotted the trail of discarded stalks, following and then surrounding the Castro force. This was the disastrous result of guerrilla inexperience in new terrain

(Rural Guard soldiers, of course, instantly knew they had found their quarry), and it was a lesson Fidel would never forget.

Now, he, Faustino, and Universo carefully reached for stalks near them to suck the sugar syrup for nourishment. But during the three days and nights they would stay in that canefield (and two more days and nights in one farther east) because the Batista forces would not go away, the rebels developed mouth sores from sucking the stalks, lacking knives to slash them open. Every dawn, to quench their thirst, they licked the dew from the leaves, but this also caused mouth and tongue infections because of the leaf roughness. Occasional night rain served only to soak the ground—and the three men.

Just as exasperating, certainly for the ever-restless Fidel, was their immobility. He decided that they should remain on their backs day and night so that no movement in the canefield could be detected from the outside or from above. To relieve himself, a man urinated away from his body; by not eating, he could avoid bowel movements. And aside from vast physical discomfort, the nervous tension from the need to remain endlessly motionless gripped them.

The first night in the canefield, Castro placed his rifle vertically atop his body: He fitted the barrel against his throat and lodged the butt against his feet. He released the safety catch, curling his fingers around the trigger. "I shall not—never—be taken alive by the soldiers of tyranny while I sleep!" he announced in a dramatic whisper. "If I am found, I'll just squeeze the trigger and die."

His two companions looked at each other with incredulity. "Fidel, you are crazy," Universo Sánchez told him. "I don't want to get caught, either, but what you're doing is suicidal. There are lots of land crabs here, and a crab could trip the trigger, you know." Castro, who dislikes being contradicted, answered in a sulking murmur: "Fine. You do what you want. *I* am going to sleep like this." And that was the way he slept every night under the sugarcane *paja,* the rifle barrel at his throat. Universo settled in, sleeping with his rifle cradled in his arms. Faustino had no weapon.

The only sacrifice of which Fidel Castro was not capable was absolute muteness. He simply was unable to refrain from talking. And it was not only his normal compulsiveness to hold forth on any subject (he is not a man of small talk, so he is serious almost every time he opens his mouth). Pinned down in the canefield, he was furiously and excitedly thinking ahead, planning their escape, organizing in his mind a guerrilla army, preparing the victory, and outlining revolutionary laws and measures. He spoke day and night in a controlled whisper, not really expecting answers or comments—it was like a monologue or a murmured speech. This is how

Faustino Pérez, the liberal and sophisticated physician, remembers Fidel's speeches-under-the-leaves:

"Discussing the modalities of continuing on to the Sierra Maestra, Fidel was already convinced that we would meet with our companions. We would go that same afternoon, he was saying, or the next day in the morning. Personally, I was thinking at that moment that perhaps it would be possible for us to arrange a truce, that is to say to get out of there, try to organize ourselves again, and try to return.

"These were my personal thoughts that I was unable to articulate because Fidel had already begun to speak of reuniting the companions and proceeding. And it wasn't just a reunion in the days to come. He was already taking it for granted that there would be a reunion, and he was talking about the small combat actions in which we must engage in order to keep growing—not only with the participation of our expeditionaries that we would encounter, but also of peasants who would wish to join us. In other words, what would actually happen afterwards was already clearly seen by Fidel at a time when there were just the three of us, knowing nothing about the others.

"To be able to speak, we had to put our heads together, to talk tonelessly in whispers, because we were certain that the army had surrounded us. And in this whisper, speaking with the enthusiasm that characterizes him, Fidel told us his future plans. But not only plans for the future. For the first time, I could hear him speak amply about other things, about the meaning of life, about our struggle, about history, about all these things. And I can say to you that it was there that my deepest understanding of Fidel and my absolute confidence in Fidel became crystallized. Because there, in the canefield, he spoke about what glory signified.

"I remember that for the first time I heard him repeat the phrase of José Martí that 'all the glory of the world fits inside a kernel of corn,' I'd known the expression of Martí, but not in this context, not in the context in which Fidel now spoke of what the struggle means for a revolutionary, what meaning does the struggle have, what meaning does life have for a revolutionary, and how one may not fight for personal ambitions, not even for ambitions of glory. . . . He spoke of the necessity and the satisfaction, at the same time, that a revolutionary has in fighting for others, in fighting for his people, in fighting for the humble ones . . .

"In the order of ideas, what I'm telling you is what struck me the most [in Fidel]. But there were many other things of which he spoke. About organizing the country, about the people of Cuba, the history of Cuba, the future of Cuba. And about the necessity of launching a revolution, a real revolution. We didn't speak of Marxism and communism in those days, but of a social revolution, of a true revolution, of the role of imperialism in our country.

"I can tell you that a moment would come when one would say to oneself, 'But gentlemen, if there are only three of us, what sense does it all have, how can one talk about the struggle and a future victory? Fidel must have gone crazy!' Then, we would meditate, and Fidel with his explanations would make us think hard about the significance of it all.

"You see, Fidel didn't really believe that our group—the eighty-two expeditionaries—or even the entire Twenty-sixth of July Movement that was being organized throughout the country would lead the people to victory. Those who were going to conquer victory, those who Cubans needed, would be a group of the vanguard, a group that would be the shining path. And having a group that would show the way, give the example, the people would do what they had to do in order to win victory. This was the significance of Fidel discussing it among the three of us, wanting to gather ten more, and fifteen more, and fifty more fighters—this was what he wanted to do in the Sierra, giving this example, lighting this flame, because the nation would certainly respond with its own acts of struggle. It would be the decisive struggle, the popular struggle, the struggle of the masses. And this is the great lesson, the great lesson of faith and optimism—and at the same time of realism—that Fidel preached in those days.

"In this sense, Fidel had absolute faith. Yet, how does one explain this faith, which is really more than faith, because there is idealist faith, blind faith, faith without any real basis—but he had faith and confidence with a base of reality."

And this is how Universo Sánchez, the peasant from Matanzas who shifted his lifelong allegiance from the Communist party to the 26th of July Movement because he thought the Communists were not doing enough against Batista, recollects Fidel Castro's whispered oratory under the sugarcane *paja:*

"At one point, Fidel starts discussing—apparently to give Faustino and me some courage—what will be the revolution and the future. He talked of the revolutionary program, he raised our spirits. And at no time did Fidel consider himself to be defeated. It was always his thing with him about regrouping our people. I began to believe at one stage that Fidel was crazy. I said, 'Shit, he's gone crazy.' You will see that my rifle has my name engraved on it with the tip of the bayonet because I thought that when they would kill me, my family would know it was I who was killed, that I didn't disappear. At that point, I didn't think I'd get out alive of the Sierra Maestra, so I put my name on my rifle. . . . I was thinking then that Fidel had to be crazy, how can he beat Batista with these few people? Fidel always predicted that the Sierra would some day fill up with people—with fighters—and that moment came. And I would say, 'Shit, look how Fidel predicted it.' And I had been thinking of dying, of how I would die."

Fidel's loquacity is a legend, the episode in the canefield being a case in point. He holds forth anywhere, anytime. Once after an American television-interview taping session with his big brother at the Palace of the Revolution in Havana in 1985, Raúl Castro was asked whether he had watched the full five hours before the cameras.

"Oh, my God, no, no . . . ," Raúl, who has much more of a sense of humor than Fidel (the big brother does not laugh at himself unless *he* makes the joke), replied in mock annoyance. "I think I've heard Fidel talk enough to last me for the rest of my life. You know, when I was moved into Fidel's prison cell where he had been in isolation for about a year—this was at the Isle of Pines Presidio late in 1954, when we were serving our sentences for the Moncada barracks assault—he didn't let me sleep for weeks. Having been alone all that time, he just talked day and night, day and night . . ."

Moreover, Fidel Castro demands undivided attention when he addresses a visitor on a one-to-one basis. He often prefers to stand rather than sit down during conversations—he also tends to pace rapidly when he is excited by an idea or an indignity he believes he has suffered in some context of world politics—and if the visitor is not riveted by his words, he might suffer a punch in the arm or the chest.

Castro can also be a superb listener when he is interested in the subject or the speaker. And he is a great questioner, centering swiftly on the heart of the matter under discussion. One often does feel that he may be dying to speak out, but his courtesy and curiosity usually prevail—and Fidel will remain silent for very long minutes, fiddling with his cigar, lit or unlit (before he abruptly quit smoking late in 1985), or twisting his beard with his fingers between his chin and lower lip in a characteristic gesture of thoughtfulness.

CHAPTER

3

That within just over two years of the canefield drama the Castro revolutionary war in the Sierra Maestra and the urban struggles by the 26th of July Movement and militant university student organizations would force the collapse of the Batista regime is a matter of history. It is similarly part of the history of the twentieth century that Fidel Castro, the uncompromising bearded *guerrillero,* has led Cuba through the greatest social revolution since Mao Zedong imposed communism on China a decade earlier, immeasurably improving the human condition of millions of Cubans—there were six million in 1959 and over ten million in 1986. As a belatedly self-anointed Marxist-Leninist, Castro has organized Cuba as the first (but thus far the only) Communist country in the Western Hemisphere, allying himself politically, economically, and militarily with the Soviet Union. He has pursued these policies and alliances in defiance of seven successive United States presidents (Ronald Reagan being the latest in this period of nearly three decades), defeating an American-supported invasion attempt and, in conspiracy with Nikita Khrushchev, bringing the world to the brink of nuclear conflagration in 1962.

Fidel Castro has held power longer than any other important head of government except North Korea's Kim Il Sung and Jordan's King Hussein (in Communist Bulgaria, Todor Zhivkov has been in power since 1954), and remains a highly active and influential player in international affairs. His health appears excellent, and all the signs are that, barring assassination, he will be present in the world arena for a very long time as a senior statesman (he is five years older than the Soviet Union's Mikhail Gorbachev) whose actions must be considered with utmost seriousness in the context of changing conditions and policies. Though in the future he may turn over to his closest associates some of the real responsibilities for the day-to-day conduct of Cuban internal matters, he will never abandon his determination to retain a major impact on the battles and controversies over the fate of humanity.

It has been said that Castro exaggerates his own importance, and it may be so, but he believes this is the only way a small nation can exact attention and respect from the superpowers as well as from other lesser countries. Evidently Castro's Cuba commands sufficient attention from the Soviet Union to receive around $4 billion annually in economic assistance. This is enormous for a population of ten million in terms of traditional foreign-aid programs of the superpowers. There is an additional half-billion dollars in sophisticated military hardware, from computerized control-and-command systems to MiG-23 jets. To the United States, which after twenty-five years is still committed to the idea that Castro must vanish, Cuba is a permanent nightmare as Washington worries about its influence in Africa, Central America, the Caribbean, and even our "own" Puerto Rico.

Castro knows how to play his adversaries and his allies. He has been negotiating some form of settlement with the United States on and off for over a quarter-century, and he has been feuding openly and covertly with the Soviet Union for just as long. Paradoxically, the United States cannot afford to settle with Castro because it might mean full-fledged acceptance of him, and the Soviet Union cannot afford to break with him because it would mean an awesome defeat in the contest with the United States and China for influence in the Third World.

To much of the Third World, which made him chairman of the Non-aligned Movement for the 1979–1983 term, Fidel Castro is a hero, and not only because Cuban troops are in Angola and Ethiopia to defend them from "imperialism," and Cuban advisers are in Nicaragua for the same ostensible reason. The Third World perceives Castro as its advocate and at times its conscience. He thinks other peoples in the Third World deserve the kind of dignity as nations and individuals that the revolution has granted the Cubans. This is what he likes to talk about most, hour after hour, with foreign visitors.

He insists that Cuban "internationalism" transcends the estimated forty thousand combat troops and advisers involved in wars from Angola and

Ethiopia to nearby Nicaragua, because, in the Castro logic, a much greater and more lasting impact is made by tens of thousands of Cuban doctors, nurses, teachers, and technicians assigned to countries on three continents. He says that some 1,500 Cuban physicians are assigned to Third World countries. Increasingly consumed by what he sees as Christian and Marxist parallels, he argues heatedly that "if the church has missionaries, we have the internationalists." Castro makes a point of recalling that the Christian-Marxist nexus he proposes is not a sudden inspiration, because in meetings with Chilean Christian leaders in 1971 and Jamaican churchmen in 1977, he had already urged a "strategic alliance" between the two forces "to achieve the necessary social changes in our countries." He is taken sufficiently seriously for a delegation of United States Roman Catholic bishops to have gone to meet with him in Havana early in 1985—Castro had a marvelous time dazzling the bishops with his familiarity with theology and liturgy—and to confer with the Cuban episcopate late that year, the first such meeting in the twenty-six years of the revolution.

In sum, Fidel Castro is a fascinating phenomenon in our century's politics: to the increasingly gray and dull Western world, a man of panache, a romantic figure, an ever-defiant, dizzyingly imaginative, and unpredictable rebel, a marvelous actor, a spectacular teacher and preacher of the many credos he says he embraces. But Castro makes other impressions as well. Though his personal popularity in Cuba is immense, there is a segment of Cubans to whom Castro looms as a ruthless and cunning dictator, a cynical betrayer of liberal democracy in whose name he first rallied millions of Cubans to his cause and banner, a servile satellite of the Soviet Union, the idolized object of a personality cult he needs as he needs the air he breathes, and a cavalier creator and perpetrator of fundamental economic-policy errors at home.

Such are the complexity and the dimensions of Fidel Castro as a personality and a statesman that he may indeed be *all* of the above things—the hero of humble mankind and, at the same time, the repressive Communist dictator in the eyes of many Cubans. He himself does not mind appearing devious, or worse, if in his opinion this is "historically justifiable," the most spectacular example being Castro's proud public admission that he had deliberately concealed his Marxist-Leninist orientation during the guerrilla war to avoid antagonizing bourgeois and other potential supporters. What is the real and final truth about Fidel's Marxism may never be known—it is beyond prediction what he may say about it and himself in the future—but for a long time his Marxist performance had been ideologically vague, uncertain, and ill-formed. Castro doubtless realizes that people have short memories, and that in Cuba there is already a whole generation now moving into positions of power and responsibility, educated by and under the revolution, and to whom the past does not exist historically or intellectually—except as a time of shame and scorn. That is how, demonologically, Castro has defined the Cuban past, the political past in which he grew.

In seeking to portray Fidel Castro precisely, an imprudent proposition under the best of conditions, a crucial element is that he thrives on contradiction and paradox. It is an alluring intellectual exercise for Castro to reconcile most logically, when challenged by events or by an interviewer's questions, whatever of a contradictory nature he may have said publicly over nearly three decades about Marxism, democracy, and Christianity, the Soviet Union and the United States, the future of sugar in the Cuban economy (produce less or produce more), the real-life progress of his Revolution (a word that is always capitalized in Cuba in print and in speech), and just about any subject springing from his fertile imagination and computerlike memory. With a mischievous sparkle in his brown eyes, he has engaged in virtuoso performances with American television correspondents whose homework on Castro left much to be desired. In most instances he has the upper hand, rhetorically and intellectually, and he takes advantage of it delightedly.

Inasmuch as Castro conducts domestic government and Cuban world policies mainly by frequent public speeches or endless interviews (secret negotiations and decisions are reserved only for the most delicate situations), it is truly impossible to keep track of what he has said and when he said it. Even the historical department of the Council of State was unable to provide the precise number of public speeches Castro has delivered as "Maximum Leader" since January 1, 1959, the day he assumed effective power. An educated guess is that they exceed 2,500 (some of them running five hours or more, and the record being around nine hours in 1959), but not all of them have actually been transcribed by the Council of State's teams of stenographers, not all have been published, broadcast or telecast, and it is impossible to locate the text of every speech Castro has made. For example, between January 1, 1966, and October 1984, he delivered *130* very lengthy public addresses on his favorite topic of public health and medicine alone.

Indeed, Fidel Castro's revolution—or, at least, the selling of this revolution to Cubans—might not have succeeded without the medium of television. From the first day, in fact, Castro has governed through television, the first such massive use of this technology in the craft of government as distinct from campaign politics. While he does have a natural rapport with his audiences and he used this symbiotic emotional relationship in the first years of his rule by addressing crowds as large as one million, television was vital in carrying the face, the voice, and the message beyond the meeting plaza to Cubans in their homes. Later, television became the regular channel of communication between Castro and the population.

By Latin American and even United States standards, Cuban television was quite advanced technically early in 1959 when Castro forced Batista's ouster, and the number of sets in the country was relatively high, especially in the cities. But what mattered the most was that Castro, whose revolutionary concept was always built on communication with the masses, instantly understood that he and television were made for each other. Actually, Cuba had a tradition of the use of the radio in politics, and Fidel had been very effective with the microphone on the limited occasions when he was allowed to get near one. In the second and last year of the guerrilla war, Castro installed a radio station—Radio Rebelde—at his headquarters atop the Sierra Maestra, rapidly turning it into a superb instrument of propaganda and the dissemination of coded operational orders. He often addressed Cuba over Radio Rebelde.

The switch to television was thus natural, and Castro's ideal on-camera presence and his rich dramatic gifts did the rest. The Cuban propaganda apparatus is so well honed that the nation may be treated to a Castro speech carried live (always in its entirety) as well as to a number of taped rebroadcasts over the two national channels, sometimes over a period of days. Additionally, every public appearance by Castro is either carried live in special reports or as part of the regular television news programs (the radio, of course, carries the Fidel sound as well).

It may be difficult to believe that Castro, who seems to adore public speaking, actually fears it before the first words are out. He told the Cuban magazine *Bohemia* that "I confess . . . I suffer from stage fright when I speak in Revolution Square. . . . It is not at all easy for me." As a young man, he forced himself to deliver speeches in front of a mirror in his room until he was satisfied they were adequate to encourage him to pursue a career in law and politics. On most occasions, Fidel begins his speeches in a low, almost uncertain voice, talking quite slowly—until he has that sudden feeling of having established rapport with his audience. From there on, it is Fidel Castro, the great orator. Other famous orators never conquered the initial fear, among them Gladstone and Winston Churchill.

Castro is fascinated by the art of public speech. He has reminisced that when he was a high-school student he became friendly during summer vacations with a well-educated Spaniard in Oriente who told him that in order to overcome his speech difficulties, Demosthenes used to place a pebble under his tongue. This led Fidel to recount that, still in high school, he began collecting speeches by other great classical orators, but that he subsequently concluded that he disliked their oratory because "it was too rhetorical and grandiloquent, depending too much on wordplay." Moreover, he says, today Demosthenes and Cicero would "have great problems if they had to face concrete realities and explain their society," Castro's point being that he ceased to admire Athens' democracy when he understood that it meant that a "tiny group of aristocrats met in public place to make decisions." Fidel's favorite speaker, it turns out, was Emilio Castelar, the brilliant Spanish statesman and thinker who headed Spain's short-lived First Republic in 1873, but as "marvelous" as Castelar's parliamentary speeches had been, "today he would have been a complete fiasco in any parliament." In the end, Castro decided to practice the exact opposite of what all the great orators in history had done, creating his own fiery yet chatty style. It is unlikely that there is another Communist ruler in the world nowadays who delights in dissecting classical oratory, or is capable of it.

There have also been "secret" Castro speeches, unknown in number, delivered before Communist party or armed-forces leadership groups, and unpublished *charlas* (chats) at meetings of, say, the Cuban Women's Federation or the Committees for the Defense of the Revolution. Additionally, all the top revolutionary leaders—Castroites and "old" and "new" Communists alike—engage in a permanent flood of oratory to keep rallying the nation behind Fidel, to demand new efforts, and confess past errors. It is like living in an echo chamber, and inevitably words lose meaning and coherence.

But it is untrue that nowadays Castro's speeches turn the Cubans off. First, there still is a sense of fascination with him and his oratory; second, nobody in a society as tightly and rigidly organized ideologically as Cuba— it is more so than Eastern European Communist countries—can afford *not* to know what the President of the Republic is saying. Ideological indoctrination (*Fidelismo,* Cuban history, and Marxism-Leninism skillfully blended together by Castro) is so important that soldiers, workers or students must study his speeches as promptly as possible after delivery, and be able to explain, preferably in his words and slogans, the views he holds on domestic and foreign problems.

It would be difficult to imagine genuine indifference toward Fidel Castro—in Cuba or abroad. He may be loved or he may be hated, but there are no neutral sentiments about him, only strong emotions. This is why Castro seems to attract so magnetically every kind of admiring or pejorative adjective to his name.

Naturally Castro, who is long on vanity, basks in this attention, and the only time I saw Fidel deeply disturbed and angered by criticism was early in 1985, when he read a description of himself as "cruel" in an article by a leading Spanish journalist who had recently interviewed him at great length. "How does he know I am cruel?" Castro fumed, furiously pacing up and down his office. "Has he ever seen me commit cruel acts? Has he ever heard me order an execution?" Fidel was deeply wounded on his personal scale of moral values, and would not drop the subject for long minutes. Actually he *has* ordered executions in the name of "revolutionary justice," but he resented the implication that it was done wantonly.

Political criticism—or rivalry—is not welcome. After Peru's president-elect, Alan García (a thirty-six-year-old left-of-center politician of great personal appeal whom Fidel saw as a potential junior partner in Latin America), in July 1985 dared to question the Castro view of the world, the Cuban exploded in wrath. García opposed Fidel's notion that Latin American debtor countries should refuse to pay their huge debts to U.S. banks altogether, noting that while the Western financial institutions were "imperialist," so were the Communist Warsaw Pact military alliance and Comecon, the Communist common market, to which Cuba has belonged since 1972. On the day of García's inauguration, Fidel responded with an incredible message of formal congratulations, listing all the ills of Peru, from illiteracy to "misery of all types," and adding grandly that "if you really decide to struggle seriously, firmly and consistently against the Dantesque picture of social calamity and free your nation, as you have publicly promised, from the domination and dependence of imperialism, the only cause of that tragedy, you may count on the support of Cuba." It was a message probably unmatched in the annals of contemporary diplomatic insults, and it may not have been entirely irrelevant that García is the age of Fidel's son, Fidelito; no mature leader likes an upstart. Subsequently, the two governments established a degree of cordiality, but Castro avoids mentioning García publicly.

Castro has a limitless capacity for indignation, anger, and fury, both real and contrived, and does not hesitate to display it privately or publicly. He would have a tantrum in the Sierra Maestra when a single bullet was wasted by a careless guerrilla (and he would threaten the man with direst punishments), and even as president of Cuba, he will throw a fit—complete with the foulest language imaginable in Spanish—when he learns of a bureaucratic stupidity, not an uncommon occurrence after more than a quarter-century of his great revolution. His most senior associates fear the "Fidel furies."

On another level there is little doubt that Castro ordered the exodus from Mariel of over one hundred thousand Cubans to the United States in the spring of 1980, as a gesture of supreme personal rage against President Jimmy Carter for the encouraging attitude he had taken toward a wave of asylum-seekers in foreign embassies in Havana. Emotion remains a powerful factor in the Castro decision-making, and he suspended an immigration agreement that had been signed with the United States in 1985 because the Reagan administration put into operation a hostile radio station named "Radio Martí." Castro told friends that what he had resented was the use of the hallowed Martí name against the Cuban revolution; he did not care about the broadcasts themselves.

Castro's rages (and his brooding, petulant moods—sometimes in lieu of rages) are part of his uncompromising personality, his unshakable sense of righteousness, his toughness, courage, risk-taking instincts, and pride. He demands instant response to his slightest whims, and his entourage is fully tuned to it. These whims, which are frequent, range from the sublime, requesting at once a rare literary opus, to the ridiculous, insisting that a top adviser in attendance provide his (Castro's) combat-boot size to a friend who had offered to find him a special pair in Texas.

Although it is not generally known, Fidel Castro has been at the brink of death an incredible number of times. He nearly died of peritonitis at the age of ten (peritonitis followed appendicitis in those days before antibiotics and penicillin). As a university student in Havana in the mid-1940s, Castro, in the midst of the wave of political gangsterism of that period, was always carrying a gun and being shot at. He participated in an abortive invasion of the Dominican Republic (swimming to the Cuban coast from a vessel bringing back the expedition because he feared assassination), and less than a year later he found himself in the center of a bloody uprising in Bogotá, the capital of Colombia, where he was helping organize an "anti-imperialist" student congress.

On July 26, 1953, now a political leader, he led the assault on the Moncada barracks in Santiago, and miraculously escaped death on at least two occasions—when the army beat back the attack and when he was captured days later in the mountains. In the courtroom and prison, he daily defied the Batista authorities to the point where it was generally assumed that he would be quietly murdered. Amnestied two years later, he almost forced his assassination by the Batista police in Havana, before fleeing to Mexico, announcing that he would soon return to oust the dictator. In Mexico he was arrested by the federal police and nearly killed by Cuban agents (a spy penetrated his clandestine organization in Mexico City). The yacht *Granma,* carrying Fidel and his expeditionaries to Cuba, was so insanely overloaded it practically sank in a storm during the crossing (she reached the Oriente coast in the wrong spot in what Che Guevara described as a "shipwreck," not a landing).

Castro's Sierra Maestra campaign began with the Alegría de Pío disaster, and until the final victory he insisted on personally leading every march and every attack. After the revolution there were thirty or more assassination plots against him—most engineered by the Central Intelligence Agency—and during the CIA's Bay of Pigs invasion, Castro was on the battlefield with his forces until the attackers surrendered. Eighteen months after the invasion, he faced a full-fledged American assault at the time of the missile crisis. In 1963, Castro came close to death when the Soviet airliner bringing him on his first visit to Russia was within seconds of crashing as it landed in a thick fog in Murmansk (an occurrence that was not publicized at the time).

It seems, in truth, that Castro tends to court death. In 1981 he decided to sail to the Mexican port of Cozúmel aboard a high-speed launch, rather than fly to a secret meeting with the president of Mexico, in order to test personally the degree of vigilance of the U.S. Navy in the Yucatán Channel. The navy was patrolling the Gulf of Mexico to determine whether the Cubans were shipping arms to Nicaragua, and Fidel's idea of fun was to see whether it would catch *him*. He then flew home, having made his point, though this incident was not mentioned in the Cuban media. It is an intriguing thought what a U.S. Navy destroyer skipper would have done if he had discovered Fidel Castro aboard a heavily armed craft, escorted by two missile-carrying patrol boats, in international waters between western Cuba's San Antonio Cape and Cozúmel. The *guerrillero* laughs about it.

Moreover, Castro's absorbing hobby is undersea fishing: Weighed down with two pounds of lead at the belt, he stays deep under water for more than two minutes (which is long without oxygen tanks) off a Caribbean key, aiming his spear gun at *pargo* fish and lobster. He frequently flies aboard his personal, Soviet-built executive helicopter, even at night, and, on at least one occasion, he lost his bearings while piloting a twin-engine launch in a powerful storm off the Oriente coast.

Ernesto Guevara Lynch, the octogenarian father of Che Guevara, once remarked to me at his house in Havana that "Fidel must have a pact with God or the devil" because there can be no other explanation for his charmed life. Don Ernesto may be near some sort of truth.

CHAPTER

4

Fidel Castro's success in war and peace, apart from his qualities of leadership and iron determination, is due in great measure to the supreme loyalty he has always been able to command among his fellow revolutionaries, relatives, and friends as well as among the Cuban people after the victory. Three decades later, the loyalty of the first companions of the *Fidelista* revolution remains as absolute as it was in the youthful ardent days of these men and women. In this, the old *Fidelistas* resemble a medieval religious and military order, like the Knights Templars in the time of the Crusades. Inevitably many have grown apart from each other, and ideological differences—sometimes very deep ones over the issue of communism—have developed among others. Still, they share a seemingly unbreakable bond of loyalty to Castro as their historical leader. To them, Fidel can do no wrong. But why is this so?

There is a character in the play *Galileo* by Bertolt Brecht, the German Communist playwright, who declares, "Unhappy the land that has no heroes," to which Galileo replies, "Unhappy the land that has need of heroes." In Cuba's case both reflections are true: The country never had a triumphant hero (José Martí was killed before independence was attained), and its depressing history made it hunger for one. Clearly, Castro was the answer.

And like Maximilien Robespierre, the "Incorruptible," Fidel Castro speaks the language of The Revolution, he is the voice of the deepest revolutionary principles of his time. His proven commitment to live up to these ideals assured him of his following. In a corrupt nation under the Batista dictatorship, Fidel represented the values of honesty, political legality, and social justice—and he was believed because of his impressive oratory and his willingness to challenge the system at great personal risk.

Aside from pure politics, Castro inspires widespread loyalty on the basis of human chemistry. An immensely attractive and contagiously energetic man, he is an unmatched persuader. He can and has convinced scores of men and women of varying backgrounds and temperaments to participate in military actions about which they are told nothing specific until the last moment (the assault on Moncada and the voyage of the *Granma* are cases in point). The army lieutenant who captured him in the mountains after Moncada was convinced by Castro's courage to spare his life, and years later he joined Castro's victorious army. Cubans and foreigners without number have found it impossible to say no to Fidel.

Another inspiring trait has been Castro's personal heroism. His courage is so great that it borders on the insane. At one point in the Sierra Maestra, all his officers signed a petition asking him to stop exposing himself to hostile fire by insisting on being in the front line in every skirmish and battle.

Most vitally, Castro was always concerned by the welfare of his people. He refused to take men who had wives and children (though he himself already was a husband and father) on the Moncada attack and endlessly inquired about the personal problems of his Sierra fighters and their families, meticulously supervised food allocations to assure that all shared equally (once he berated Universo Sánchez for being unable to account for every piece of hard candy with which he was entrusted), and established an iron rule that the *guerrilleros* must pay the peasants in cash for every ounce of rice and every chicken they took.

The peasants' loyalty to Castro had made his survival possible during this revolution—there was a $100,000 prize on his head for informing the army of his whereabouts—and this loyalty grew even more when the rebels helped with the harvest and started rudimentary schools and clinics for Sierra children. A priest who joined the guerrillas as a self-appointed chaplain, Father Guillermo Sardiñas, spent much of his time christening children of peasant families in the Sierra, a greatly appreciated gesture inasmuch as, according to Castro, there were neither churches nor priests in the mountains. As a result, Fidel recalls that numerous families "wanted me to be godfather of their children, and Father Sardiñas christened there scores and scores of peasant kids . . . which in Cuba is like being a second father." He says that "I have masses of godchildren in the Sierra Maestra, and many of them perhaps already are army officers or university graduates."

Castro believes that the presence of Father Sardiñas in the Sierra and the christening of children served "more to link the families with the Revolution, the families with the guerrillas, and to make even strong ties between these populations and the guerrilla leadership." He thinks that the priest, even though he supported the revolution, refrained from political preaching, and his work among the peasants was "more of a religious kind." Still, Castro recognizes that "indirectly" this helped the revolution. Fidel told a Brazilian friar that he wore a cross on a chain around his neck during part of the war because a little girl had sent it to him from Santiago with a "tender message"; he said: "If you ask me whether this was a question of faith, I'd tell you, 'no,' it wouldn't be honest if I told you it was a question of faith; it was a gesture toward that girl."

Evidently Fidel understood how politically essential the peasant support was, but the very poor people of the Sierra Maestra, as they retell their war stories twenty-five years later, always choose to regard the guerrillas as friends and heroes, not as men simply in quest for political power. This loyalty to Fidel Castro has stood the test of time, in truth, growing with it. After the 1983 United States invasion of Grenada, for example, Castro made a point of visiting almost daily at a Havana hospital the Cuban soldiers and construction workers who were wounded in combat on the tiny Caribbean island. He brought them books (among them, Tolstoy's *War and Peace* in a handsome Spanish edition), videocassettes, and plenty of conversation. These visits were not publicized, but the word got around that the Commander in Chief cares about his men.

Armando Hart Dávalos, one of the first organizers of the 26th of July Movement after the Moncada attack, says that of the group with which Castro met secretly in Havana four weeks before leaving for Mexico in mid-1955 to prepare the invasion, "every single person either remains with the Revolution today or is a martyr of the Revolution." This meeting had been called in the harbor district home of two elderly ladies to select the Movement's National Directorate, which would head the underground struggle in Cuban cities in support of the planned Sierra guerrilla war, but Hart himself was soon caught by the Batista police and remained imprisoned until the revolution's triumph. Castro immediately named him education minister (Hart was twenty-nine years old at the time, four years younger than Fidel), and over the decades he has been one of the Maximum Leader's most trusted advisers.

Coming from the moderate political background that characterized the urban 26th of July organization, Armando Hart (who earlier had belonged to a right-of-center nationalist group) unhesitatingly followed Castro in the shift to communism, and is currently a member of the party's ruling Political Bureau as well as minister of culture. Most likely he would again change ideological allegiances if Fidel so ordered.

Faustino Pérez, Fidel's canefield companion of Alegría de Pío, represents another dimension of Castroist loyalty. He had joined the Movement after Moncada (about the same time as Hart, proceeding from the same moderate anti-Batista faction), and he, too, had attended the clandestine Havana meeting prior to Castro's departure. He then followed him to Mexico, landed from the *Granma* as a deputy military commander, survived with Fidel the first month in the Sierra, and was dispatched to Havana to establish the contacts between the guerrillas and the urban 26th of July Movement.

As a top city underground leader, Faustino Pérez participated in planning the bitterly disastrous general strike in April 1958, a milestone event that triggered the fundamental break between pro-Communists and anti-Communists (and various intermediate factions) within the revolutionary movement. The revolution never fully recovered from it, protestations of total unity notwithstanding.

The strike was opposed by the orthodox Communists, still very much at arm's length from Castro, and it was chiefly advocated by the liberal-minded groups in the 26th of July Movement that thought it would quicken Batista's fall. Fidel himself had first publicly supported the general strike (though grudgingly) before siding with Che Guevara, who was with him in the Sierra, and with the Communists in the cities to denounce the failed stoppage. Guevara, ideologically the most radical rebel commander, was from the outset the chief foe of the liberal wing of the Movement, and an excitable leader in the internal revolutionary struggles. The Communists, who traditionally favored political strike strategy, were against the April stoppage because they were being almost entirely left out of it, and they feared the adverse effects on their future influence in Cuba. Allegations of actual Communist sabotage of the strike cannot be proved.

In the Florentine climate of intrigues rising within the revolution, Castro chose to turn against the stoppage organizers, but only after the strike had run its course. He had come to believe that the urban wing of the 26th of July Movement—known as the *llano,* which means lowlands—had been withholding arms and money from the Sierra Maestra guerrillas seeking to use the general strike to capture the overall control of the revolution. Fidel, who still becomes agitated discussing these events despite the passage of so many years, further believed (and probably accurately) that the urban organizations, with their "bourgeois" origins, would try to block the profound social revolution he was planning for after the victory, and that they would settle on Batista's overthrow without allowing the overhaul of the entire Cuban system.

That Castro was planning from the beginning to dismantle the old social order, established by the Spaniards and continued under United States supervision after Cuban independence in 1902, was not made clear to most people on the island (and beyond it) when he was organizing and fighting in the Sierra. Ex post facto, certain historians and commentators claim that Castro's real intentions—the implantation of "secular salvation" for the downtrodden—were expressed in his brief at his trial after the Moncada assault, and were quite deliberately concealed until he took power. Later, Fidel had no compunctions about admitting this publicly, explaining that the people simply were not yet ready for the "real" revolution. The American historian James H. Billington, who has written extensively about revolutionary phenomena, uses the Castro enterprise as the latest example of modern revolutions that were not understood at the time. He writes that "previous political upheavals—even when called revolutions—generally sought a new leader rather than a new order." Billington suggests that "the norm was revolt rather than revolution," and that, in effect, Castro was inspired by the French Revolution in the sense that "never before was the word *revolution* related to the creation of a totally new and entirely man-made order."

To effect this transformation Fidel Castro required trustworthy and experienced allies, and the Communists entered the picture during the last months of the national insurrection because of the convergence of a series of political situations.

To the extent that it is possible to reconstruct precisely the revolution's internal battles, it is certain that Castro's growing suspicions of the 26th of July Movement in the cities, encouraged by Che Guevara as his Sierra letters to Fidel make clear, led him to form an alliance with the old Communists and to create his own new Communists. It is Castro who would later first use the expression "new Communists" publicly. By the same token, Castro is known to have concluded that among his most loyal Sierra fighters there were few with any sound political or administrative background or experience, required to operate a future government or handle the politics of the revolution. And since Castro had decided not to rely on liberal "bourgeois" managers in the conduct of civilian affairs, except in the initial transition period, he was determined to build a brand-new Cuban revolutionary army from the *barbudos* (bearded ones) of the rebel contingents, a great many of them illiterate. It became a matter of revolutionary principle that the Batista armed forces be entirely destroyed (though preserving a handful of professional officers ready to join the cause), and he was proved right when, within hours of the dictator's escape from Cuba on January 1, 1959, an attempt was made in Havana to create a military junta that would have effectively barred Castro from power. His theory was that the destruction of the old army was a precondition for the establishment of a new national order.

Cuba's "old" Communists had been running labor unions, sitting in parliament, infiltrating the university, and publishing newspapers for nearly forty years, providing a pool of both dependability and experience for Castro to draw on. Castro was especially attracted by the Communists' sense of discipline and their organization skills. Although he had never belonged to the "old" Communist party, some of his closest university friends did, and this may have influenced Fidel's thinking during his time in the Sierra. Further, he felt that the new revolutionary armed forces must from the outset be controlled by these "old" Communists, and transformed into the principal "new Communist" power base in the revolution. Castro later claimed, presumably for the benefit of public opinion at home and abroad, that he had had no practical alternative to the absorption of Communists in his victorious Movement, though he omitted to explain how he used them to take over the Rebel Army.

Considering that the Communists had opposed and vilified him as an "adventurer" after Moncada and even when he set foot in the mountains, it remains astonishing that Castro would suddenly regard them as dependable partners and mentors. Being Fidel, he doubtless assumed he could control them. The Communist "sectarian" conspiracy against him in 1962, four years after they made their deal, would later come as a shock. Moreover, the professional quality of the Communists mobilized by the revolution to run Cuba was generally abominable, and whether this also came as a surprise raises questions about Castro's administrative judgment.

The overall conclusion to be drawn from all these events is that the historical decision that the revolution should lead to the establishment of socialism and then communism in Cuba was reached by Castro alone with utter finality in the late spring of 1958—probably following the series of crucial political meetings held in the Sierra during May and June.

Numerous Cuba scholars and specialists abroad have argued that Fidel Castro was a secret Communist since Moncada or earlier; or, obversely, that a year or two after he won power, he had been pushed into communism by the hostility of the United States. Both views appear invalid in the light of careful analysis of existing materials as well as in-depth discussion with key Cuban personalities who participated directly in the entire revolutionary process. Fidel always knew where he was going, adjusting strategy and tactics according to changing political situations; he dreamed of a sweeping revolution, but not of a Communist revolution as defined by *the* Cuban Communist party.

Castro's bitterness toward the United States, dating back to his student days when he was active in a variety of "anti-imperialist" organizations in Havana, and aggravated by American deliveries of bombs and ammunition to the Batista aircraft at the U.S. naval base in Guantánamo, just below the Sierra, must have been an extremely important factor in his decision to go the Communist route to fulfill his broad revolutionary program. He must also have understood from the outset that this course would assuredly antagonize Washington to an infinite degree, and that, sooner or later, he would be forced to obtain Soviet assistance and support for his revolution to survive. He had calculated correctly that the Russians, facing at the time the split with China, would help. In this manner, Castro was able to define beforehand—from the Sierra—the relationships that would involve Cuba, the United States, and the Soviet Union before either superpower suspected what this Caribbean rebel had in mind.

There are those who believe that Castro has always held love-and-hate sentiments toward the United States, secretly hoping for the Americans' approval of him, and this may be true in a highly subjective way. His avidity for personal contacts with Americans of every description—from congressmen and journalists to marine biologists, churchmen, and musicians like the trumpet virtuoso Dizzy Gillespie—may support this idea. But it does not alter Fidel's basic political attitude, best expressed in a private message to Celia Sánchez, his closest Sierra companion, on June 5, 1958, right after Batista aircraft had hit the rebels with U.S.-supplied bombs:

"I have sworn that the Americans will pay very dearly for what they are doing. When this war has ended, a much bigger and greater war will start for me, a war I shall launch against them. I realize that this will be my true destiny." Castro's insights, of course, proved to be entirely correct, though not even he could have suspected that as early as March 10, 1959, only two months after he entered Havana, the National Security Council (NSC) in Eisenhower's administration had already reviewed modalities for bringing "another government to power in Cuba." The record of this review exists in the classified NSC archives, a generally unknown—but fundamental—element in the broad Cuban-American tragedy.

Meanwhile, the immediate upshot of the April general-strike fiasco was that Castro was able to establish undisputed sway over all the revolutionary factions in Cuba, and—most specifically—to subordinate the National Directorate of the 26th of July Movement to his political and operational control. In this sense, the final phase of the Sierra war did for Castro what the Long March in China in 1935 did for Mao Zedong: It proclaimed his absolute primacy among his country's revolutionary leaders. Inevitably Fidel's relationships with his fellow revolutionaries began to change.

Thus Faustino Pérez, who would tell friends many years later that "rightly or wrongly, I was considered to be part of the right wing of the Twenty-sixth of July Movement," was replaced in Havana by a commander picked by Castro, and recalled to the mountains. But Fidel was not about to ditch Faustino. Disregarding Che Guevara's violent criticism of Pérez (the two finally had a savage confrontation at a meeting organized by Castro in the Sierra), Fidel named him chief of civil administration in the liberated territories. Faustino's office shack was inside the rebel headquarters compound high up at La Plata, a few hundred feet from the tree-hidden, ramshackle command-post structure that Castro shared with Celia Sánchez, and the two men were in permanent contact.

Castro has a strong sense of loyalty toward old companions when, even in disagreement with him, they are careful not to engage in what he considers betrayal. He also has his own concepts of loyalty, betrayal, and justice: He is merciless if he thinks he is being betrayed by those he trusts, yet often he goes out of his way to try to salvage a friendship and a relationship, and to convince himself that there had been no real breach of loyalty to him.

In the case of Faustino Pérez, a strange minuet developed between him and Castro. When the revolution triumphed, Pérez, along with Armando Hart and many other figures then perceived as moderates, was named to the first revolutionary government as minister for the recovery of stolen property. He was never invited, however, to join the inner circle, where Fidel was secretly preparing to assume total power in Cuba and veer leftward. This was during the brief period in 1959 when Cuba lived under an "official" government headed by President Manuel Urrutia Lleó (whom Castro, still in the Sierra Maestra, had designated to the presidency) and Fidel's hidden but de facto government, secretly negotiating with the "old" Communists for the joint takeover of the republic in the name of the "real" revolution.

After the first open crisis over the issue of communism in the revolutionary regime erupted in 1959, and after Castro ordered on October 21 the arrest on charges of treason of Major Huber Matos, a highly popular but outspokenly anti-Communist guerrilla commander, two of the moderate ministers resigned at a stormy cabinet meeting with Fidel on November 26. One of them was Public Works Minister Manuel Ray, an engineer and urban underground leader in Havana (but not a Castro "companion"), who soon thereafter began conspiring against Castro and presently fled Cuba. (Ironically, Ray was regarded as too "liberal" by the CIA to allow his anti-

Castro movement to participate in the Bay of Pigs invasion in 1961.) The second minister was Faustino Pérez, but he did so without audible protest, publicly denying that he was leaving over the Matos affair. Castro responded by personally guaranteeing his safety, as a demonstration of loyalty to a comrade-in-arms. In the ensuing years, Pérez fought alongside Castro at the Bay of Pigs, worked in obscure government posts (in 1969 he supervised the construction of a hydroelectrical power plant in the Stalinist tradition of relegating undesirable politicians to obscurity), and notwithstanding his distaste for communism, he joined the party when Castro organized it as the "highest leading force of the society and of the state."

In 1980, Faustino Pérez was a member of the party's Central Committee, the chairman of a National Assembly committee, and coordinator of a mass-membership organization in Cuba's local self-government system. He was relaxed and philosophical as he recounted the crises of revolutionary ideologies, neither repentant nor accusatory, and it was quietly evident that Faustino Pérez retains sincere admiration and loyalty for Fidel Castro. Without being hailed as a "hero of the revolution" by the official propaganda, Pérez is a highly respected figure in Cuba.

In July 1960 came the resignation of Communications Minister Enrique Oltuski, an engineer, a former provincial coordinator of the 26th of July Movement, and the youngest member of the cabinet. He was the last moderate to go, criticizing Communist inroads in the revolution (the overwhelming majority in the first cabinet were highly experienced, talented, and motivated men and women, all of whom had fought against Batista, but to Fidel their political moderation was their fatal flaw). Castro had Oltuski arrested on vague charges, and he spent several years in prison. Having undergone ideological "re-education," he reappeared in the government in the early 1980s as a deputy minister of fisheries, obviously chosen by Fidel, who prefers to preserve talent, if at all possible.

Finally, Universo Sánchez, the other member of the Alegría de Pío canefield trio, faced no ideological soul-searching when the revolution acquired its Communist identity. Finishing the war with the rank of *commandante* (major), then the highest in the Rebel Army, Universo faithfully performed over the years a variety of military and civilian tasks. Though never qualifying personally for Castro's hard-core political circle, he wound up his revolutionary career in his late sixties as head of environment-protection programs. He resumed his membership in the Communist party when Castro reorganized it, presumably because all of Fidel's followers who wished to stay with the revolution were expected to be members. However, unlike many others, Universo Sánchez invariably makes a point of telling visitors that the "old" Communist party did not fight Batista the way it should.

*   *   *

Loyalty, of course, is as elusive a political notion as treason, and Fidel Castro reserves for himself the right of defining both, naturally quite subjectively in terms of the persons involved. Castro, his face and voice exploding in fury, delivered an incredible seven-hour prosecutorial summation before the revolutionary court in the trial of Major Huber Matos, claiming Matos had committed treason because he had "conspired" with some of his officers in the Camaguey province command to resign in protest against Communist infiltration. Matos, one of the best Sierra commanders, was sentenced to twenty years in prison. That first he had written Castro a private letter, loyally beseeching him to act against communism in order to protect democracy in Cuba, was not an attenuating circumstance. In Fidel's eyes Matos's actions constituted a treasonable conspiracy because they threatened to split wide open his revolutionary regime and armed forces, playing into the hands of the United States and other enemies of the revolution. In this sense, Castro's definition of loyalty versus treason was purely practical and political on the grounds that the defense of *his* revolution overshadowed all other considerations. It is impossible to say fairly whether Matos would have been treated more leniently had he been an old companion since the days of Moncada—he joined Castro only in the Sierra, bringing a planeload of desperately needed arms from Costa Rica—but, conversely, Castro could have demanded the death penalty for him. Thus, while Castro argued publicly at the time that political resignations were vicious blows at the revolution, he chose not to apply this condemnatory standard to Faustino Pérez and many others on lower levels who behaved discreetly.

As noted earlier, Fidel is merciless with those he considers traitors, and to him any "counterrevolutionary" is a traitor, an appallingly broad definition. Thus in the mid-1960s, there were at least fifteen thousand so-called counterrevolutionaries serving terms in Cuban prisons by Castro's own admission, a figure reduced to three thousand in 1977, and to around five hundred in 1985, some of them resentenced for unclear reasons. In conversations with foreign visitors Castro defines counterrevolutionaries as those who have in some fashion risen against the revolution, but the regime includes even potential cultural and political dissenters in this category. He says that there are no "prisoners of conscience" in Cuba, but international organizations attempting to monitor the observance of human rights there report, with names, that a number of persons are imprisoned under an article of the Cuban penal code which punishes those who "incite against the social order, international solidarity or the socialist State by means of oral or written propaganda, or any other form; make, distribute, or possess propaganda of the character mentioned in (this) clause."

Incitement by "oral . . . propaganda" is such an incredibly arbitrary concept that its rigorous (or capricious) application in Cuba nearly guarantees the absence of organized dissent in the Eastern European or even Soviet sense. And, indeed, no such dissent exists in Cuba as far as it can be ascertained on the island. Castro occasionally releases political prisoners as goodwill gestures toward foreign governments, groups or individuals (such as the group released in 1984 to the Reverend Jesse Jackson), but he has never defined the criteria under which he selects persons to be freed. Final decisions concerning crime and punishment in Cuba are Fidel Castro's personal province.

Strangely, for example, Castro has sought on at least two occasions to give a last chance to men who conspired to kill him—perhaps out of a sense of loyalty to former guerrilla fighters who played important roles in the war, or out of his private notion of justice. Castro very seldom explains the motivations for his actions or makes them public unless it is absolutely necessary. These two stories have never before been told.

The first one concerns Comandante Humberto Sori-Marín, a lawyer who was Castro's judge advocate at the Sierra headquarters and, simultaneously, was in charge of economic planning for the future. Sori helped to draft the revolution's first agrarian-reform law, signed by Castro in the mountains on October 10, 1958, while the fighting was still going on, and he became agriculture minister after the victory. Along with Universo Sánchez, Sori was a member of the revolutionary tribunal that sentenced to death Jesús Sosa Blanco, a Batista officer charged with multiple murders, at the first great "war criminals" trials at Havana's sports stadium. Sori was not invited to participate in the drafting of a more radical agrarian-reform law promulgated by Castro on May 17, 1959, and immediately came into conflict with Che Guevara, who accused the agriculture minister of being exceedingly moderate. This was part of the early power and ideological struggles within the revolution, and, rather than battle the unbending Che, Sori submitted his resignation from the cabinet. Castro tried hard to dissuade him (though he would not promise to keep Che off his back), but the lawyer left on June 14, 1959.

During the summer, Sori began to plot against the regime and fled to the United States. In 1960 he returned clandestinely, apparently with the aid of the CIA, to link up with anti-Castro armed groups in the Escambray Mountains in central Cuba and to try to assassinate Fidel. Sori was captured, suffering a bullet wound in a shoot-out with state security agents, but his brothers Raúl and Mariano (Raúl had stayed in Cuba as a supporter of the revolution, and Mariano, who had left, returned from his Miami exile) succeeded in arranging a meeting with Castro to plea for Humberto's life. Castro proposed that Mariano accompany him at a prison encounter with Humberto (Raúl would have been regarded with suspicion by Humberto), but the Miami brother refused, fearing a violent confrontation. His fears proved right.

As Mariano Sori-Marín would learn from a witness later, Castro went to see Humberto in prison and told him, "Humberto, you have betrayed us, and anyone but you would pay with his life." This was meant as encouragement to Humberto to ask Fidel for clemency that he was prepared to grant, wanting to be asked. Sori, however, reacted violently, insulting Castro and telling him, "*You* are the traitor of the revolution!" Fidel departed, and presently Sori was sentenced to be shot. He was executed on April 20, 1961, while Castro was leading his militias in defeating the invaders at the Bay of Pigs.

The other story was that of Rolando Cubela Secades, a physician who led the Students' Revolutionary Directorate guerrilla forces in the mountains of central Cuba in the war against Batista. But Cubela was recruited by the CIA in 1963 under the top-secret AM/LASH project (which was even kept secret from President Kennedy) to assassinate Castro and overthrow the regime. Cubela enjoyed Castro's confidence (despite a confrontation between Castro and the directorate that nearly led to an armed clash the first week after victory), and in the mid-1960s was named envoy to UNESCO in Paris.

Late in 1963, Cubela was contacted by the CIA in Paris and Madrid and advised that special weapons had been smuggled to him in Havana to assassinate Castro, which he had earlier agreed to do. But the Cuban secret service, one of the best in the world, learned of the plot. When Cubela arrived in Havana on a routine trip from Paris, Castro summoned him to a meeting at the Palace. According to witnesses, Fidel asked Cubela: "Is there anything special that you want to tell me?" but the doctor replied in the negative. He was arrested as he was leaving the Palace, and he testified at his trial that he had planned to "shoot Premier Castro with a high-powered telescopic rifle and later share in top posts of a counterrevolutionary regime." Had Cubela confessed the plot to Castro, the trial might have been averted. As it was, Cubela was sentenced to fifteen years in prison—a relatively mild sentence by Cuban revolutionary justice standards, which often led to executions, and even that sentence was reduced. Cubela now lives in Spain. Why did he receive such leniency?

Apart from questions of betrayal and counterrevolution, Castro quietly practices personal loyalty toward old companions, making sure they live comfortably. In a great many instances, nice-sounding jobs were created for aging or less than highly qualified individuals because Fidel felt they must be given a sense of self-respect and of knowing that the revolution was eternally grateful for their deeds. This has caused no resentments: Most

Cubans understand it. In many cases, on the other hand, Castro allowed his judgment to be clouded, naming old companions, for example, to ambassadorial posts where they wound up embarrassing Cuba before being recalled.

On one occasion, Fidel demoted an "old" Communist from an important industrial post for revealing that the general manager of the establishment, who was a friend of Castro's from the days of Moncada, was "sleeping at his office, drunk as usual." Castro knew that the friend was an alcoholic who had to be gently removed from higher positions, and he had stopped by at the man's office to see how he was coming along. Hence his violent anger at the Communist, not an anti-Batista fighter, for his contemptuous slur on the old revolutionary.

Even within the Castro brotherhood there are profound nuances among the knights, flowing from the length of their service, and, very humanly, from the nature of their direct relationship with Fidel and their own personalities. These factors define the present power constellation in revolutionary Cuba as they would elsewhere, except that among Cubans the personal element is particularly significant.

As Armando Hart had emphasized, none of the *Fidelista* knights who gathered around Castro before his departure for Mexico in 1955 ever abandoned the faith. Those who survived the war stood by him when the new revolutionary socialist society was being given birth. But, as it happens, Hart is the only member of that first post-Batista cabinet—the "visible" government of the first year that drafted a short-lived "democratic" constitution for Cuba in February 1959—to remain in a ministerial position in the regime and be close to Castro. Most of the other ministers have fled the country and some have died.

Aside from Fidel Castro, the revolutionary names best known outside of Cuba are Che Guevara and Raúl Castro. Guevara was killed in 1967, leading his own guerrilla force in Bolivia, and even before he embarked on this unwinnable enterprise, he had severed—apparently voluntarily—all his ties to Castro and Cuba. The eternal critic and ideological maximalist, Guevara really had no place in Castro's scheme to institutionalize the Cuban revolution in full collaboration with the Soviet Union, which Che increasingly regarded equally as rapacious as the "capitalists" in dealing with the Third World. In fact, Che's last public speech, delivered in Algiers in February 1965, bitterly accused the Soviets for imposing unjust trade relations on underdeveloped nations.

Sadly but cynically, it seems that today the romantic and oddly attractive Che has more political value to Cuba and world revolutionary causes as a dead martyr than as an aging rebel with nothing but his "heroic *guerrillero"* past to live on. As Guevara's father tells it to friends, Che and Fidel were entirely different human beings, though intellectually and politically they complemented each other. Che was one of the very few people in Cuba capable of stimulating Castro's intellect, and it is a grievous loss that no record of their wartime conversations exists (there are only the Sierra letters and messages). Fidel does not talk about it. In any event, Che's face stares down at Havana's Revolution Plaza from an immense billboard opposite the statue of José Martí and Castro's massive Palace of the Revolution beyond it.

The present power structure under Castro and the relative importance of the figures in it are generally little known and understood, even in Cuba. First come the old companions. Among them, Raúl Castro is number one, having been at his older brother's side since before Moncada. His special responsibility is the armed forces. He is general of the army and defense minister in addition to all his titles as Fidel's deputy for everything—and he is his designated successor and second secretary of the Communist party. He conducts much of the day-to-day business of governing Cuba as his brother devotes more time to his global and ideological concerns. With a clipped mustache and a roundish face, Raúl resembles a self-satisfied Spanish grocer, but he is most respected for his toughness and his skills. The few foreigners from the non-Communist world who meet Raúl generally find him quite charming and interesting.

Raúl divides his government leadership functions (he is first vice-president of the Council of Ministers) with Osmany Cienfuegos Gorriarán, a member since 1986 of the party's Politburo, long a vice-president of the council and secretary of its Executive Committee, who has gradually become one of Cuba's most powerful though little-known men. An architect by training, the mustachioed, bespectacled Osmany Cienfuegos is *not* a knight of the Castro companions order; he participated in no revolutionary combat in Cuba, but chose to sit out the Sierra war with a Communist group in Mexico. However, he was the elder brother of the wildly popular Comandante Camilo Cienfuegos, an old companion whom Castro named Chief of Staff of the Rebel Army early in 1959. Camilo Cienfuegos disappeared in a mysterious aircraft accident in October of that year, days after he had gone to Camagüey on Fidel's orders to arrest Huber Matos. A month later, Osmany Cienfuegos was appointed public works minister in the place of Manuel Ray who had resigned over the Matos case, while Camilo was entering the pantheon of martyred revolutionary heroes.

As a university student, Osmany belonged to the Youth of the Popular Socialist party (as did Raúl Castro), which now is called simply Communist Youth, and he is a hard-line Communist. Not well liked in Cuba, Osmany appears very rarely in public; he is the classic quiet, gray "inside man," and he is an example of the fact that it is no longer absolutely necessary to be a *Fidelista* knight to wield power on the island. Still he does not enjoy personal closeness to Castro or easy camaraderie.

Until 1986 the next most powerful man was Comandante Ramiro Valdés Menéndez, who until then had served as interior minister and was a member of the party's Political Bureau, and vice-president of the Council of State and the Council of Ministers. His departure from the ministry (he had served twice in that post) and the Politburo was a major surprise. Castro offered no explanation during the Communist party's Third Congress in February 1986 except that the Politburo had to be "rejuvenated." But it seems that Valdés, who was kept on the party's Central Committee and given a vague technical assignment, lost a power struggle with Raúl Castro, who would not tolerate a rival power center in the police-intelligence area of the regime. Valdés was the only top leader besides Fidel Castro to customarily wear the olive-green uniform, and he has kept his Sierra pointed beard, giving him a slightly satanic countenance. Raúl Castro often appears in casual clothes, such as a black leather jacket over shirt and necktie. The ex-interior minister is one of three revolutionary veterans to carry the title of *Comandante de la Revolución,* a great distinction that even Raúl Castro does not possess; the others are Juan Almeida Bosque, the black army Chief of Staff, and Guillermo García Frías, the first Sierra peasant to join the rebels, though neither has much political clout (García also lost his Politburo seat and his post as transport minister, mainly because he was incompetent.)

Valdés's successor as interior minister, General José Abrahantes Fernández, directs the State Security Services (the political secret police), the elite Special Forces of State Security (a militarized shock force of some five thousand men with armor and aircraft), the National Police, the overseas intelligence services linked to State Security, and the tremendous network of neighborhood security watchers and informers of the Committees for the Defense of the Revolution (CDR). The Security Services have an impressive record in dealing with the regime's enemies everywhere, and they easily infiltrate Cuban exile communities in the United States, Canada, and Western Europe. However, Abrahantes, who was Valdés's deputy, has no political clout of his own—he used to be the man in charge of Castro's personal security—and most likely Raúl Castro now controls the police network along with the armed forces.

As a revolutionary, Valdés had participated in every action from the Moncada attack in 1953 to the autumn 1958 invasion of the lowlands by rebel columns from the Sierra (he was Che Guevara's deputy in command of Column 8, which conquered the central provinces), and his influence with Castro was enormous. He was one of the most pro-Soviet figures in the top leadership, and his ties with Soviet and Eastern European intelligence services were most intimate, particularly after 1968 when Moscow imposed the primacy of the KGB (the Soviet secret service) in Cuba.

The rest of the power structure includes *Fidelista* knights as well as men who joined the revolution at subsequent stages. Among the former are Pedro Miret Prieto, an engineering student who trained the Moncada plotters in the use of arms and was at Castro's side during the attack; he later served as a distinguished military commander both in the Sierra and at the Bay of Pigs. Miret, now the vice-president of the Council of Ministers in charge of industrial development and a member of the party's Political Bureau, is a pleasant and deceptively harmless-looking short man with a moustache. Once one of the toughest revolutionaries, Miret is a greatly trusted adviser in the Castro inner circle.

Armando Hart and Comandante Almeida both belong to this enchanted circle, though they are not among the principal decision-makers. Jesús Montané Oropesa, an accountant who fought both at Moncada and in the Sierra, was removed from the Political Bureau at the 1986 Party Congress, but remains a Central Committee member and is still very close to Castro.

Sergio del Valle Jiménez, a physician who was a Sierra *guerrillero* of outstanding courage and later a leading military commander, is part of the inner group. He was minister of health, an extremely important assignment given Castro's public-health priorities, Until December 1985, when Fidel drafted him for a full-time but discreet political role outside the Politburo. José Ramón Machado Ventura, also a former guerrilla physician (the Rebel Army had an inordinate number of fighting doctors, starting with Che Guevara and Faustino Peréz, who came aboard the *Granma*), has become the new Communist party's chief ideologue, carrying great weight with Fidel. He is on the Politburo.

Another old-time *Fidelista* with recently growing power is Jorge Risquet Valdés-Saldaña who at the 1986 Party Congress became a member of the party's Politburo, in addition to the key post of a secretary of the Central Committee he has held for a decade. Relatively unknown to the public, Risquet is the chief executor of Castro's military and political operations in Africa as well as an enforcer of labor policies at home.

Vílma Espín de Castro, who is Raúl Castro's wife, was in the Sierra with him (they were married in Santiago shortly after victory, but Fidel was too busy to attend the wedding), and she is now a member of the Council of State and (since February 1986) a full member of the party's Political Bureau (the only woman in this ruling body). As president of the Cuban Women's Federation, Vílma Espín (who once studied architecture at the Massachusetts Institute of Technology) holds very considerable political power. A woman of charm, Vílma occasionally acts as informal hostess on social occasions at the Palace of the Revolution, standing next to Fidel in the reception line.

Besides old *Fidelistas,* two men of diverse backgrounds count powerfully in the Castro power edifice, principally because he has immense personal respect for each of them. They are Carlos Rafael Rodríguez Rodríguez, the intellectual and political genius of the "old" Communists, and José Ramón Fernández Álvarez, a former Batista army officer who was imprisoned during the Sierra war for antiregime conspiracies, and who now serves Castro as a vice-president of the Council of Ministers, education minister, and worldwide diplomatic troubleshooter.

Rodríguez, his white goatee and his manner giving him a touch of Old World *boulevardier,* is by far the most experienced politician in Cuba: His career stretches back to the 1930s, when Castro was a child in Oriente. A most affable man of considerable learning and sophistication as well as a prolific writer, he has been Fidel's most valuable collaborator, politically and intellectually, since the 1959 victory. They remain extremely close, conversing every day. Rodríguez is thirteen years older than Castro, and is one of the few Cubans who address him as *tu* in the Spanish familiar usage of friendship.

As vice-president of the Council of State and the Council of Ministers and member of the party's Political Bureau, Rodríguez is the third member of the trio, along with Raúl Castro and Osmany Cienfuegos, that forms the top level of the power structure directly under Fidel. Foreign policy and economics are his principal personal responsibilities (Foreign Minister Isidoro Malmierca Peoli, an old-line Communist, executes foreign policy but does not plan it), and Castro makes very few decisions of any kind without consulting the septuagenarian Rodríguez.

It is principally thanks to Rodríguez's perspicacity that the *Fidelista-*Communist fusion could occur, and that the Castro-led new Communist party emerged from all the Sierra and post-Sierra maneuvering. Rodríguez had served as a minister in the Batista cabinet during World War II when the Communists shifted to "popular front" policies (this ministerial phase in his career is now tactfully omitted from official Cuban biographies), and he was the first of the otherwise unimaginative leadership of the illegal Communist party during the insurrection to understand that Castro would dislodge Batista—and that the Communists had better get aboard.

Even as late as the failed general strike of April 1958, Communist leaders thought that, at best, a victorious Castro would turn into a Cuban version of Egypt's Gamal Abdel Nasser, a radical nationalist, and that moves toward socialism would remain confined to the Communists. The idea of coming to power themselves as a result of the anti-Batista revolution had never occurred to the essentially conservative Communists (that idea had not occurred to the Soviets at the time either), and it took the flexible and fertile mind of Carlos Rafael Rodríguez to see new possibilities.

After the general strike, the Communist party authorized its members to join the guerrillas if they so desired, and quite a few of them went up the mountain. A number of them chose (or happened) to serve in Raúl Castro's new autonomous command northeast of the Sierra Maestra possibly because Raúl, unlike Fidel, had been a Communist Youth member at Havana University. In June 1958, Rodríguez himself made the ritual trek to the Sierra Maestra headquarters, where Fidel, whose notion of asserting his importance is often to keep people waiting endlessly to see him, had the Communist emissary cool his heels for several days.

There is no record of their lengthy conversations, but Rodríguez remained in the Sierra through August and the start of the final rebel offensive against Batista, indicating that the foundations for a future Communist Cuba were laid then. Castro had already decided on this course in principle, and the discussions with Rodríguez must have centered on the modalities of creating a Communist apparatus that Fidel would control. In the ensuing months, the "old" Communists fielded a small guerrilla operation in central Cuba, conveniently joining forces with the column coming down from the Sierra under the command of Che Guevara and Ramiro Valdés.

If, indeed, Rodríguez played a crucial role in the forming of the initial Communist alliance with Castro—including the secret negotiations in Fidel's hideaway house outside of Havana within weeks of victory—he was equally important in preventing breakdowns. In 1962, when the new Communist party was still in its organizational phase, Rodríguez sided with Castro to put down a dangerous attempt by old-line Communists, perhaps encouraged by ever-greedy Moscow, and to assert their domination over it. The crisis was so grave that Castro vanished from public view for weeks, preparing a counterattack. A similar situation developed in 1968 when problems within the Cuban Communist party were immensely aggravated by Castro's open quarrels with the Soviet Union over military, security, and economic cooperation policies. Here again Rodríguez saved the day as the chief mediator. Castro once more beat down domestic orthodox Communist conspiracies—imprisoning thirty-seven "old" Communist leaders (some of whom were still incarcerated in 1985) to show that he would not countenance Communist "counterrevolutionaries" either—but he had to make far-reaching international concessions to the Kremlin. Subsequently, Carlos Rafael Rodríguez assumed the specific responsibility for all aspects of relations with the Soviets.

Unlike most other top Cuban Communists, Rodríguez thrives on good conversation and every form of art and literature, being himself the author of remarkably readable books of essays and memoirs. A onetime Havana University professor and newspaper editor, he enjoys attending art gallery openings, interesting social and diplomatic occasions, and intellectually satisfying small dinners—always shining with wit and courtesy. But sad to say, Carlos Rafael Rodríguez seems to be the last of his breed: They no longer make politicians of his brains and charm, Communist or not, in Cuba.

José Ramón Fernández Álvarez, the vice-president and troubleshooter, is a Cuban revolutionary politician of still another type, but also a most valuable one to Fidel Castro and increasingly important. Tall, ramrod-straight, and white-haired, he is similarly a revolutionary charmer of foreigners. A few years older than Fidel, Fernández attended the same high school in Santiago and shares a family background in rural Oriente very similar to Castro's own. They met first, however, after the insurrection when the former army officer was released from the Batista prison on the Isle of Pines (now the Isle of Youth), the same prison where Castro was locked up five years earlier. Also of Spanish parentage, he is known as "Gallego," though his parents came from Asturias and not Galicia.

Fernández, who at the time had no ideological links of any kind, was a professional soldier (and a graduate of the U.S. Army's artillery school at Fort Sill, Oklahoma) and anxious to enjoy civilian life now that the hated dictator had been overthrown. The day of Batista's fall, he assumed the military command of the prison, presently returning to Havana to look for a job. Castro, however, had heard about him and summoned Fernández for a late-night conversation shortly after arriving in the capital from Oriente.

The ragtag Rebel Army had a desperate need for professional officers in order to expand and modernize, and Castro urged Fernández, the officer "with clean hands," to get back into uniform, ignoring that he had just become a well-paid sugar-mill manager. For an hour Fernández resisted, and finally Fidel cried with mock petulance: "Okay, Gallego. . . . That's just great. . . . You go and run your sugar mill, I shall retire to write books, and fuck the Revolution. . . . Is that what you want?" The great persuader won, and Fernández spent the next two years rebuilding the Rebel Army and purchasing arms overseas. He helped to buy the first shipment of Belgian FAL automatic rifles that were Cuba's best weapons until the Soviets became the arms supplier. As the field commander of Cuban forces during the Bay of Pigs invasion, Fernández deserves a great deal of credit for the rapid victory on the ground while Castro coordinated the overall strategy.

Later, Castro used the much-liked Fernández whenever he had hard problems: in the armed forces, in the vast operations of the Education Ministry, and, in the latest phase, in establishing diplomatic and political relations with Latin American governments. Authoritative and efficient, a rare combination in socialist Cuba, Fernández is one of the revolution's best assets (in his spare time he graduated from the Communist party's superior studies' school in Havana to become a bona fide Marxist-Leninist as well). At the 1986 Communist Party Congress, he was named Alternate Member of the Politburo.

While the top power structure in Cuba is reasonably well coordinated in terms of the overall management of the country, the fundamental problem remains Fidel Castro's psychological inability, rather than conscious refusal, to let go of any power. The result is that all authority and responsibility continue concentrated in his hands, a state of affairs that paralyzes all initiative on lower levels. Castro's compulsive dedication to detail, and the conviction that no matter what the subject he knows more about it than anyone else, have combined to make him an obstacle to an efficient development of the economy and the society. Most Cuban managers in the totally state-owned economy simply have no courage to make decisions within their purview, fearing displeasure on high and, most likely, punishment. therefore, a mutually protective association of bureaucrats has come into being, and the bitter Havana joke is that Cuba does have a two-party system after all: the Communist party and the bureaucratic party. The waste of resources and talent is staggering.

Castro, of course, bristles at any suggestion that he is a dictator and that all the decisions are made by him. In 1977 he told an American interviewer that "I'm a leader, but I am very distant from having unipersonal power or absolute power." He went on to say that though "my personal power was very great" during the war, almost immediately afterward, the revolution moved to "establish a collective leadership . . . a leadership group from among the most capable leaders."

In 1985, Castro still insisted on the collective character of the Cuban leadership and on the fact that the process of institutionalizing the revolution has been completed. This, he said, was achieved with the 1975 national referendum approval of the new Cuban constitution (which became effective in 1976). This created novel mechanisms such as the "Popular Power" local government system under the National Assembly, which votes on proposed laws and in theory supervises their implementation. Still, it taxes the imagination to visualize a rejection by the assembly (which meets twice a year for two-day sessions) of a Castro-proposed law or the removal of a high official on the grounds of, say, inefficiency. Officials may be recalled by municipal or provincial assemblies on the local government level, but policies are not questioned. Castro told me that in the Communist party's Political Bureau he held only one vote out of fourteen and had been overruled on occasion, though he cited no examples.

The question is whether Castro accepts that in the light of Cuban realities these claims are not wholly credible. On the other hand, it must be recognized that he is caught in a political-image trap because he cannot publicly acknowledge that he does actually hold "unipersonal power." Such an admission would undermine the integrity of the institutions he has created, removing whatever semblance of independence or autonomy they may have. The real point, of course, is whether these institutions can survive Castro's death or incapacitation, and how the issue of succession and Cuba's future is to be resolved. To be sure, it was Castro who named his brother Raúl his successor, thus in effect imposing succession.

The official media portray the Council of State and the Council of Ministers (both presided over by Castro) as the decision-making organs of the republic. But in November 1984, for example, an infuriated Castro went before the National Assembly to denounce the shortcomings of an economic-development plan for the following year prepared by the Central Planning Board, and its parallel five-year plan. Overnight he established a "Central Group" in the Council of Ministers, headed by Osmany Cienfuegos, to develop at once a new plan for 1985—hardly the best way of running the economy. Not long thereafter, the Planning Board president, Humberto Peréz Gonzáles, once considered the most promising new-generation technocrat, was "liberated from his functions," the general euphemism for being fired.

In general, it is Osmany Cienfuego's Executive Committee of the Council of Ministers that administers Cuba on a day-to-day basis, and Castro does not always attend the meetings held in the building adjoining the Palace of the Revolution. Carlos Rafael Rodríguez felt in mid-1985 that the Executive Committee had "freed" Castro from concern with most current problems, it being agreed that he should have the time to concentrate on great global issues. Only decisions affecting Cuba's relations with the United States and Latin America must be personally approved by Fidel before action is taken, Rodríguez said.

In real life, however, Castro insists on being posted on just about everything, and consequently even relatively minor decisions may be delayed or postponed until the Commander in Chief can catch up with the problems in the midst of his other interests and commitments. His frequent speeches confirm that he is abreast of all aspects of every problem in Cuba: He touches upon all of them as he preaches the virtues of hard work and the tremendous need to save resources. The printing of the Communist party's official daily newspaper, *Granma,* the voice of the regime, may be held up until dawn hours while Castro personally edits a lengthy policy speech he had delivered extemporaneously the previous day, or a major policy editorial (he writes some of the page-one unsigned editorials, his style being unmistakable for its color and subtle invective).

Fidel Castro's style of government is based on what he calls dialogue or rapport with the population. In practice, it means oratorical hard sell of new policies or an insistence on the fulfillment of old ones. This is usually done through televised speeches delivered before large audiences that applaud Castro and reply affirmatively when he asks whether they approve of what he proposes. (In this fashion, Cubans since 1959 have "approved" executions of Batista-regime torturers, the military presence in Angola, multifarious social and economic sacrifices, and even the transfer of the port town of Moa from Holguín province to Guantánamo province in eastern Cuba.)

Crowds have never said no to Castro, and this technique of popular "consultation" was refined by Fidel at the outset of his rule. He says it is "direct democracy," preferable to old-fashioned elections, and it remains his most powerful political weapon in any crisis—the recourse to the masses. Chants of "Fidel, Fidel!" punctuate the mass rallies as he whips the audiences into a frenzy of enthusiasm.

Mass-circulation newspapers and magazines—as well as specially printed booklets of interviews granted by Castro to foreign radio and television networks and publications—are another facet of government-by-verbiage. After *Playboy* printed a long, boring Castro interview in August 1985, an expanded text in Spanish was published as a special pullout section in *Granma,* but it carefully avoided any reference to *Playboy* as the magazine in which it had appeared. Instead, *Granma* provided only the names of the two American interviewers. Fidel Castro is still prudish, at least concerning his image at home, and image is what counts the most.

CHAPTER

5

Fidel Castro's personal life and activities are normally kept well out of public view in Cuba for reasons of privacy as well as security. Only his very frequent official performances are reported in the press and on the radio and television, and most Cubans are not even aware of how Castro works or of whom his immediate entourage is composed.

Extraordinarily busy as he is at the Palace of the Revolution and elsewhere, with occasional around-the-clock schedules, nearly always surrounded by people, Castro nevertheless projects an overwhelming impression of loneliness. A bachelor for the thirty years since his divorce, with many of his closest friends and companions dead or simply no longer around, Fidel does not appear to have anyone near him to whom he can turn in full trust and with whom he can share great and small victories and defeats. Not even his brother Raúl qualifies for such intimacy. His ex-wife, Mirta, had tried to be a friend and companion, but Fidel never included her in his plans and ambitions. They are said to have been in love. She visited him in prison immediately after Moncada, and then they briefly exchanged letters. But in 1954 he decided to divorce her for political reasons: Her parents and brother were too close to Batista, and she was on the official payroll.

By nature, Castro is secretive and tends to keep his counsel, but a few human beings in the past were extremely close to him. The most important of them, as every Cuban *does* know, was Celia Sánchez, the doctor's daughter from an Oriente sugar mill who for twenty-three years until her death from cancer was Fidel's unconditionally devoted helper and adviser in war and peace—in fact, his conscience and alter ego. She also held great power.

Celia lived in a cramped and shabby apartment, at number 1007, a small building on a short block of Eleventh Street in a middle-class section of Havana's residential district of Vedado. It was the center of Fidel's and her life and work; he frequently slept in the untidy apartment and Celia worked there, as his chief aide as well as preparing hot meals in the minuscule kitchen to be sent to him wherever he happened to be in Havana during the lunch or dinner hours. Even after she died in 1980, the apartment with chromos on the walls is still home to him. The street block continues to be cordoned off with iron chains and guarded by armed State Security troopers in olive-green.

Celia, who was Fidel's senior by five years, never married. With Castro, she was the "first lady" of Cuba, quite beloved and respected on the island. Since her death, she has been virtually canonized by the revolution, with hospitals and schools named after her. While she was the very warm, very human, and very Cuban symbol of the revolution, she was also the firm *compañera* who protected Castro from too much outside pressure and from himself. She was probably the only one in Cuba to tell him to his face that he was making a bad decision, although she told others that "Fidel is always right." Celia kept Fidel's days and nights from turning into total chaos, found time to help other Cubans with problems that could only be solved at the top of the government, designed a spectacular public recreation park and restaurant in suburban Havana (Lenin Park), preserved antiques and museums, and organized an oral-history project of the revolution up to victory in 1959.

At her death, Celia was secretary of the Council of State, with ministerial rank, and a member of the party's Central Committee. Strikingly, a large number of outstanding Cuban women of various backgrounds, including some from the highest circles of prerevolutionary Havana society, came forth to play invaluable roles in helping and supporting Fidel from the very outset. Many risked their lives for his cause and to this day are fiercely protective of Fidel and his good name, even if they have not seen him in years.

Then there was Dr. René Vallejo, a distinguished surgeon who had served with the American Army in Europe in World War II, who joined Castro in the Sierra, remaining with him afterward as personal physician, *aide-de-camp,* and around-the-clock friend and companion. Vallejo, one of the most *simpático* early figures of the revolution, died at forty-nine from a sudden illness in 1969, leaving behind a crushed Fidel Castro. Like Celia, Dr. Vallejo could never quite be replaced in Castro's life.

In a very special fashion, there was Castro's friendship with Che Guevara, two years his junior, an unmatched relationship. Though Guevara, whom Fidel first met in Mexico during the conspiratorial period, came on the *Granma* expedition as a physician (he was listed on the roster as "Lieutenant Ernesto Guevara, Chief of Health"), he quickly turned into one of the principal guerrilla commanders. In the absence of radio or telephone links during most of the mountain war, Castro and his officers communicated through written messages, carried by couriers (often women). From these papers, most of them preserved, it appears that apart from communications with Celia, the single greatest volume of exchanges was between Fidel and Che.

The two men were the only intellectual equals in the Sierra, and in addition to operational orders and reports, they wrote each other long political letters that illustrated the ideological evolution in the guerrilla, with Che emphasizing his radical leanings, and Fidel being more practical and pragmatic about the politics of the war. there were also personal touches. As the Batista offensive began in May 1958, Fidel wrote Che: "It's been too many days since we've talked, and that's a matter of necessity between us." On another occasion, complaining that he was not receiving promised ammunition from the cities, he opened a letter by saying, "This is a complete fuck-up." Admitting that Castro had been correct in warning him about an enemy attack, Che wrote: "As so many other times, your excellency (isn't there a junior lieutenant colonel rank?) was right and the army got as far as our beards."

In the first years of the revolution, Castro and Guevara were inseparable, not only as the top leaders in Havana engaged secretly in moving Cuba toward socialism, but also as friends continuing to need each other. Conchita Fernández, who was Castro's personal secretary, aiding Celia Sánchez during the initial period, recalls that Fidel and Che lunched together alone almost every day, sharing the hot meal sent by Celia Sánchez from the apartment.

It will probably never be known what exactly caused Che Guevara's mysterious leavetaking of Cuba in 1965; was there, for example, a deep personal break between them, as some mutual friends think? In any event, no one has taken the mystical Argentine's place as Castro's intellectual partner. Carlos Rafael Rodríguez comes the closest to it in the present power structure, but they share no experiences of conspiracy, war, and danger as Fidel and Che had.

Castro once told a visitor that "I detest loneliness, total loneliness, maybe because of the need man has for company." He added: "Aristotle said that man was a social being, and it seems I belong to that species," remarking in a reminiscence about the months spent in solitary confinement in the Batista prison that "the fact that I detest loneliness does not mean that I am not capable of standing it." In the absolute sense of physical loneliness, Castro clearly does detest it because he is so seldom alone—the acolytes are always on hand. But of course the company he keeps nowadays tells a lot about the quality of his daily life, particularly his personal life.

First there is the palace entourage, the court over which Fidel Castro presides in a manner sometimes bordering on the royal, being in demeanor something of a Spanish royal personage himself, Marxism-Leninism and olive-green military garb notwithstanding. Castro is much more of an imposing and electrifying personality than Communist leaders elsewhere in the world, certainly more so than the Soviet Union's Mikhail Gorbachev.

The entourage at the Palace of the Revolution, a sprawling structure with a broad front staircase built by Batista for the Cuban Supreme Court, is rather undistinguished though blindly loyal to the Commander in Chief. Basically, the staff is composed of officials attached to the Council of State, which functions as Cuba's principal government organ. Castro is president of Cuba because he is chairman of the Council of State, and all his and his staff's offices are in the Palace of the Revolution, located in the center of a closely guarded and well-landscaped government complex on the Plaza de la Revolución. The palace adjoins the building of the Central Committee of the Communist party. Raúl Castro's Revolutionary Armed Forces (FAR) Ministry, the Interior Ministry (MININT), and the party newspaper, *Granma,* are nearby.

The man who has taken over Celia Sánchez's responsibilities as the council's secretary is Dr. José M. Miyar Barrueco, commonly known by his nickname, "Chomy," and he must be the most overworked individual in Cuba. A physician who met Castro in the Sierra, Chomy was chancellor of the University of Havana for a time after the revolution, then he was summoned to the palace to replace Celia; he later became a member of the Communist party's Central Committee.

Chomy has his own office on the ground floor of the palace, but he is at Castro's beck and call night and day, attending most of the official meetings as well as sessions with foreign visitors that often stretch into the dawn. Every time Fidel has a fresh idea, a question or a request—which is all the time—Chomy writes it down, then passes on pertinent instructions to appropriate officials. Only when Castro goes to sleep is Chomy—who has only a six-person staff—free to attend to his paperwork. He is responsible for organizing and reorganizing the schedule of the *Jefe,* and the direction of the council's historical division. He is also the most opposed personally to the publication of serious historical material about Castro and the revolution, and he is a most effective watchdog of the archives. A harried, superficially affable man who is obsessed with photographing Fidel all the time with everybody, Chomy has limited intellectual and political input, but he wields the power of the doorkeeper.

Heading Castro's personal "Coordination and Support Group" at the palace is Government Minister José A. Naranjo Morales, known as "Pepín," who fought with the Student Revolutionary Directorate guerrillas and later became governor of Havana province. Naranjo is a bureaucratic jack-of-all-trades, running political errands for Fidel and coordinating the preparation of background studies and information for him on all imaginable subjects. This support group—ten men and ten women, all carefully picked, free to cross all bureaucratic lines in the party and government, and equipped with up-to-date computers—is informally called the "Fidel staff," and its relatively young members are groomed for promotion later to important administration positions. The new head of the state radio and television network was recruited from the Fidel staff when Castro strongly (and deservedly) criticized the quality of Cuban television in an off-the-record chat with leaders of the Women's Federation early in 1985.

The most interesting men around Castro, however, are the less visible ones, and all come from State Security Services. General José Abrahantes Fernández, who replaced Ramiro Valdés in December 1985 as interior minister, is still directly responsible for Castro's security, his Special Forces of State Security being the Praetorian Guard handpicked to defend Fidel and the regime in the direst of dangers. The Special Forces detachments are organized along the lines of the Soviet KGB's uniformed security forces. Whereas Abrahantes, even as minister, plays a limited political role, two Security-linked figures in particular were long exceedingly important in the foreign-policy field.

One was José Luis Padrón González, whose title was president of the national Tourism Institute (which operates all the Cuban hotels), but who served Castro extensively as a discreet international emissary and negotiator, bypassing the Foreign Ministry. A former colonel in State Security (there are more Security-linked officials in Cuba than meet the eye), Padrón was part of the new revolutionary generation, seemingly being groomed for high leadership. Padrón first impressed Castro in helping to set in motion the Cuban military intervention in Angola in 1975—he was rushed to Luanda by State Security to prevent the collapse of the Marxist Popular Movement for the Liberation of Angola (MPLA) in the civil war that followed independence from Portuguese colonial rule—and he has been an important "inside" player ever since.

Over the years, Padrón (assisted by Tony de la Guardia, another Security colonel at the palace) had met secretly with senior United States Department of State officials in Miami, New York, Washington, Atlanta, and Cuernavaca, Mexico, to explore the possibilities of agreements between the two countries. The 1979 accord on the American acceptance of thousands of Cuban political prisoners and on Cuban exiles' visits home from the United States was worked out by Padrón, who reported only to Castro during the long and secret negotiations. Early in 1986, however, Padrón mysteriously vanished from sight, losing his job and palace access, becoming a nonperson. It happens in Fidel's Cuba.

A notable Security figure at the palace is Comandante Manuel Piñeiro Losada, long known in Cuba as "Barba Roja" (for his red beard, now white). Piñeiro, a member of the Communist party's Central Committee, is head of its American Department, and, as such, he is the chief coordinator of all Cuban operations in the hemisphere, from Nicaragua and El Salvador to Panama, Peru, Argentina, and the United States.

Given Castro's ambitions for Latin American leadership, Piñeiro's position is extremely important. His article in the Autumn 1982 issue of the Communist party's theoretical journal *Cuba Socialista* on "The Present Crisis of Imperialism and Revolutionary Processes in Latin America and the Caribbean" offers thoughtful proposals for the launching of "democratic, popular and anti-imperialist revolutions," insisting on the unit of Communists with other political parties and leftist organizations. This is, of course, the lesson of the Cuban revolution as conducted by Castro and his close collaborators, among whom Piñeiro was a most active one.

Piñeiro acquired his anti-Americanism (and his first wife, an American) when he studied at Columbia University in New York in the early 1950s, and was humbled and "radicalized" by his defeat by a "rich South American kid" in the elections for president of a student association. He was in the anti-Batista underground back in Havana, where his apartment was an arms depository before the failed and horribly bloody attack on the presidential palace by young revolutionaries in 1957, before joining Raúl Castro's command in the Sierra. He wound up as military governor in Santiago.

When Ramiro Valdés, Che Guevara's deputy, became chief of the Rebel Army's G-2 Section (security and intelligence), Piñeiro was named number two. He presided over the revolutionary tribunal that sentenced Batista aviators to severe prison terms in 1959 after they had been acquitted under another judge, then retried on Castro's orders. He subsequently took over as head of G-2. He was Cuba's top political policeman under Valdés until

1968, when a Soviet-imposed reorganization of the Security Services forced him out—and Castro put him in charge of Latin American affairs.

When Castro toured Africa, Eastern Europe, and the Soviet Union for over two months in 1972, Piñeiro was a senior member of the entourage, his name appearing near the top of the Cuban list in all the joint communiqués with foreign governments. It is rare for Piñeiro not to be present at social functions at the Palace of the Revolution (usually standing near Castro with a small group of the top leadership), and it is not uncommon to find him in the middle of the night having a milkshake with friends at Chomy's downstairs office. Naturally, he has permanent access to Castro's third-floor office.

Dr. Antonio Nuñez Jiménez, a geographer, explorer, historian, ardent Communist, and vice-minister of culture also has full access to Castro, whose story he is slowly writing in multiple volumes. Nuñez Jiménez (whose daughter is married to José Luis Padrón, the ex-Security colonel turned nonperson) is regarded primarily as a personal companion—he accompanies Fidel on most of his trips around Cuba and shares vacations with him, but he does not fit (nor chooses to fit) into the palace entourage. Finally, there is Jorge Enrique Mendoza, the editor of *Granma,* a close ideological and propaganda adviser; he is an unpleasant, irascible, hardline, dogmatic Communist who knew Castro in the Sierra, serving as the Rebel Radio announcer in the final phase of the war.

Castro enjoys the friendship of many people who are personally and even ideologically attuned to him but exercise no political influence. Foremost among them is Gabriel García Márquez, the mustachioed Colombian Nobel Prize laureate in literature. Currently the greatest Latin American novelist, his worship of Castro is evident in an early brief portrait he wrote titled "My Brother Fidel," based on conversations with Castro's sister Emma. Castro's and García Márquez's friendship is so close that when the Colombian comes on one of his visits, the two men often converse continuously for eight or ten hours, then repeat it over several days and nights. Former Colombian President Alfonso López Michelsen, who was brought to Castro's island by García Márquez in 1984 and was with them much of the time, says that, among other things, the novelist "recommends books" to Fidel. López Michelsen says that "Fidel is a reader of extraordinary avidity. . . . Gabito [García Márquez's first-name diminutive] brings him five books and he stays for ten days, and the day he leaves, Fidel comments on the books one by one. They are not necessarily serious books, they are agreeable books that statesmen use to rest."

The Commander in Chief takes pleasure in the company of creative individuals who usually reciprocate his feelings; he fascinates the intelligentsia.

Among Castro's visitors of this type—and there have been hundreds, ranging from great thinkers to a Brazilian soap-opera celebrity—were the French philosophers Jean-Paul Sartre and Simone de Beauvoir, the American historian Arthur M. Schlesinger, Jr., the British novelist Graham Greene, the British actor Alec Guinness, who along with Noel Coward had gone to Cuba a few weeks after the victory in 1959 to film *Our Man in Havana*, and the American actor Jack Lemmon.

Soviet poet Yevgeny Yevtushenko met Castro in Havana in the early days of the revolution (he had actually made a point of learning Spanish beforehand), and his political impressions of revolutionary Cuba were most vivid. In his autobiography Yevtushenko tells the story of two Cuban plotters, one of them a realist and the other an abstract artist, who argued furiously while waiting for orders to attack the Batista palace, then "went to fight for the future of their country, and both were killed." Yevtushenko added: "I very much wish this story were known to those dogmatists who write off all modern artists as lackeys of bourgeois ideology. One must be more patient and more discriminating." This was, of course, during the relative liberalization in the Soviet Union that accompanied the de-Stalinization process, and Yevtushenko found it useful to cite revolutionary Cuba as an example of new freedoms.

Elsewhere in the autobiography, he writes that it was the same evening in Moscow that he spoke publicly about Cuba and read for the first time his famous poem "Babi Yar" about the place where the Nazis had massacred thousands of Jews during the war. Then Yevtushenko reported intriguingly that a white-haired old man, leaning on a stick, came up to him afterward to say: "What you've said about Cuba and what you've written about Babi Yar are one and the same. Both are the Revolution. The Revolution we once made, and which was afterward so betrayed, yet which still lives and will live on. I spent fifteen years in one of Stalin's concentration camps, but I am happy that our cause, I mean the cause of the Bolsheviks, is still alive." Yevtushenko has neither returned nor written about Cuba since those early days.

Contacts with artists and intellectuals are among Castro's great pleasures and stimulations. But on another level, he has also expressed interest in meeting Henry Kissinger (who has his own curiosity about Fidel) as well as David Rockefeller, and in 1982 he greatly enjoyed a secret conference with Lieutenant General Vernon A. Walters, a former Central Intelligence Agency deputy director and an ambassador-at-large when President Reagan sent him to Havana.

Perhaps the greatest danger facing Fidel Castro after all these years in power is precisely the intellectual and political isolation of the environment in which he lives. In this sense, the deaths of Che Guevara and Celia Sánchez were terrible blows to Castro (for one thing, now nobody dares to contradict him). This is harshly demonstrated by the caliber of today's immediate entourage, essentially fawning and sycophantic, and the debatable quality of most of his top advisers in the government (no more than three or four are first-rate).

The personality cult around Castro not only does not abate, but it is continually enhanced. This is a very touchy subject with him, and Fidel angrily denies that such a cult even exists, stressing that the first decision made by the revolutionary government was to forbid the naming of localities of streets after living leaders or to erect monuments to them.

This, of course, is true in the strict sense inasmuch as only the names of dead heroes—Che Guevara, Camilo Cienfuegos, Celia Sánchez or others—are given to schools, hospitals, factories, and so on. In practice, however, Fidel Castro lives bathed in the absolute adulation orchestrated by the propaganda organs of the regime. There are very few huge posters or billboards with Castro's face adorning Cuba's streets and highways, but his portraits are in every government office (as portraits of democratic leaders appear in government offices in their countries) and in a great many homes—sometimes in shrinelike fashion.

The adulation is expressed through the use of all his titles—Commander in Chief, president of the Councils of State and Ministers, and first secretary of the Communist party—in every single printed or broadcast reference to him, sometimes every few paragraphs. Editorials and speeches invariably speak of his wisdom and genius as "the guide of Cuba," and the 1976 Cuban constitution proclaims the decision "to carry forward the triumphant Revolution . . . under the leadership of Fidel Castro." Quotations from Fidel are printed everywhere, including at the bottom of every page in the Havana telephone directory.

Every public act by Castro, no matter how routine, is printed on the first pages of newspapers and is the lead item on the evening television news. A shot of Fidel waving benignly from a balcony is part of the introduction to the news. Virtually every speech he delivers is published in full, sometimes as a special supplement. Books and articles about the history of the revolution are adoring in every mention of Fidel. The first volume of Antonio Nuñez Jiménez's planned history of the revolution is a hymn to Fidel Castro. He is usually called *Comandante-en-Jefe* in normal conversation, and nobody in his right mind, even in private, dares to criticize him. It is hard to believe that a man of Castro's immense intelligence does not realize that he *is* surrounded by such an incredible cult of personality. (The Castro cult is in interesting contrast with the Soviet Union's Mikhail Gorbachev, who has forcefully banned even the kind of veneration given Leonid Brezhnev, one of his recent predecessors. Gorbachev has ordered editors to use his name as little as possible, and has forbidden public praise.)

It has been remarked that Napoleon succeeded in ruling France because he kept intact his group of outspoken but loyal marshals. Death, naturally, took away many of Fidel's great old companions, but for political reasons he has rid himself of others—the nonpersons or nearly nonpersons of contemporary Cuba—who were valuable and honest revolutionaries.

What has developed therefore is government by courtiers, apparatchiks, bureaucrats, and yes men who tend to tell Castro what he wants to hear but which is not always the truth. A man who knows him well says that Fidel is a polemicist and needs to be contradicted, but cannot tolerate it. Constructive discussion is therefore often ruled out, and notwithstanding Castro's protestations about collective leadership, he makes command decisions all the time.

The progressive militarization of Cuban society which resumed in earnest in the early 1980s emphasizes even more Castro's role as Commander in Chief, and the slogans, painted on walls and repeated in countless public pronouncements every day in speeches and on radio and television, are "Commander in Chief! Give the Orders!" and "Commander in Chief! The Rearguard is Secured!" The slogan of "Nobody Surrenders Here!," taken from the defiant cry by Juan Almeida Bosque at the Alegría de Pío disaster, is being revived.

Not surprisingly, Cubans regard themselves as a besieged society, constantly under the threat of an armed attack by the United States. The Bay of Pigs, the U.S. military intervention in the Dominican Republic in 1965, the Grenada invasion in 1983, the CIA's "contra" operations against Nicaragua, and the steady flow of warnings by the Reagan administration that it will go to "the source" of all Central American affrays—i.e., Cuba—have given Castro sufficient reason to emphasize defense as the nation's first task. Inevitably, however, this phenomenon leads to a permanent situation of national emergency in which the *military* leader of the ten million Cubans must be supreme. And being supreme, he cannot tolerate being contradicted.

When contradicted, Castro may explode in anger or just pout. Even small things matter. Being as genuinely generous (another strong character trait) as he can be vindictive, he once sent lamb chops and legs of lamb from his suburban weekend home, where he also has a vegetable garden, to a couple of American friends spending some time in Havana. Visiting their house one afternoon, Fidel instructed the friends to bread the thin lamb slices and then fry them in oil, and grew quite annoyed when the wife gently suggested that lamb can also be good broiled. Saying brusquely, "So make them however you want," Fidel strode out of the kitchen.

Castro has immense access to all forms of data and information, but in the absence of useful interlocutors, he lacks the opportunity to discuss or analyze them seriously with anyone else. When he draws correct conclusions, it is because of his superb political instincts. Castro's favorite expression is "Let us analyze it," and he engages aloud in the analytical process on a given subject, sometimes for hours, in front of a visitor or his staff. If Fidel comes up with a strong opinion, even foreign visitors refrain from expressing disagreement, out of courtesy to the president of Cuba and because he is intimidating in dialogue.

Castro's hunger for information is gargantuan. He carefully reads Cuban newspapers and magazines daily, and he receives around-the-clock copy from U.S. and European news agencies tickers as well as from Cuba's own *Prensa Latina*, perusing it quickly, and setting aside what interests him. Dish antennas at the palace pick up U.S. radio and television news broadcasts from the United States day and night; and in 1985, for a half-million dollars annually, the government purchased the extensive computerized financial service of the Reuters news agency of Britain (U.S. restrictions on any commercial dealings with Cuba prevented buying a similar American service).

Castro also goes over the daily dispatches from Cuban embassies abroad and reports from what he calls with a wink, "our special services." From overseas he receives a steady stream of clippings, special publications, reports, and books. Some are translated or summarized for him, others he reads in full (Castro has a good command of written English but hesitates to speak it). In the summer of 1985, he read and virtually memorized a study of U.S. protectionist trade practices, prepared by the Japanese chamber of commerce. His interests are so wide and intense that almost any topic can fascinate and inspire him to learn more about it, especially if it deals with economic development, agriculture, public health or education.

(Castro amuses himself extrapolating startling conclusions from data available to him, mainly to make a dramatic debating point. Talking about Latin America's huge external debt on one occasion, he informed his audience that he had figured out "with pencil and paper" not only what citizens in a region owe per capita, but also how much of the debt corresponds to an acre of arable land. Such unusual statistics tend to impress his listeners, and Castro has also worked out how many pounds of sugar a Caribbean nation has to export to import a tractor from the "capitalists," emphasizing that the cost of the tractor in terms of sugar exports keeps growing astronomically. This is the sort of language that Latin Americans understand inasmuch as balance-of-payment statistics are abstractions to the hemisphere's millions of poor people, and it is very effective foreign policy for Cuba.)

When Castro is away from Havana, which is frequently, helicopters deliver to him twice a day batches of publications, news agency cables, diplomatic reports, and everything else required to keep him fully informed at all times. Even when he spends a few days at his fishing hideaway in the Caribbean, he religiously follows the routine of absorbing all these materials.

Conversing with a friend from Washington, Castro showed curiosity about the briefing practices at the White House, inquiring how much information President Reagan was being given and how often. He offered no comment upon being told that American presidents usually have only one daily, and relatively short, foreign-policy and intelligence briefing, but he seemed surprised by this.

Visitors from abroad (and chiefly from the United States), whom Castro sees in astounding numbers and at astounding lengths, often find themselves under third-degree interrogation. When a Texas oilman and self-made millionaire was brought by a Texas congressman, who is a friend of Castro's, to dine at Fidel's tiny fishing retreat on a key directly south of the Bay of Pigs, he was questioned in detail about his Horatio Alger-like history—and about offshore drilling. A wealthy rice broker from Arkansas, accompanying an Arkansas congressman, was debriefed about planting techniques, Cuba still being a rice importer and hoping to augment domestic production. Another American was quizzed about Reagan's tax policies (and told by Castro that sales taxes would be more acceptable to Americans than higher income taxes). A newswoman who had just visited Mexico was asked about Mexican poppy-plant eradication (Cuba forbids all drugs, but Castro admits that during the war he tolerated the planting of marijuana by some Sierra peasants because it was their only cash crop; now it is prohibited). A Texas pilot was asked to advise Fidel on the best private jet aircraft.

Castro's informality can startle Americans. When the Texas oilman (who has a beard) alighted on the fishing key from the Cuban air-force helicopter late at night, he almost collided with a tall bearded figure all in dark blue: sweatsuit, windbreaker, sailing cap, and sneakers. The man in blue said, "*Bienvenido* to Cayo Piedra," and when the congressman began to introduce the oilman, Fidel broke in, saying, "We've just met, we had a collision softened by our beards."

Fidel's curiosity about people, especially Americans, is so boundless that during a recent six-month period he met with a delegation of U.S. Roman Catholic bishops, a score of congressmen, the daughters (on separate occasions) of Robert F. Kennedy and Nelson Rockefeller, a half-dozen book publishers, two television network correspondents (and crews), interviewers from a leading U.S. newspaper and from a men's mass-circulation magazine (in the latter case for some thirty hours of taped conversation), a middle-level and pronouncedly hostile State Department official (who did not expect to be received by Castro during his business visit), a famous jazz musician, a number of businessmen, and several marine biologists.

In the same period he received the presidents of Algeria and Ecuador; the secretary-general of the United Nations; numerous cabinet ministers from all over the world; leaders of Latin American political, labor, press, and journalism organizations (these groups came to Havana to attend conferences on hemisphere economies, with Castro present at the daylong sessions); and Japanese and Mexican businessmen interested in trade with Cuba. At one point, Carlos Rafael Rodríguez said to a visitor, "You know, this is Fidel reborn, this is the crusading Fidel of the Nineteen-Sixties."

With his prodigious memory, Castro appears to absorb and remember everything he has read, heard, and seen in the past half-century. It is more than *memoria technica,* as the Romans called the memorizing technique of ancient orators, because he can play infinite variations on his themes before his audiences, never forgetting facts and figures. As a law-school student, Castro completed the last two years in one year, studying day and night, and when it came to memorizing he would destroy all the materials he had learned by heart so that he was forced to depend on his memory.

A voracious reader whose chief occupation during the nearly two years he spent in prison was reading books, Fidel has accumulated knowledge and erudition that are stunning. In relaxed conversation or improvised speeches, he ranges easily from references to obscure Roman laws on debt moratoria to Victor Hugo's critique of Louis Bonaparte, the "Little Napoleon," from tales of the Spanish conquest to quotations from Abraham Lincoln and José Martí, from a forgotten passage from Lenin to a line from Curzio Malaparte. In a written battlefield order in the Sierra, he once lapsed into Latin to urge a *manu militari* solution in a tight spot.

Presently, Castro maintains his furious pace although he talks about relaxing it. Asked by American interviewers why he had failed to attend the funeral of Soviet leader Konstantin Chernenko in Moscow early in 1985, Castro replied that on the day of Chernenko's death he had worked "for forty-two consecutive hours . . . no rest or sleep," and, in effect, was too exhausted to fly the great distance. But the word in Havana was that a petulant Castro did not wish to spend days in Moscow waiting his turn to be received by a new Kremlin leader for the third time in a year (he had attended the funerals of Leonid Brezhnev and Yuri Andropov, and he already knew Mikhail Gorbachev). However, Castro shone at the Soviet Communist Party Congress in February 1986—Gorbachev's first congress—at a time when Cuba's economic dependence on Moscow was again desperate. Then, restlessly, he went to North Korea, probably the world's toughest Communist dictatorship.

At home or abroad, Castro keeps incredible hours. The punishment is self-inflicted, but he is so compulsive that he thrives on unpredictable schedules. His only concession to orderly behavior is public punctuality, which he recently and surprisingly began to practice with his characteristic obsessiveness. Whereas in the past he would be hours late arriving at a meeting, rally or reception—or delivering a scheduled speech—he is now shockingly and exactly on time. At the Communist Party Congress in February 1986, Castro berated delegates for being a few minutes late for a 9:00 A.M. session, charging that if Communists are unable to get to a conference on time, they are probably incapable of running the country well. The delegates hung their heads in shame, like chastized children before the schoolmaster.

Castro seems to get by on very little sleep. Even under the best of circumstances he does not go to sleep before three or four o'clock in the morning, yet he looks fresh and rested at nine o'clock that same morning at an international conference or another public function. When in Havana, Fidel may sleep at the Eleventh Street apartment (at dawn or for a daytime nap), in the small bedroom behind his office on the third floor of the Palace of the Revolution, at his new and very private villa in the suburbs west of Havana (he experiments with enormous hydroponic tomatoes in the garden there), or at almost any location, such as a friend's home.

In the first year of the revolution, Castro used the twenty-third floor of the Habana Libre Hotel (formerly the Havana Hilton) as home and office in addition to the apartment, kept by Celia Sánchez on Eleventh Street, a spacious house overlooking the sea on a hill in the fishing village of Cojímar, five miles east of the city, which a rich prerevolution politician lent him, and a house next to the old Charlie Chaplin (now the Carlos Marx) theater in the Miramar residential section. This was why it was so hard for everybody, sometimes even for Celia Sánchez, to locate Fidel in those days. Then the established offices at the building of INRA, the National Agrarian Reform Institute. Conchita Fernández, his former secretary, recalls that he raced all the time from one place to another in Havana and beyond in his motorcade of Oldsmobiles (the black Mercedes-Benzes came later) with gun-toting *barbudo* guards, carrying with him in the car papers, reports, and notes in a briefcase. He usually worked while being driven around, reading or dictating memoranda or ideas to Conchita, who came along. "He never rested, not in the car, not anywhere else," she recalls.

Conchita also remembers days when Castro would arrive at his INRA office at eight o'clock in the morning on some occasions, "and he would leave only three or four days later . . . not stopping in the morning, not in the afternoon, not at night, not at dawn. . . . At best, he'd say, 'I'll rest on the couch for three or four hours, and you wake me up at such and such a time,' and ten minutes later he was back in my office to read the correspondence.''

Twenty-five years later, Castro's life is more orderly and more elegant, but he has not changed very much in behavior and attitudes. He has de luxe Soviet helicopters with inside wood paneling at his disposal as well as a fleet of Mercedes-Benz limousines (he usually travels in a three-limousine motorcade, two of them carrying security guards), but he still has a weakness for jeeplike vehicles. He himself enjoys driving the Soviet *Gazik* vehicles during countryside outings, and to be driven in them in Havana suburbs; the drives remind him of the Sierra, which he often describes as the "happiest time" of his life. Occasionally, he switches from *Gazik* to limousine or the other way around (all the cars move in an armed convoy). On these drives he does not seem overly concerned for his security; occasionally he even stops at red lights.

As head of state, Castro enjoys a degree of luxury and privilege, yet on a modest scale compared with rulers elsewhere, including Communist countries. The most spectacular display of luxury is the receptions he offers for visiting dignitaries on the ground floor of the palace, where as many as a thousand guests are invited to eat and drink in a huge area decorated with living green plants and rare ferns from the Sierra Maestra. As a rule, Castro strolls with the guest of honor, introducing him or her to his friends. The best meats, fish, and lobster are served along with aged Cuban rum (Isla de Tesoro) and occasionally Chivas Regal scotch whiskey is available. All of this is a great treat in a country still plagued with food shortages, particularly quality food. But these are not caviar-and-champagne feasts, they end early without drunkenness, and Cubans do not seem to begrudge their Maximum Leader this form of official entertainment (the press reports the holding of these receptions).

On such occasions, as well as at major conferences or National Assembly sessions, the fastidious Castro wears the brown formal uniform of Commander in Chief, with his single-star-on-a-black-and-red-diamond rank insignia and laurel leaves. The shirt is white and the tie is black. The rest of the time, he prefers his perfectly tailored whipcord olive-green campaign uniform. Though open-necked, the uniform is quite heavy, and Castro wears an undershirt under the front-zippered tunic. To feel comfortable in this attire, he has the palace air-conditioning system turned up to the point where his aides, in *guayaberas,* or sports shirts in the Cuban style, seem to be freezing.

Sometimes Castro chooses to wear olive-green fatigues, keeping his cap on over well-trimmed hair even when sitting in his own office or at a friend's house. He always wears black combat boots, and army orderlies on duty rush over to tuck the trousers inside the boots if they slip out, an instance of Castro's royal behavior with his staff. The orderlies also clean mud or dirt from the boots when he returns to the office.

Because of the Sierra Maestra legend and his dedication to olive-green uniforms, Fidel's image is still that of the *guerrillero,* careless about dress. But he had always been extremely conscious of his appearance: As a university student leader and budding politician running for elective office, Castro preferred dark suits and ties to the *guayaberas* worn by most Cuban men, and it was in a good suit that he returned from prison to Havana, left for exile in Mexico, and planned the voyage of the *Granma.* Photographs of the period show Castro looking quite tall and elegant, a kerchief in his jacket pocket, wearing a pencil-line moustache. He looked for all the world like the scion of a Cuban millionaire family (even when, after prison, he owned only one suit), knowing that it had more dramatic political effect than being one more youth in a casual shirt. A picture taken with his son, Fidelito, a few hours before leaving for the Moncada assault in 1953, shows him looking dapper and fashionable. The tailored campaign uniforms of today are consistent with Castro's prerevolutionary dress code.

Fidel is quite nearsighted, but his vanity does not keep him from wearing horn-rimmed glasses when he wants to see well. He wore them in the Sierra and at the Bay of Pigs, and one of the few billboard pictures of Fidel portrays him during the battle wearing a brown beret and his glasses. A superb marksman, Castro wears his glasses when shooting, having devised a system allowing him to aim with *both* eyes open, instead of just one as most marksmen do because he was more accurate using the glasses.

Fidel Castro is a curious combination of *hidalgo* courtliness and innate beautiful manners, especially toward women, and of outright rudeness and peremptory treatment of subordinates. He may spit on the floor and use astoundingly filthy language when only men are present. He swears easily, when playing chess or dominoes, being a master at both. He has a regal carriage, but he allows morsels of food to be caught in his beard when he eats, damaging the great image.

Castro is a perfectionist to the point of pedantry. He spends hours correcting and editing his speeches and other writings to achieve stylistic excellence in the Spanish language he so beautifully and elegantly deploys his pages are a labyrinth of inserts, arrows, interlines, and squiggles in tiny script. Sometimes he does his editing in a fast-moving car, which drives his secretaries to sheer despair, much as they are used to his handwriting. Gallego Fernández, the vice-president, recalls the occasion when Fidel began drafting in his limousine a letter to a foreign head of state, and had the driver go around in circles for four hours until he finished his meticulous missive, as he did not wish to have his concentration broken by stopping to go up to his office.

Fidel still improvises many of his speeches, but he admits that now he writes the ones he considers the most important—for example, before the United Nations in New York in 1979 and at the Nonaligned Movement's summit in New Delhi in 1983. Unquestionably his ad-libbed discourses are far better, but his sense of perfection must prevail in all he does. Having a drink one afternoon at a friend's house in Havana, Castro was visibly uncomfortable until he finally reached for the whiskey bottle to screw on the top properly; the host had left it a bit askew.

The rare times Castro really appears to relax is with a few companions or visitors at Cayo Piedra, a small volcanic key in the Caribbean, ten miles south of the Cuban coast, where he flies in his helicopter to engage in his favorite sport of underwater fishing. The key, once the site of a lighthouse, has a four-room rustic caretaker's house with a veranda and a pergola, and this is Fidel's real home. Visitors are put up in a modern guesthouse on the other side of the key (there is a collection of José Martí's works, but no Marx or Lenin, at the guesthouse), and all the meals are served aboard a barge tied to the wharf. Cayo Piedra has a pool where Castro swims every morning against the clock. A helicopter pad is an essential facility.

Most of the day is spent fishing off one of the two large powerboats (always escorted by two naval missile launches), with Fidel in his wet suit diving deep with his spear gun. A champion diver, he invited Jacques-Yves Cousteau, the famous explorer, to join him off Cayo Piedra during the Frenchman's stay in Cuba to study marine life. After his dives, Fidel is given eyedrops and nosedrops by his physician. At dinner fresh turtle soup (from turtles bred off the key) is served before the baked red snapper and lobster speared personally by Castro.

Meals with Castro usually follow the same pattern—at Cayo Piedra, at the seaside house on the Isle of Youth where he sometimes fishes, or at a military reservation in western Cuba where, wearing a camouflage suit (he calls it a "mercenary" uniform), he hunts wild ducks on a vast lagoon partly covered with forest and mangrove. First, there is a long cocktail hour with light banter—usually about the day's fishing or hunting—and occasional serious conversation, then dinner in a very small group, lasting sometimes until long after midnight. That is one of Fidel's favorite settings for a good discussion, at which he most brilliantly explains revolutionary Cuba to foreign visitors, dazzling even right-wing Republican congressmen from the mainland. He tends to grant interviews in his office, a relatively sparse L-shaped room with bookcases behind his desk, and he usually sits

on a sofa under a modern Cuban mural while conversing. Camilo Cienfuegos looks down from a portrait on another wall. Fidel's reasonably orderly desk includes a large transistor radio (telephones are on an adjoining
small table), cassette tapes, piles of documents, a jar containing his favorite
hard candy, and until late 1985, boxes with long and short cigars. He
increasingly favored the short ones until he suddenly decided to quit smoking sometime in October of 1985. Castro announced this event in a pre-
Christmas interview on Brazilian television, and evidently such remains the
fascination with him that the story made TV newscasts from the United
States to Japan, and was printed prominently in newspapers and magazines
worldwide. In his announcement he said that "I reached the conclusion
long ago that the one last sacrifice I must make for public health is to stop
smoking; I haven't really missed it that much." Given his proven
willpower—and given his dedication to physical fitness—chances are that
Castro, who began smoking at the age of fifteen when he was in high
school, will stick to his decision. He explained that "if someone had forced
me to quit, I would have suffered . . . but since I forced myself to halt
smoking, without making any solemn promise, it worked."

Years before, Castro had launched a vast antismoking campaign to persuade Cubans that tobacco, which made the island so famous, was a danger
to their health. Television, radio, billboards, magazines, and newspapers
were mobilized for the campaign (for example, pregnant women were
shown on televised sketches, indicating how smoking can affect the fetus),
and the price of cigarettes was raised to nearly two dollars per pack. Anyone returning to Cuba after a long absence is immediately struck by the
extent to which Cubans ceased to be a nation of chain-smokers. Castro's
own "last sacrifice" was the ultimate weapon in the campaign, adding the
virtue that he now practices what he preaches.

If Fidel did away with his personal trademark of the cigar, he made it
absolutely clear that the other symbol—the beard—stays on. He explained, as he had done in the past, that he and his companions grew their
beards in the Sierra simply because shaving was too much trouble. Subsequently the *barbudo* cult grew, and Castro acknowledged that "the beard
became a symbol of the guerrilla." But typically he also noted that beards
"have a practical advantage" because "if you calculate fifteen minutes a day
to shave, that is five thousand minutes a year spent shaving," and the time
can be spent better reading or exercising. While the notion of a beardless
Fidel is naturally politically absurd thirty years later, he has strongly discouraged other Cubans, even in the highest ranks, from wearing guerrilla
beards. It is really his personal mark of distinction.

Occasionally, Castro will talk about his foreign travels and the people he
has met on every continent. The United States is the country he knows

best. He went there three times as a young man: on his honeymoon in 1948, the second time a year later to escape political gangsters' death threats in Havana, and in 1955, to collect funds for the revolution he was planning to launch in Cuba from Mexico. He went back in April 1959, as the chief of the victorious rebellion (he met Vice-President Richard Nixon, and charmed guests at the Cuban embassy in Washington, asking them to "help me to help my country"), and twice more to give angry speeches before the United Nations.

Castro toured Panama, Colombia, and Venezuela as a student leader, and most of South America as the Maximum Leader immediately after the revolution. In 1971 he visited his friend, Chile's elected Marxist president, Salvador Allende Gossens, advising him not to antagonize the United States. In the 1980s he made trips to revolutionary Nicaragua, imposing advice on the *Sandinistas* he helped win and now supports.

He visited the Soviet Union nearly a dozen times (not all his trips are necessarily publicized), Eastern Europe twice, and Vietnam once. He stayed away from China mainly because Cuba was firmly on the Soviet side in the Sino-Soviet feud, and he acquired deep contempt for Mao Zedong, whom he never met (he criticizes Mao for having allowed himself "to become a God," but of the Soviet Union he says only that "during Stalin's time, a personality cult developed and abuses of power did take place").

Fidel, the best-traveled Communist chief of state in the world, has been to Africa several times, mainly to Algeria, whose independence the Cubans championed after their revolution, and to Angola and Ethiopia, where Castro dispatched combat troops in the mid-1970s—and where they remain a decade later. He has been to India, but the only time he touched the ground in Western Europe was during a one-hour stopover at the Madrid airport.

With all his responsibilities and obligations, Castro still tries to be a free soul, to act on the spur of the moment, to do the unexpected. His decision to attend the inauguration of Nicaraguan President Daniel Ortega Saavedra, his revolutionary protégé, in January 1985 was taken at the last moment when he discovered that no other head of government would be present (he took along Gabo García Márquez, telling him aboard the plane that they were flying to Santiago in Oriente). In Managua Fidel inevitably upstaged the short and uncharismatic Ortega, but he meant well. When a Cuban youth delegation sailed to the Soviet Union in 1985, Castro brought the entire 399-member National Assembly to the Havana dockside (the assembly was having its biannual session that week) to see the youngsters off.

The Commander in Chief as a rule no longer attends diplomatic receptions (except at the Soviet embassy, now a huge modernistic compound on the sea in uptown Havana), but he may suddenly turn up at dinner at an ambassador's residence, invited or self-invited. The first embassy Fidel ever visited was that of Brazil in 1959, and he was still carrying his rifle (he checked it at the door) wherever he went.

Appearing unexpectedly one night at the French embassy and staying until four o'clock in the morning in conversation with visiting parliamentarians, Fidel sent over the next day a case of Cuban scotch-type whiskey, which is not the greatest contribution to the pleasures of drinking. However, Castro also plans to produce a Camembert-type cheese and *pâté de fois gras* in Cuba, and the enthusiastic chef has already become a theoretical expert on force-feeding geese. He is the dilettante extraordinare in all esoteric pursuits.

A revolutionary and a *guerrillero,* Castro is far from an ascetic as his liking for Chivas Regal and his pâté interests indicate. He had always been partial to good food and cooking; and in May 1958, as the great Batista offensive was opening, he dispatched a desperate note to Celia Sánchez in the Sierra headquarters, reporting that "I have no tobacco, I have no wine, I have nothing. A bottle of rosé wine, sweet and Spanish, was left in Bismarck's house, in the refrigerator. Where is it?" In Mexican exile he and a friend decided to splurge on caviar on a day the churchmouse-poor revolutionaries received an unexpected donation from abroad.

Cuisine in many forms has preoccupied Fidel since youth, and, curiously, spaghetti was always among his favorite dishes. Manuel Moreno Frajinals, Cuba's leading historian who befriended the young Fidel in the late 1940s while he was still at the university, remembers his frequent visits at the family's apartment to discuss politics and grab a meal. On one occasion, Castro arrived just as the maid was frying plantains. Smelling the aroma, he rushed to the kitchen, telling the girl, "Let me show you how to fry them properly." When Moreno Frajinals's wife, who is an architect, asked, "Do you think you know everything?" Fidel replied, "Almost everything."

In prison on the Isle of Pines (called the Isle of Youth after the revolution), Fidel continually attempted to cook spaghetti on a small electric plate in his solitary cell; guiding visitors through the prison, which he does occasionally, he never fails to recount how many hours it took for the spaghetti to be ready. His sister Emma says he went on cooking spaghetti in the Sierra for his fellow fighters. Conchita Fernández tells that Castro often dined in the kitchen of the Habana Libre Hotel (where he also granted all-night interviews), and that he frequently tried to show the cooks how to prepare the red snapper properly. As mentioned earlier, he has definite views on preparing lamb chops. He believes that the *confit* of duck should be made in a *bain-marie,* and he is partial to grilled fish and lean meat. At a casual dinner in the countryside with close friends, he will settle for fish, chicken, and salad, often with a Bulgarian white wine and an Algerian red.

Castro also holds powerful opinions on intellectual aspects of sports: Having played both basketball and baseball of almost professional quality, he once provided a visitor with a learned explanation of why basketball is the thinking man's game (There is no telling where a conversation with Fidel may lead). His theory is that whereas basketball requires strategic and tactical planning as well as speed and agility—thus preparing a man for guerrilla war—baseball poses no such needs (the subject came up when Castro forcefully denied a rumor then circulating abroad that he had once hoped to play for the majors in mainland baseball).

It is probably inevitable that comparisons are made between spectacular figures of our age, notably between great *guerrilleros,* and a comparison between Castro and Yugoslavia's late Marshal Josip Broz, Tito, instantly comes to mind. Both were guerrilla chiefs, both fought forms of Fascism, both seemingly became Marxists in the name of social justice, and both had world-arena aspirations. Yet, there are more differences than similarities between them. Tito was an avowed Communist, under Comintern discipline, long before the German invasion and the creation of his partisans' guerrilla army—whereas Fidel (whatever his real inner thoughts might have been when he launched *his* insurrections) certainly did not start out as a loyal member of the Moscow-oriented Communist movement. And though neither man has made fresh contributions to Marxist thought, Castro has much more intellectual depth than Tito had—the point is made in Fidel's writings and speeches before and after the revolution. It is idle to compare Tito's courage in the face of the Nazis and then Stalin with Castro's courage in the face of Batista and the United States; the circumstances are extremely different. The two leaders dealt with distinct and separate problems in modernizing their nations. Tito succumbed to the temptation of great vanity, making himself a marshal in a white gold-braided uniform, sailing in his private yacht, and taking refuge in the luxury of Brioni Island. Castro, with his own bent of indulgence, has basically remained a guerrilla leader at sixty years of age.

To be sure, the two men hated and despised each other, precisely because of the surface similarities and the contrasts; there was a long rivalry between them. They fought bitterly in Tito's final years of life for the control of the Nonaligned Movement (Castro having succeeded Tito as chairman in 1979), and the old marshal battled the Cuban's attempts at the Havana summit conference to bring the Nonaligned Movement too close to the Soviet Union.

Here, perhaps, Tito had tried to teach Castro a useful lesson: Once the Kremlin's devoted wartime ally, the Yugoslav broke with the Russians over their determination to dictate his country's future, preferring to become a "neutral" Communist state. Castro, despite many painful experiences with Moscow, went on faithfully espousing in public every Soviet foreign-policy position; Tito seemed to be warning the younger *guerrillero* not to mortgage away his independence forever.

The complexities of Fidel Castro's personality are immense, and therefore no future change of course by him can ever be ruled out if he becomes convinced that it is in the interest of Cuba, the revolution, and his own. Naturally, these three sets of interests blend and overlap as long as Castro dominates the Cuban scene. And one is tempted to think that his personal destiny was to achieve this domination.

II

THE YOUNG YEARS

(1926–1952)

CHAPTER

1

Slightly more than thirty years after José Martí, the "Apostle" of Cuban independence, was killed in combat with Spanish troops in 1895, Fidel Alejandro Castro Ruz was born at a *finca* in the province of Oriente, barely twenty-five miles from the Dos Ríos battlefield. In terms of history, three decades are a short segment—it already roughly equals the span of Castro's own revolution—and in this sense, his life has been intertwined from the outset with the struggles and symbolisms of Cuba's past.

As the island's greatest thinker and patriotic hero, Martí was always Castro's role model, and in landing with his rebels on the shores of Cuba to dislodge tyranny, Fidel was fulfilling the Martían code. Martí, the leader of the Cuban Revolutionary Party, had launched the final war of independence with a proclamation from his New York headquarters on January 29, 1895, and came ashore in a rowboat on the Oriente coast two months later to join the guerrillas fighting the Spaniards. The Apostle was killed on May 19, astride a white horse, weeks after coming home to Cuba from exile. He was only forty-two, a slim, sad-faced man with a bushy, pointed moustache and a half goatee under his lower lip, and almost constantly in poor health.

It was therefore only natural for Fidel Castro to seek the most complete personal identification with Martí's martyrdom, and, quite predictably, soon after the victorious revolution, he made a pilgrimage to Playitas, the beach where the historic landing occurred on April 2, 1895. The pilgrimage produced an hour-long color documentary, exhibited in movie theaters and on television, showing Castro in battledress standing dramatically alone on the small horseshoe-shaped stretch of white sand, narrating the tale of Martí's sacrifice. Then, the cameras followed him to a nearby shack where Fidel interrogated the only surviving witness, a spry nonagenarian, about all he could remember of that stirring event.

The parallels between 1895 and 1956 are many, and they have their provenance in the many emotional, intellectual, and political traits shared by Castro and Martí . . . Martí after countless attempts to overthrow Spanish rule, concluded that a revolution in Cuba could only succeed from an expanding guerrilla warfare; Castro came to the same conclusion after the failure of his assault on the Moncada army barracks in 1953. Martí likewise understood the immense personal risks involved in leading a revolution, writing a friend on the eve of his death that "every day I am in danger of giving my life for my country and for my duty"; Castro would pledge before embarking on his invasion that "we shall be free or martyrs." Both men had quite early formative political experiences. Martí was imprisoned by the Spaniards at the age of seventeen for opposing colonial rule, sentenced to hard labor, and exiled. Castro's own political rebellion took shape at Havana University before he was twenty-one. Fundamental principles motivated both men, and Fidel would remark many years later that "the sense of personal honor is held by almost every Spaniard." This was the heritage of the rebel sons of Spain.

The tragic poet was convinced that even if he died, the liberating revolution would triumph (some biographers insist that Martí deliberately sought death to create an aura of martyrdom in the war of independence), and events proved him right—up to a point. Castro had demonstrated on numerous occasions—notably at Moncada and Alegría de Pío—that he was ready to die for his cause, but his bravura stemmed principally from his temperament as well as his belief that revolutionary chiefs must lead their men. Sons of Spaniards, both Martí and Castro represent a very special strain of Iberian mysticism and romanticism combined with a powerful dose of New World nationalism. Moreover, the real Cuban issue to both Marti and Castro was a revolution in depth, a social revolution and not just changes in the political status quo. Familiar with his own country and with much of Latin America's poverty, Martí advocated the grant of land to its tillers (but without taking it away from others) and a better distribution of national wealth: "The nation where there are a few rich men is not rich: it [is] rich where everyone has a little of the wealth."

Martí wrote that the government has the duty of providing needed education to the people because "to read is to walk." One of Castro's first major revolutionary undertakings after 1959 would be an islandwide crash literacy campaign. The Cuban philosopher opposed, however, Marxist radical social transformations. His biographer, M. Isidoro Méndez, a prerevolutionary historian, found Martí to be a "social republican," believing in "prudent socialism" without extremism.

Appearing decades later on the Cuban political scene and in a totally changed world, Castro's extremism in social change, his rejection of direct-vote elections as the keystone of political life that Martí had urged in the nineteenth century seem a stark and telling indication of the difference in their political platforms for revolution. This ideological difference surely stems from their distinct temperamental and psychological makeups. Martí, the classical democrat, essayist in three languages, and lyrical patriotic poet, believed in civilian government with the consent of the governed. Castro is the quintessential Spanish military *caudillo,* wrapped today in a Marxist-Leninist mantle of convenience, offering the intellectual rationalization that "real" revolution is impossible under an elective system, and that Communist authoritarianism is the necessary instrument for its implementation. He is thereby proving the sad theorem that "without power, ideals cannot be realized; with power, they seldom survive."

In a speech in the mid-1960s, when his regime was caught up in complex internal ideological struggles, Castro went a historically curious step further to identify his revolution with the Cuban "fathers of independence." He proclaimed that Carlos Manuel de Céspedes, a wealthy landowner who led the first patriotic insurrection in 1868, and José Martí were not Marxist-Leninists simply "because in the epoch in which they lived and in the historical conditions in which their magnificent struggles developed, they could not be." He then added in a phrase instantly made into an official slogan: "Then, we would have been like them—today, they would have been like us!" Castro understood that without an independent and Martían foundation, a Marxist system would be unacceptable to Cubans.

While Martí and Castro would disagree on the means of achieving their ends, Castro is truly Martí's direct philosophical and political heir in his views on radicalism, agrarian reform, racial equality, and social justice. Castro and Martí also share fears and suspicions of the United States and its intentions toward Cuba. North American aspirations to annex or even buy Cuba (as Louisiana was purchased) go back to the early days of the last century. The American consul in Havana wrote in 1833 that "in the fullness of time, when Cuba and Spain and we should all be of one mind—without discussion, or revolution, or war—Cuba would doubtless be added to the Union."

Martí, having spent long years in the United States in forced exile from

Spanish Cuba, had nightmares over an American grab of his island from Spain as the outcome of the independence war then under way. In the letter he penned the day before he died, Martí said that he was battling the Spaniards for Cuban independence in order to impede "the extension of the United States through the Antilles." He wrote that in the United States he had lived "inside the entrails of the monster," and he was concerned with American "economic imperialism," remarking that "the disdain of a formidable neighbor who does not really know us is the worst danger to our America."

Fidel Castro had, of course, absorbed all these sentiments long before his own experiences and confrontations with the "formidable neighbor."

José Martí's days coincided with the time of America's Manifest Destiny. Even twenty years before the independence war, the United States economic presence in Cuba was already weighty. Cuban trade with the United States was six times larger than with Spain and, as Martí saw it, there seemed an economic and geographic inevitability about the island coming under complete American political domination.

José Antonio Saco, a philosopher and one of the first great Cuban patriots, warned in 1847 that Cuba "is so important that her possession is well worth a war . . . her possession would give the United States power so immense that England and France not only would see the existence of their colonies in America threatened, but would also experience the weakening of the powerful influence they exercise in other parts of the world."

No annexation of Cuba ever occurred formally, yet the fears of the nineteenth-century patriots were fully realized when the United States declared war on Spain in 1898 after the battleship *Maine* blew up in the Havana harbor from unknown causes. This conveniently provided the casus belli for an open conflict with Madrid—for which Americans were spoiling anyway. As Theodore Roosevelt and his Rangers charged up San Juan Hill in Santiago and other American forces were landing elsewhere along the Cuban coasts, the exhausted Spaniards were quickly defeated (they had been fighting the merciless Cuban guerrillas for three years). Later that same year, the United States and Spain signed the Paris peace treaty, transferring *control* over Cuba to Washington. At the same time, the United States also acquired Puerto Rico, the Philippines, and Guam.

The Cubans had begun struggling for their independence in 1868, when Carlos Manuel de Céspedes, the Oriente landowner with a doctorate of law from the University of Madrid, sounded his *Grito de Yara* (the call to freedom issued from the town of Yara his rebels had seized) to proclaim the revolutionary war against Spain's "decrepit and worm-eaten" power. A revolutionary junta, then a provisional government were formed, a constitution was drafted, and a congress established with veto powers over presidential decisions. But the rebels, led by President Céspedes, the congressional leader Ignacio Agramonte, and the military chiefs General Máximo Gómez and General Antonio Maceo, could not dislodge the Spaniards from the island. As much as anything else, the Cubans were too deeply divided among themselves to plan and conduct a successful liberation war: Céspedes fired the Dominican-born General Gómez, then was deposed in turn, finally being killed by Spanish troops in 1874. Gómez was brought back, then removed again. In the end the first independence war was terminated ten years after it began when the two sides signed the Peace of Zanjón. Maceo, the black general, kept fighting for another year. Shortly thereafter, Maceo and General Calixto García launched what became known as "The Little War," lasting through 1880, again with a Cuban defeat. The final independence war came with Martí and General Gómez in 1895—and the American intervention and Spain's expulsion from the New World in 1898. For Fidel Castro the history student, the divisions caused by the independence wars emphasized the need for revolutionary unity concentrated in the hands of one leader, and for freedom from Yankee control.

When Cuba came under outright United States military occupation, the island had been ravaged by thirty years of continuing warfare. The independence wars seemed in vain as the Cuban economy was made wholly dependent on the northern neighbor, and even the educational system was thoroughly Americanized in flagrant ignorance of the local culture and language. As with the inhabitants of Puerto Rico, the idea was to prepare Cubans to become good Americans someday. Inevitably, the four-year occupation established the foundations for turning Cuba into at least a de facto United States protectorate in the Caribbean for the next sixty years.

This protectorate status was engineered through the forcible insertion of the so-called Platt Amendment in the Cuban constitution, drafted with American blessings as a prelude to the grant of independence, as well as through the enforcement of trade and investment arrangements allowing United States interests a completely free run of Cuba. The amendment, devised by Senator Orville H. Platt as part of a United States Army appropriations bill, authorized the president to "leave the government" of Cuba to its own people, with the proviso that the Cuban constitution recognizes that "the United States may exercise the right to intervene for the preservation of Cuban independence, the maintenance of a government for the protection of life, property, and individual liberty . . ." The Cubans, given the choice of accepting the Platt Amendment in their new constitution or possibly remaining forever under military occupation, engaged in fervent debate before capitulating.

On May 20, 1902, Cuba was proclaimed an independent republic, the last colony in the Americas to achieve such status. Even this independence was farce and fiction; Leonard Wood, the last governor-general, wrote President William McKinley that "there is, of course, little or no independence left in Cuba under the Platt Amendment." A year later, Senator Chauncey Depew declared that "the day is not far distant when Cuba, resembling the United States in its constitution, laws and liberties . . . will have from five to six million people who are educated upon American lines and worthy of all the rights of American citizenship. Then, with the initiative from Cuba, we can welcome another star to our flag."

No such initiative ever came; instead, continuous internal political unrest resulted in a second United States military occupation lasting from 1906 to 1909, the landing of U.S. Marines to protect American interests and citizens in 1910, and another landing in 1917 to persuade Cuba to enter World War I, given the island's strategic importance astride the sea-lanes. To protect American properties in Oriente from labor unrest and sabotage, the marines remained in Cuba until 1923—three years before the birth of Fidel Castro. Throughout this period it was apparent that the American government held Cubans in deep disdain. Small wonder then that the young republic grew up with a paralyzing inferiority complex, and an anti-American sentiment to which Castro and his generation became heirs. The Mexican novelist Carlos Fuentes would sum up the American attitudes toward Latin America a half-century later, writing that the destiny of the U.S. was "to be strong with the weak."

Though the Platt Amendment was removed by Franklin D. Roosevelt in 1934, under the new Good Neighbor policy for Latin America, the United States' political and economic stranglehold over Cuba would not vanish until the great Castro revolution. The ouster of the Batista dictatorship in 1959 would allow Fidel and his *barbudos* the true achievement of Cuban independence, independence for which Jose Martí had died sixty-four years earlier, and which was refused Cuba by the Americans in 1898 and 1902. This historical dimension of the revolution would elude the United States, however, along with the comprehension that Cubans had lived all these years in the shame of being, as Castro called it, a "pseudo-republic."

A large sugarcane and cattle estate, the Manacas *finca* is located in the municipality of Birán in the Mayarí region of the northern Oriente coast. It is about twenty-five miles south of the Bay of Nipe and roughly the same distance east of dos Ríos, the spot where José Martí was killed in a Spanish ambush in 1895.

It was at Manacas that Fidel Alejandro Castro Ruz was born on August 13, 1926, and by the time he was ready to attend the local elementary school, he was already imbued with the Martí legend. This was part of every Cuban childhood, especially in the proud and ever-rebellious Oriente province. The revolutionary tradition touching the area of Fidel's birth would again be emphasized when his brother Raúl Castro, leading his Rebel Army column from the Sierra Maestra to the Sierra Cristal in the northeast to establish the guerrilla war's Second Front, marched past the Birán house in April 1958. Raúl would make a point of mentioning this fact in his lengthy report to Fidel on the progress of his operations in the Mayarí region. The youngest brother was probably the most family-minded member of the vast Castro clan.

The head of this clan was Ángel Castro y Argiz, an émigré to Cuba from his native village of Áncara, near the town of Lugo in the northwestern Galicia region of Spain. He was a destitute thirteen-year-old orphan when he arrived in Cuba. Born around 1874, as a child he had lived with an uncle in the Galicia *pueblo* in Spain's poorest and bleakest corner. Seven or eight years before the last Cuban war of independence, Ángel, increasingly maltreated at home, left Spain to join another uncle, one who had settled on the faraway Caribbean island. Castro has claimed on at least one occasion that his father had been sent to Cuba to fight as a Spanish soldier when the independence war erupted in 1895, that he was repatriated after the war but, having liked the island, came back as a penniless emigrant in the first years of this century. This account is vague, and probably inaccurate; at least two of Fidel's sisters admit they have never heard about Ángel Castro's military experiences.

Significantly, Fidel Castro seems and professes to know astonishingly little about his father's background, and this must be a conscious or subconscious expression of his negative attitude toward Don Ángel, for reasons that may be deeply personal, political, or both. In a 1985 interview with the Brazilian Dominican friar Frei Betto, Castro admitted that "I do not know much what were [his father's] first years, because when I had the opportunity to ask all this, I didn't feel the curiosity that I may feel today to know what were his first steps . . ." Considering that Fidel was thirty years old when his father died, he would surely have had ample occasion to ask questions—had he cared. Elsewhere in that interview, Castro volunteers the comment that although his father had the political ideas of a "landowner," he was a "most noble" man because he never turned down an appeal for help. This, Fidel said, "is very interesting." Compared with his frequent warm and personal references to his mother, the comment about his father seems forced and supercilious.

Ángel Castro's uncle lived in the town of Santa Clara in central Cuba where he had a brickmaster's business. Ángel was put to work there (naturally there was no time or opportunity for him to go to school), but after some five years he evidently tired of the uncle's bricks and struck out on his own. He moved east, probably walking most of the way, and for reasons no longer remembered in the family, he chose the Mayarí zone in Oriente to try his luck. He must have arrived in the area just as the end of Spanish rule was approaching and the American era was about to begin. He never talked much about his youth; he died in 1956 at the age of eighty-two, his past shrouded in oblivion.

Mayarí is a fertile region, ideal for planting sugarcane and tobacco and for raising cattle. Its principal town, also named Mayarí, lies on the river of the same name, and the beaches and fishermen of the Bay of Nipe are only four miles away. Sometime after young Ángel Castro reached Mayarí, the once sleepy town of wooden houses dating back to the early nineteenth century, was transformed into a commercial center of activity fueled by American capital.

At the end of the overlapping Cuban independence war and the Spanish-American War, the devastated country had been thrown wide open to United States investments whose safety from troublemaking Cubans was guaranteed by the military occupation. Throughout Cuba, these investments more than tripled in eight years, from $50 million in 1898 to $160 million in 1906, chiefly in land. And lush and rich Oriente was the preferred province. In 1899, for example, the Cuban-American Sugar Company bought seventy thousand acres in Chaparra on the north coast, and a year later the property accounted for 10 percent of Cuba's sugar harvest. At the same time, the United Fruit Company and its subsidiary, the Nipe Bay Company, purchased 240,000 acres in the Mayarí area—a veritable private fief carved out of Oriente.

Speaking with enormous indignation about the economic consequences of the Cuban "independence," Fidel Castro noted in a bitter anniversary speech in 1968, one century after the first insurrection against Spain, that "someone named Preston bought in 1901, 75,000 hectares [185,250 acres] of land in the Bay of Nipe zone for 400,000 dollars, that is, for less than six dollars per hectare of this land." He added that "the forests that covered all these hectares with precious woods and that were burned in the furnaces of the sugar mills, were worth many times, incomparably many times, this sum of money . . . they came with bulging pockets to a nation impoverished by thirty years of war to buy the best land of this country for less than six dollars the hectare." Until the revolution nationalized them in 1959, the United Fruit Company's Preston and Boston properties in Mayarí remained Cuba's main foreign-owned sugar mills and estates.

By the time Castro was born in 1926, American investments in Cuba exceeded $1.6 billion. Today, this total investment would be equivalent to $3 billion. With the collapse of world sugar prices in 1920 (following the "Dance of the Millions" of the previous years when prices were ten times higher, and fantastic fortunes were made in Cuba), United States interests could and did pick up the pieces cheap. Foreign banks controlled 80 percent of the sugar production; American companies gained monopolies in all the Cuban railroads, electric power supplies, and telephones; and Cuban deposits in United States-owned banks on the island soared from 20 percent in 1920 to 69 percent in 1921, as most Cuban banks disappeared because they could not compete with the political power and resources of Yankee bankers.

Cuba's president in 1926 was Gerardo Machado y Morales, an American-supported friend of big business (Washington forced him out years later, however, when he turned into a despotic dictator and the country's economic stability was threatened by the rising rebellion against him). He was as corrupt as his predecessors who ran the "pseudo-republic" in cahoots with the "better classes" of Cuban society.

But among the new Cuban generation, a new anti-American nationalism was beginning to develop. Not only were the Cubans saddled with the Platt Amendment and the United States economic hegemony, but they could also watch American military interventions by the U.S. Marines in Mexico and Nicaragua. The Cuban Communist party was created clandestinely in Havana just a year before Fidel Castro was born. Cuba had begun to stir.

The Mayarí region where Fidel grew up featured probably greater American presence and control than any other place in Cuba. The United Fruit Company, a Boston-based corporate giant with operations throughout Latin America, maintained special housing in Mayarí for its American (and a few Cuban) employees, hospitals, schools (for the children of the sugar-producing elite), stores stocked with American foodstuffs, a post office, and, later, swimming pools and a polo club. In addition to the Rural Guard, a United States-trained Cuban gendarmerie, the company was protected by its own armed police force that assured order and kept out undesirable Cubans.

The United Fruit Company also exercised great political power in Cuba, even more than other American companies and banks. The first sizable acquisition of sugar land in Cuba was made by an American investor from Boston named E. F. Atkins in 1882 (although the first American-owned sugar mill dates back to 1818). Atkins was followed by the Boston Fruit Company, originally concentrating on bananas, which changed its name to United Fruit Company in 1898 when it began its massive sugar-plantation purchases. United Fruit and the other companies had key Cuban politicians in their pockets (or on their payrolls): The Cuban-American Sugar Company, founded by a Texas congressman at the turn of the century, was represented on the island by Mario G. Menocal, who served as the U.S.-blessed president of Cuba from 1912 to 1920. The United Fruit Company was saved from the nationalization of one half of its holdings by the revolutionary government of President Ramón Grau San Martín in 1934 when Agriculture Minister Carlos Hevia, at odds with the rest of the ministers, talked Grau out of doing it. Hevia, who later served a brief presidential term, conducted the defense of United Fruit as a negotiator between his own government and the United States embassy in Havana. Until the revolution, United Fruit was untouchable: Succeeding governments protected it from labor unrest, excessive taxation, or any interference with its privileges. The company also had immense fruit-plantation holdings, mainly banana, in most Central American countries where it likewise was the dominant political force. In 1954, United Fruit worked hand in hand with the CIA to overthrow the leftist regime of Colonel Jacobo Arbenz Guzmán in Guatemala, after the company's economic domination was challenged by the local government.

In Cuba thousands upon thousands of canecutters and mill workers lived with their families in miserable *bohíos* (shacks) on the estates during the four months of the annual *zafra* (harvest), usually earning less than a dollar a day (sometimes only forty or fifty cents, without food). In the year's remaining months—the sinister "dead time" in Cuba—there was simply no work, and the *guajiro* (peasant) families tried to survive the best they could. This was the social environment that Fidel Castro remembers from his childhood, and it awakened him politically as he matured.

When Ángel Castro first came to Mayarí, there were occasional jobs available on the new railway the United Fruit Company had built between its mills and the port of Antilla on the Bay of Nipe, and he was briefly employed as a laborer laying down track as one of the many menial tasks he performed to stay alive. He was probably around twenty-five years old when he decided to start his own business as an itinerant peddler among canecutters and woodsmen up and down Mayarí.

It was beautiful country, the classical Oriente landscape with clusters of tall palm trees rising proudly amid green canefields and meadows, then deep woods extending far beyond in the direction of the sierras in the south. Rivers faithfully irrigated the fields. As the war ended and foreign capital poured in, chimneys of the new sugar mills being erected all over Oriente began to punctuate the skyline. More and more, there were cattle grazing in the pastures, with *guajiros* in big straw hats, mounted on their tough little horses, guarding the herds.

Selling lemonade he prepared every morning and transported in small barrels and tankards was Ángel Castro's first mercantile enterprise. With a donkey cart, he toured the fields and the woods of Mayarí, serving the lemonade to the thirsty men. With his first tiny profits, he started buying wholesale a variety of merchandise, peddling it from *finca* to *finca* in the ranch countryside. Fidel says he remembers hearing that his father then organized a group of local workers, whom he paid to cut trees for new sugar-planting fields and for burning wood in the furnaces of the big mills. He apparently had a work contract with an American sugar company. As Spanish immigrants, and especially the *gallegos,* the poorest and the most determined of all, have always done overseas, Ángel Castro worked incessantly to earn and save as much as possible. Somewhere along the line, he learned how to read and write.

Probably around 1910, when he was thirty-five or so, Ángel began leasing land from the United Fruit Company in the Birán area, thirty-six kilometers southwest of the town of Mayarí, putting the proceeds from his sugar sales into the acquisition of parcels of land. Thus he became a *colono,* planting sugarcane for sale to the company's mills, a practice the corporation encouraged because it tied the small farmers closer to it. He employed his first farmhands, and gradually became Don Ángel Castro, an increasingly affluent landowner in Oriente.

About that time, Ángel Castro married his first wife, María Argota. She is thought to have been an elementary-school teacher in the Mayarí area, though virtually nothing else is known about her. They had two children, · Pedro Emilio and Lidia, the latter born in 1915; in 1985 both of them lived in Havana, rather aged and rarely seeing their famous half-brother but in quiet comfort assured by him. Lidia, who eloped as a very young woman to marry an army officer and became widowed within a few years, devoted the rest of her life to Fidel, helping him immeasurably during his clandestine periods, the imprisonment, and the Sierra war. Pedro Emilio was a minor politician prior to the 1952 Batista coup, then reverted to his real interest—Greek and Latin studies.

There is something of a mystery about the first Señora de Castro, and about the circumstances of Ángel Castro's second marriage. All the published accounts about the Castro family are extremely sketchy (Fidel likes to keep it that way), but they coincide in affirming that María Argota de Castro died shortly after her second child was born. Juana Castro, Fidel's younger sister, insists however that her father either divorced or simply left María (this point is entirely unclear inasmuch as actual divorces in rural Catholic families in Cuba in the 1920s were most uncommon). Juana Castro also says that this first wife lived very long, dying well after the revolution.

Ángel Castro's second wife, the mother of Fidel and his six sisters and brothers, was Lina Ruz González, a woman easily twenty-five years younger than the Birán landowner. She appears to have been born in the westernmost province of Pinar del Río, and her daughter Emma once described her as "a Cuban for a long time," presumably meaning that her parents were not first-generation immigrants from Spain. Juana says her mother was from "the most humble origins," but it is unknown when and why she had come to Oriente. According to most published versions, Lina worked as a cook or a maid in the Castro household while María Argota de Castro was still in residence.

Fidel says that his maternal grandparents moved "one thousand kilometers in a cart" from Pinar del Río to Oriente at the start of the century, with Lina and their other children. The grandparents were extremely poor and, according to Fidel, Lina's father and his two brothers drove oxcarts transporting cane from fields to mills. It is unknown what happened to Grandfather Ruz, but Fidel recalls that his maternal grandmother lived about one kilometer from the Birán house and that she had even gone to Havana with Lina after the revolution in 1959.

Castro, who has rich memories of his mother, has often told of her being "practically illiterate" until, as an adult, she taught herself to read and write. Both his mother and his grandmother, he says, were deeply religious, "the religiosity coming from some family tradition." Because there were no churches or priests in the Birán region, Fidel's mother's devotions were at home, and during the Sierra war, both women had made endless promises to God and the saints for the lives and safety of Fidel and Raúl. The day the revolution triumphed, Señora de Castro, her head covered by a black mantilla, knelt at the altar in the Santiago cathedral to thank God for her sons' survival and victory. Castro recalls that when his mother and grandmother told him of the promises and their faith, he listened to them with interest and respect. "Although I had a different concept of the world, I never discussed these problems with them, because I saw the strength, the encouragement, the consolation they derived from their religious sentiments and their beliefs." Castro later remarked in all seriousness about his mother: "The fact that we completed our struggle alive must have, doubtlessly, expanded her faith." Of his father, he says in a peculiarly detached fashion, "I saw him more preoccupied with other subjects, with the political thing, the daily struggle . . . rarely, almost never, I heard him expound on religion. Perhaps he was skeptical in matters of religion. That was my father." It almost seems as if Fidel resented his father for not sharing his mother's religious faith in his own destiny.

Some accounts by foreign writers claim that Ángel's and Lina's first three offspring—Ángela, Ramón, and Fidel—were born out of wedlock during the period when Lina worked as a maid or cook for the household. All three have chosen, unsurprisingly, not to discuss this allegation publicly. It is entirely possible that the first wife simply decided to walk out and that the family subsequently let her slip into oblivion (although Pedro Emilio and Lidia went to live with her somewhere, at least for a time).

In any event, Ángel Castro and Lina were married in church after Fidel's birth in a ceremony arranged by the priest Enrique Pérez Serantes, a friend of the groom, who as bishop many years later would be instrumental in saving Fidel's life from Batista soldiers. Ángela, Ramón, and Fidel were baptized in church later on with the proper surnames of Castro Ruz. There is nothing to suggest that Fidel's alleged illegitimacy had ever caused him the slightest discomfort or problem in the tolerant Cuban society.

At the time of Ángel Castro's first marriage, the Birán *finca*'s two-story hilltop frame house on wooden piles had already been partly built. It was quite large with most of the bedroom windows facing the sierras in the south, and the cattle and the dairy barn under the building. Upon Ángel's arrival in 1899, the village of Birán had some 530 registered inhabitants and was growing rapidly when the Castro children were being born. Marcané was the nearest town of any importance having both a school and a doctor. All of this territory lay within the confines of the United Fruit Company empire.

Castro believes that his father built the house on wooden piles with space underneath for cattle and fowl because it was in the architectural style of well-off landowners in Galicia; Ángel Castro, born in a humble one-story Spanish stone house, was eager to enjoy the prosperity that he had earned through his hard work in the new country. Fidel has saved a photograph of the Galicia house, and he shows it to visitors to underscore his family's early poverty.

When Castro was a child, the house was expanded to include an office for his father, and later a cow barn was erected some one hundred yards from the main building, followed by a small slaughterhouse and a repair shop. In time, Ángel Castro built a store and a bakery. Eventually, Castro claims, the tiny post office and the small rural school were the only structures in Birán not belonging to his father. Near the house there was a cockpit where every Sunday during the harvest cockfights were held; Castro tells that "many humble people spent there their scarce earnings [on betting]; if they lost, nothing was left to them and if they won, they spent it immediately on rum and *fiestas*."

Manacas, the Castro *finca*, became in time a 26,000-acre domain (1,920 acres belonged to Don Ángel and the rest was rented permanently) with some 300 families living and working on the property. Many of these people were indigent Haitian canecutters brought to Cuba across the narrow body of sea from the island of Hispaniola to work the sugarcane fields. The cane was sold by the *finca* to the United Fruit Company's nearby Miranda mill. Don Ángel also grew fruit, raised cattle, and owned forests in Pinares de Mayarí where his sawmill processed lumber for sale in big volume. A small nickel mine belonged to him as well (the Bay of Nipe zone is very rich in nickel and other minerals).

Ángel Castro played with gusto the role of the landed Spanish-Cuban patriarch, an imposing figure almost six feet tall in a wide-brimmed hat covering his completely shaved head. Either his wife or one of the daughters would use a hand clipper to keep his head hairless and shiny. Until the age of forty, he had a beard, then shaved it off, too.

Don Ángel was an incredibly hard worker who, even as a rich man, rose at dawn every day to take breakfast personally to canecutters and planters in his fields. On the eves of Christmas and other major holidays, he would sit at an outdoor table in front of the warehouse abutting the main residence to distribute vouchers to the workers for feast food as presents from the master. Despite this beneficent aspect of Don Ángel, the Castro children also remember him as a man of an extraordinarily violent temperament, given to unpredictable explosions—traits he passed on in full to his son Fidel.

CHAPTER

2

Everything about Fidel Castro seems to be controversial, even the exact date of his birth. There has been for years a lively disagreement as to whether he was born in 1926 or 1927. The Commander in Chief himself swears that 1926 is the correct year—he once said laughingly that he actually wished he had been born a year later so that "I would have been an even younger chief of government, thirty-one years old and not thirty-two, when we won the Revolution." Even the Soviet press gave the wrong date in a lengthy biography published in 1963, and the error was maintained in Cuban newspapers reprinting it. This confusion seems to have stemmed from a change made in his school records when he lost three months of classes because of postappendectomy complications. In any case, when he was born at 2:00 A.M. on August 13, weighing ten pounds, Castro was named after Fidel Pino Santos, a very wealthy Oriente politician and friend of Don Ángel. The origin of his middle name, Alejandro, has been forgotten, but Fidel has used it as his nom de plume in anti-Batista newspaper articles after the 1952 coup, as his code name in clandestinity, and his nom de guerre in the Sierra (it may have reminded the historically minded Fidel of Alexander the Great).

Fidel is both dramatic and mystical about the circumstances of his birth.

He told Frei Betto, the Brazilian Dominican friar, that "I was born a *guerrillero*, because I was born at night, around two o'clock of the dawn. . . . It seems that night might have had an influence in my *guerrilla* spirit, in the revolutionary activity." By the same token, he attaches great importance to the number 26 in a very Cuban tradition of superstition and spiritism. The year of his birth was 1926, he points out that he was twenty-six years old when he launched his conspiracy against Batista in 1952 (Fidel also notes that fifty-two is the double of twenty-six). The assault on Moncada, the date having been chosen by Castro, was July 26, 1953, and his revolutionary movement became known as the 26th of July Movement. The *Granma* landing in Cuba was in 1956 (this date includes at least the numeral 6). His friends say that Castro often picks the twenty-sixth day of the month for major decisions and acts: In 1962, for example, he chose March 26 to deliver a crucial speech that served to destroy the challenge against him by the so-called Sectarians of the Communist party; on that day he emerged from self-imposed isolation (he had not been seen publicly for over a month) to strike the decisive counterblow. As Fidel once remarked, "There may be a mystery around [the numeral] twenty-six."

The choice of first name for the future leader of the revolution was prophetically and politically felicitous: Fidel comes from the Latin word for "faithful," and has a good solid ring to it. In a nation where popular leaders are often known and called by their first names, and with Castro always orchestrating huge mass rallies, it is hard to imagine crowds chanting rhythmically the name of, say, Felisberto, Dagoberto, or even Ernesto. With *Fidel* he had a phenomenal political head start.

His name is a matter of pride to him, and he remarks that "I'm entirely in agreement with my name, for faithfulness and for faith. . . . Some have a religious faith and others another [faith]; but I have been a man of faith, confidence, and optimism." Castro also observes that April 24 is the day of his saint—the day of San Fidel—and he says that this was the date "of my saint, because there is a saint called San Fidel; before, there was another saint, I want you to know it."

He has gone to great lengths to make clear that while he was named after Fidel Pino Santos, this local millionaire never actually became his godfather. Interestingly, Castro says that the reason he was not christened until he was five or six years old (he is not certain which) was that it had been impossible to bring to Birán both the priest assigned to the region and the very busy Fidel Pino Santos. The question of whether his parents were married when he was born is not mentioned in Castro's rare and incomplete versions of his childhood, and he says that he was named Fidel because "they could wait six years to christen me, but they could not wait six years to give me a name."

Castro has said that because he had not been christened, other children in Birán called him "the Jew." Though he did not know at the time what "Jew" meant, he realized it had a pejorative connotation for his not having undergone baptism. There is a black-beaked bird in Oriente known as a *Judío* (Jew), and at one point Fidel thought that for some reason he was being called *that*. With his selective memory, it is interesting that he remembers so well his trauma as an unchristened child. He was finally baptized at the cathedral in Santiago, where he was then living and going to school. His godparents were Luis Hibbert, the consul of Haiti, and his wife, Belén Feliú, both mulattoes. Belén, the sister of his primary-school teacher, taught piano. It is not even certain that Fidel's parents were actually present at the christening.

His birth date makes Castro a Leo, and, interestingly, Simón Bolívar—the great South American Liberator whom Fidel is determined to emulate—also was a Leo, born on July 24, 1783. This fact is important in superstition-prone Cuba, where astrology, the occult, and even forms of black magic in Afro-Cuban *santería* rites are very much a part of life, scientific Marxist-Leninism or not, and Castro remains most aware of it. Finally, if one searches for historical analogies, it may be noted that, like China's Mao Zedong, Castro was the son of a very rich peasant, receiving the superb education denied to his father, and becoming a revolutionary and a Marxist.

Myth-making official propaganda rejects a portrayal of the Castro family as nouveau riche and of Fidel's father as an uneducated man. The truth would not seem opprobrium in the Cuban rags-to-riches tradition of generations of immigrants, particularly in the new revolutionary age of the common man. But the Castro biography, republished in the official newspaper *Revolución* from the Soviet press, which had to obtain its data from Cuba, describes a young Fidel "spending hours in the company of his father, who tells him the tales of Independence and epic narrations of Troy and other ancient wars and their legendary heroes." Russian readers may well accept this, but it is entirely improbable. Don Ángel was a man of natural intelligence, interested in politics and public affairs, a careful reader of Havana newspapers to which he subscribed at the *finca*, a devoted radio listener, and, in his old age, a breathless fan of televised wrestling. Still, he was a person of few words and no-nonsense mien.

Fidel's childhood appears to have been very pleasant and basically happy, certainly a privileged childhood even by the standards of affluent Cuban landowners of the day. The Castro children seem to have received much love from the parents, despite Ángel's outbursts of violence, and clearly they were spoiled. The seven children of Lina were close despite age gaps, and Juana Castro, the fifth child, says that Raúl was their mother's favorite (as well as hers) because he was "tender and loving." But there is no question that Fidel was the most assertive and combative, the one who always knew how to get his way. It is difficult to attribute his sense of rebellion to any childhood rejection or a hostile home environment.

Family album photographs show Fidel at the age of three, looking most serious and composed in an elegant little boy's suit with short pants and a jacket with a large round collar. His hair is carefully parted on the right side, and he holds a book in his left hand. His big brown eyes stare hard at the camera. In another picture, Fidel sits atop a wall, between his older sister Ángela and brother Ramón who are standing on the ground. Looming over them, he dominates the scene.

At the age of four, Fidel entered the public grammar school in Marcané that Ángela and Ramón had already been attending. He learned to read and write before he was five years old, remaining at the local school until after his fifth birthday. There were fifteen or twenty pupils at what Castro called a "kindergarten." His parents decided at that point that he should transfer to Santiago, the capital of Oriente province, to study under the very disciplinarian Marist brothers. As Castro recalls it, his parents simply lost their patience with him over his disorderly behavior at the Marcané school. It appears that from a very young age Castro had to have his way and that he rejected all form of authority—though responding to acts of kindness and special attention directed at him. When he could not have his way, he struck back with violence—against his parents, teachers, siblings, and playmates.

Fidel's happiest moments were spent outdoors, climbing hills, swimming in the Birán River, riding horses, and, when he grew a little older, hunting with a shotgun and a pack of four dogs. He was a natural athlete. His passion for active physical life—while still living at the *finca* and then during summer vacations at home from schools in Santiago and Havana—conditioned him from childhood to the future hardships of guerrilla life in the Sierra.

He learned to shoot with his U-type shotgun and delighted in firing at almost any target he spotted. According to the brief section on his youth in the biography published in *Revolución,* Fidel liked to practice his aim on the *finca*'s hens, "and if one of his sisters threatened to tell their mother on him, he would convince them to shoot, too, so that they couldn't say anything." No matter what he did, the *Revolución* biography reports—play, swim, study, or work on projects at home—"he never wants to lose and almost always arranges to win." This biography also reports Castro "constantly changes things from one place to another," and that at the dinner table, he "does thousands of combinations with a drinking glass," suggesting permanent restlessness, easy boredom, pedantry, and perfectionism—he seemed to want the glass to be placed just so.

His sisters recall that Fidel once organized a baseball team in Birán, his father having let him order bats, gloves, and other equipment (baseball is Cuba's national sport). Characteristically, Fidel's preferred position was that of pitcher, though his very fast ball had little control. However, he was no sportsman; when his side was not winning, Fidel would simply halt the game and go home. He may have learned this sort of reaction from his father: The sisters remember that Don Ángel's favorite game was dominoes (now it is Fidel's too), which he played every night with one of his employees or with his wife, but when an argument developed or he was losing, he would grab the dominoes board and hurl it to the ground below the veranda where they sat. Then, there would be no dominoes game for a week or so.

Fidel Castro himself is the foremost authority on the fact that he has always been violent, given to tantrums, devious, manipulative, and defiant of all authority. In a most remarkable conversation he held in 1959 with Carlos Franqui, then the editor of *Revolución,* Castro revealed his complicated personality at great length in one of the very few autobiographical interviews he has ever granted (but which was never published in Cuba).

Fidel says of the Birán school that "I spent most of my time being fresh. . . . I remember that whenever I disagreed with something the teacher said to me, or whenever I got mad, I would swear at her and immediately leave school, running as fast as I could. . . . One day, I had just sworn at the teacher, and was racing down the rear corridor. I took a leap and landed on a board from a guava-jelly box with a nail in it. As I fell, the nail stuck in my tongue. When I got back home, my mother said to me: 'God punished you for swearing at the teacher.' I didn't have the slightest doubt that it was really true." In Castro's self-portrait between the ages of four and six, he acknowledges that "I had one teacher after another, and my behavior was different with each one." He adds: "With the teacher who treated us well and brought us toys, I remember being well behaved. But when pressure, force or punishment was used, my conduct was entirely different."

Fidel's behavior was presumably tolerated because his father was wealthy and influential in the area, and Castro recalls that in general "everyone lavished attention on me, flattered, and treated me differently from the other boys we played with when we were children." These other children, he remarks, "went barefoot while we wore shoes; they were often hungry; at our house there was always a squabble at table to get us to eat."

He is not certain whether he was dispatched to Santiago "because I caused too much trouble at home or because my teacher convinced my family that it would be a good thing to send me away to school." In any case, Fidel was five or six years old (he says he does not remember the exact age) when he and his older sister Ángela traveled by train to Santiago, clear across Oriente to the south coast, and a new phase began in his life. He remembers how "extraordinary" the big city appeared to him—"the station with its wooden arches, the hubbub, the people"—and that they went to stay that evening at the home of the sister of the schoolteacher from Birán and her Haitian husband who soon thereafter would become Fidel's godparents. Fidel recalls: "I remember that I wet the bed on the first night."

In Castro's version of his young years, as given to Carlos Franqui in 1959, he was dispatched to Santiago to be enrolled in the Marist brothers' La Salle school, a private establishment for boys from affluent families. But in interviews granted in 1985, a wholly different story emerges. Thus Fidel claims that he did not attend school at all during the first two years in Santiago when he lived with his godparents, the Hibberts. Instead, according to this account, his godmother taught him at home, and the studies were confined to memorizing the four arithmetic operations from the back cover of a booklet (he says that he memorized them so well that now he can add, subtract, multiply, and divide as rapidly "as a computer") and to improving his orthography and handwriting; there were no books in the Hibbert household.

It is hard to understand why Fidel's parents would have allowed such a state of affairs to exist, and it is just as odd that, as he tells it, it took his family a full year to realize how badly he was treated in Santiago. Castro now appears to feel very resentful about that entire period, portraying himself as a victim of a situation mysteriously concocted, it would seem, by his parents and the Hibberts. He uses expressions like "When they sent me away to Santiago—I was very small—I suffered such need and had so much work . . . ," and a year later, "They sent me away again to the house in Santiago." There, Castro says, "I went hungry and I was the target of injustice," and it was a waste of two years "of a hard life, work, sacrifice." He goes on to complain "that I was the victim of some exploitation by this family that was paid by my parents to have us there."

Without further explanation, Fidel tells of having entered La Salle in his third year in Santiago as a first-grade day student (in the Franqui version, he says he was immediately in "the first grade" at that school) and of having to make up for the lost years. He still lived with the Hibberts, returning home for lunch ("then, there was no more hunger") and enjoying "having professors, classes, companions with whom to play, and many other activities that weren't available to me when I was a solitary student learning arithmetics from the back of a booklet."

But Fidel was soon unhappy again. For example, while boarders at La Salle were taken to the beach or out for walks on Thursdays and Sundays, Fidel, being a day student, was left behind and his life was "very dull." He began to despise the Hibberts. As he tells it, he was six or seven years old when he took matters into his own hands, "engaging in my first rebellion" and applying guerrilla tactics of sorts to compel his godfather into letting him be a boarder, too. One day, when Hibbert spanked him "on my rear end" for some infraction, Fidel proceeded to "rebel and insult everybody, disobey all orders, shout, and say all the words that were forbidden." Castro says that "I behaved so terribly that they took me straight back to school and enrolled me as a boarder; it was a great victory for me." He was one of thirty boarders; two hundred boys were day students. It cost his father thirty dollars per month.

At La Salle the boys had to observe a strict dress code, wearing suits and ties. There is a group photograph of the school's second grade with Fidel sitting in the front row, tie loosened, and a look of bored contempt painted on his face. There is nothing in his recollections to indicate that he felt unhappy being away from his parents at such a young age; he was obviously content taking care of himself. He does say that "on our first holidays, we went home for a three-month vacation; I don't think I've ever been happier; we hunted with slingshots, rode horseback, swam in the rivers, and had complete freedom during those months." But not a word about the family.

When Fidel completed the third grade, Ramón and Raúl joined him at La Salle, where a special grade was created for the three of them to be together, seemingly incongruous because Ramón was ten, Fidel was eight, and Raúl was four. Castro explains that it was made possible because the family was rich.

His stay at the Marist school was punctuated by these battles for his rights, and Castro goes out of his way to draw a picture of himself as an uncompromising and violent boy. He recalls, for example, beating up a schoolmate—the teacher-priest's pet—in a fight that followed a boat-ride argument. That evening, the priest summoned Fidel away during a solemn chapel service to ask him what had happened and, without awaiting an explanation, he "gave me a slap that just about stunned one side of my face. . . . I spun around, and he slapped my other cheek. . . . When he let me go, I was in a complete daze, I felt painfully humiliated." Castro continues: "Another time . . . we were marching in single file and he struck me again, this time on the head. Then, I promised myself not to let it happen again. We were playing ball one day. The kid who was at the head of the line always had the best position, and I was half arguing over first place with somebody else, when the priest came up to me from behind and hit me on the head. This time I turned on him, right then and there, threw a piece of bread at his head and started to hit him with my fists and bite him. I don't think I hurt the priest much, but the daring outburst became a historic event in school."

In what Fidel calls "a decisive moment in my life," Ángel Castro decided during the boys' summer holidays after the fourth grade that they would not go back to school. The father had not only been receiving reports from La Salle that his three sons did not study and "were the three biggest bullies who had ever gone there" (Fidel says it was "an unfair report but they believed it at home"), but he also discovered that they were cheating, obtaining solutions to math problems given them by a tutor at the *finca,* from an answer book they had procured at school.

Ramón Castro was delighted to end his education because he preferred the life in Birán, the fields, the animals, the machines. The little Raúl, unable to defend himself, "was packed off to a military school run by a country teacher, a sergeant, who also gave him a hard time." But Fidel was determined to return to school. As he tells the story, "I remember going to Mother and explaining that I wanted to go on studying; it wasn't fair not to let me go to school. I appealed to her and told her I wanted to stay in school and that if I wasn't sent back, I'd set fire to the house. . . . So they decided to send me back. I'm not sure if they were afraid or just sorry for me, but my mother pleaded my case."

Fidel was learning quickly that absolute and uncompromising stubbornness was a powerful weapon. This may have been the most important lesson he had drawn from his young years at the *finca* and at the Santiago schools, and he never forgot it. Now, having in effect blackmailed himself back to school, Castro was enrolled in the fifth grade of the much more demanding and much better Dolores boys' school in the center of the city. There, at the age of nine, he began his Jesuit education, one of the most significant intellectual influences in his life. Fidel says that Dolores "was a school that set very high standards; I had trouble keeping up with the others." Castro claims that he changed schools at his own demand, refusing to stay at La Salle after the teacher had slapped him.

But again he found himself a day student, staying with a merchant's family he deeply resented; his sister Ángela, attending a girls' school in Santiago, was his only friend at the hostile house. Fidel was receiving from Birán a twenty-cent weekly allowance, spending ten cents on Sunday movies, five cents for ice cream afterward, and five cents for the comic book *El Gorrión* (The Sparrow) he bought on Thursdays. However, Fidel recalls, his allowance was cut off if he failed to get the highest marks. Therefore, "I decided to take steps to protect my interests."

He did so by informing his teachers that he had lost the report card with his marks, and he was given another one. "From then on," Fidel says, "I would put my grades in the new book and take that one home to be signed—with very good grades in it, of course. The other notebook, the one they put the real marks in at school, I signed myself and returned to school." At the time, he seemed to have been cultivating a more angelic demeanor, so he must have been above all suspicion. A photograph shows him, a half-smile on his face, sitting on a wooden bench at the long dining table at the school. He wears the Dolores uniform of white trousers, dark blue jacket, shirt and tie, and a white Sam Browne belt; the Jesuits favored a military school atmosphere, and there even was a band with which, as another photograph shows, Fidel marched under a Cuban flag and a Dolores pennant.

Right after turning ten, Castro developed appendicitis, spending three months at the Colonia Española Hospital in Santiago because the incision would not heal properly. But, as usual, he used the time well, and he enjoys recounting the experience: "I was practically alone, and I made friends with all the other patients. I am telling this because I think it shows I already had an ability to relate to other people; I had a streak of the politician. When I wasn't reading comic books, I spent my time visiting other patients. . . . Some people thought I might make a good doctor, because I used to play with lizards and a Gillette razor blade. I had been impressed by operations like the one I'd been through. . . . Few sanitary measures were taken, and that's why the wound opened up and I had to stay at the hospital three months. After that, I would 'operate' on lizards— lizards that usually died, of course. Then I would enjoy watching how the ants carried them off, how hundreds of ants working together could carry the lizard and move it to their heap."

The illness prevented Castro from skipping a year at school as he had been encouraged to do by a black woman teacher known as "Professor Danger." She had been tutoring his sister and saw great potential in him. Fidel says that he never had "a preceptor or a guide who would help me" in his youthful rebellions when his character was being formed, but "that black professor in Santiago was the nearest to being a preceptor." She was, Castro adds, "the first person I knew who stimulated me, who gave me a goal, who was able to make me enthusiastic about studying at such an early age."

Back at the guardian's home, Fidel, now in the sixth grade, was increasingly resentful. For one thing, he was angry that when he came home from school, he was shut up in a room for hours to study "when all any boy wants is . . . do nothing, listen to the radio or go out." But he refused to study, instead letting his imagination "fly off to places and events in history, and to wars." Castro says that he liked history "very much, and particularly the stories of battles . . . I even used to invent battles." The hours he was locked up, he recalls, "were a kind of military training. . . .

I'd start off by taking a lot of little scraps and tiny balls of paper, arranging them on a playing board, and setting up an obstacle to see how many would pass, and how many wouldn't. There were losses, casualties. I played this game of wars for hours at a time." When he could not stand the guardian's home any longer, he told the merchant's family "all to go to the devil," and became a boarder at Dolores that same afternoon. Fidel does not explain whether his parents had any say in it, but stresses that time after time "I had to take it upon myself to get out of what I considered an unpleasant situation," listing the fights over his years at La Salle and the confrontation at Birán over the continuation of his education.

Castro, who then was barely eleven years old, says that "from then on I definitely became my own master and took charge of all my own problems without advice from anyone. . . . I played soccer, basketball, jai alai, all kinds of sports. All my energy went into them." He explains that "I personally suffered from the lack of the most elementary perception about teaching and about the psychology of educating boys," but that "I'm not blaming my parents, who were ignorant people without a proper education; they left us in the hands of others they believed were treating us properly, but we had a hard time of it." It would appear that Castro's view of his parents was principally one of contempt. Still, he went on for years using them to his advantage, accepting financial aid from them even as late as the preparations for the *Granma* invasion in 1956. His sister Juana believes that Fidel respected his father, but very clearly there was no warmth between the two proud and stubborn Spaniards. Fidel's threat at the age of nine to burn down the house if he were not allowed to return to school was his first major confrontation with Don Ángel. At the age of thirteen, while his father paid the Dolores bills, Fidel used the summer vacations not only to drive the *finca* tractors (one of his preferred pastimes), but also to try to organize the sugar workers against Don Ángel. At eighteen, while studying at an expensive Jesuit college in Havana, he fought with his father repeatedly over the family's "capitalism," accusing him of "abusing" the workers "with false promises."

Other relatives say that at the same time Fidel would criticize Don Ángel for administering the estate badly, insist on examining the books, and protest that a peasant working for the family was permitted to go on owing it six thousand pesos. When Fidel was in prison after the Moncada attack, his father (who had just finished financing his university studies) sent money monthly for the support of the rebel's wife, Mirta, and the little Fidelito. According to Juana Castro, Fidel found no time to see his parents at Birán during the seven weeks he spent in Havana between amnesty from prison and his departure for Mexico and the revolution; Raúl, released from prison on the same day, did manage to visit the *finca* for a week or so. The last time Fidel saw his father was early in 1953, before Moncada. He was in Mexico when Don Ángel died from a ruptured hernia in October 1956. A person who has known Fidel Castro from childhood says that he had "no tenderness for anybody, not even for his wife," and that he is "a passionate but not a caring or tender man, living at the margin of all human problems, except his own."

This is a very harsh judgment, and it is not necessarily borne out by other accounts of Fidel's behavior. Fidel had undeniably become his own master, however, something he seemed to have worked on quite conscientiously while at school. In the sixth grade at the Dolores school, Fidel was joined by Ramón and Raúl, their father having again changed his mind about educating all his sons. Raúl recalls that during one Sunday outing at the Siboney beach, near Santiago, the priest in charge punished two students by forbidding them to go swimming. Hearing this, Fidel approached the priest and asked him, "Father, if I dive down from the ten-meter [thirty feet] board, will you let them swim?" The priest answered, "My son, it is too high. . . . You'll never dare!" Fidel repeated his question, and the Jesuit said, "Well, yes, if you dive." Fidel jumped, mustering all his courage, winning the swim for his companions—but also proving to himself that he could always overcome fear.

Raúl, who hated school, remembers: "For me, it was a prison. School for me, it was prayer, the necktie, the fear of God. But what really killed me was the prayer. We prayed from morning till night. But Fidel, that was different. He dominated the situation. He succeeded in everything. In sports, in studies. And, every day, he would fight. He had a very explosive character. He challenged the biggest and the strongest ones, and when he was beaten, he started it all over again the next day. He would never quit." Again, Fidel's goal was to hone his courage.

Juan Rovira, who was Fidel's classmate at Dolores (and is now an exile in Miami), recalls him as a sports hero and a student with a phenomenal memory. Rovira says: "Everybody was very enthusiastic about Fidel when there was a basketball game with La Salle, or when there was a track meet, because he ran so well and his sports qualities were fantastic. He was outstanding in all the sports, and for this reason he enjoyed general sympathy. When it came to studies, Fidel didn't stand out as much, but when the exams came, he studied a lot. Boarders were allowed to rise early, at four o'clock in the morning, to study, and he had a prodigious memory. He wrote everything exactly the way he read it, and it looked as if he had copied it, but he had it all engraved in his memory. And he got good grades for his fantastic memory."

Castro's familiarity with mountains first came when he was at Dolores. The boys were taken on outings by school bus to climb mountains, sometimes to El Cobre, to Gran Piedra, or even the foothills of the Sierra Maestra. Fidel recalls that "I also loved to take off along the rivers when they were swollen, cross them, and hike a while before coming back. The bus always had to wait for me. . . . I did not imagine that mountains would one day play such an important role in my life!"

When he was fifteen, Castro graduated from Dolores (that year, he says, "I was one of the best in my class"), but he still had an arrogant attitude. Asked at final exams to name a reptile, he answered, "A *majá*" (a large, nonpoisonous Cuban snake), and when told to name one more, he replied, "Another *majá*."

His determination to excel and to distinguish himself knew no limits. When a Santiago radio station sponsored a poetry contest, with parents requested to vote for the best entry, Castro entered several poems. José Martí, his object of admiration, had been a great poet, but this was a gift Fidel totally lacked. He admits that in the contest his poems "weren't the best, but I had made friends with all the boys, which I think again reveals perhaps a political streak in me. . . . Almost all the kids asked their parents to vote for me; as a result, letters were sent in . . . that went something like this, 'Elpidio's poem to mothers is very beautiful and very touching, but our vote goes to Fidel . . .'"

After his revolution triumphed, Fidel Castro has often said that his sense of social consciousness was born at the little country school in Marcané and at the Birán *finca* where he studied and played with the children of the poor. This is substantiated in a lengthy personal letter he wrote to a woman friend from his prison cell in January 24, 1954, where he was serving the sentence for the Moncada attack: "My classmates, sons of humble peasants, went to school barefoot and, in general, wore miserable clothes. *They were very poor.* They learned poorly the first letters, and they soon left school, even if they were more intelligent than was necessary. They drowned, then, in a bottomless and hopeless sea of ignorance and poverty, without any of them ever escaping the inevitable disaster. Today, their children will follow in their steps, shouldering the burden of social fatalism. I, on the contrary, *could* study, I continued to study. . . . Nothing has changed in twenty years. . . . It is probable that it has been like this since the day the Republic was born, and that it will continue without anyone seriously taking in hand such a state of things. . . . Nothing that could be done in the domain of the technique and organization of education can lead anywhere unless one reshuffles from top to bottom the economic status of the nation . . . because that is where the real root of the tragedy lies. . . . Admitting that, with the aid of the state, a young man reaches an enviable technical level, even there he would drown with his diploma, like a paper boat, in the terrible narrows of our present economic and social status . . ."

In his autobiographical interview with Carlos Franqui in 1959, Castro notes that "all the circumstances surrounding my life and childhood, everything I saw, would have made it logical to suppose I would develop the habits, the ideas, and the sentiments natural to a social class with certain privileges and selfish motives that make it indifferent to the problems of others." Yet, he says, "one circumstance in the middle of all this helped us develop a certain human spirit: It was the fact that all our friends, our companions, were the sons of local peasants."

It was with this sort of background that Fidel Castro was exposed to the teachings of the Jesuits, first in Santiago, then in Havana. After his first experiences with Jesuit schools, he concluded that "the educators were better prepared than those in other schools . . . there was a spirit of discipline . . . creating habits of discipline and study was good. I am not against that kind of life, Spartan to some degree. And I think that, as a rule, the Jesuits formed people of character."

Castro had no way of knowing it, but an astonishing number of young Cuban boys, boys destined to become his closest revolutionary companions, were at that same time studying at Jesuit schools in Santiago, Havana, and other cities. As for Fidel, he was now almost sixteen and ready for the next great change in his life.

CHAPTER

3

On an October morning in 1941, Fidel Castro's knees were shaking, and he was perspiring profusely from nervous tension. He was standing before Father José Rubinos, the director of the Avellaneda Literary Academy, to deliver from memory a ten-minute speech that would mark the birth of his professional political life.

If it pleased the demanding Father Rubinos, the speech would mean acceptance in the academy, the school of oratory at Belén College, an exclusive Jesuit high and preparatory school in Havana where Fidel was sent by his father to pursue his studies after completing the four years at Dolores. He had turned sixteen in August, and while vacationing in Birán, he had persuaded his parents to let him attend Belén because he regarded it as the best school in the country even though it was also "the center of great prestige of the cream of the cream of Cuban aristocracy and bourgeoisie." Castro always knew what was good for him.

Studying in Havana and leaving behind the provinciality of Oriente was a tremendous step toward a proud career, Fidel being the first in his family to be given such an opportunity. Belén, above all, was the road to the university, Castro's next planned move. The bustling, cosmopolitan, vital, sensual, and explosively loud capital city on the north coast, ever mysterious in its promise of ideas and experiences, was a breathless new world to the gangling and still rough-cut youth from the sugarcane countryside. It was his first time in Havana, and Fidel arrived there by train from Santiago with "a lot of money to buy clothes and other articles . . . and to pay tuition, purchase books . . . and for other expenses." Tuition and board cost fifty dollars monthly, which, Castro says, was "very cheap" at the time, considering Belén's ample facilities, but out of reach for the children of a schoolteacher, for example, whose salary was seventy-five dollars a month.

Fidel did not know a soul in Havana, but naturally he was determined from the first moment to make his mark as soon as possible, overcoming obstacles as he encountered them. José Ignácio Rasco, his schoolmate at Belén and later at Havana University, recalls that on that October day at Avellaneda Academy, Castro "was desperate because he worried that his nerves wouldn't let him pass the all-important test." In the end, Castro was able to satisfy Father Rubinos and was accepted into the academy. Nobody remembers what he actually said, but the test (it was elective, not compulsory) was another victory over himself. In Cuba one cannot succeed politically without being a first-rate public speaker, and at the age of sixteen, Fidel was already overwhelmingly attracted by the craft, art, and glamour of politics and power—little as he knew about them.

Castro is recognized today as one of the great orators of his time, yet the adolescent timidity never really left him. Rasco (who has lived in exile in Miami since 1980, having broken with Castro soon after the revolution over the issue of communism) remembers that before delivering his first *public* speech as a university student five years after the agony of Avellaneda, Fidel had spent a week at Rasco's house writing, rewriting, and memorizing the address before practicing his delivery in front of a mirror. Again, timidity was the great problem, and again, he overcame it. His confession to Havana's *Bohemia* magazine in 1985 that he suffers from fear as he faces an audience was greeted with disbelief and derision by readers at home and abroad, but it was no invention. He can be a timid (or shy) man, and there is always that moment of uncertainty when he begins to speak, usually in tentative tones. When he becomes enraptured with his theme, the timidity vanishes for the long hours he remains before the cameras and microphones.

In Fidel Castro astounding tenacity accompanied the timidity. Rasco emphasizes that as a student Fidel had immense powers of concentration as well as his phenomenal memory working for him. Frequently, distracted by other matters that interested him more, he would fall back in his studies—then recover spectacularly. In his last year at Belén, Rasco says Castro was suspended after the first semester from French-language and logic classes for failure to maintain his grades, which, in turn, prevented him from being allowed to take special final exams before Education Ministry inspectors (because Belén was a private school, the law required that students be certified by the ministry in addition to normal graduation). Under the circumstances, Fidel bet Father Larrucea, the supervising teacher, that if he obtained a 100 grade (the highest possible) in both French and logic in the second semester, he would be permitted to present himself at the ministry exams. The priest agreed and, of course, Castro won the bet.

His memory was so stunning that, as Rasco remembers, his schoolmates would playfully ask him, "Fidel, what does page forty-three of the sociology text say?" and Castro would recite it with a straight face, and if the word at the end of the page was hyphenated, he would stress that, too.

Castro's tenacity and determination were physical as well. Fidel had made up his mind to be Belén's best baseball pitcher, but because he had a muscle problem in his throwing arm, he practiced sometimes until eight o'clock in the evening at the school's sports grounds. Long after the catcher got tired and left, Castro would go on throwing the ball against a wall. Rasco tells the story of Fidel bragging one day that he could succeed in anything he wanted to do, and when challenged by a student named Cabella, he bet him that he could hurl himself on a bicycle, head first, against a brick wall. Not surprisingly, he fainted from the impact, and he had to spend three days recovering at the school's infirmary.

During the four years Castro spent at Belén, he was by far the outstanding athlete, both because he enjoyed sports and because it was a question of principle for him to excel at everything. As several of his former schoolmates point out, however, Fidel was principally a solo player, not a team player, which was a natural reflection of his character.

He was a track star, a Ping-Pong champion, a pitcher in baseball, and, as he explains it now, a "thinker" in basketball—the de facto team captain. Enrique Ovares, who knew Castro very well at the university because of their many common political involvements, says that the Belén basketball team "was famous because Fidel made famous everything he touched." Ovares, an architect who now lives in exile in Florida, cannot be accused of being a Castro propagandist: He spent seven years in prison in Havana after 1960 for antirevolutionary plots. He reminisces that Castro's dedication was so great that "he went to all the practice sessions, and if it was required to shoot fifty baskets, Fidel would shoot one hundred of them."

Fidel himself tells proudly that "I succeeded in becoming outstanding in basketball, football, baseball, track, and almost all the other sports from the moment I arrived there." He maintained his love for mountain-hiking, and he says that after the first outings by the Belén Boy Scout troop, "the teachers determined that I was outstanding, and they kept promoting me until one day they made me the chief of the school's Boy Scouts, 'the general of the explorers,' as they called it." He remembers organizing and leading an expedition to climb the tallest mountain in western Cuba, taking five days instead of the scheduled three days because the ascent of Guajaibón Peak was so difficult, and alarming the Jesuits who feared that Fidel and his group had suffered an accident. He says that "I didn't know then that I was preparing myself for the revolutionary struggle, nor could I imagine it at that moment."

In the context of Cuban society of the 1940s, the politically conservative rich families as well as those who had just acquired wealth—such as Spanish immigrants—were expected to send their sons to Jesuit schools. This was not because Catholicism was so deeply engrained in them but because these were the best educational establishments. To attend a Jesuit school carried a social and snobbish cachet as well; girls from "good families" went to convent schools like Sacred Heart or the Ursuline Sisters. The goal was solid religiously oriented upbringing, at least on the surface.

Affluent but liberal "bourgeois" parents preferred nonreligious private schooling for their offspring, often military academies for the boys. Middle-class or even the poorer families' children attended public schools known as *institutos,* or technical and vocational schools. Peasants and other truly poor Cubans seldom could send their sons (much less daughters) to study beyond the local grade school. In this sense, social stratification in Cuba was implemented from childhood, and children on all levels knew instantly their place in society. For the now-rich Castro family it was logical for the sons to go first to Dolores in Santiago, then to Belén in Havana (Ramón, the oldest son of Ángel and Lina, returned from Dolores to work at the *finca,* which he liked best, and was married at nineteen; Raúl was kept at Dolores until it was time for him to enter Belén, too; Ángela, their oldest daughter, went to the Ursuline Sisters' convent in Havana to be followed there by Juana).

Fidel Castro is extremely keen, however, to avoid any impression that he came from an aristocratic or upper-class bourgeois background even though his family had accumulated wealth. He makes the accurate observation that in "privileged" schools like Belén, the students were divided into two groups, "not so much by money, although money was the basic fact, but by social category, the homes where they lived, the traditions."

He remarks that while the Castros perhaps had adequate resources to rise socially, they never did so because they lived in the countryside: "We lived there among the people, among very humble workers . . . where animals were under the house, the cows, the pigs, the chickens, and all that." Fidel may be exaggerating the simplicity of his origins, particularly because there is no question that at least his father undertook to create a social and political position for himself in the region (contributing to political campaigns and participating in them on behalf of friends and associates), but he is correct in portraying the status of Cuban rural-class culture.

"I was not a grandson of a landowner nor a great-grandson of a landowner," he says. "Sometimes the great-grandson of a landowner no longer had money, but he conserved a full culture of the aristocratic, rich, oligarchic class. Inasmuch as my mother had been a very poor peasant and my father a very poor peasant, who succeeded in accumulating certain wealth, the culture of the rich, of the landlords . . . I think that if I had been a grandson of a landlord or a great-grandson of a landlord, I might possibly have had the misfortune of receiving that culture of class, that spirit of class, that conscience of class, and I would not have had the privilege of escaping bourgeois ideology."

It is striking that so many years after his victorious socialist revolution and his supposed evolution into a full-fledged Marxist-Leninist, Castro still finds it necessary to engage in essentially gratuitous protestations about his social origins. He is defensive about his parents' affluence, offering the disclaimers that they "worked every day under hard conditions" and had "no social life," even though he willingly accepted parental financial support well into adulthood. Because Fidel's youthful reminiscences, limited as they are, always seem designed to construct most carefully the official myth of Fidel Castro (as he wishes to be perceived at a given time), he is determined to erase any suspicions that the bourgeois taint lurks somewhere in his background. He may also want to exorcise the temptation existing among so many bourgeois who became Communists to remain, even subconsciously, more bourgeois than Communist. In any event, "bourgeois" is an immensely obscene word in Castro's lexicon, a word he almost fears, and it is a theme to which he constantly returns.

As the leading educational center of the Cuban establishment elite, Belén was located in its own imposing building, built in the 1930s on a large tract of land in the mainly residential Alturas de Belén (Bethlehem Heights) district of Havana, off Fifty-first Avenue. The school had originally functioned in cramped quarters in Old Havana, but the new building had comfortable accommodations for its two hundred boarders (out of a total of a thousand students), several baseball and basketball fields and courts, a track oval, and even a swimming pool. Fidel enjoyed very much this privileged environment and facilities of the Cuban aristocracy and bourgeoisie; after the revolution Belén was turned into the Military Technical Institute, a university-level technological center for the armed forces (Fidel's old room remains unchanged as a shrinelike showplace maintained by the Museum of the Revolution).

Discipline at Belén was strict, but Castro seemed to have no problems with it. The boys wore uniforms, were awakened at 6:30 A.M. to attend mass at 7:00 A.M. (mass and prayer and periodic religious three-day retreats were the only aspect of Belén life Fidel now says he disliked), then breakfast and classroom. Castro was given the responsibility for the main reading room where pupils studied between dinner and bedtime: He had to make sure that windows and doors were locked and lights turned off after the study period, but he often stayed on alone for hours to read for exams.

Castro tells that he made "many friendships" with fellow students, and that "without realizing it, or trying for it, I began to acquire popularity among them as a sportsman, athlete, explorer, mountain-climber, and as a person who, after all, had good grades." He says that "perhaps during that time there, some unconscious political qualities were emerging" in him. They must have been very unconscious indeed, because Fidel never developed any political or other following at the school, and no Belén companion of his ever participated in Castro's subsequent revolutionary activities.

Fidel and other boarders were allowed to go out on weekends if they had relatives in Havana or were invited to friends' homes. But despite his sports-rooted fame, Castro seems to have rated few such invitations. His father was relatively rich, yet in class-conscious Cuban society this did not automatically grant him entrée in old-money Havana circles, and behind his back he was occasionally called *guajiro* (peasant). Enrique Ovares recalls first meeting Fidel during a weekend at the home of Carlos Remedios, a Belén basketball player whose father was a powerful politician and member of the Chamber of Deputies. This, however, was only a "basketball friendship." Ovares, who knew Castro as well as anybody during his young years, goes still further in defining the latter's situation. "I think that the worst damage Fidel's parents did him," he says, "was to put him in a school of wealthy boys without Fidel being *really* rich . . . and more than that, without having a social position. . . . With Fidel's kind of maturity, when he grew from child into an adult, I think that this influenced him and he had hatred against society people and moneyed people."

He recalls that Fulgencio Batista, even as Cuba's president, could not be elected member of the super-exclusive Havana Biltmore Yacht Club, being blackballed each time his name came up. This is a sad commentary on how the privileged in Cuba behaved in those days, and Fidel is the first to point it out. It is difficult to judge whether Ovares is correct in concluding that his inferior social status pushed Fidel into "hating the rich"—Castro has not spoken of these years in those terms, but it is doubtful that he could remain indifferent in a situation where his personal popularity was in question. In any event, Fidel was not exceedingly sociable, then or later. Ovares says that "we used to like going to parties and so on, but he did not. He was introverted." Juan Rovira, a fellow boarder, remembers Fidel as a person of "a somewhat difficult character . . . he wasn't very open, he wasn't very constant," and "being happy if things worked well for him, and depressed if they went badly." He was prone to violence, and one of his former schoolmates says that on at least one occasion he got into a fistfight with other players over a basketball referee's decision he questioned.

In his third year at Belén, Fidel, then age eighteen, was proclaimed Cuba's "outstanding collegiate athlete," but in the classroom he concentrated only on the subjects that interested him: Spanish, history, geography, and agriculture (which is probably why he let French and logic lapse). Interestingly, Fidel favored sacred history as well because of its "fabulous content . . . it was marvelous for the mind of a child or adolescent to know all that had occurred since the creation of the world until the universal deluge."

Castro was fascinated by the Bible: the story of Moses, the crossing of the Red Sea, the Promised Land, and "all the wars and battles." He says that "I think it was in sacred history that I first heard about war, that is, I acquired a certain interest in martial arts . . . it interested me fabulously, from the destruction of the walls of Jericho by Joshua . . . to Samson and his Herculean strength capable of tearing down a temple with his own hands. . . . All that period, which one might call of the Old Testament, Jonas, the whale that swallowed him, the punishment of Babylon, the Prophet Daniel, they were marvelous stories." Then, he says, came the New Testament, where "the whole process of the death and crucifixion of Christ . . . produced an impact on the child and the youth."

At the same time, however, Castro also remembered his education in sinister terms: In a 1961 speech, he said that "I was formed in the midst of the worst reaction and I lost many years of my life in obscurantism, superstition, and lies," evidently not seeing a contradiction between this view and his recollections of the fascination sacred history held for him at Belén. Again, this is how Castro goes on weaving and re-creating his own myths about himself.

Fidel Castro's stay at Belén—autumn of 1941 to spring of 1945—straddled the Second World War and coincided with Fulgencio Batista's first presidency in Cuba. This former army sergeant, for seven years the power behind the throne in Cuba as the country's military chief, now emerged as a democratic leader.

Batista's political career began on September 4, 1933, when he led a coup by noncommissioned army officers to establish the armed forces as the conservative arbiter in Cuban politics, although at the outset he allied himself with a radical regime.

The crisis of the 1930s in Cuba stemmed from the move by President Machado, initially supported by the United States, to prolong the four-year term in office by five years. First elected in 1926, Machado organized two years later a phony election with himself as the only candidate, receiving a new term to run from 1929 to 1935. Machado's maneuver triggered a wave of opposition bringing together revolution-minded students, the young Communist party, and moderate traditional political leaders, and Cuba lived for five years in deep unrest and violence. The United States blissfully ignored this state of affairs almost until the end even though the Platt Amendment was still in force and Washington could have intervened legally for a good cause. But so long as American economic interests were not in jeopardy, the United States paid no attention.

Starting a tradition that Fidel Castro would resurrect a quarter-century later, revolutionary students and young professionals along with worker and peasant leaders formed the spearhead of the antidictatorial movement. From the university sprang the Students' Directorate, which would be reborn during the Castro revolution, and the militant youth of the day became known as the "Generation of 1930," profoundly nationalistic and social-justice conscious. The anti-Machado struggle produced its heroes and martyrs: In January 1929, Julio Antonio Mella, a student leader and secretary-general of the illegal Cuban Communist party, was assassinated in Mexico by agents of the dictatorship; in September 1930 the police killed Students' Directorate leader Rafael Trejo during an anti-Machado street demonstration. From thereon Cuba was plunged in virtually permanent violence.

Only in 1933, when American businessmen and investors became concerned about the Cuban economy and their stake in it, did the United States government awake to the island crisis. President Franklin D. Roosevelt dispatched Benjamin Sumner Welles, his top diplomat in Latin America, to mediate between the Machado regime and the opposition groups. Protesting against the belated Welles effort, however, was the Generation of 1930 leadership. The Directorate and its allies saw in the anti-Machado battle the opportunity to bestow on Cuba "real" independence, free her of American influence, and put into practice Martí's teachings of freedom and justice. Thus Martí was the model and hero of the revolutionaries in the Thirties, a fact that goes far to explain subsequent Cuban history.

Machado was forced to resign on August 12, 1933, under the pressure of a revolutionary general strike on one hand and United States demands on the other. The strike was successful despite a last-minute decision by the Communist party to withdraw its support on the unconvincing grounds that it might lead to armed United States intervention (it was no longer necessary anyway), and an agreement between the party and Machado. The

Communists, who seemed to have forgotten Mella's murder, could not save Machado, but this incident proved the first example of the extraordinary political flexibility of the Cuban Communist party—if not always necessarily of its wisdom.

Acting on Welles's advice, the traditional political parties and the army joined in handing the presidency to Carlos Manuel de Céspedes whose father led the first independence war in 1868. But the Students' Directorate and other radical groups were not satisfied with the simple disappearance of Machado, desiring a full-fledged revolution. This is where Fulgencio Batista, the stenographer-sergeant, entered the scene, and Cuban politics. Noncommissioned officers headed by Batista rebelled against the army command on the night of September 4, taking over the power in order to hand it to a five-man civilian commission named by the Students' Directorate. Céspedes had lasted three weeks.

The new president was Dr. Ramón Grau San Martín, a professor of physiology and an idol of the students after the Machado regime arrested him along with other university professors in 1931; they had been charged with sedition. Grau lacked political and administrative experience, but he was fully in tune with the rising wave of nationalism and radicalism among the Cuban youth. As Jaime Suchlicki, an exiled Cuban historian, wrote about those intoxicating days, "With Grau, the Generation of 1930 was catapulted into power" and the "students held Cuba's destiny in their hands." The guiding spirit in this revolution was Antonio Guiteras, the twenty-five-year-old interior minister, who pushed for social and economic reform, and the abrogation of the Platt Amendment. While the Communist party now backed the government, it did not dominate it in any significant way, and neither Grau nor Guiteras approved of the establishment of soviets in the sugar mills by Communist workers.

Nevertheless, this was an overall situation the United States would not tolerate, and the Cubans were reminded once more that they were victims of "historical fatalism," in Martí's words, which meant that nothing could be set in motion on the island without American blessings. This went for the Grau reforms. In discreet contact with Fulgencio Batista and his military commanders, the United States pushed for Grau's overthrow. Washington had never recognized the Grau regime, and early in January 1934, thirty navy warships encircled Cuba, sending a clear signal that no more Cuban nonsense would be permitted—and that the marines were ready to land. On January 14, Batista's army ousted Grau, and Carlos Mendieta became the provisional president, instantly recognized by the United States. The Grau revolution lived for one hundred days, and from then on, Cuba was run by a succession of five stooge presidents manipulated by Batista—until he was ready to run for president in 1940. Two years after

Grau's fall, Antonio Guiteras was killed by the police when he attempted to flee Cuba after failing to launch a fresh revolutionary movement. Together with the Communist leader Julio Antonio Mella, Guiteras came to symbolize the Generation of 1930 for the generation of Fidel Castro. And Batista symbolized the evils of the past.

Batista's presidency was a crucial period for Cuba, but Fidel was a totally apolitical student up to the time of his graduation from Belén shortly before his nineteenth birthday. As a teenager, he had written a letter to President Franklin D. Roosevelt congratulating him on his 1940 reelection, declaring his support for democracy and his opposition to Nazism, and asking for a twenty-dollar bill. Fidel may have simply hoped for a signed reply from FDR, but all he received was a note of thanks from the Department of State and an expression of regret that the money could not be sent.

Castro has said repeatedly that his social-justice sentiments were awakened during his youth among the poor peasantry of Birán, but there was nothing in the environment of his Jesuit college to strengthen them. In fact, Belén was more an intellectual center for the preparation of Cuba's future right-wing leaders. Most of the teachers were Spanish priests of extreme rightist persuasion, having come to Cuba after the end of the Spanish Civil War in 1939 and the victory of the Franco nationalists. Often they represented the strain of Spanish anti-American nationalism that neither forgave nor forgot that the United States had wrested the island away from Spain in 1898. And, of course, right-wing and left-wing nationalisms and populisms tend to blur.

By far the most militant in these views was Father Alberto de Castro, a young Spaniard who taught sociology and history, and certainly the most influential teacher at Belén. Father de Castro propounded the thesis of *Hispanidad,* which attributes historical superiority to the political and cultural influence of Spain and Spanish thought, and he harangued his pupils with the notion that Latin America's independence had been frustrated because there had been no social reforms and because Anglo-Saxon values had dislodged the Spanish cultural impact. He had formed a small student association, called *Convivio,* to propagate his ideas, but it made no headway.

Juan Rovira remembers Father de Castro predicting that the Americas would be the future scene of world wars, that a war between the United States and Latin America was inevitable, and that all the small nations of Central America and the Caribbean should unite with South America to face the threat from the north. Rovira says the priest had "a very expressive face" and was a "fantastic orator."

It is unclear if Father de Castro's preachings had much—or any—direct influence on Fidel Castro, but, coincidentally or not, Fidel does share the Jesuit's opinions on the fundamental incompatibility between the United States and Latin America as well as the conclusion that liberal democracy is "decadent." Many of Fidel's companions at the university say they remember his intense interest in the writings of José Antonio Primo de Rivera, the founder of the Fascist Falange party in Spain prior to the civil war (and the son of a dictatorial Spanish military prime minister). Still, it would be unjustified to ascribe fascistic tendencies to Castro. In Mediterranean philosophical climates, rightist and leftist extremes meet easier than elsewhere, having both social populist characteristics. Benito Mussolini marched on Rome as a socialist before seizing ultimate power as a Fascist.

Castro himself has little to say about the Belén faculty politically apart from observing that without exception the teachers' ideology was "rightist, Francoist, reactionary," and that at that time there were no "leftist" Jesuits in Cuba. He remembers that at Belén communism was considered "a very bad thing," but still, his political activities were nil. Castro has told Frei Betto that he "observed" the rightist philosophy of his professors, though "I didn't question it much; I was in sports . . . I tried to advance my studies."

Coexisting with rigidly rightist Spanish Jesuit teachings at Belén were ideologies of groups favoring liberal Christian social trends in Cuban politics (now they would be described as Christian Democrats), and serious efforts were made to attract the apolitical Fidel Castro. José Ignácio Rasco, who launched a full-fledged Christian Democratic movement in Cuba in the 1950s, continues to believe that while Fidel stayed away from such groups he still harbored his Roman Catholic faith. He thinks Castro lost his faith suddenly at the university and became an atheist.

Rasco's version is naïve, and Fidel must be believed when he says that he never had religious faith and that at school "they were unable to inculcate such values in me." By the same token, it is fully credible when Castro declares that "later, I had another type of values: a political belief, a political faith that I had to forge alone through my experiences, my reasoning and my sentiments." But it sounds more like self-aggrandizement when he says that "unhappily, I had to be my own preceptor all my life" and says also that he would have been grateful for a mentor in politics and "revolutionary ideas" when he was in his teens.

Castro's Belén schoolmates who remember him the best are convinced that from the outset the Jesuits had an eye on him as a leader they could form to have a great destiny in Cuban politics. Father Rubinos, the head of the oratory academy who was regarded as the "ideologue" of Belén, centered his attention on Fidel as the most intelligent student in the college, the best athlete, and the leading outdoorsman. In the end the Jesuits failed to mold him; Castro is grateful for their intellectual influence but contemptuous of their efforts to bring him into the fold.

Twenty years earlier, the Jesuits had the same plans for Eduardo "Eddy" Chibás, the son of an Oriente millionaire, and a Dolores and Belén student of extraordinary promise. Instead, Chibás took a revolutionary role in Cuban politics, first against the Machado dictatorship, then against the corrupt Batista-manipulated puppet presidents of the 1930s and Cuba's quasi-colonial status. By the time Fidel Castro reached Belén, Eddy Chibás had become a famous opposition politician and the nearest figure in the young Cuban history to the incorruptible José Martí. In the years to come, Chibás would be Castro's political mentor and protector—and Castro his worthy successor. In its own fashion, Belén was the school of great Cuban leaders.

Though Cuba enjoyed the first period of full-fledged representative democracy in its history, vested economic interests and corruption continued to exist, and the country still could not risk to antagonize the United States and American investors. The Platt Amendment had been abridged after the fall of the Grau regime in 1934, but the "empire," as Martí called it, had not relinquished its decisive voice in Cuban affairs.

In any case, Fulgencio Batista was a constitutional president, governing with the backing of the armed forces, the conservatives, and the Communists, an unprecedented coalition. There was an elected congress where all voices could be heard, and there was a free press. In 1940 elections were entirely clean, and Cubans hoped that finally a new era had dawned. It was Batista himself who proposed the free elections, preferring to be president in name as well as in fact after seven years of ruling Cuba from behind the scenes in Camp Columbia, the Havana military headquarters. Unquestionably, the struggle for the survival of democracy in the world, even though the United States was not yet formally in the war, played a crucial role in Batista's decision to act in a democratic fashion.

To win the presidency, Batista defeated Ramón Grau San Martín whom he had overthrown in 1934. But this time there was a reversal of roles. Grau was no longer the radical professor of the past, reverting instead to the classical model of Cuban presidents and politicians whose overwhelming interest was pomp and wealth. Batista, on the other hand, chose to present himself as a candidate with advanced social and economic ideas, earning the leftist support Grau once had. In the meantime, the former sergeant quietly succeeded in becoming a very rich man.

Batista's election to the presidency followed the drafting of a new Cuban constitution early in 1940. The constituent assembly that produced it had been freely chosen in probably the first such exercise of the vote in the history of the republic. This constitution was remarkably progressive by Cuban and Latin American standards of the day, including clauses proscribing *latifundia* and thus opening doors to agrarian reform (which Fidel Castro would institute twenty years later according to this model), establishing new social-welfare provisions, and limiting the presidency based on democratic elections to a single four-year term.

The behavior of the Communist party at this juncture is extremely important not only historically, but with respect to future revolutionary developments in Cuba and Fidel Castro's complex relations with the Communists. In fact, the seeds of Communist Cuba were planted in 1940 when Castro was still at Dolores in Santiago, and the men with whom he would form his alliance after the 1959 revolution were already seasoned organizers and politicians.

Cuban communism has deep roots. A Communist Republic of Cuban Soviets was formed in Havana on August 1920 by a handful of admirers of the Russian Revolution of 1917, but obviously it was no more than a gesture. In 1922 a Communist Association was organized in Havana by Carlos Baliño, then seventy-four years old and a remarkable figure in Cuban history. Baliño, who had been an associate of José Martí in the Cuban Revolutionary Party in New York (he lived as a worker in the United States much of his life), was the founder of the Socialist Workers' Party in 1904, and was probably the first serious Cuban Communist militant.

Other "Communist Associations" appeared throughout Cuba, and in August 1925, Baliño and several fellow believers called a congress to give birth to the Cuban Communist party. The party was founded by the seventeen delegates attending the congress, including Baliño, Julio Antonio Mella (a student leader), and Fábio (Abraham) Grobart, a Polish-born apprentice tailor in his early twenties who had arrived in Cuba three years earlier—speaking only Yiddish—to escape the persecution of Communists in his homeland. In 1986, Grobart, well in his eighties, was the only surviving founder of the party, a superbly lucid politician and a treasure trove of Cuban Communist lore. He had tears in his eyes when Fidel Castro awarded him a medal on the sixtieth anniversary of the party's foundation; at the party's Third Congress in February 1986, it was Grobart who introduced Castro to the delegates.

Though illegal, the Cuban Communist party functioned in a brilliant and disciplined fashion, its influence vastly exceeding its numbers. Intellectuals, artists, and labor leaders formed the backbone of the party, making it possible for the Communists to occupy key positions in national life. Rubén Martínez Villena, one of the leading Cuban poets of this century, was an early party leader; there was widespread mourning when he died after an illness while still young. Among party militants and sympathizers was the outstanding painter, the late Wilfredo Lam; poets and writers Nicolás Guillén (in 1986 still president of the official Writers' Union though in his eighties), Alejo Carpentier (a distinguished Cuban novelist who died after the revolution), Juan Marinello (the party's president in the 1940s and a member of Castro's Politburo in the 1970s), Raúl Roa García (who until his death served as Castro's fire-breathing foreign minister), Pablo de la Torriente Brau (Castro's university friend), Emilio Roig de Leushenring; and the leading Cuban economists Jacinto Torras and Raúl Cepero Bonilla. In truth, there were few creative personalities in Cuba since the 1930s who were not of the left, or the extreme left.

Communist leaders, not always confessing their political persuasion, had dominated the powerful Cuban labor unions since the early 1930s, organizing political strikes and exercising a considerable influence on the economy. Following the directive issued in Moscow by the Seventh Conference of the Communist International (Comintern) the Cuban party adopted the policy of cooperation with non-Communist parties in organizing "popular fronts." In Cuba this meant collaboration with Fulgencio Batista.

In a move that Fidel Castro may have found inspiring many years later, early in 1938 the Communists first created the Revolutionary Union Party (PUR) to serve as a legal vehicle for some of their activities while the illegal party continued its own operations. The PUR then helped organize the Popular Revolutionary Block (BRP) with other opposition parties arrayed against the candidacy of ex-President Grau, promising their support to Batista in the 1940 elections. Straight-faced, the Communist party's Central Committee announced in mid-July of 1938 that Batista was no longer "the center of reaction" and that all efforts should be made to force him to keep "progressive" promises.

This was a neat piece of political footwork, and the government of President Federico Laredo Brú (a Batista puppet), aware of the Communist strength in the labor unions, rushed to nail down the party's support by legalizing it in September 1938, thereby ending its thirteen years of ostensibly illegal existence. The Communists stayed with Batista, continuing to build up their power. In January 1939 they merged the PUR with the real Communist party to produce the Communist Revolutionary Union Party (PURC), and launched the Cuban Workers' Confederation, representing a half-million workers, under Communist leadership. In 1959 and 1960, Fidel Castro, working with the same Communist leaders, imitated this twin maneuver when he proceeded to unite all the revolutionary groups and to join with the Communists in controlling the labor confederation.

When Cubans voted for the Constituent Assembly in 1940, the Communist party had only ninety thousand members, and was able to win only six seats. The six Communist constituents succeeded, however, in attracting an extravagant amount of attention by submitting their own draft constitution emphasizing the "anti-imperialist" struggle for propaganda purposes, though never expecting their proposals to be adopted, and by persuading the assembly to allow radio broadcasts of their sessions. Thus, Communists won access to a nationwide audience, another lesson that would not be lost on Fidel Castro.

In the presidential elections, the Communists basked in Batista's triumph, and celebrated their own victories in winning ten seats in the Chamber of Deputies, eighty municipal assemblymen throughout the island, and the mayoralty of the town of Manzanillo in Oriente—the first time Cuba had a Communist mayor. The party justified its support for Batista by stressing the importance of his commitment to a liberal constitution and his program of building schools, hospitals, and roads. But old-line Communist leaders acknowledge privately even today that the toughest test they had faced with the party membership was the effort to explain Stalin's wisdom in signing the nonaggression pact with Hitler's Germany in 1939. Still, party discipline prevailed—as it would prevail again and again in the future.

Batista naturally welcomed Communist support because of the party's power in organized labor, and this alliance became even smoother after the Nazis invaded the Soviet Union in 1941 and the Russians became the allies of the Western democracies. Batista established diplomatic relations with Moscow and gave Juan Marinello, the party president, a seat in the cabinet; subsequently Carlos Rafael Rodríguez, the party's chief intellectual and now Fidel Castro's closest adviser, took that cabinet seat.

Marinello's and Rodríguez's membership in the Batista cabinet does not appear in official postrevolutionary histories of the Communist party (though Fábio Grobart, the octogenarian party founder and a man of great intellectual honesty, mentions it in a monograph on Marinello). Fidel Castro says in one of his accounts of contemporary Cuban history only that the Communist party "had a certain influence" in the Batista government. Cuban political memories are indeed selective when it comes to the Communist trajectory. Both the party's pro-Batista phase or its role in the revolution when it refused for a very long time to go along with the *Fidelista* vision of history are very lightly passed over.

Acting on Moscow instructions in 1944, the Communists engaged in still another "tactical maneuver." With the end of the Batista term and the rise in anti-Communist sentiment, the party changed its name to the Popular Socialist Party (PSP), doing away with the uncomfortable word "Communist." Although the PSP delivered twice as many votes in 1944 as in 1940, it could not prevent Ramón Grau's election. Marinello, however, was elected to the Senate and the party retained its hold on the labor confederation through the presidency of Lázaro Peña, a member of the PSP leadership. Generally, however, Cuban workers elected Communists as their leaders not because of Peña's and the others' ideology, but because of the high quality of their unionist performance. But after 1944 in general Communists came under growing pressures in Cuba.

As part of the war effort, Cuba under Batista had granted military bases to the United States, including the big Guantánamo naval base in Oriente, reaffirming once more the island's immense strategic importance in the Western Hemisphere. Batista also agreed to sell Cuba's entire 1941 sugar crop to the Americans for less than three cents a pound, a very low price indeed. Like so many thoughtless American policies toward Cuba, this was taking advantage of the little country, and was one more reason for rising resentments among the new generation. At Havana University, for example, an "Anti-Imperialist League" was organized even before the war ended, not because of any specific American acts, but on general principle.

The economic relationship between the two countries was regulated by the 1934 Reciprocal Trade Agreement, which gave the United States total control over the insular market. This agreement was the quid pro quo for the Roosevelt administration to abrogate the already embarrassing Platt Amendment of 1902. Both moves stemmed directly from the overthrow of the threateningly radical first Grau regime by the Batista army, but as far as most thinking Cubans were concerned, Cuba still had to live with its "Platt mentality inferiority complex." This meant that Washington could go on doing what it pleased in Cuba. The corresponding psychological phenomenon in Cuba was "historical fatalism," a feeling that Cubans could never do anything with their own country without North American approval. History and Castro would defeat this fatalism once and for all.

The extent to which the adolescent Fidel Castro, completing his studies with the Jesuits at Belén during the war years, was aware of all these political pressures is most uncertain. There is nothing in the record and nothing in the recollections of his former teachers and fellow students to suggest the slightest curiosity on Fidel's part in Cuban and world politics. In fact, Castro would say later that he was a political illiterate until he entered the university. Yet, as one of his contemporaries has remarked, "Fidel has fantastic political radar," and it is entirely possible that events in Cuba during his time at Belén had not gone wholly unnoticed by this "political animal" from Birán.

In any case, Belén gave Fidel Castro an extraordinary sendoff. His basketball coach and teacher, Father Francisco Barbeito, wrote in the school's year book: "Always Fidel distinguished himself in all the subjects related to letters. He was *excelencia* [in the top ten of the graduating class] and *congregante* [a student who regularly attends prayers and religious activities] and a true athlete, always defending the banner of the *Colégio* with pride and valor. He has known how to win the admiration and affection of all. He will make law his career, and we do not doubt that he will fill with brilliant pages the book of his life. He has excellent qualities. . . . He has the timber, and the artist will not be absent."

Castro was *not* number one in the graduating class, but it is remembered that he received the biggest and warmest ovation from his fellow students when he was summoned to be handed his diploma. The life of Fidel Castro would progress from ovation to ovation.

CHAPTER

4

If, as he says, Fidel Castro was politically illiterate before he entered the university, his education in this field developed with lightning speed. Enrolling in Havana University's law school in October 1945 (students then entered directly the faculties from which they hoped to graduate in their chosen professions), Castro almost immediately plunged into politics. He had just turned nineteen during the summer vacations in Birán, and his political innocence indeed seemed nearly total, but the savage environment he found at the university forced him into active involvement.

The political situation at Havana University reflected the overall state of affairs in Cuba, though it was inevitably accentuated and aggravated by youthful passions and their manipulation by professional politicians. Ramón Grau San Martín, the physician who headed the short-lived radical junta in 1933–1934, was elected to the presidency in 1944 as the candidate of the opposition *Auténtico* party (claiming it stood for the ideals of José Martí), principally because Batista let him win. In the postwar democratic climate, Batista chose not to put up a military-supported successor and retired to Daytona Beach, Florida, to continue building his personal fortune while keeping an eye on Cuban politics. Grau, increasingly moving to the right and forgetting his dedication to social reforms, allowed his regime to wallow deeper and deeper in corruption and chaos.

Corruption and brutal political rivalries led to a generalized violence throughout the island that Grau could not or would not stem. Havana University turned into a battlefield among armed gangs whose origins went back to the anti-Machado "action groups" of the 1930s, and whose leaders now sought a new generation of recruits. It was impossible for politically conscious students to stay away from these confrontations; in Cuba the university was the steppingstone to national political careers. Still, it was extremely difficult to define the programs or ideologies of the different factions roaming the university "hill" in the center of Havana and the broad expanse of stairs leading down from the faculty buildings—the famous *escalinata* where most of the political rallies were held.

Under the circumstances, Fidel Castro, the new law student, had a wide choice of allegiances both ideologically attractive ones or those merely a political convenience. There are reasons to believe that he did a certain amount of shopping around at the outset, avoiding for a time becoming clearly identified with any of the factions. His instinct for elusiveness was already asserting itself, allowing him a way of maintaining his freedom of action as long as possible.

Political identities and loyalties could be highly volatile in the crucible of the university battles, and radical shifts were part of the scene. Only the Communists on the campus, those admitting to membership in the Popular Socialist Party (PSP) or in the Socialist Youth (JS) and those concealing it, were rock-solid and disciplined in the face of rising attacks on the party by the Grau government.

Castro was not a member of the party, and it is very hard to reconstruct precisely his political profile during his university years in the light of his diverse activities and positions and variety of personal friendships and relationships. Essentially, he was engaged in creating a personal reputation and myth as rapidly and as dramatically as possible, exhibiting both his flair for the spectacular and his attraction to the limelight. Castro has said almost nothing publicly about his university period, apart from accounts of his conversion to Marxism. However, on a visit to his alma mater immediately after his revolutionary victory in 1959, he described Havana University in his day as having been more dangerous than the Sierra Maestra. Like most activist students then, Fidel never ventured anywhere without a gun on him.

From the first day on campus, Castro is remembered as a serious and intense young man, over six feet tall, powerfully built, and very excitable and violence prone. Though his face was still boyish, he had a presence that could not be ignored. While most students wore *guayaberas* or sports shirts appropriate to Havana's heat and humidity, Fidel often made a point of wearing a dark wool suit and a necktie as if to set himself apart from the crowd by an aristocratic elegance. It was all part of his myth-building, and he would say in 1959 that "I was the Quixote of the university, always the target of cudgel blows and gunfire" in the midst of the waves of gangsterism.

Aside from myth creation, Fidel Castro was an extremely attractive person, appealing to men and women alike. His almost Greek profile, his tall Spanish *hidalgo* carriage, his piercing brown eyes, his physical courage in university melees, and his powers of persuasion quickly pushed him to the front ranks of popularity. Fidel's athletic prowess, particularly as a high jumper and 400-meter track runner, added to his growing reputation. Now he had to address himself to active involvement in university politics, and to become known beyond the students' *escalinata* in Havana. Social-background considerations did not matter at the university, as they did at Belén, which allowed Castro to take full advantage of his talents.

Havana University was composed at the time of thirteen schools, ranging from law to medicine and architecture, and each school annually elected its president. The University Students' Federation (FEU) was the mainstay of student political activity and had a great deal of influence in Cuban politics. The president of the FEU and other top officials were elected by the presidents of the thirteen schools. The presidents of the individual schools were elected by delegates from each class year (i.e., the four years of the law school or the six years of the architecture school), and the year delegates were chosen by delegates from each academic course who had to be selected by the students attending those courses. Basically this was a very democratic procedure, but all these elections were part of the larger Cuban political process by which everybody, from the government to the Communists and the gangster groups, tried to exert influence with votes, money, and muscle.

Because the university was autonomous and self-governing—neither the police nor the army could enter the campus—this "sacred hill" in the heart of Havana was a sanctuary for politicians and political gangsters of all persuasions. Even those gangsters who were not students had the run of the university grounds. Students who were not part of the FEU leadership or even federation members might be closely associated either with gangster groups, which inevitably called themselves "revolutionary" to improve their image, or with established political parties. Shoot-outs on and off the campus and brutal beatings were routine occurrences in which rival police units participated under one pretext or another, often gunning for each other. It was truly impossible to establish anyone's political identity with complete assurance, and this was particularly true of Fidel Castro as he battled his way through the jungle of university and Havana politics.

Castro fought hard for an elective post in the FEU, but the only time he succeeded was in 1945 when he was chosen a delegate from one of the courses in the first year of law school, shortly after he entered the university. This was the lowest elective position in the FEU structure, and apparently Fidel could never muster support for the presidency of the law school. If he had, it would have made him a member of the federation's governing body and opened the way to the chairmanship of the FEU. The student-elected president of the law school that year was Baudilio Castellanos, a childhood friend from Birán and a personal friend to this day.

The most plausible explanation of Castro's failure to be elected to any significant university post is that he was constitutionally unable to be a team player—in politics as in sports—and therefore none of the politically organized student groups wanted to risk supporting him. He was too unreliable and unpredictable. Ironically, even the Communists at the university, including his best friends, refused to back him in elections, notwithstanding their personal admiration for him.

In a speech after the revolution, Castro recognized that at the university "my impetuosity and my determination to stand out led me to combat" and that "my frank character made me enter rapidly into conflict with the milieu, the venal authorities, the corruption, and the system of bands that dominated the universitary environment." This view of himself is corroborated by his friends and contemporaries of all political denominations. Castro's savagely independent attitudes during his entire university career suggest that his basic passion for independence has never been extinguished and that despite today's formal ideological commitments he will always remain his own man.

Enrique Ovares, who served five years as president of the architecture school and three terms as FEU president, says that "Fidel could never be elected president of the [law] school" because he would not work with others. In a lengthy interview in Miami in 1984, Ovares, meticulously objective about Castro even after the years he spent in his prisons, said that on the surface it seemed "inexplicable" that Fidel could not win a major election. "At that time," he said, "Fidel indisputably had great dedication, had a series of leadership traits that could already be seen, and he had political ideas even if they were a bit anarchist. . . . He did not know well what he wanted, but he expressed himself well, and he had people who followed him." Ovares is also convinced that the Communists, well entrenched in the university in the mid-1940s though weak elsewhere in Cuba, refused to support Castro for office because they knew they could not control him. "Anybody who followed closely Fidel's political process as a student knows that Fidel at no time maintained relations with the Communists. Fidel was an individual who, because of his ideological formation, was a negative type for the [Communist] party because he was an individual who would say 'white' today, and 'black' tomorrow, and 'gray' the day after. He was totally independent, he could not be controlled."

In 1947, when Ovares was running for his second term as FEU president, he won in a 7–6 vote by the presidents of the individual schools, with the Communist leader of the school of philosophy and letters casting the decisive vote. On the defeated slate, Fidel Castro ran for the post of FEU secretary, and Humberto Ruíz Leiro, a Catholic student leader, was the presidential candidate. According to Ovares, who had Communists on his slate, Castro went with the Catholic faction when he realized that he would not have Communist backing if he joined the Ovares group, and he campaigned intensely for the Ruíz Leiro ticket. Ovares insists that Castro never had Communist support at the university, and that the Communists allied themselves with the victorious presidential ticket because "it was convenient for them to have me there." He says that although he was opposed to communism, he believed that there should be room for all tendencies in the FEU, and the Communists knew "they would get in. Fidel would have never let them come in, and that is why the Communists never supported Fidel."

That Fidel Castro was indeed too independent for the Communists is confirmed in considerable detail by Alfredo Guevara who is one of the most intriguing figures in Cuban revolutionary politics, and one of the men most trusted by Castro over the forty years of their friendship. Alfredo Guevara (no kin to Argentina's Che Guevara), presently the Cuban ambassador to UNESCO in Paris, has spoken freely of his university time with Castro.

Guevara entered Havana University at the same time as Castro, but enrolled in the school of philosophy and letters because he and his friends Lionel Soto Prieto and the late Mario García Inchaústegui arrived with the extraordinary idea of "capturing the FEU." Guevara and soto were of humble social origins (Guevara's father was a Havana railroad locomotive engineer) and had graduated from a public school where they directed an anarchist-minded student association. But at the university they began to shift their ideas to what Guevara calls "socialism," and soon both became members of Communist Youth (JC).

Guevara picked the philosophy school because women were in majority among the students, and he calculated correctly that he could be elected president there more easily as part of the takeover plan. The law school adjoined the philosophy school, and almost immediately he began hearing of an interesting student named Fidel Castro next door. Marching over to the law school, he found Fidel standing in the patio surrounded by students listening to him. Guevara's interest was in "leadership" affairs, and he introduced himself to Castro who impressed him greatly. But Guevara has said, "At that moment I had prejudices against Fidel, because I came from a public school, a poor student from the Havana institutes, and he came from a religious school. . . . from Catholic priests. For me in those days, a religious school and a man of religion were the same thing . . ." Despite this, Guevara concluded that Castro was "a volcano," that he could block his Communist group, "and therefore we have to conquer or vanquish him." He says that he felt that "here I have found a boy who will be José Martí or the worst of the gangsters because I saw in him a man of action, and for me the image of men of action was one of gangsters fighting. . . . But I wanted to conquer him. But the problem was that we tried to organize political meetings the way we did at public school, having one person speak, then another, and another, so that we could capture the meeting in the end. But Fidel, who had no arrangements with us, whom we did not know, would start speaking on his own after everything had been organized, and he changed everything."

This experience was evidently the first indication to the Guevara Communist group that Castro could not be trusted and thus should not be supported politically. In the 1947 FEU elections, it was Guevara who beat Castro for the post of federation secretary when Fidel ran with the Catholics against the Ovares ticket. Moreover, Guevara admits that for a period he was not certain whether Castro would even become a "socialist." He says that "I knew there was honesty in him, and a nationalist, anti-imperialist, revolutionary, and radical position, but I was not sure he would attain socialism."

Guevara may have had this preliminary assessment of Castro, and the Communists may have refused to support him, but this did not interfere with the formation of personal friendships. In Cuba, as in much of Latin America, deep political differences do not preclude friendly relations. Havana University was, after all, a small place, and outstanding students tended to gravitate toward each other even if they differed ideologically. Enrique Ovares, an anti-Communist, says that in the 1940s "it was the era of hatred against Nazism and falangism, and more or less all the Cuban political leaders were people of the left. . . . To be on the left, then, was a normal and logical thing, apart from the age of the kids at the university. They were full of ideals, and one really believed that all these things proposed in theory by the socialists and the Communists could be achieved . . ."

A strong friendship developed between Guevara and Castro, and they took part together in a series of political confrontations at the university. In 1959, Alfredo Guevara was part of a secret task force that allowed Fidel Castro to assume full control of the new revolutionary government, still filled with independents, and then to move Cuba toward communism (he also ran the cinema industry and served as vice-minister of culture). Despite their close relationship, it does not follow that Guevara was necessarily responsible—and this is true of other leading Communists who knew Fidel well at the university—for Castro's conversion to Marxism.

Having succeeded in controlling the Communists after the revolution's victory in 1959, instead of being controlled by them (a powerful party faction would attempt again to impose itself in 1962 and 1968 before being smashed), Castro made full use of his Communist university friends. Although none of them participated in any of the great *Fidelista* revolutionary actions—the party did not believe in them at the time—afterward they were given important assignments.

Knowing his Communist friends had no other place to go after 1959, and believing in their organizational skills, Castro recruited those at the university to set in motion, at first secretly, Cuba's transition to communism once he made the basic decision to do so for overwhelming strategic reasons. Before the revolution he had had virtually no contact with the top party leadership, and these older men were brought into the 1959 political picture in a different fashion and in a different power context. With a few exceptions, they were merely decorative.

Meanwhile, Alfredo Guevara instantly entered the most intimate circle, and he remains there. Lionel Soto, Raúl Valdés Vivó, and Flavio Bravo organized in Havana the first Marxist instruction school for the new Communist leadership emerging from the revolution, a vital mechanism in the transition. In 1986, Lionel Soto became a member of the Communist party's Secretariat, a key policy post, Flavio Bravo Pardo was chairman of the National Assembly and a member of the Council of State, and Valdés Vivó was a member of the party's Central Committee. Alfredo Guevara commutes between Havana and Paris, enjoying full access to Fidel and Raúl Castro. None of these men, however, is the regime's ideologue: This is Fidel's domain.

To the extent that it can ever be fully proved, available evidence indicates strongly that Castro became a convert to Marxism in his own way and in his own good time. Castro says that he became familiar with the ideas of Marx, Engels, and Lenin in his third year at the university, which would have been in 1948 or 1949, having first been a "utopian Communist."

Since the rate of ideological evolution cannot be measured or defined scientifically, there is no reason to doubt his version of his Marxist catechism progress or his phrase that "if Ulysses was captivated by the songs of the siren, I was captivated by the incontestible verities of the Marxist literature." He acknowledges that he patronized the Communist party's bookstore on Havana's Carlos III Street (today Avenida Salvador Allende), and it is probably true that he read or borrowed Marxist texts at Alfredo Guevara's house in Old Havana.

Much more important is how Castro's Marxism and socialism evolved in his full-fledged revolutionary period and when he assumed power. At the university he carefully refrained from even using words like "Marxism" or "socialism" in public speeches; conversely, he never attacked socialism or communism in his political pronouncements—though over the years he has strongly criticized the "old" Communist party.

Alfredo Guevara is right in emphasizing that Castro is a man of action, and there was plenty of it for Fidel outside the confines of the highly structured university politics. Despite his efforts at the university, Fidel wound up as a political loner, a role much more compatible with his personality and which would become his hallmark, and his strength in the greater political arena of Cuban politics.

The lethal realities of Havana politics, in and out of the university, were two large powerful gangster groups whose origins were in the violence surrounding the Machado dictatorship in the early 1930s. One was the Socialist Revolutionary Movement (MSR), founded in 1945 by Rolando Masferrer. Masferrer, a Spanish civil war veteran on the Republican side, had broken away with several friends from the Cuban Communist party. The other was the Insurrectional Revolutionary Union (UIR), headed by Emilio Tró, a veteran of both the Spanish war (where he fought alongside the anarchists) and World War II, during which he served with the U.S. Army in the Pacific, participating in the Guadalcanal campaign. Because the only thing the MSR and the UIR had in common was greed for power and political (and economic) influence, they were natural enemies, permanently involved in mutual killings. They each opposed the government of President Grau because it was the popular position of the day, and Grau—unwilling or unable to put an end to this political gangsterism—preferred to temporize and attempted at times to buy off the two organizations. Obviously it did not work.

The MSR claimed it stood for an anti-Communist "revolutionary socialism," whatever that meant, which would also oppose United States "imperialism" and Grau's *Auténtico* party. Its other stated objective was the overthrow of the bloody dictatorship of Rafael Leónidas Trujillo Molina in the Dominican Republic (a 1930 creature of the United States and of the marines who who were occupying the country at the time). The anti-Trujillo stance gave the MSR a certain respectability, attracting to its leadership such men as the exiled writer Juan Bosch who was later elected president of the Dominican Republic in 1926 after Trujillo's assassination. Basically, however, the MSR was an association of gunmen, and much of its power was concentrated at the university where Manolo Castro, an engineering student and an associate of Masferrer, was FEU president for five years. The UIR's official objective was to rid Havana's streets of "assassins," presumably meaning the MSR

The great conflict between the two organizations erupted early in 1947 when President Grau committed the folly of naming the UIR's Emilio Tró and the MSR's Mario Salabarría as majors in the National Police, a desperate effort to neutralize the gangs. Tró was also made chief of the National Police Academy, and Salabarría was appointed head of the Investigations Department of the National Police. Thus Grau managed to instigate a war within the police apparatus as well. Grau's other idea was to name Manolo Castro to be the national sports director. This post prevented Manolo Castro from running for reelection as FEU president in 1947, opening the federation to a succession struggle (won by Enrique Ovares and lost by the slate that included Fidel Castro) in which the MSR and the UIR were naturally deeply involved.

The war between the MSR and the UIR directly affected Fidel Castro's political interests and destiny because in order to survive, politically as well as physically, he had to maneuver skillfully between the two gangster organizations. This was his first taste of real infighting, and Fidel instantly demonstrated his talent for concealing his true positions and playing both sides of the street, a practice known by his subsequent admirers as the exercise of "political genius."

Even today it is impossible to determine accurately Castro's relationships with the MSR and the UIR, possibly because they were of such shifting nature, especially in terms of personal ties with leaders on both sides. Fidel is known to have developed a good relationship with MSR's Manolo Castro when the latter was winding up his presidency of the Students' Federation, but he was never seriously identified with that movement. In 1946, Fidel was accused, but without proof, by some political leaders of wounding with a gunshot an obscure UIR student activist named Lionel Gómez to please Manolo Castro.

According to contemporary accounts, Castro aligned himself with the UIR while completing his first year at the law school in the spring of 1946, but it is not certain whether he actually joined it. Jesús Diegues, a UIR leader, told an exiled Cuban historian in a letter many years later that Castro "used us for his own political battles within the university without ever really identifying [publicly] with UIR." This would be par for the course for Fidel, although the reverse may also be true. The slate on which he would run in the July 1947 FEU elections was supported by the UIR, while Ovares and his Communist allies had the MSR's backing. It has also been said that Fidel went to the UIR when the MSR rebuffed him. A published report claims that Fidel Castro was present when President Grau swore in Emilio Tró, the UIR chief, as director of the National Police Academy, but there is no further corroboration. In the end, it is virtually impossible to ascertain the truth.

Meanwhile, Fidel's first outside political performance came in the spring of 1946, before he had turned twenty. The occasion was a meeting at the home of Carlos Miguel de Céspedes, a rightist politician and grandson of the 1868 independence leader, running for mayor of Havana with ties to the Machado dictatorship and hoping for the support of the Students' Federation. He invited Manolo Castro, then FEU president, to negotiate student backing, but Manolo Castro insisted on bringing along three colleagues from the university, including Fidel Castro, who in previous weeks had led law-school students in a violent attack on a group of youths, described as "Nazi-Fascists," trying to hold a meeting on the campus.

Céspedes, according to an article published in *Bohemia* magazine in June 1946, outlined his campaign plans and asked for comments. When Fidel's turn came, he started out by saying that he would support the candidate— and here the hosts broke out in smiles—but on three conditions. He made the classical Fidel pause, and said that the first condition was for all the young revolutionary leaders killed by rightist regimes, including Communist party cofounder Julio Antonio Mella, to be brought back to life, that Céspedes and his friends return to the national treasury all the money they had "stolen from the people," and that history be set back a century. Dramatically, Fidel announced that "if these three conditions are met, I shall immediately sell myself as a slave to the colony into which you want to turn Cuba." Presently, he rose and marched out of the mansion. This little-known episode is a milestone of sorts in Castro's ideological history: The powerful Cuban nationalism and his sense of betrayal of Cuba by one and all since the wars of independence, and the thundering protest against the "exploitation" of the poor by the rich continue to be Fidel's fundamental themes forty years later.

At the university Castro joined the newly formed Anti-Imperialist League and the FEU's Committee for the Independence of Puerto Rico as a great many other students did. Given the deep nationalist sentiments surfacing among young Cubans during and after the war, and the concurrent exacerbation of resentments against the United States for the role it had played in Cuba since 1898, it was not particularly surprising that a man like Fidel would have become a member of such organizations. It did not at all define him as a Marxist. When it comes to "anti-imperialism," he has been consistent in this attitude from his student days as a *Cuban* perhaps more than as a Marxist, which is a subtle point not often appreciated in the United States.

Fidel crowned the first full year at the university with his debut as a public speaker on November 27, 1946. He was in his second year at law school (academic years straddle calendar years), he had just turned twenty, and was already sufficiently well known to rate front-page treatment in the next day's newspapers. The patriotic occasion was the seventy-fifth anniversary of the execution of eight medical students by Spanish colonial authorities as punishment for their pro-independence activities. Traditionally, the ceremony was organized by Havana University and held before the martyrs' pantheon in the sprawling Colón cemetery in the Nuevo Vedado district.

Castro naturally paid tribute to the memory of the medical students, but instantly launched into a tirade against the Grau government, denouncing the president's unconstitutional plans to try for reelection in 1948, accusing the regime of exploitation, and appealing to Cubans to abandon their apathy and rise against those who allowed them to starve to death.

This was the speech Fidel had so carefully rehearsed at the home of his fellow student, José Ignácio Rasco, and it evidently made a great impression, especially when he attacked "the presidential tolerance for some ministers who steal public funds and for the gangs that invade the inner circles of the government." In rhetoric that would become familiar to Cubans in the years and decades to come, Fidel proclaimed that "if Machado and Batista assassinated and persecuted decent persons and honorable revolutionaries, Doctor Grau has killed all hopes of the Cuban people, transforming himself into a stigma for the nation."

Castro had hit on the great populist formula of Cuban speechmaking, and he would never really deviate from it. Though he was the last speaker at the lengthy ceremony, he was quoted in the newspaper *El Mundo* in the fourth paragraph of its lead article on page one, ahead of much better-known political speakers. The newspapers, referring to him as Fidel *de* Castro, failed to explain on whose behalf he had spoken—it may have been in the name of the law school—but it did not truly matter because for all practical purposes a new star had ascended the firmament of the contentious Cuban politics. And he was not yet legally of age.

CHAPTER

5

Nineteen forty-seven was for Fidel Castro the year of definitive political commitment, the year he launched his political career in earnest, and the year of great romantic political adventure and immense personal peril. Following up on his cemetery speech the previous November, Castro was one of thirty-four signers of a declaration against President Grau's reelection by the central committee of the FEU, which he helped draft as a delegate of the law school. Among other signers was the federation president, Enrique Ovares, the law-school president, Baudilio Castellanos (Fidel's childhood friend), and his future brother-in-law, Rafael Díaz-Balart.

The declaration, issued on January 20, 1947, had a *Fidelista* ring to it (though his companions had a similar penchant for magniloquent oratory), affirming that "the ideas of reelection, extension of the period in power, or even the imposition of candidates can be found only in the sick minds of traitors, opportunists, and the constantly insincere." It pledged "to fight Grau's reelection even if the price we have to pay in the struggle is our own death—it is better to die on your feet than to live on your knees." As Cuban scholars have noted, Castro would often use this slogan about life and death, a phrase appropriated from Mexico's revolutionary leader Emiliano Zapata.

Now a believer in maximal public exposure and confrontation, and full of ideas, Fidel shortly thereafter organized a trip by law students to the Isle of Pines to inspect a new prison. Built according to blueprints for a high-security prison in Illinois, it became known in Cuba as "the model penitentiary," but Castro found there execrable food and brutal treatment of inmates, and nearly started a free-for-all with the guards. Returning to Havana, he publicly chastised the prison administration, and, of course, made the newspapers. Ironically, it was the "model penitentiary" at the Isle of Pines where Batista would incarcerate Fidel and his rebel companions seven years later.

Castro realized from the very beginning of his political life that in order to succeed he had to operate on a variety of levels, often simultaneously. He had already become visible in politics in and out around the university—from the FEU to the "revolutionary" gangs—and he had learned the value of well-managed confrontation. In the spring of 1947, he decided to penetrate the world of traditional politics as well.

The opportunity to do so came when Senator Eduardo "Eddy" Chibás, the voice of anti-Grau opposition and the immensely popular champion of the Cuban common man against governmental corruption and exploitation by the rich, moved to form his own political party. Chibás, whose battle cry was his famous slogan "Shame of Money," was regarded as a future president and as the most honest and idealistic Cuban leader since José Martí. He was elected to the Senate at the age of thirty-seven, and now, as he was launching his Cuban Peoples' Party (PPC), he was just forty years old. Grau's determination to seek reelection, which Castro denounced in his January speech, triggered Chibás into leaving the official *Auténtico* party.

Fidel, who knew the Chibás legend from Belén College, was a sufficiently important "name" in Cuban politics to be among one hundred or so outstanding citizens who were invited to the historic gathering on May 15, 1947, when the PPC was officially born. The record shows that this meeting was attended by six senators, ten congressmen, numerous mayors, experienced politicians, academics, and businessmen and industrialists.

Fidel Castro, not yet twenty-one years old, was the only university student leader asked to come that afternoon to the headquarters of the Youth Section of the *Auténtico* party from which Chibás was breaking away. It is an exaggeration to say that Castro was a founder of the PPC, but he certainly was present at the creation, and his association with Chibás would be invaluable in the years to come. The PPC quickly became known to the public as the *Ortodoxo* party because it claimed to stand for orthodoxy in its loyalty to the principles of Martí, a notion that fitted perfectly with Castro's own sense of Cuba and personal destiny.

What Castro achieved by joining the *Ortodoxo* party was to give himself the option of pursuing his long-term political ambitions through establishment politics, and to position himself to take the best advantage of it. Being an *Ortodoxo* and dedicating much time to the *Ortodoxo* Youth Section, Fidel committed himself full time to a political life and involved himself in its chaos and violence.

There is no real contradiction between Castro's very practical decision to play politics from the inside through the new party, and what he has described as his revolutionary instincts and his evolution toward Marxism. Even at this young age, Fidel had enough sense and political radar to know that revolutions are not accomplished overnight and that the proper climate must exist for them to occur (his intense study of the French Revolution and his readings of Marx surely helped him to understand this). Obviously he was unable to imagine and predict that Fulgencio Batista would launch a coup d'état five years later and thereby create a revolutionary climate. At this time Castro had the notion that he could propagate the ideas of a social revolution through the media (whose darling he was becoming) and through a seat in the congress to which he already aspired. Should some kind of revolutionary conditions develop, however, as was always possible in Cuba, Fidel says that he would have instantly shifted to the vanguard of a revolution.

Moreover, Castro was not compromising his image and principles by joining the *Ortodoxos*. Chibás and the new party were not only the most powerful opposition instrument in Cuba, but they projected liberal and socially "progressive" views, far to the left of President Grau's ossified *Auténtico* politicians. Chibás himself was a romantic figure, full of panache and excitement, a man who challenged political opponents to sword duels when he felt aggrieved. Most of Cuba came to a standstill on Sunday afternoon when Eddy Chibás made his fiery weekly radio broadcast.

It suited young Castro in 1947 to give the deliberate impression, still persisting among many Cubans, that he was Chibás's favorite disciple, and, in time, his political successor. He campaigned hard for Chibás in the senator's unsuccessful bid for the presidency in 1948, and it is to Fidel's credit that he publicly warned him that the young people would abandon him if he continued to seek alliances (which he did) with rich landowners in Oriente. Castro was important enough for Chibás to answer from the rostrum in Santiago that "No, *Compañero* Fidel Castro, you may dissipate your doubts . . . Chibás would be incapable to defraud the devotion of the masses. . . . The day Chibás senses an extinction in the citizens' love, he will shoot himself in the heart."

Privately, as fresh evidence from interviews and written materials shows, Chibás and Castro rather resented each other; the former feared Fidel's ultimate rivalry, and the latter saw the senator as an obstacle to his future advancement. Raúl Chibás, Eddy's brother, who briefly replaced him as the *Ortodoxo* chief when Eddy committed suicide in 1951, says that "I do not recall my brother ever talking to me about Fidel" and that at the outset there were people in the new party who did not wish Fidel to join it "because they considered him a negative element . . . he was a fourth-rate figure." Raúl Chibás was in the Sierra with Castro and joined him in drafting the first guerrilla manifesto to the nation, making him a credible witness and not a Fidel detractor (though he is now in exile in the United States).

By the same token, it suited Castro in that early period to recognize Chibás as his guide and mentor, particularly when Fidel decided to run for congress in 1952; identity with Eddy enhanced his public standing. As for Chibás, he quickly recognized Fidel's political value, so long as he could be kept in check. But Raúl Chibás's revelations, and the fact that Castro stopped mentioning Eddy after the victory of the revolution, suggest strongly that the two men simply used each other, with no love lost between them. Fidel would use Eddy most spectacularly in the aftermath of the senator's self-inflicted death.

To Castro, the association with Chibás and the *Ortodoxos* must have been exceedingly important, and except for the clash during the 1948 presidential campaign, he accepted the senator's leadership. The *Ortodoxos* was the only political party Fidel joined before the revolution. The explanation for his never contemplating membership in the PSP—the Popular Socialist Party, as the Communists were known then—is that he simply saw no future for himself in submission to party discipline. Besides, the Communists were at their lowest point in national influence, except in the university and some labor unions, and Castro has never believed in alliances with losers.

Fidel's pragmatic disposition is further demonstrated in that despite his personal friendship with Communists at the university and his growing attraction to Marxism-Leninism, he did not let Eddy Chibás's strong anti-communism and refusal to enter into an electoral coalition with the PSP affect his alignment with the *Ortodoxos*. Castro remained a member of this party until 1956, just before the yacht *Granma*'s voyage to Cuba to initiate the guerrilla war. By then, he already had his own 26th of July Movement.

In an extremely candid conversation in 1981 with a Colombian writer, Castro said that even when he acquired "my Marxist-Leninist formation," he did not enroll in the Communist party, preferring to work within his own organization. This was not because "I had prejudices against the Communist party," he said, "but because I understood that the Communist party was very isolated and that it was very difficult to forge ahead from the ranks of the Communist party with the revolutionary plan I had conceived. . . . I had to opt between turning into a disciplined Communist militant or building a revolutionary organization that could act under Cuban conditions."

By the spring of 1947, Fidel Castro had become a full-fledged politician, operating on every possible level, and on the threshold of astounding adventures.

So completely was Fidel Castro engaged in his political and revolutionary endeavors in 1947 that he simply had no time to study at the law school or to enjoy even a modicum of social life that glittering Havana, with its bars, restaurants, hotels, casinos, theaters, cinemas, nightclubs, brothels, beaches, and swimming pools, offered its student population.

Castro passed his exams at the end of his first year of law school with little problem in the spring of 1946, but by 1947 he was so involved in so many extracurricular activities that he did not even present himself for exams after the second year. As he tells it, he audited third- and fourth-year courses without being a regular student, which complicated his standing as a law-school delegate in the FFU.

Fidel's single-mindedness about politics affected his social and personal life as well. He simply never went out evenings or weekends except to attend political meetings or visit fellow students or other young people he wanted to convince of his ideas. No sooner had he joined the *Ortodoxo* party than he began to build a personal following within the Youth Section through a group called Orthodox Radical Action (ARO), which, under Fidel's influence, preferred to seek power through revolutionary rather than electoral means. ARO and Castro published a mimeographed pamphlet, *Acción Universitaria,* with a rather limited impact. But ARO and Castro's network of young political friends would be the embryo of the *Fidelista* movement: Castro was always planning ahead for all conceivable contingencies.

As for his social life, it was also a question of time and interest, certainly not of money (he received enough from home for recreation) or of social standing. Max Lesnick, who was head of the *Ortodoxo* Youth Section and a friend of Castro's at the university (he is now exiled in Miami but has not lost contact with Havana), says that he never saw Fidel at any of the dancing and drinking spots patronized by students in the capital. "I've never seen Fidel dance, and I don't know anybody who has seen him dance," Lesnick says, adding this was very unusual in Havana where young people in the 1940s were as devoted to dance and music as they were to politics. "The idea of Fidel dancing is inconceivable." Lesnick also recalls Fidel's awkwardness and timidity with women. On one occasion, he says, he and

Fidel were at the *Ortodoxo* party headquarters on Prado Avenue in Havana when three "very pretty, very well-dressed young women" walked in to ask a question. According to Lesnick, Castro was most popular among women students, yet "behaved with incredible shyness." Lesnick says that "the man who was capable of discussion with a youth, an old man, a politician or a student, froze in front of these girls."

At the university, Lesnick says, Fidel had no girl friends, except for Mirta Díaz-Balart, a philosophy student whom he knew from Oriente and whom he married in 1948. Lesnick remembers that "at that time, politics were Fidel's obsession, and he would never miss a meeting to take a girl out or to go to a dance." But, his friends say, he was not practicing chastity either.

Fidel shared meals and evening conversations with young political friends at their homes and even at boardinghouses. In 1947 one such favorite spot was a boardinghouse on "I" Street in Vedado, near the university, where several students lived. It was run by La Gallega, a Republican woman refugee from the Spanish civil war, who was married to a Cuban architect and, in the words of a former boarder, was "the political brain behind the Students' Federation."

Castro often came to the boardinghouse with his friend Alfredo Guevara, and they sat until late at night at a dining room table, talking about politics, laughing and bantering. Fidel, invariably wearing his dark suit though with the necktie loosened, puffed on a cigar and sometimes played with the revolver he usually carried. One evening, he amused himself charging and recharging the chamber with the gun pointing up. When a boarder who had served in the Royal Air Force during the war told him sharply that a gun should always be pointed down while being charged to avoid an accident ("In the army, you would've gone to the brig," the ex-RAF man said to him), Fidel put it back in his pocket without a word. His real experiences with weapons would come soon.

Violence went on rising in Havana in the spring and early summer of 1947; on May 26, for example, an MSR leader named Orlando León Lemus was wounded by a gunshot, and rumors spread that he would be "finished off" for using his alleged "revolutionary" credentials for personal gain, which was true of most of the gang *pistoleros*. The MSR's action squad responded by machine-gunning members of the rival UIR organization in the streets.

In July Enrique Ovares, backed by the MSR and the Communists, was elected FEU president (to replace Manolo Castro, the new national sports director), beating the slate on which Fidel ran for federation secretary-general. As it happened, the UIR was supporting this ticket, leading the

MSR leadership to conclude that Castro was beyond any doubt identified with their enemy. From that moment on, Fidel was persuaded that the MSR was out to murder him, and he compounded this risk by public attacks on political gangsters.

He wrote denunciatory antigang articles in the student newspaper, *Saeta* (The Arrow), he helped to launch with Communists friends in 1946. This was the first regular publication to print Castro's editorials. He kept up his attacks as one of the principal speakers at the inaugural session of the University Constituent Assembly on July 16. Castro was among the main movers behind the effort to provide Havana University with a charter guaranteeing its freedoms and modernizing its educational methods, and he was at the end of the speakers' table in the university's auditorium in the company of the chancellor, several deans, and the FEU president to address the 891 delegates. This was Fidel's first full-fledged political speech (longer and more formal than at his cemetery appearance the past November). Also it is believed to have been the first time his photograph appeared in Havana newspapers and he was identified in the caption.

The text of this Castro speech has vanished (as have so many Cuban historical materials), but his unmistakable tone emerges clearly from excerpts published in Cuban newspapers in which he is still called "Fidel *de* Castro." Fidel started by paying tribute to the "pleiad" (classical education and allusions were a must in Cuban political speechmaking) of student martyrs, such as Communist party cofounder Julio Antonio Mella, murdered in the defense of a "progressist university movement." Praising dead heroes, Castro knew then as he knows now, is the best way to strike the emotional chord in the audience, preparing it for the real message the speaker wishes to convey.

He charged that "false leaders"—he had Fulgencio Batista and President Grau in mind—had been leading students in recent years toward "indifference and pessimism." The university, Castro said, must not be a place where "ideas are traded as if they were merchandise," and "a shameful environment of collective cowardice." He took on the gangs, especially the MSR, when he urged students to "unmask the merchants who profit from the blood of martyrs," and he was inviting even more hostility by describing the Grau government "as a tyranny that has descended over our nation." This was the classic structure of a Castro speech, manipulating and shifting audience moods, awing the crowd with flashy oratorical imagery, and establishing absolute authority of the speaker whose words and thoughts must be obeyed and followed. Fidel has not altered this basic technique over the decades, obviously because it works. In the university auditorium, he received tremendous applause after his speech, and his credentials as an orator were now firmly set. He spoke to students and politicians whenever and wherever he could find them—Max Lesnick says that "Fidel was the only one who could instantly mobilize fifty persons to follow him, and when he could not get students, he went out to chase followers in the street."

Castro had drafted Martí as a historical and inspirational ally from his first days as a campus politician, and he never again let go of the Apostle. As a university student, Castro recorded Martí speeches on wire, and listened to them to hone his own style. He has a rhetorical monopoly on Martí quotations, which are so numerous (Martí's collected works in a current Cuban edition run to nineteen volumes) that he must be the only one to remember them—and to know when a Martí phrase fits the best. In today's Cuba, Martí is regarded as a prophet whose words are sacred, and Castro is gaining a similar prophet status through the massive dissemination of his every public (or written) expression.

In the meantime, the Grau government and the MSR reached the conclusion that they had had enough of Fidel Castro, now the most outspoken and increasingly popular critic of the regime and its friends. Since he could be neither co-opted nor corrupted, he was given an ultimatum: to abandon his antigovernment and antigangster stance or to leave the university altogether. The warning was sent by Mario Salabarría, the top associate of MSR's founder Rolando Masferrer, who had been named secret police chief by Grau earlier in 1947. Salabarría had an assassin's reputation, and Castro, who referred to him as the "owner of the capital," feared that reprisals against him would go beyond being removed from the university.

Fidel decided to go away alone to a beach near Havana to ponder his situation, and to make up his mind what to do in the light of the threats. As he recalled it later, "This was the moment of great decision. The conflict hit me like a cyclone. Alone, on the beach, facing the sea, I examined the situation. If I returned to the university, I would face personal danger, physical risk. . . . But not to return would be to give in to the threats, to admit my defeat by some killer, to abandon my own ideals and aspirations. I decided to return, and I returned—armed."

This was one of the first significant turning points in Castro's life. He knew that if he capitulated, his career as a politician and the revolutionary leader he was determined to be would be instantly finished. In a country where masculine qualities like physical courage have an exaggerated weight, a coward had no place as a leadership figure. It was in Fidel's nature, as his entire life demonstrates, to accept challenges and take high risks in the name of principle. But it was also in his introvert's nature to isolate himself at the start of a crisis and to make the great decisions in solitude. He would do it on many future occasions, and emerge renewed and strengthened from these retreats, ready to do battle. Moreover, Fidel would discover that disappearances from public view also had the value of keeping his foes off balance, wondering where he was and what he would do.

Fidel Castro could never leave well enough alone, and as soon as he emerged from the Salabarría confrontation, he volunteered for the next adventure, an invasion of the Dominican Republic to oust the Trujillo dictatorship. The expedition was being organized by a group of Dominican exiles led by Juan Rodríguez García, a millionaire, and the writer Juan Bosch, the future Dominican president, supported and financed by both the Grau government's top officials and Masferrer's and Salabarría's MSR. All in all, it was a blend of political and economic greed and opportunism, touched by a very respectable idealism, that was recurrent in the Caribbean in those days.

To recruit Cuban idealists for the invasion, the planners had to turn to Havana University where the MSR, always seeking an idealistic and revolutionary image, held sway. In practice, this meant that Masferrer, Salabarría, and Manolo Castro controlled the recruitment. If Fidel were to join the Dominican adventure, he would have to be accepted by the MSR chiefs and be given guarantees that their gunmen would not assassinate him in the training camps. He remained convinced that the MSR was out to get him. Therefore, he needed a truce, or a deal, with his enemies.

Fidel was so anxious to participate in the invasion because it was both a matter of revolutionary honor, as he saw it, and of political limelight, which he craved. To help overthrow the hated Trujillo would have been a badge of honor for a young leader like Fidel. Contemporaries like Max Lesnick confirm that most students of that generation were powerfully motivated against Trujillo, and that it was "normal" for Castro to volunteer.

Early in July, Castro still had to complete several exams to graduate from the third year of law school, but when he learned of the Dominican project, he says, "I considered that my first duty . . . was to enroll as a soldier in the expedition and I did so." The implication is that this was the reason he failed to take the exams and thus lost both his standing as a regular student along with his freshly gained position as president of the law school. This is one of the many convoluted episodes in Castro's history. He explains that while someone else had been elected earlier in 1947 to the presidency of the law school—it was a student named Aramís Taboada (who fought at Moncada and was imprisoned in the early 1980s in Cuba for allegedly defending a counterrevolutionary at a trial)—the student majority rejected this president and named Castro instead. This version is not borne out elsewhere, but at least one Havana newspaper late in 1947 referred to Fidel as the law-school president.

Fidel recounts that at the time he was chairman of the university's Committee for Democracy in the Dominican Republic, and while he was not among the organizers of the expedition, he was close to exiled Dominican leaders and was duty-bound to sail with them. He does not tell, however, the troubles he had in being permitted to join the anti-Trujillo force.

Enrique Ovares, the new FEU president and an organizer of the invasion with the rank of *comandante,* says that he negotiated Castro's participation with the MSR chieftains. Ovares and Castro had friendly relations from high school days, their university political differences notwithstanding. "Fidel came one day to my house in Vedado, and we sat in the garden. 'It's not possible that they will deny me the opportunity of offering my life to do away with the Trujillo dictatorship. . . . But I cannot go because they will kill me there, because you know that Masferrer will kill me,' he said. So I asked Fidel if I, as president of the FEU, all of whose students must be accepted [in the expedition], insist that he goes, would he go? Fidel said: 'If you guarantee my life, I shall go.' So I went to see Manolo Castro [the outgoing FEU president and MSR leader], who was a friend of mine, and I told him there was a problem with Fidel Castro. Manolo Castro replied: 'Fidel is a shit, but he is right. I am going to speak to all these people and Fidel will go to the camps.'"

Because Manolo Castro was one of the chief advocates of the invasion, along with the incomparably corrupt education minister, José M. Alemán, and the Army Chief of Staff, General Genovevo Pérez Dámera (the latter two saw profit and power in the Dominican Republic after Trujillo was overthrown), Fidel received his guarantees. Late in July, he was sent to Holguín in the north of Oriente province to receive his first experience of basic military training at the local polytechnic institute. On July 29, Fidel and his companions were driven to the port of Antilla in Nipe Bay (near the Birán family home), and placed aboard four vessels to sail to Cayo Confites, an islet north of the coast of Camagüey province, adjacent to Oriente.

The expeditionary force totaled around 1,200 men who spent fifty-nine days on Cayo Confites under a blistering sun and a permanent mosquito assault, undergoing additional military training, but basically doing nothing because the invasion leaders could not make up their minds to proceed with it. Fidel says that he was named lieutenant in charge of a squad, then promoted to company commander when word was received from Havana late in September that the whole campaign was being called off. It is not entirely clear why this happened. Castro says simply that "contradictions between the civilian government and the army" forced the cancellation. But there is evidence that Cuban and international politics intervened to dismantle the expedition.

Soon thereafter, Emilio Tró Rivera, the head of the UIR, the political-action group with which Fidel had strong ties, was assassinated in Havana on September 15 by MSR hit men as a new wave of violence swept the capital. Tró also had been the director of the National Police Academy, and police agents under the secret-police chief, Salabarría, had failed to kill him in a previous attempt on September 2. UIR gunmen murdered a MSR-connected policeman on September 12, and Salabarría ordered Tró's arrest. Three days later, Salabarría's men located Tró dining at the home of a suburban police chief. At the end of a three-hour fusillade during which several bystanders were killed, Tró was dead, riddled by bullets. The Cuban press reported that there were sixty-four political assassinations and one hundred assassination attempts in Cuba during Grau's 1944–1948 term.

The Grau government's official version links, not very credibly, Tró's murder and the liquidation of the Cayo Confites expedition. In a press briefing on September 29, the army spokesman explained that while investigating Tró's death, military investigators found clues that led them to the América ranch near Havana, belonging to Education Minister Alemán, where "fantastic quantities" of arms and ammunition were discovered along with documents concerning plans for the Dominican invasion. The army then learned that the expedition's headquarters were at the Hotel Sevilla on Havana's Prado Avenue, close to the presidential palace. With this knowledge in hand, the army acted to prevent the invasion.

As the invasion preparations were widely known from the outset, with the education minister and the army Chief of Staff behind it, this connection seems unlikely. The truth appears to be that Trujillo complained to Washington that he was about to be invaded, and the United States quietly convinced the Grau regime to halt the invasion. In those days, Washington exercised this sort of influence in the Caribbean. After the Cuban army and navy rounded up most of the expeditionaries (General Pérez Dámera now being committed to regional law and order), the State Department expressed its satisfaction that a "threat to peace" had been removed.

Castro was on Cayo Confites when Tró was killed, though he was probably then unaware of it. He says that when orders came to cancel the invasion, some men deserted, but his battalion sailed anyway. They were twenty-four hours away from landing, he recalls, when "we were intercepted and everybody was arrested."

Most likely, the vessel was boarded by a Cuban navy unit and ordered to turn back (other expeditionaries were arrested by the army on Cayo Confites), but Fidel escaped detention. "I did not let myself be arrested, more than anything else, for a question of honor: It shamed me that this expedition ended by being arrested." Therefore, when the small coastal freighter *Caridad*, which the men nicknamed *Fantasma* (Ghost), sailed back west, Fidel jumped from the deck in front of the Oriente fishing port of Gibara, and swam southwest along the coast for eight or nine miles, in waters supposedly full of sharks, to reach Saetía at the mouth of Nipe Bay. According to one published version, Fidel actually went in a small boat that he lowered in the night from the ship, but he says flatly that he swam all the way, and there is no reason to doubt it.

Enrique Ovares is certain that Fidel swam ashore—"you can never say Fidel was a coward"—but he suggests another reason for it. Masferrer and his men were aboard the vessel with Castro who, according to Ovares, now feared that they would try to kill him and decided that he could best save his life by swimming to shore. Ovares says that he thought Fidel was right: "I could guarantee his life while he was in the camps, but not after the invasion was aborted."

From Saetía Castro went on to Havana (it is not known whether he stopped at home in Birán to rest and change clothes) in an immense hurry to return to his political battles and the university. He says he wasted August, September, and October on Cayo Confites, again missing his law-school exams. Once back in the capital, Fidel went to the apartment of his sister Juana—he had no other place to stay—and instantly plunged into the Havana infighting. He had reached maturity, turning twenty-one, on that desolate and frustrating Caribbean key, and now he craved action.

Castro wasted no time. He swam from the ship to Saetía at dawn of September 28, and on September 30 he was already delivering an anti-government speech at the university. The occasion was the anniversary of the killing of a student during the Machado dictatorship, but Fidel used it to blame the Grau government for having betrayed the cause of Dominican liberation and to urge, once more, the resignation of the president. As a child, Fidel had learned that permanent attack is the only route toward victory—even if he had not yet defined his ultimate objective—and he now adopted the principle of one of his French Revolution heroes, Danton, acting "with audacity, always audacity . . ." Audacity has been Fidel Castro's most overwhelming and distinctive trait.

His intense political involvement interfered with systematic studies, and he chose not to enroll officially for the continuation of the third year of law school because he wanted to avoid failing his exams. Instead, he signed up to audit third- and fourth-year classes. Despite the change in his student and political status, Castro claims that "at a certain moment, without seeking it, I became the center of that struggle against the Grau government."

Fresh from the Cayo Confites experience, Castro used every opportunity to harass the regime and its principal figures, and Havana was turning into a permanent battlefield in the closing months of 1947. At the September 30 demonstration, Fidel's principal target was Education Minister Alemán for his role in the Cayo Confites episode as well as for his corrupt practices and private corps of armed thugs. In the Senate the opposition submitted a censure motion against him. Alemán's response was the unhappy idea of having his followers organize a public manifestation of "adhesion" on October 9, but a scuffle developed in the crowd, and a high-school student named Carlos Martínez Junco was shot dead by one of the education minister's bodyguards.

The boy's death triggered a near-revolt, particularly when Alemán proceeded with his plans to hold the self-congratulatory meeting in front of the presidential palace where Grau was foolish enough to come out on the balcony to praise his education minister. Within hours, thousands of students, carrying the coffin of Carlos Martínez Junco, marched on the palace, shouting demands for Grau's and Alemán's resignations, shaking their fists at the balcony as they filed past. Fidel Castro was in the first row of the demonstrating students.

When the student mass reached the *escalinata,* the vast stretch of stairs on the university campus, Castro addressed them in an outpouring of emotion. He blamed Grau for the boy's death, saying that "there is no culprit of these tears and this grief other than President Grau." Noting that the following day, October 10, Cuba would be commemorating the anniversary of the first independence war in 1868, the history-conscious Castro accused Grau of celebrating it "with the criminals of this government . . . as a feast of joy with lights and champagne while the students cannot commemorate this date because we must bring here, to bury him, the cadaver of one of ours, of a student assassinated by the new thugs . . ."

A forty-eight-hour general strike by students, supported by labor unions, followed to demand Alemán's resignation, and demonstrations, went on for several weeks with Fidel Castro always present, always visible, always audible, and always in charge. Senator Chibás, the chief of the *Ortodoxo* party, chose to leave the street demonstrations to Fidel, his youth leader, himself working in the Senate for the approval of the anti-Alemán motion. In his weekly radio address on October 12, he lashed out at the regime's murderers. Now there was no question that Chibás and the twenty-one-year-old Castro were Cuba's most important opposition leaders, each concentrating on his public.

Inevitably, the MSR and the regime's police were out gunning for Castro. Salabarría had been arrested by the army after the Cayo Confites fiasco (ten $1,000 bills were found inside his shoes), but Rolando Masferrer remained firmly in control of his gangsters. Numerous attempts to ambush Fidel were made at the university, but each time he escaped. One day, Evaristo Venéreo, a lieutenant of the campus police, tried to disarm him. Castro pointed his gun at the policeman, saying coldly, "If you want it, try to grab it by the barrel." Venéreo surprised Fidel by challenging him to a pistol duel in a deserted corner of the university sports stadium. Castro accepted at once, his honor being at stake, but he took the precaution of bringing along a group of armed friends. This was fortunate because the lieutenant had posted policemen in the bleachers to ambush Fidel; they were spotted and fled together with Venéreo as the students shouted insults. Castro said later, "It was a miracle I came out from that alive."

A feverish imagination was part of Fidel Castro's political arsenal—as he would demonstrate to the nation in the first days of November with the October disturbances barely over. He had devised a plan he thought could lead to a mass popular uprising and Grau's overthrow, a premature expectation perhaps but one that was successful in creating a national scandal of vast proportions. In this, again Fidel had a genius for summoning Cuban history to inspire the masses as he attempted to stage his coup. His chosen instrument in this case was the La Demajagua bell, the Cuban equivalent of the U.S. Liberty Bell, which Carlos Manuel de Céspedes rang at his estate called Demajagua, near the Oriente port of Manzanillo to mark the opening shot in the 1868 independence war. For many decades, the bell had been entrusted to Manzanillo as part of a national shrine.

Actually, it was the Grau government that first became interested in the bell for a mixture of historical and political reasons, a common Cuban phenomenon. Grau's idea was to bring the Demajagua bell to Havana to make it ring at the next year's anniversary commemorations as it had done eighty years earlier; Grau was still thinking about reelection in 1948, and it struck him the bell would toll well for him.

Unexpectedly, however, the Manzanillo municipal council not only refused to let the bell go to Havana, but virtually expelled Grau's emissary. Most likely the reason was that Manzanillo had a radical political tradition based on the sugar-mill workers in the area and industrial workers in town, and that it was a center of opposition to Grau. Moreover, Manzanillo had Cuba's first Communist mayor in history in the person of Francisco "Paquito" Rosales, elected in 1940. Batista's agents murdered him in 1958.

Hearing that Manzanillo had turned Grau down on the bell, the ever-inventive Fidel Castro conceived the idea of having Havana University students bring the bell to the capital—he knew that the municipal council would agree to this—and organize a mass gathering at which the bell would be rung, after which crowds would descend on the presidential palace to demand Grau's resignation.

Fidel, outlining his concept to his Communist friend Alfredo Guevara, was fully confident that it would work and that Grau's overthrow would be achieved. The next person brought into the scheme was Lionel Soto, also a Communist at the university, and a friend of both Guevara and Castro.

Finally, FEU President Enrique Ovares was informed (it was necessary because of the role the university was to play in Fidel's scenario), and he, too, liked the idea.

Ovares confirmed that the plan, presented to him by Fidel and Alfredo Guevara, aimed at a confrontation with the government. He says that he agreed to accompany them to Manzanillo, but the plotters decided that Lional Soto would go instead of Guevara. Max Lesnick recalls that several party leaders, including Senator Chibás, contributed around $300 to cover the expenses of the Manzanillo trip.

While Fidel, Lionel Soto, and Enrique Ovares traveled by train to Manzanillo on November 1, Alfredo Guevara in Havana had the task of procuring arms with which the rebel students would presumably face government forces. He remembers that his first contact was a gang leader named Jesús Gonzáles Cartas, known as *El Extraño* (The Strange One), who received him sitting on a throne, surrounded by flags, and bathed in indirect lighting. *El Extraño* promised to sell Guevara arms but never did; the students, Guevara says, obtained them through other channels. In Havana in the Forties, it was not too hard to buy arms if one had the right contacts.

To extract maximum political effect from their enterprise, Fidel and his companions spread the word that they were bringing the venerable bell, and thousands of students awaited the train's arrival in Havana on November 5 (two Manzanillo citizens came along to keep an eye on their treasure). A large convertible car carried the 300-pound bell from the railroad station to the university in a triumphal parade lasting over two and a half hours. There is a contemporary photograph of a very youthful-looking Fidel Castro in his dark striped suit and a flowery necktie, his right arm around the bell, and his left hand clutching a ceremonial candle holder.

At the university Castro addressed the cheering crowd from the convertible, declaring that Manzanillo patriots had refused to surrender the symbol of Cuban independence to "puppets at the orders of foreigners." But, he said, "the liberators of yesterday have faith in the student youth of today to continue their labor of independence." The bell was placed in the Gallery of Martyrs next to the office of the chancellor of Havana University for safekeeping while the students spent much of the night on the campus, planning the great anti-Grau demonstration the next day. The following morning, when the chancellor's office was opened, the students discovered that the bell had been mysteriously removed. The police, which had surrounded the campus all night (together with a hundred MSR gunmen), denied any knowledge of the theft.

Enrique Ovares recalls that Fidel, Alfredo Guevara, and several other student leaders appeared at his house early in the morning to apprise him that the bell was gone. Ovares said that as FEU president he would deliver a written accusation to the university chancellor on the grounds that *he* was responsible for the Demajagua relic (the elderly chancellor later challenged Ovares to a duel because he felt insulted). Fidel rushed to a radio station to charge the regime with stealing the bell while thousands of students began gathering at the university. Around noon, Castro was on the campus, grabbing the microphone to shout, "Let the rats stay here; we are going to denounce this robbery," and to lead thousands of students to the nearest police station to make a formal complaint.

In the early afternoon, Fidel returned to Ovares's house, but several police officers he had named as culpable in his broadcast followed him, brandishing their guns. Ovares and his mother convinced the policemen to leave. Fidel was subsequently accused of cowardice for hiding in the house, but Ovares says that "Fidel is an intelligent man; why would he want to get out of the house? To commit suicide? . . . Fidel always did it that way. In the Sierra as well. He would not come down from the mountains to get himself killed. He had to preserve himself for the end."

At night a mass rally was held at the campus (though the chancellor had decreed a seventy-two-hour suspension of classes to prevent disorders), and Fidel wasted no time in delivering a slashing attack on President Grau. He accused him of breaking his promises to look after the restoration of "national dignity," after the neglected peasants and the hungry children, saying that "the faith has been lost." Castro warned Grau that the students for whom "the deception was the most terrible" were there to proclaim that "a young nation can never say, 'We surrender.'"

In a passage that would become a Castro theme in the years to come, he spoke of a "betrayed revolution," the nationalist revolution Grau had promised Cuba, with peasants still without land, and "the country's wealth in foreign hands." Then he moved on to what would be another *Fidelista* rhetorical discovery and weapon: the power of statistics. He had already mastered the need for homework, his prodigious memory providing the means for conveying his knowledge comprehensively to his audiences. In this instance, Fidel was able to inform his listeners that in the three years or so that Grau had been in power, his government had been given 256 million pesos (the peso being equal to the dollar), but public health was awarded only 14 million pesos and public works 112 million pesos, while defense—meaning the armed forces—were given 116 million pesos. Castro was always obsessed with the need for huge investments in public health, which he was able to assign when he obtained power, but this was the first time he raised the issue publicly.

The threat of "militarism" was another Castro concern, and he warned his university audience of the growing power of the military—more than five years before Batista's coup. One of Fidel's special gifts has been his ability, both instinctive and analytically intellectual, to predict the future moves of his adversaries. He displayed it that November evening in 1947. Meanwhile, he urged students to become militant in the battle for national unity of the people "to obtain its true independence, its economic liberation, its political sovereignty, its political liberties . . . the definitive emancipation of our nation."

Some scholars of the Castro revolution consider the November 6, 1947, speech at the university as Fidel's moment of maturity as a political thinker, and as his first coherent attack on the status quo from a leftist perspective. This is debatable inasmuch as Castro himself concedes that at that time his process of evolution toward Marxism was still developing. There is no question, however, that at this juncture Fidel had created the distinctive political style that he would nurture for the rest of his life.

As to whether Castro was alone on that occasion in expressing leftist and nationalist views, the centrist newspaper *El Mundo* reported in its front-page article on the university rally that this was not so at all. Practically all the student speakers, it said, "made special references to 'Yankee Imperialism.'" This is an important point: In 1959 the American government and public opinion were, by and large, under the impression that Cuban nationalism and anti-Yankee sentiment were essentially Castro inventions when he captured power. When it came to Cuba, the island always taken for granted by the United States, almost nobody paid attention or did his homework. Hence the surprise of the true meaning of the *Fidelista* revolution when it finally came.

Several days after the Havana disturbances, the Demajagua bell was delivered to President Grau by parties unknown, and immediately sent back to Manzanillo. It marked the end of this particular incident, but young Fidel Castro had achieved new fame as Cuba's most promising rising political star whose increasingly controversial ascent would continue in the new year.

CHAPTER

6

For Fidel Castro, 1948 was a breathless year, politically and personally. He was asserting his political identity, engaging in immense activity, and acquiring new responsibilities.

Around him, Cuba was disintegrating and decomposing socially and politically. The Cuban society was more and more polarized, the island having lived in a revolutionary environment since the eruption of the insurrection against the Machado dictatorship in September 1930.

Cuba had the outward trappings of a democracy, advanced social legislation, and a progressive social and political instrument in the form of the 1940 constitution. In reality, the frustrations of a nation sharply divided between the huge wealth of a minority and the quite abysmal poverty of the majority—peasants and urban and rural salaried workers—had created a state of affairs in which the country was essentially ungovernable. The urban middle class was too small and divided to provide a solid political center.

Tolerating political violence and corruption on a grand scale, President Grau had abdicated not only national leadership, but his elementary responsibilities for the maintenance of day-to-day law and order. He compounded all the resentments and divisions by his unconstitutional decision

to seek reelection to a second term in 1948. The advent of a new generation, represented by the leadership and the rank-and-file in the University
of Havana (and even in secondary-education schools), was a direct challenge
to the putrid status quo, and the conditions were perfect for the emergence
of a challenger like Fidel Castro, a believer in a "true" Cuban revolution.

Elsewhere in postwar Latin America, similar pressures were gathering. In
Argentina Juan D. Perón had seized power in 1945 to launch a populist
nationalist movement in the name of social justice, producing a militaristic
and fascistic dictatorship while remaining a hero of the masses. Peronism
was a phenomenon in which Fidel Castro at the northern end of Latin
America was developing a special interest. In 1948 in Peru, the army had
smashed a revolt by the nationalist and socially committed APRA movement, one of the few wholly original revolutionary movements in Latin
America, eschewing classical socialism and Marxism. During the same year
in Venezuela, a nation belonging emotionally as much to the Caribbean as
to South America, a tyranny had fallen and been replaced by a democratic
and socially radical government. And next door in Colombia, a vicious civil
war with deep social undertones was in progress.

In Cuba the political and social tensions of the era were accentuated and
magnified by the island's physical smallness, and revolution as a way out of
the fundamental national crisis was the talk in 1948. The unhealthy and
disturbing relationship with the United States was an added dimension in
the bitter Cuban search for identity.

For years, unbridled political gangsterism flourished in the guise of "revolutionary" movements and organizations, rendering the word "revolution"
meaningless. By 1948, however, the whole question of the unfinished
Cuban revolution, its partial independence, became the subject of most
serious public debate by spokesmen of mainstream political and ideological
currents.

The rightist *Diario de la Marina* in Havana agreed editorially that Cuban
youth must find ways to channel its demands, but protested against attempts to bring together "workers, peasants, and students" because it
smelled of "Communist" inspiration. Summing up the ruling establishment's total incomprehension of the new realities, the editorial remarked
that "among student, proletarian, and agrarian youth there is not necessarily any community of origins and interests." Francisco Ichaso, a well-
known rightist polemicist, wrote that a "minority" in Cuba had convinced
itself that one of the safest and most profitable endeavors is "what they call
revolution," using the youth to activate it.

On the other end of the spectrum, Raúl Roa García, a professor at
Havana University (who would become Fidel Castro's ideologically radical
foreign minister and relatively elderly adviser in 1959), argued that it was

wrong to claim, as the rightists did, that the 1940 constitution "had concluded the revolution" in Cuba. Though the "colonial structure" had been removed, Roa wrote, the constitution was "a road, not a goal." Rafael García Barcena, also a university professor and later an anti-Batista revolutionary leader, put it best, saying that "a nation that enters into revolution, as José Martí said it, does not leave it until its extinction or until it crowns the revolution" with victory.

As for Fidel Castro, he was now obsessed by revolution. Manuel Moreno Frajinals, a noted Marxist historian who befriended young Fidel at the university that year, recalls that every conversation with him dealt with "deep revolution." Castro frequently visited the Moreno Frajinals household, and the historian says, "Even then, Fidel was determined to carry out a revolution, he was convinced it would happen, and all he did was in preparation for it. . . . He talked incessantly about the revolution." Castro, he says, realized early that access, if not control, of the communications media was vital, and while still at the university, he concentrated on those media—newspapers, magazines, and the radio—that were highly developed in Cuba. In fact, Fidel lost no opportunity to be as visible as possible.

On January 22, 1948, Jesús Menéndez Larrondo, a black union leader of sugar workers, a Communist, and a member of the Chamber of Deputies was shot to death by an army captain in Manzanillo, the Communist labor stronghold in Oriente province. Invoking his parliamentary, he had refused to submit to arrest and was summarily shot. Menéndez had been briefly arrested the previous October along with hundreds of leftist labor leaders as part of the campaign by Labor Minister Carlos Prío Socarrás to rid the unions of Communist control. But this time Menéndez had been warned that he was on the death list.

General Pérez Dámera, the army Chief of Staff, publicly commended the captain for his actions as a model example to the army, "so that every time a similar situation occurs, action be taken in identical form." The Grau government evidently no longer had any contact with political reality, as the Menéndez assassination came as a tremendous shock to the public opinion. The labor leader had been very popular, and his casket was placed on display at the National Capitol; tens of thousands of Cubans filed past the bier to pay him homage.

Among the mourners was Fidel Castro. Mario Kuchilán, a noted Cuban journalist, recorded in his memoirs that Fidel, who stood next to him at the cemetery, suddenly asked angrily, "What would you say if I climbed on a tomb to summon the people to march on the presidential palace?"

Two weeks later, Castro was back in the limelight. On February 11 a student demonstration was held in the center of Havana to protest police brutality toward students in Guantánamo in Oriente, and a streetcar was burned during the riot. The police charged the students, chasing them back to the university, but Commander José Caramés, the police chief of the university district, personally raced up the *escalinata,* pistol in hand. Before his men convinced him to leave the campus, he pistol-whipped a student who had a limp, and a serious confrontation was narrowly averted with the armed students.

Nevertheless, a violation of the university's autonomy had occurred when the police entered the grounds, and Fidel Castro called for a peaceful protest demonstration the following day, February 12. In Havana, however, there was no such thing as a peaceful protest. While a group of students deployed a .50-caliber machine gun atop the *escalinata* in the event of a new police invasion, Fidel and a fellow student led a march into town, carrying a huge Cuban flag and signs proclaiming "We Protest the Violation of University Autonomy!" Singing the national anthem, the students reached police barricades at an intersection, where they began shouting, "Out with Caramés, down with Grau—the assassins!"

Riot policemen moved on the crowd with clubs in their hands, and Fidel was among the first to be hit. Headlines in the next day's newspapers announced that he had been "injured" (this was the first time he made the headlines), and the news stories said he had "suffered a grave contusion" on the head and was taken to the Calixto García Hospital to be X-rayed. Castro's injury turned out to be superficial and he refused to remain at the hospital. Nonetheless, he had shed his first blood for the revolution, and that was all that counted, that and the publicity. Besides, the police released students who had been arrested, and Caramés was suspended pending an investigation. It was a most successful day for Castro and his cause.

An event of extraordinary gravity was the assassination in Havana on February 22, ten days after the university riots, of Manolo Castro, the national sports director and a founder of the MSR organization. Castro was shot in front of a cinema, of which he was a co-owner, by unknown assailants after being called out to the street under some pretext.

The first assumption was that Manolo Castro was killed in retribution for the death in September 1947 of Emilio Tró Rivera, chief of the rival UIR political gang and head of the National Police Academy, but the real assassins were never found. Manolo Castro was a very important politician, and he had received numerous death threats in the previous weeks and months. He was uncertain who exactly wanted him dead, and he went to the extreme of asking help from his former Communist friends from the university days. There was nothing they could do for him.

It was Fidel Castro, however, who was immediately accused by the MSR of murdering Manolo Castro, possibly because of Fidel's early ties with the UIR. He had ended them when he joined the new *Ortodoxo* party, but the MSR group under Rolando Masferrer (from whom Manolo Castro had protected Fidel on Cayo Confites) was obviously determined to do away with Fidel as his most dangerous political rival.

A few weeks before the Manolo Castro death, *Tiempo en Cuba,* a publication belonging to Masferrer, happened to have printed an article seeking to link Fidel with university gangsters. The day of the assassination, a nephew of Masferrer publicly charged Fidel. Within three days, on February 25, Fidel and three fellow student leaders were arrested in a car on the Havana seaside boulevard at 11:00 P.M. by policemen in a cruiser. The reason given was the investigation of the Manolo Castro death. Fidel was formerly identified as the law-school president.

The four student leaders indignantly denied any involvement with the murder. They produced the *Tiempo en Cuba* article about their alleged contacts with campus gangsters, and Fidel testified that he had spent the afternoon of the assassination day at the El Dorado café with friends, whom he named, and that he had spent the night at the Plaza Hotel. He told the investigating judge that when he saw his name the next day in the newspapers mentioned in connection with Manolo Castro, he went instantly to the nearest police station to offer testimony, but the officer in charge requested him to leave because there were no orders for his arrest.

The latter is an important point because over the years numerous published reports had created the impression that Cuban authorities actually issued a detention order against him, and the myth has survived. The arrest on the evening of February 25 was on the initiative of one of the officers in the police cruiser. The four men were submitted that night to a paraffin test to determine whether any of them had recently fired a weapon, and the investigating judge ordered them released on "conditional liberty" at 2:00 A.M.—whereupon Fidel immediately held a press conference at the police station.

He charged that Masferrer "wants to take over the leadership of the university to make it serve his personal interests," but "we have not allowed him to do so, in spite of the coercion and violence practiced against us for quite some time." Fidel said that Masferrer "wishes to incite action against us, using Manolo Castro as a pretext; in other words, he wishes to profit from the death of a friend. . . . If we had known beforehand what was going to happen, we would have prevented it." Fidel Castro's reputation in this context is also defended by the FEU ex-president, Enrique Ovares, now in exile in Miami. Ovares says that "Fidel had absolutely nothing to do with the Manolo Castro thing," and "I have no reason to lie to you." He adds: "If you have to attack Fidel, attack him with the truth.

Fidel has done terrible things. But why is it necessary to invent? This is the problem that bothers me about people who write [about Fidel]. If there is sufficient truth, why should one lie?"

Fidel's innocence in this crime was no guarantee, however, that Masferrer or others would not try to kill him, and so he decided it would be wise to vanish from sight for a time. His sister Lidia, Alfredo Guevara, and Mario García Inchaústegui helped him to hide and lead a semiclandestine life. But, as usual, a new project suddenly materialized to capture Castro's attention and, most conveniently, provide him with an opportunity to leave Cuba for a time. It would quickly turn into Fidel's greatest adventure to date.

The new project was an "anti-imperialist" association of Latin American students, a concept that was initially proposed and prompted by the Perón regime in Argentina. Castro embraced it with total enthusiasm. The group was designed to organize a Latin American students' association, heavy on nationalism and anti-Yankeeism, and a preparatory session of hemisphere student leaders was to be held in the Colombian capital of Bogotá early in April.

As happens so frequently with events in the life of Fidel Castro, there are contradictory versions concerning his exact role in the preparations for the Bogotá meeting and the entire background of the effort to create the Latin American student organization. Specifically, it is unclear where and how the idea of the "anti-imperialist" association was really hatched and developed: by the Peronists aided by Fidel Castro, or by Fidel Castro aided by the Peronists?

It is a matter of record that Perón had been actively seeking to spread the influence of Argentina under the guise of his *Justicialismo* (social justice) politics throughout Latin America. Usually covertly, Peronist funds flowed to labor unions, journalists and publications, and student groups to convert them to pro-Argentine sentiments and to the Perón doctrine: The general saw himself as a continental and world figure and his movement as a "Third Force" in international affairs.

This effort had been much less than successful, but in 1948 the Argentines were still investing in it. The notion of an "anti-imperialist" students' association was probably born in Buenos Aires, but it is unknown why the Peronists turned almost exclusively to Cuban students to set the operation in motion. Senator Diego Molinari, who was chairman of the Argentine Senate's foreign-affairs committee, at least one cabinet minister, and several lesser lights appeared in Havana at the start of 1948 to meet with Cuban student leaders and persuade them to help organize the association. By most accounts, Argentina agreed to pay all the expenses.

The Argentine proposal was well received in Havana by all student segments: the FEU president, Enrique Ovares, was for it as was Fidel Castro, the leaders of most of the Havana University schools, and militants of the Socialist Youth branch of the Communist party. In retrospect, this positive response makes sense: The Cuban resentment against the United States and "Yankee imperialism" was a fact of political life for the new generation—and Perón seemed to offer them a way out of isolation and into a community of nationalist solidarity. That Perón had nationalized British and American utilities, that he was at sword's point with the United States (which had futilely tried to undermine and oust him), and that he was portraying Argentina as a victim of British colonialism for retaining the Falkland Islands in the South Atlantic served to enhance his prestige. Young Cubans were prepared to overlook that Perón was a military dictator, and the Socialist Youth studiously ignored the anti-Communist aspects of *Justicialismo*. Nationalism and anti-Americanism were the common denominators. Just as the law student from Havana, Fidel Castro, was coordinating plans for the congress with Perón's envoys, Ernesto Guevara de la Serna, later the "Che" of the Cuban revolution, a nineteen-year-old second-year student at the medical school of Buenos Aires University in 1948, was already a Perón supporter.

Discussing the Latin American situation prevailing then, Fidel Castro explained in a 1981 interview what he and other young Cubans saw at that juncture: "There already existed strong contradictions between Perón and the United States. Our position in this [Latin American] movement was confined to these points: democracy in the Dominican Republic, the struggle against Trujillo, the independence of Puerto Rico, the devolution of the Panama Canal, the disappearance of colonies surviving in Latin America. These were four fundamental points, leading us to establish certain contacts, let us say tactical, with the Peronists who were also interested in their struggle against the United States . . . and demanding the Malvinas."

Castro went on to say that Peronists were busy sending delegations to various countries, meeting with students, "and from this convergence between the Peronists and us emerged a tactical approachment with them." He said that the Argentine student delegation came to Havana to coordinate with the Cubans how each of them would work in different areas "so that the forces of the Left in Latin America would organize this congress of Latin American students." Castro still regards Perón as something of a Latin American hero because of his commitment to social justice. Castro has indicated his contempt for the Argentine military who overthrew Perón in 1955, ushering seven years of democratic government, then restoring the dictatorship of the armed forces until the aging general was reelected to the presidency in the middle 1970s. The Cubans supplied arms and funds to the *Montoneros,* the radical leftist guerrillas who spun off from the Peronist movement during the "dirty war" with the Argentine military between 1976 and 1983. And in 1985, Castro went out of his way to receive in Havana Peronist senators and labor leaders who opposed the newly elected government of President Raúl Alfonsín.

As a student leader, Castro said, he took it upon himself to represent Cuban students despite his "conflicts" with the leadership of the FEU and the fact that not being officially enrolled in the university at that stage, he could not act as a federation official. In this Castro version, he became the central figure in setting up the congress because "I represented the great majority of the students who followed me regarding me as the leader." He said that "the idea of organizing the congress was mine," and that "I conceived the idea" that this congress be held in Bogotá simultaneously with the planned conference of Western Hemisphere foreign ministers, "called by the United States to consolidate its system of domination here in Latin America." The students were to meet, he explained, on the basis of anti-imperialist principles.

In this fashion, Fidel Castro planned to create an open confrontation with the United States and the Organization of American States in a Latin American capital, an astoundingly ambitious undertaking on the part of a twenty-one-year-old Caribbean revolutionary unknown beyond the confines of Cuba. Fidel never considered any of his ideas to be unattainable, though in this instance he could not possibly have foreseen that Bogotá would turn into a catastrophic explosion for reasons that were purely coincidental with his purposes. Still, Castro's political career is built from historical coincidences—events which cleared the way for his success and advancement.

The way Fidel tells the story, both the Peronists and the FEU were almost marginal in the preparation of the Bogotá congress, with him looming as the main inspirational and organizational presence. Fidel does tend to interpret history in this manner, never one to show modesty over his role in the continuing process of myth-making. Some of his companions, however, have slightly different recollections. Enrique Ovares was one of the four Cuban student leaders who went to Bogotá for the congress; others were Alfredo Guevara, the FEU's Communist secretary-general and Fidel's friend; Rafael del Pino, also his friend; and Fidel himself. While Castro claims that he paid for his travel to Venezuela, Panama, and then Colombia, maintaining that "we had very little money, just for the tickets"— Ovares says that Miguel Ángel Quevedo, the publisher of the Havana weekly magazine *Bohemia,* had given Fidel $500 for the trip to Caracas on the first leg of his precongress trip, and introduction letters to Venezuela's new democratic leaders. Castro was traveling with Rafael del Pino, and, according to Ovares, the "FEU sent plenty of its money to Caracas" for Fidel and his companion. Contradicting Castro's claim that he was the chief Cuban student representative, Ovares says that he sent FEU credentials for Fidel and Rafael del Pino in Caracas before they flew to Panama and then Bogotá. Ovares and Alfredo Guevara then went directly to Bogotá where they met the two others. Guevara, a slim young man who always wore dark glasses, was a bit on the spot: He was both Fidel's closest Communist friend and Ovares's deputy.

While Castro may overstate his personal commanding role in preparing the congress—the Cuban organizing committee had nine members, including himself, Ovares, Guevara, and Del Pino—but he was unquestionably its spokesman. On March 15, three weeks after the problems resulting from Manolo Castro's assassination, Fidel issued a declaration from his hiding place in Havana outlining the plans for the congress.

"We hope that this act will initiate a movement of major proportions that will find support in all Latin America, especially among university students, united under the banner of anti-imperialist struggle," he said. Castro announced that preparatory sessions for the congress would be held in Bogotá early in April, during the inter-American ministerial conference, "in order to support accusations against colonialism that various Latin American countries will present." Always the strategist, Fidel noted that "it will be easier to make such accusations if we launch a wave of protests."

On March 19, Castro drove to Havana's Rancho Boyeros (today José Martí) Airport to catch a plane for Caracas, his first stop en route to Colombia. But the police detained him before boarding on the grounds that he was attempting to violate the "conditional liberty" on which he had been released four weeks earlier during the murder investigation and he was taken before a judge. Ready to capitalize on the incident, Fidel informed the judge that he was carrying out a mission designed to "strengthen the bonds of friendship" among Latin American students. Taking the offensive, he demanded that the authorities issue a public statement concerning plans by armed thugs in Havana to assassinate him. The judge promptly let Castro go, erasing all charges, but Fidel first told newsmen that he had been victimized by those determined to "obstruct" his student activities and "to create for me an unfavorable situation before the public opinion." He left for Venezuela the next day, full of indignation and surrounded by fresh publicity.

This was Castro's first voyage abroad, and inevitably it started with a touch of adventure. Fidel recalls that his plane stopped in Santo Domingo, and that he committed the "imprudence" of getting off and risking being recognized by Dominican officials as having been a Cayo Confites expeditionary the year before. But he says, "With luck . . . I got back on the plane and nothing happened."

In Caracas, Castro and Rafael del Pino met with university students, who agreed to send a delegation to the Bogotá congress, visited the editors of the government newspaper, and made a courtesy call on the home of Venezuela's president-elect Rómulo Gallegos, a noted poet and novelist. The military dictatorship had been overthrown by a revolution led by young intellectuals and officers of a left-of-center persuasion. From there, the two Cubans arrived in Panama in the wake of demonstrations against United States control of the Canal Zone, and Castro remembers visiting a Panamanian student who had suffered a permanent injury as a result of the riot. Students in Panama similarly agreed to send representatives to Bogotá.

Even from afar, he kept an eye on political developments in Cuba. Presidential elections were to be held on June 1, and Castro, as a member of the oppositionist *Ortodoxo* party, supported the candidacy of Senator Eddy Chibás. But he disagreed with Chibás over a pact with the Communists. Although the Popular Socialist Party, the Communist organization's current name, had its own candidate in Juan Marinello (a former Batista cabinet minister), it offered to throw its backing to Chibás to help prevent a government victory. On March 31, Castro issued a statement to the Cuban press approving this move by the Communists perhaps as a means of bolstering Chibás's chances, as he was then last in the polls. Chibás, however, rejected any alliance with the Communists, even in congressional races.

Castro's public support for a pact with the Communists raises again the eternal question of when he actually became a Marxist or a Communist— and if his membership in the *Ortodoxo* party was a cover. There is no absolute proof one way or another, but even the anti-Communists among his fellow university students reject the notion that Fidel was a "hidden" Communist. Speaking in 1981 about his political stance when he went to Bogotá, Castro described it this way: "I had already entered into contact with Marxist literature. . . . I felt attracted by the fundamental ideas of Marxism, and I was acquiring a socialist conscience. . . . At that time, there were some Communist students at the University of Havana and I had friendly relations with them, but I was not in the Socialist Youth, I was not a militant in the Communist party. My activities had absolutely nothing to do with the Communist party of that period. . . . Neither the Communist party of Cuba nor the Communist youth [organization] had absolutely anything to do with the organization of that Bogotá congress. . . . I was then acquiring a revolutionary conscience, I was active, I struggled, but let us say I was an independent fighter."

During the events of Bogotá and subsequently, Castro was widely charged with being part of a Communist conspiracy designed to torpedo the foreign ministers' conference. Fidel would be deeply involved in the violence that surged in Bogotá, but this was more a display of adventuresome youthful nationalism than proof of any Communist militancy. Again it is Ovares who rises in Castro's defense, insisting that "I was there and I have not the slightest notion that he was a Communist."

"These are lies," Ovares says about reports that Castro had met with Communists and had received at the time a letter from the Cuban Communist leader Blás Roca Calderío. "The Communist is the one who was with me, and this was Alfredo Guevara." At a meeting with Colombian students immediately after his arrival in Bogotá, Fidel was interrupted by Jesús Villegas, a Colombian Communist party leader, who demanded that the Cuban display his credentials because "provocateurs frequently use revolutionary phraseology." This episode was reported in the Cuban press many years after the 1959 revolution when Castro no longer had interest in concealing his Marxist persuasion, though insisting that his conversion to communism was evolutionary, not having yet fully occurred at the time of Bogotá.

In the bizarre manner in which *Fidelista* politics operate, the official emphasis today, including by Castro himself, is to establish as a matter of history that he was a convinced Marxist by the time he reached his third year at Havana University in 1948—mainly, he says, because the *Communist Manifesto* had a "tremendous impact" on him—whereas, until 1961, it was considered reactionary and counterrevolutionary even to suggest that Fidel had any such ideological stirrings.

Castro arrived in Bogotá with Rafael del Pino on March 31, staying at the small, three-story Claridge Hotel downtown. Enrique Ovares and Alfredo Guevara flew in from Havana the next day, going to the San José boardinghouse, near the Claridge, because it was even cheaper. On April 1 the Cuban group met with Colombian and foreign student delegates at the university to organize the work of their congress. But there is a dispute as to who presided over the meetings. Ovares insists that he was named chairman because he was the president of the Cuban FEU, which was the organizer of the Bogotá congress. The meeting was very small because the Argentines never came, and there was only a scattering of Venezuelans, Colombians, a few Mexicans, and a Guatemalan. Ovares says that to "tranquilize him," Fidel was named emissary to Jorge Eliécer Gaitán, the immensely popular leader of the "progressive" wing of the Colombian Liberal party, to invite him to attend the student congress.

Castro has a diametrically different version of this, although the chance to meet the fifty-year-old Gaitán was a stroke of luck. He says with considerable exaggeration that "different progressive and leftist forces of Latin America" were present at the student sessions and that "the congress was organized" because of steps taken by him. He recognizes that he could not officially represent Cuban students because he was no longer a FEU leader, "but I spoke with considerable vehemence, I explained all that I had done, how I had done it, and why . . . I must say that in a practically unanimous manner the students supported me, after I made my presentation, a bit passionately as was to be expected at that time and at that age." Fidel went on to explain: "I was presiding over that meeting. I said that . . . what interested me were the struggle and the objectives of this struggle. . . . The students applauded very much when I spoke, and they supported the idea that I remain the organizer of this event." The students then passed a resolution condemning the conference of American foreign ministers that was being inaugurated in Bogotá on April 3.

This was the Ninth Inter-American Conference, and the United States delegation was headed by Secretary of State George C. Marshall, underscoring the importance of the occasion for Washington. The previous year the foreign ministers had drafted the so-called Rio Pact (they had met near Rio de Janeiro in Brazil) of mutual defense; now in Bogotá they were to debate and sign the charter of the Organization of American States, meant to replace the old Pan-American Union as the hemispheric instrument of collective policies.

The Cold War was already well under way—President Truman had already proclaimed his "doctrine" to defend Greece and Turkey from communism, and the Marshall Plan to reconstruct Western Europe had been launched in 1947. To young Latin Americans, the Inter-American Conference loomed as a United States scheme to assert its total domination over the region. There was little awareness that the previous month, March 1948, Communists had liquidated representative democracy in Czechoslovakia through a bloodless political coup d'état.

In Bogotá the students set out to harass the foreign ministers, and soon Fidel Castro found himself arrested by the Colombian police. One evening, Fidel, a Cuban companion he does not identify, and a Colombian student scattered "anti-imperialist" leaflets at a Bogotá theater during a gala performance for the visiting dignitaries. Castro says that "we were a little immature," and that they did not realize they were violating laws in propagandizing their congress; anyway, he remarks, "this was not an infraction against the Colombian state, but against the United States.

A pervasive anti-American sentiment comes across in most of Castro's statements, declarations, interviews, and speeches from the earliest days of his political life to the present time, particularly in his retrospective musings. It is as if he must analyze every situation affecting Cuba, and even the world, through the prism of anti-Americanism. This personal distortion may be the most tragic heritage the United States left behind among so many Cubans.

Castro suggests that his arrest by the Colombian secret police had resulted from their surveillance of the student group; they "knew something about our activities" in organizing the congress. Fidel recalls that they were arrested at their hotel, taken to "tenebrous" offices in "sordid" buildings, and interrogated by detectives. Evidently he was able to talk their way out of this new predicament, his talent for persuasion already being well developed, because they were released within a few hours. Castro says that he explained the ideas and ideals of the proposed "anti-imperialist" congress, and that he must be right in concluding that "I gained the impression that whoever was in charge there, did like what we were expounding. . . . We had been persuasive with them." Talking himself out of tight spots, including situations when his life was at stake, is an art Fidel had mastered from childhood.

On April 7, Castro visited Jorge Gaitán with Rafael del Pino, being taken to his office by Colombian "Liberal (party) students." At that stage, Colombia for over two years had been immersed in a savage civil war between the traditionally rival Conservative and Liberal parties, with thousands of dead in cities, towns, and villages. In 1948 Colombia's president was Mariano Ospina Pérez of the Conservative party, and the country was on the brink of a complete fratricidal catastrophe with political factions only able to squabble among themselves.

A few days before the conference of the foreign ministers whom the Colombian government had so imprudently invited to its tinderbox capital (and before the arrival of the Cuban students), Gaitán had led 100,000 persons in a March of Silence to protest police violence and brutality and delivered a "Speech in Favor of Peace." Receiving the Cubans, Gaitán gave them the text of the speech, explained the Colombian political crisis to them, and, according to Castro, agreed to close the student congress with a mass rally and an address by him to the delegates.

The next day, Castro went to a Bogotá court to observe Gaitán in action as a lawyer, defending a police lieutenant charged with killing a conservative politician. This was a major *cause célèbre* in Colombia, and the proceedings were broadcast over a radio network. Fidel found Gaitán was "brilliant" in the courtroom. He says he had "a really good impression of Gaitán . . . because he was a virtuoso orator, precise in language and eloquent . . . because he was identified with the most progressive positions in the country against the conservative government." Gaitán invited Castro to meet with him again on Friday, April 9. In terms of Fidel's political maturation, knowing Gaitán was crucial for him; again, events favored him.

Meanwhile, Fidel had another curious encounter in Bogotá. Alfredo Guevara recalls that at a meeting with students at the National University, he and Castro were introduced to a youth whose name was Camilo Torres. At the time, Torres meant nothing to the Cubans. But this was the young Colombian revolutionary who first became a Roman Catholic priest and then the famous chief of a guerrilla force that for years fought the army in the Andes. Father Torres was killed in the 1960s, and he now belongs to the pantheon of Latin American revolutionary martyrs and heroes. The ever-loyal Alfredo Guevara says that "it is sad to think that human beings do not carry a star on the forehead because one would have understood at that moment that it was *the* Fidel Castro and *the* Camilo Torres who stood there."

Fidel's appointment with Gaitán on April 9 was for two o'clock in the afternoon. He says he had lunch at the hotel, then went out on the street to take the short walk to Gaitán's office. But, Castro says, suddenly "there appeared people, running frantically in all directions . . . people who seemed crazed. . . . People shouting, 'They killed Gaitán!' 'They killed Gaitán!' . . . Angry people, indignant people, people reflecting a dramatic situation . . . telling what had happened, word that began to spread like gunpowder."

Gaitán had just been shot and killed on the sidewalk in front of his office building, and although his assassin, identified only as one Juan Roa, was instantly lynched by the crowd, Bogotá and Colombia blew up like a revolutionary volcano. The presence of the American foreign ministers' conference at that moment fueled suspicions that the assassination was planned to trigger a revolution and deal a terrible blow to the Organization of American States, but nothing was ever uncovered to corroborate it. Castro, who witnessed and participated in the Bogotá street battles says flatly that "nobody organized [the events of] April 9. . . . I can assure you that it was a completely spontaneous explosion that nobody had or could have organized. . . . What April 9 lacked was organization . . ."

The *Bogotázo,* as that urban revolt is now known, became Fidel Castro's real baptism as a revolutionary, and had an enormous impact on him, his thinking, and his future planning. It was the most important single event of his life up until then, and unquestionably one of his major experiences, providing him with a unique opportunity to observe an unfolding revolution—and to learn from it.

His own account of his activities during the five days of the *Bogotázo* is the most accurate and complete in existence, and those who were then in contact with Castro accept it as such. Fidel told his Bogotá story over long hours of conversation with the Colombian journalist Arturo Alape in 1981, cautioning him that after thirty-three years, he might have forgotten certain details. Still, his memory is so fabulous that the transcript of the interview reads like a marvelously vivid adventure tale.

The first thing Castro recalled after the news of Gaitán's death fired up the city was the sight of a man in a little downtown park trying to smash with his hands a typewriter he had somehow obtained, but having a terrible time of it. Fidel says he told the man, "Hey, give it to me," and "I helped him, grabbed the typewriter, threw it up in the air, and let it fall. Seeing this desperate man, nothing else occurred to me."

Castro decided to go to the National Capitol, where the ministerial conference was being held, but increasingly he came across people smashing "windows and things." This, he said, "began to preoccupy me because even then I had very clear and precise ideas of what is a revolution and what things should not happen. . . . I began to see manifestations of anarchy, to tell the truth . . . and I wondered what Liberal party leaders were doing."

He saw the congress building invaded by a furious crowd, some people carrying clubs, others with weapons, and watched office furniture being hurled out of the windows into the square. Castro, who was with Rafael del Pino, then went to the boardinghouse, where they found Ovares and Guevara. From there, Fidel saw a huge crowd rushing down one of the main avenues in the direction of a police station. It was at this instant that Castro, a round-faced young man improbably attired in light trousers, shirt and tie, and a leather jacket, made up his mind to join the revolution in Colombia.

"I join the first ranks of this crowd," he said. "I see there is a revolution erupting, and I decide to be part of it as one more person. . . . I had no doubt that the people were oppressed, that the people who were rising were right, and that the death of Gaitán was a great crime." At the police station, policemen were aiming their rifles at the crowd but not firing, and Castro says that many of them were joining the rebel throng. After entering the police station, the only weapon Fidel could still find there was a tear-gas shotgun, so he grabbed it along with twenty or thirty gas cartridges.

"I didn't have a rifle," Castro continued, "but I had something that at least can fire, the shotgun with the big barrel. But here I am in a suit, not dressed for a war. I find a cap without a visor, and I put it on. And my street shoes are not fit for a war. . . . I climb to the second floor, and enter the officers' room. There I start looking for clothes, and more weapons, and I start putting on a pair of boots. An officer runs in, and, I'll never forget it, in the middle of all this chaos, he wails, 'Oh, not my boots, not my boots . . .'"

In the courtyard Castro came upon an officer trying to organize a police squad, and he swapped his tear-gas shotgun for the officer's regular rifle and bullets. He replaced his cap with a beret and put on a policeman's jacket: "This was my uniform." Now the armed crowd, with policemen and soldiers in its midst, rushed like a raging torrent in another direction with Castro in the vanguard. Several cars with students he knew from the university drove by, and Fidel learned that a student group had taken over a radio station in the city. Hearing that the radio station was being attacked, Castro led a group to the aid of the besieged students, but the confusion, the firing, and the rioting were so overwhelming that he no longer knew what to do. And, Fidel remarked critically, "there were people who had been drinking, arriving with bottles of rum."

Castro said that it was unclear at that stage whether the army, like much of the police, had joined the uprising, but he suddenly came across a battalion in front of the War Ministry. "Possessed by revolutionary fever myself," Fidel recalled, "and trying to attract the greatest number of people to the revolutionary movement, I jump on a bench to harangue the soldiers there to join the revolution. Everybody listens, nobody does anything, and there I am on the bench with my rifle, delivering my harangue."

From the War Ministry, Castro and several companions resumed their walk to the radio station (he remembers that his wallet with all the money he had was stolen from him at that moment), but suddenly they came under heavy rifle fire, barely managing to hide behind some benches; Fidel said that "miraculously, they didn't kill us all." Unable to reach the radio station or the National University, he resolved to take over a nearby police station with his fellow students.

"It was assumed that I was the one to take the police station because I was the only one who had a rifle," Castro recounted. "This really was suicidal. . . . But luckily that station had already been taken in a [police] uprising . . . and they received us in a friendly fashion." He sought out the police-station commander, who also was the leader of the rebel policemen, and explained that he was a Cuban student organizing a congress here. Thereupon, Fidel says, "the police commander names me as his aide."

The two of them got into a jeep to go to the headquarters of the Liberal party, and Castro said that he was delighted because he had been concerned all day over the chaos and the lack of any organization. He added that "everything I'm telling you is rigorously exact about the incredible things that happened that day." At the party headquarters, the police chief procured a second jeep, and they returned to their station. At night Fidel and the police chief drove again to the party headquarters, each in his jeep, but the officer's vehicle broke down, and "I carried out the quixotic act" of giving him "my own jeep." Castro and a few students were left behind in the street, finally meeting a squad of rebel policemen with submachine guns and making their way to another police station now in revolutionary hands. But, he said, he did not have a cent for a cup of coffee.

This was the Fifth Police Division station, and Castro says it had some four hundred armed policemen and civilians. However, there was enormous confusion in organizing the station's defenses, and he was assigned a post on the second floor. What bothered Fidel the most that night was the looting in Bogotá streets, with "people resembling ants and carrying on their backs a refrigerator, or a piano," and he says that "unhappily, because of lack of organization, because of a problem of culture, because of a great situation of poverty . . . many people carried away all they could. . . . Through lack of political preparation and other factors, the city was looted. . . . I was very preoccupied that instead of seeking a political solution, many people chose to loot."

Seeing the rebel force was being kept inside the police station, Castro took it upon himself to tell the division chief and his officers that "the entire historical experience demonstrates that a force that remains in its barracks is lost." He cited Cuban military experiences to urge the police chiefs to dispatch their forces into the streets, assigning them an attack mission against government positions. Fidel was heard out amiably, but no decisions were made, although he kept insisting that "a revolutionary force kept inside is lost."

"I had some military ideas that emerged from my studies of the history of revolutionary situations," Castro said, "including of the movements that occurred during the French Revolution, the taking of the Bastille, of the Cuban experience—and I saw with full clarity that this was an insanity. . . . They were waiting for an attack by government forces." Castro also criticized the rebel policemen for beating up progovernment policemen they had captured: "This disgusted me."

It had occurred to him, Fidel said, that he did not really know what he was doing there alone "in a mousetrap," foolishly awaiting an attack instead of going out to strike the enemy. He wondered whether he should stay at the police station, but decided to remain "because then I had an internationalist thought, and I reasoned that, well, the people here are just like the people in Cuba, people are the same everywhere, and this is an oppressed, exploited people. I had to convince myself: They had their principal leader assassinated, this uprising is absolutely just, I am going to die here, but I am staying."

Castro finally persuaded the police chief to assign seven or eight men to him to patrol the hill behind the station from where the army could attack them. At one point, the patrol ordered a civilian car near the station to stop, suspecting that it was driven by a government spy. But, as the prudish Fidel Castro said with immense indignation, the man was with two prostitutes, taking them to have sex. "Can you imagine," he asked, "the city burning, the war erupting, and this man driving around Bogotá with two prostitutes?"

On the morning of April 11, a Sunday, word circulated that an agreement was being reached between the government and the Liberal opposition. Fidel recalls that he still had on his improvised uniform with a beret, his rifle with nine bullets, and a saber. Within hours, an accord was announced, and the rebels were asked to surrender their weapons. Castro had wanted to keep the saber, but he was not allowed. He believed that the peace agreement was a "betrayal" (one of his favorite words) of the people because after the rebels gave up their arms, government forces "began to hunt the revolutionaries all over the city."

Fidel returned downtown, finding Ovares and Guevara at the boardinghouse where they had sat out the rebellion. But the owner was a Conservative, and he started saying "horrors" about Gaitán and the Liberals. Castro said he lost his patience, became "exalted," and contradicted the man by defending the Liberals. This was a half hour before the 6:00 P.M. curfew, and the owner threw the Cubans out. They reached the safety of a downtown hotel with five minutes to spare. And Fidel added, "It was immature to commit the error of engaging in polemics with the owner at twenty-five minutes before six o'clock." No lesson is ever lost on Castro.

At the hotel the Cubans ran into an Argentine diplomat they knew and persuaded him to drive them to the Cuban embassy (diplomatic cars being exempt from the curfew). Fidel recalls that they were very well received at the embassy "because we were already famous and everybody was looking for the Cubans." The Cuban students remained at the embassy until April 13, when they were flown home to Havana aboard a Cuban aircraft that had come to Bogotá to fetch bulls. Castro thought it was pleasantly ironic that they were saved by the Cuban government they so strenuously opposed.

Summing up his Bogotá experience, Castro says that "the opportunity of seeing the spectacle of an absolutely spontaneous popular revolution had to have exercised a great influence on me," and that it was part of the ensemble of experiences he had acquired before engaging in the Cuban revolutionary struggle. And he said, "Remember that I was twenty-one years old then, and I think what I did there was really noble . . . I am proud of what I did . . . I think that my decision to stay there that night, when I was alone and it all seemed like a great tactical error, was a great proof of idealism, a great proof of quixotism in the best sense. I was loyal until the last moment . . . I was disciplined and I stayed although I knew it was suicidal . . . I behaved with principle, with correct morality, with dignity, with honor, with incredible altruism . . ."

Castro claimed that the Bogotá experience led him to "extraordinary efforts to create a conscience, a political education in Cuba . . . to assure that at the triumph of the Revolution there would be no anarchy, no looting, no disorders . . . that people would not take justice into their own hands. . . . The greatest influence was in the Cuban revolutionary strategy, in the idea of educating the people during our struggle."

Nevertheless, Fidel admitted, "my presence there was accidental, and our congress had nothing to do with what happened." In fact, the *Bogotázo* ruined the organization of the congress. The plan was never revived by the Cubans or the Argentines.

The return of Fidel Castro and his Cuban associates from Bogotá made the front pages of Havana newspapers. Despite contradictory reports about his activities in Colombia and charges that he was part of a Communist conspiracy, Fidel's image at home gained considerably. At twenty-one, he was now an international as well as a national political figure in the eyes of many Cubans. True to his life pattern, he had lucked out once more; as he said himself, he materialized "accidentally" in the midst of the Colombian civil war.

Back in Cuba, Castro immediately threw himself into the presidential electoral campaign, now in its closing phase. Though he was veering toward Marxism, he maintained his strong support for the candidacy of Senator Chibás—presumably because he felt that the *Ortodoxo* party offered the best solutions for the Cuban crisis, and because he was not ready to break with the traditional political process. His revolutionary propensities notwithstanding, he was also a practical politician, and he saw no advantage whatsoever in becoming formally identified with the weak (if loud) Communist party.

Fidel spent several weeks in May campaigning with Chibás, chiefly in their native Oriente province, and national newspapers duly reported his presence along with *Ortodoxo* congressmen and mayors. Campaigning for Chibás, however, Castro was careful to preserve his personal reputation for independence, often being more outspoken than the candidate on social issues and even being publicly critical of the senator over his friendship with very wealthy landowners. It was one more major political experience for the young Castro.

On May 31, the eve of the elections, Fidel described the contest as "a decisive battle" between Chibás's "idealism" and the "vested interests" of the candidates of the Grau government. He noted that while the Communists had their own candidate, they preferred Chibás to all the others. Eddy

Chibás may have been the conscience of young Cuba, but on June 1 he was demolished at the polls by Grau's labor minister, Carlos Prío Socarrás, with the Liberal party's Ricardo Nuñez Portuondo, a conservative, coming in second. Chibás beat only the Communists' Juan Marinello.

After this defeat, Castro turned to his long-range political interests, including his radical ARO group within the *Ortodoxo* party, driving all over Havana behind the wheel of his green second-hand Buick. But in Fidel's life the unpredictable is the normal, and less than a week after the elections he was embroiled in a new problem.

A university police sergeant named Oscar Fernández Caral was shot in front of his own house on June 6; before dying he supposedly identified Castro as his assassin, and an unnamed witness corroborated it. In this instance Fidel learned of these charges from the newspapers and proceeded to hide again, believing that once more the campus gangsters and their friends in the police were out to kill him. After Fidel's protests, the witness retracted the accusation, telling newsmen that he had been bribed by the police to name Castro. But early in July, an effort was made to reopen the case, and Fidel fired one of his broadsides at the judge, informing him that he had no intention of appearing in court "to aid in the unforgivable attempt of implicating myself in something of which I am completely innocent." Always determined to have the last word, Castro told the judge that if his arrest were ordered, "some police agents" might take advantage of the opportunity to assassinate him. Predictably, nothing further happened.

He spent a short vacation in Birán, saw his parents, and returned to Havana early in September to resume his law studies, still on an auditing basis. Simultaneously, he engaged in new research into Marxism and socialism. But on September 8, Grau's outgoing government authorized the Havana bus company to increase fares, and the next day Communist labor and student leaders were the first to protest through street rallies and speeches.

By the afternoon of September 9, the FEU and university students joined the fray, capturing eight buses, decorating them with Cuban flags, and driving them to the campus. Castro could not, of course, stay away from the latest confrontation, and joined Alfredo Guevara and Lionel Soto in warnings that serious clashes would occur if the people violated the autonomy of the university. The buses, however, vanished overnight from the campus, and this particular incident ended abruptly. The fare increases were canceled.

President Prío was inaugurated on October 10, 1948, a national holiday, opening a period of corruption and mismanagement in Cuba that would exceed even Grau's time in power. Fidel Castro, however, succeeded in ignoring this occurrence, at least for a while. Two days later, on October 12, he married Mirta Díaz-Balart, the pretty, dark-haired philosophy student he had met several years earlier through her brother, Rafael, also at the law school; the two men were friends.

Like the Castros, the Díaz-Balart family was from Oriente, and the wedding took place at their home in Banes, not far from Birán. The Díaz-Balarts were wealthy, with heavy political connections in Santiago and Havana, and, except for Rafael, less than enthusiastic about Mirta's marriage to the twenty-two-year-old Fidel. They disapproved of his politics and most likely, of his family background.

Virtually nothing is known of the circumstances of this marriage, which was fated to be destroyed by politics within five years. By all accounts, Mirta was deeply in love with Fidel, and his friends say that she was the only woman in whom he was interested as he divided his time between his studies and politics. Yet this does not explain why Fidel and Mirta were married so young, particularly when he was still in law school and all his free time was consumed by politics and revolutionary dreams. A curious parallel, or coincidence, appears here between the lives of Fidel Castro and José Martí. Both were married at a very young age, and Martí never could spare time for his wife and family. Castro, as it would turn out, had no time for his family either—despite his great emotional attachment to his son.

Fidel and Mirta traveled to the United States for their honeymoon, an interesting choice, given his anti-American sentiments. Perhaps he wanted to follow, even briefly, in the footsteps of Martí, who had lived for years in New York, or he was simply fascinated by the United States. In any event, the young Castros remained on the mainland for several weeks, including a New York stay, at which time Fidel apparently gave some thought to trying for a scholarship at Columbia University. It is said that it was in a New York bookshop that Castro purchased a number of books by Marx and Engels, including *Das Kapital*.

CHAPTER

7

A political career and a professional career, both in earnest, were Fidel Castro's most urgent priorities. He realized that his frantic activities in so many directions had interfered with his serious political and professional progress, and now he had to concentrate on systematic party work and on law courses. But his ideological preoccupations never left him.

Castro would say later that at the time of Bogotá, "I was almost a Communist, but I was not yet actually a Communist," which is one of those obfuscating phrases he delights in throwing out. He would complain that it was "a great calumny" to blame the Colombian or the Cuban Communist parties, or the international Communist movement, for the terrible uprising. As far as he was concerned, "the whole people fought" in Bogotá, including Liberals, leftists, and Communists. And after so many years, there is no reason to doubt his basic conclusions.

Back in Havana, however, he must have been shocked to discover that the Cuban Communist party sharply criticized his "adventurism" and "putschism" in participating in the Bogotá street fighting. The Communist Alfred Guevara was in Bogotá with Fidel in his capacity of FEU official, but he never stirred out of the boardinghouse during the disturbances; obviously Guevara knew the party line and impulsive Fidel did not.

Whether Castro seriously regarded himself as "almost a Communist" or not in the aftermath of Bogotá, the Communists distrusted him absolutely and would do so until the eve of his revolutionary victory *ten* years later. Looking at Castro's intricate ideological positions and maneuverings, the unfolding of his beliefs and allegiances after Bogotá suggests quite plainly that he drew a fundamental distinction between Marxism, socialism, and communism on the one hand and the traditional Cuban Communist parties on the other. This is a point, as so many other points about Fidel, that the United States and the Soviet Union alike have had difficulties in grasping for a quarter-century.

Meanwhile, Fidel Castro settled down to married life, somewhat incongruously. Their home was a room at a small inexpensive hotel at 1218 San Lázaro Street in downtown Havana, one block from the university. While convenient for Fidel, it must have been confining for Mirta. While she still attended some classes at the philosophy school, and went on seeing her friends and family, Fidel was rarely in attendance. If he was not studying or politicking at the university, he was at the *Ortodoxo* party headquarters at 109 Paseo de Prado, the lovely, broad avenue linking the seaside Malecón Boulevard at the harbor with the National Capitol, for meetings with opposition politicians. Castro always cultivated pivotal *Ortodoxo* figures to keep open his political electoral options, and at that time he was already thinking of running for office. The rest of his day and night hours were consumed by other political meetings, occasional involvement in street disturbances if they were politically profitable, and voracious reading.

Fidel and Mirta lived at the hotel for a full year, presumably because they could not afford to rent a house or an apartment. His allowance from Don Ángel in Birán was around eighty pesos monthly (eighty dollars), which was probably just enough for the hotel room and the full service that came with it, including some meals. Not working, he had no other income. Mirta's family was quite wealthy, but chances are that the proud Fidel would not let his in-laws help support him, and his wife must have been loyal enough to agree. She was accustomed to great comfort, yet she did not appear to mind the hotel life with Fidel—her love compensated for deprivation and his absences. And Mirta would endure much more sacrifice in the near future.

In his fourth year at the law school, Castro received high grades in courses on labor legislation, and only passing grades in a property and real estate course which may have been a clue to his interests.

*   *   *

In Havana political violence continued in all its dimensions, and Fidel Castro was in the thick of it. The Havana bus company again demanded a fare increase, and the Prío government approved it on January 20. The FEU at a university meeting split over whether students should return to the bus-fare battle in the streets. The FEU's Struggle Committee was for it, but the leadership opposed student action, fearing a campus invasion by the police. The profight faction led by Fidel insisted that the FEU's prestige was involved in the protest. On January 24 thousands of students gathered on the campus, preparing to march downtown. But police cruisers surrounded the university, and policemen fired their guns at the students. With Castro in the vanguard, students responded with a barrage of stones and tomatoes. Then, the student committee printed fifty thousand leaflets (this was Castro's idea) urging *Habaneros* to boycott the buses.

In March 1949 an incident occurred in Havana that added to the bitterness against the United States, an incident that Cubans in general and Fidel Castro in particular never forgot, explaining why anti-American sentiment tended to rise rather than diminish with the passage of time.

On the evening of March 11 the statue of José Martí in Havana's Central Park was desecrated by a group of inebriated United States Navy sailors on shore liberty. At least one sailor urinated at the base of the statue, and another sat on Martí's sculpted head. Venerated as Martí is in Cuba, an infuriated crowd gathered quickly, and the sailors were saved just in time by the police, who took them to a station. Presently, the U.S. naval attaché and a Shore Patrol arrived at the station to take the sailors back to their ship. There was no effort by the Cuban authorities to lodge any charges against the men.

The word of the profanation spread instantly throughout Havana, and there was immediate reaction by students led by Fidel Castro, whom a friend had located by telephone. Castro and a number of students appointed themselves as a guard of honor, guarding the Martí statue all night as a patriotic gesture, and making plans for an anti-American demonstration the following day. The student protest in front of the American embassy on March 12 was directed by Castro and his friends Baudilio Castellanos, Alfredo Guevara, and Lionel Soto, the university's principal activists.

The American ambassador in Havana, Robert Butler, who fully understood the gravity of the Martí incident, came out to speak to the students and apologize for the sailors. Just then, police riot squads commanded by Havana's new police chief, Colonel José Caramés (who as university district police commander the previous year had broken up the bus-fare demonstration during which Castro was injured by truncheons), attacked the students with extraordinary brutality. Fidel again was beaten up. Caramés's actions presumably served to show the United States how efficiently Cuban authorities could protect the embassy, but contemporary accounts indicate that Butler himself was taken aback by this violence.

The ambassador then drove to the Foreign Ministry to deliver official U.S. apologies to Foreign Minister Carlos Hevia, but the students followed him. Finally, Butler had the opportunity to try to explain the incident, reminding the students of the American friendship for Cuban in the name of which, he said, the United States had helped win independence for the island in 1898. Given the Cuban-Ame. ˜an history, starting with the military occupation and the imposition of the Platt Amendment, this was not the most felicitous way of pacifying the students; they made this clear by shouting Butler down. While the ambassador went on to Central Park to place a wreath at the Martí statue (this, too, was a very obvious and patronizing gesture, since the Cubans wanted instead that the sailors be punished, at least by the U.S. Navy), Castro, Guevara, and Soto made the rounds of Havana newspaper offices. They delivered a statement, published the next day, charging that it was "a shame for Cuba" to have a police chief who instead of preventing American sailors from desecrating Martí's monument, chose "to attack those who defended our honor." Cubans have assured that the incident will never be forgotten: In his postrevolution film on the modern history of Cuba, *Viva la República,* director Pastor Vega has included old newsreel footage of the demonstrations, with a very clear explanation of what had happened that day.

Though Castro took advantage of every conceivable public situation to assert his political engagement and his leadership aspirations, the Martí statue events being a perfect example, he devoted much of his time to regular political activities on many fronts. He had to build up a reputation beyond that of a patriotic or socially inspired agitator and revolutionary. Consequently, Fidel made an effort to become known in the impoverished working-class districts of south Havana—both on behalf of the *Ortodoxo* party and its youth branch, and on his own to prepare his candidacy in the next congressional elections—visiting homes, stores, and repair shops on the miserable, narrow streets. Castro wangled invitations to speak on the COCO radio network several times a month, preaching social justice and honesty in politics, and clearly he wished to remain identified with Eddy Chibás, who despite his defeat in the presidential elections remained very popular and planned to run again in 1952. In May, for instance, Chibás issued a documented report on Batista's personal enrichment during his presidential term, attracting great attention throughout the country. Fidel knew the senator was a political virtuoso, that his "honesty" campaign was having a growing impact on President Prío's corrupt administration and believed Chibás would win the presidency in the end. Therefore, he watched the master closely, studying his techniques, and preparing to adapt them for his own future use.

On another level, Castro was active in the University Committee for the Struggle Against Racial Discrimination. In a country as racist as Cuba in terms of the ruling white establishment (even poor whites were often against the blacks and mulattoes in the heavily mixed Cuban society), opposition to discrimination was not a popular cause, but Castro embraced it from the outset. There is every reason to believe that Castro was always personally opposed to racial discrimination, and he realized that the kind of mass political or revolutionary movement in Cuba that he was already urging could never develop without major support from nonwhite Cubans. Events would prove him right, but despite his efforts Castro has not been able even a quarter-century after the revolution to eradicate visceral racism among white Cubans. As a racially mixed society, Cuba resembles Brazil the most in Latin America, and despite the existence of severe anti-discrimination laws in both countries, the whites have not by and large shed their sense of superiority. It comes out in casual remarks, jokes, and subtle societal attitudes. Possibly, mixed friendships cannot be forced.

On September 1, 1949, Fidel Castro became a father with the birth of Fidel Castro Díaz-Balart, immediately known as Fidelito. At the time of Fidelito's arrival, his parents still lived in the hotel room on San Lázaro Street, and it was many months before they could arrange to move to a small, modest apartment on Third Street. In the affluent residential district of Vedado, the building was only one block from the Malecón and the sea, and across the street from the Fifth District police station. The new Castro residence was still very far from luxurious, as Fidel basically depended on the allowance from his father. He was too busy in politics to work after his university classes, and Mirta had to watch every peso. All the furniture for the apartment, simple as it was, had to be bought on an installment plan.

About the same time, Fidel persuaded his parents to let his brother Raúl come to Havana. Raúl had done so badly at Belén College that Don Ángel made him come back home and work with Ramón, the eldest brother, at the farm's office. Fidel thought that Raúl, being highly intelligent (his Belén problem had been his hatred of discipline and particularly of prayer), should be given another chance. Consequently, Raúl returned to Havana late in 1949, hoping to be ready to enter the university the following year.

This move would have important implications on the *Fidelista* revolutionary future.

In the meantime, corruption and gangsterism were soaring under Prío, the violence rising again in and out of the university. Justo Fuentes, a top FEU leader, was killed in April by unknown gunmen. Castro was threatened innumerable times (some of his friends think that he rented the apartment facing the police station as a security precaution for his family and himself). Finally, Eddy Chibás went on the radio late in September to charge that corrupt members of Prío's *Auténtico* political party controlled the political gangs shooting it out in Havana streets.

Prío's response was to solve the violence problem with what Cubans called the "gangs' pact," which, in effect, meant that the president bought all the gangsters. It was a much more creative idea than simply putting them on the payroll: Prío's concept was to negotiate a peace pact with the principal organizations, spreading government appointments (with attendant payments and shakedown possibilities) among them on the condition that there would be no more violence. Moreover, none of the gangsters would be charged with crimes or arrested.

This was virtually turning the Cuban government over to the gangs, and the students, supported by the main opposition political parties, were the first to set in motion a plan to denounce the secret pact before the public. To this end, the *Ortodoxo* party Youth Section and the Social Youth branch of the Communist party joined in organizing the "30th of September Committee," named after Rafael Trejo, the first student leader to be killed by the Machado dictatorship on that date in 1930. In this committee, Fidel Castro would soon play a crucial and dramatic role.

Fidel was not among its founders, and had not been considered for membership because of his earlier ties with UIR, ties that for all practical purposes ended in 1947 after the assassination of Emilio Tró, the UIR's chief. The principal leaders of the new student committee were Max Lesnick, national director of the *Ortodoxo* Youth Section, and Alfredo Guevara, still president of the philosophy school and university representative of the Communists' Socialist Youth. Between them, Lesnick and Guevara controlled student politics in 1949 because they represented the most cohesive and best-organized political entities in Cuba, and they got along extremely well personally.

This was the only instance in which Chibás and the rest of the *Ortodoxo* party leadership accepted collaboration with the Communists, and the accord was confined to the university. Earlier that year, Chibás had once more turned down a Communist offer for electoral alliances in the 1950 and 1952 congressional races. The senator was increasingly anti-Communist, and Castro, who thought his party could profit from Communist support, lacked sufficient influence to change the *Ortodoxo* position.

Even then he was an advocate of the unity of all opposition groups, believing that once the common enemy is defeated, the ultimate leadership among the new victors will sort itself out—preferably in favor of the faction Fidel personally favored. Still, Castro was careful not to go too far in pushing cooperation with the Communists; politically he did not need to be identified with them. Helping him in this regard was the party's own reluctance and reservations concerning its relationship with *him*.

Castro had a pleasant relationship with Alfredo Guevara, who had been with him in Bogotá, and the other Communist leaders in the group, Lionel Soto and Antonio Nuñez Jiménez. Although he had closer personal links with the three Communists, Fidel chose to approach Lesnick first to be invited to join the 30th of September Committee. Lesnick, now exiled in Miami, recalls how Castro instantly turned into the hero of the new movement, and what happened subsequently:

"Fidel came to see me at the university to say that he wanted to join the committee. Well, it seemed absurd to me: Fidel, who was involved with the gangs, accused of being one of those who participated in it. The Thirtieth of September Committee could not welcome him. . . . I told him, look, Fidel, I cannot decide alone on this, and he asks me to propose his name to the committee. There were ten or twelve of us who were important [in the committee], but fundamentally it was Alfredo Guevara and I, the two chiefs of the *Ortodoxo*-Communist pact at the university."

Lesnick agreed, and Fidel met with him and Alfredo Guevara at Lesnick's apartment on Morro Street, near the presidential palace, to request that he be admitted to the committee. Max Lesnick continues: "Alfredo and I then started coming up with impossible conditions. No member of the committee may be armed when he goes to the university, and Fidel always carries a pistol. Fidel says, 'Well, I won't carry it anymore.' The second condition was the endorsement of the [*Ortodoxo*-Communist] pact, which demands that all those covered by the 'gangs' pact' must be denounced, and their official posts revealed. So Fidel says, 'Well, I'll sign the document.' Then Guevara asked who should be chosen to make the denunciation before the FEU, and he answered, 'I shall do it.'"

In the closing days of November, a meeting attended by the thirteen presidents of Havana University schools and about five hundred students was held late in the afternoon in the Martyrs' Gallery on the campus. In the words of Max Lesnick, "Fidel takes the floor and delivers a demolishing denunciation of the whole gangster process, and makes the confession that he, too, had the misfortune of falling into it."

Fidel then proceeded to name all the gangsters, politicians, and student leaders profiting from Prío's secret "gangs' pact," and the effect was absolutely stunning. With his network of friends and acquaintances in the political world, he had no trouble procuring detailed information on this subject. Castro must have realized that he was embarking on an extremely dangerous venture in terms of his very life, but he had evidently made a classical risk calculation—a blend of intellectual analysis and pure instinct—that his challenge would succeed, that he would be cleared once and for all of gangsterism accusations, and that his political stature would rise significantly. It was not bravura but an act of deliberate political courage. No sooner had Castro finished speaking than automobiles with armed thugs began to materialize around the university.

"The problem now," Max Lesnick says, "was to get Fidel out of there alive. I had a red convertible, and I told him that I would drive him out. It was seven o'clock in the evening. . . . I didn't do it out of courage, but figuring that if Fidel is seen in the open car next to me, the gangsters wouldn't dare to shoot. It would be too much of a scandal to attack a well-known *Ortodoxo* leader's car."

Lesnick took Fidel to his Morro Street apartment, hiding him there for fifteen days "because he would be killed if he went out in the street." The Havana newspapers prominently displayed Castro's speech, and he was an inviting target. It is not clear what, if anything, Mirta—still in the hotel room with the three-month-old baby—was told about her husband's whereabouts and plans. As a rule, Fidel never discussed his political activities with her.

The rest of this story has never been told publicly, presumably because Castro wished it kept quiet. Having agreed with his friends that he should leave Cuba until tempers cooled and it was relatively safe for him to return, Castro decided to go to New York for several months. The decision was wise, but Castro may have feared that under the circumstances a clandestine departure from the country could be regarded as a cowardly act, and he did not desire any taint on his reputation.

While his trip was being prepared, Fidel spent the time at Lesnick's apartment reading, listening to the radio, talking, and being bored. According to Lesnick, one day Fidel grabbed a broomstick and, pointing it from his bedroom window at the north terrace of the presidential palace, said to Lesnick's grandmother: "You know, if Prío steps out on that terrace to make a speech, I could get him from here with a single bullet from a telescopic-sight rifle . . ." Other friends tell the story that in 1947 when

Castro went to the presidential palace as part of a FEU delegation conferring with President Grau, he said in all seriousness to a colleague as they
waited for the audience: "What would you say if we grab Grau when we
enter his office and we throw him off the balcony?" But Fidel does have an
unusual sense of humor.

It was Lesnick and a mutual student friend named Alfredo "Chino"
Esquível who drove Castro to the railroad station in Havana in mid-
December; there Fidel and Chino boarded a train for the city of Matanzas,
then another train to Oriente to the Castro home in Birán. Apparently,
Fidel was able to obtain enough money from his father to finance his
trip to the United States. Castro flew to Miami, then to New York,
where he remained for three or four months. Virtually nothing is
known about this trip except that he lived in a room in a brownstone
on 155 West Eighty-second Street. It is unclear whether Mirta, with
or without the baby, joined him for any part of the New York stay or
how he supported himself, though he most likely received money from
home.

It is possible that Fidel attended some classes at Columbia University,
and even that he considered enrolling there for further study after graduating from the Havana law school; he spoke subsequently of his frustrated
desire to study abroad for postgraduate degrees. He may have been attracted to the idea of living in New York because that was what Martí had
done as he prepared *his* Cuban revolution. In any case, Castro had ample
opportunity to read and think and even write. He improved his knowledge
of English, but one has the impression that somehow he was not able, not
then or later, to grasp fully the mechanics and nuances of American moods,
attitudes, and politics. This shortcoming may have had an unfortunate
effect on some of his later policy decisions.

Three months before Castro's self-imposed exile in September 1949,
Mao Zedong's Communists achieved ultimate victory in the long civil
war with the United States-supported Nationalists, a milestone in the
history of great modern revolutions. The struggle that led to the creation of the People's Republic of China that year had immense relevance to
Castro's own revolutionary interests—from ideology to guerrilla warfare—
yet he seemed to ignore it, and Mao was certainly never a military model
for him. Strangely, this lack of outward interest in Communist and revolutionary events abroad (except where Cuba is directly involved, as in Nicaragua or Angola) is part of a pattern. He rarely refers to them, and during
the 1950s he is not known to have spoken publicly of the Chinese experience, the Vietnamese victory over the French at Dienbienphu, or the antiregime uprisings in Poland and Hungary. To Castro, Cuba evidently
was and is the absolute center of his attention, and he did not seem

to need Mao or Ho Chi Minh to teach him guerrilla warfare. He always knew best.

Castro had resolved to graduate from law school during 1950, and through the spring and summer he lived day and night with his books. Rather than attend classes, he prepared himself for the exams at home by reading voraciously, absorbing in less than six months a normal two-year work load. His willpower and superb memory helped him greatly. His friends were astounded that he had so totally removed himself from all political activity, and feared that in staying away from meetings and discussions he might be forgotten. He knew he would not be. Though he had very little time for his family, he evidently doted on Fidelito, and right after his return from New York, he took photographs of the baby, sending prints to grandparents in Birán and Banes. His brother Raúl was in Havana, about to enter the university, and his youngest sister, Emma, came through the city en route to private school in Switzerland.

Finally, September 1950 saw Fidel Castro graduated from the University of Havana with the titles of Doctor of Law, Doctor of Social Sciences, and Doctor of Diplomatic Law; the Cuban system permitted a student to work for multiple degrees. Fidel says that during his crash study program he completed forty-eight out of fifty courses on his own schedule, a record that no other student had matched during a comparable period. Being short of the fifty courses, Castro was not eligible for the scholarship abroad that he had thought of pursuing. By this time, Fidel says, he was too impatient with the political "realities" at home to pursue the scholarship-abroad notion, deciding to bid farewell to academia and dedicate himself to politics as well as to a law practice.

In terms of conventional politics, Castro recalls, his ties with the *Ortodoxo* party remained "strong" as he graduated from the university, "although my ideas had advanced much more." This party, he says, was able to channel much of the discontent, irritation, and confusion of the Cuban masses over unemployment, poverty, and the lack of housing, schools, and hospitals. As an *Ortodoxo* youth leader, Fidel was not "preaching socialism as an immediate goal at that point," he says, but campaigning against "injustice, poverty, unemployment, high rents, low salaries, expulsions of peasants [from the land they tilled], and political corruption." In his opinion, looking back nearly thirty-five years, this was a program for which Cubans were much better prepared, the phase during which the start had to be made to move the people in a "veritably revolutionary direction."

Castro had never alluded publicly to his Marxist-Leninist evolution and persuasion until he declared himself a Communist in a speech in December 1961, and over the ensuing years he maintained a high degree of imprecision as to the stages of his socialist conversion, blurring considerably and deliberately his whole educational process. But he is consistent in stating with complete candor that he had concealed the socialist character of his political program until he thought Cubans were ready for it and he felt sufficiently in control of the country to reveal it. He goes to the extreme of insisting that his first political program, publicly enunciated in 1953, was really socialist in nature if one reads it a posteriori in the proper context. Cuban revolutionary ideologues have, in fact, published thick essays to substantiate this interpretation. Castro cannot escape the temptation of manipulating his own political history.

In an interview in 1977, Fidel declared that "I became a Communist on my own, and I became a Communist before reading a single book by Marx, Engels, Lenin or anyone. . . . I became a Communist by studying the capitalist political economy. . . . When I had a bit of understanding of those problems, it actually seemed to me so absurd, so irrational, so inhuman that I simply began on my own to elaborate formulas for production and distribution that would be different." This, according to Castro, occurred when he was in his third year in law school, which would have been around the time of the Bogotá uprising in 1948, and his revolutionary baptism by fire.

However, in 1981 in a discussion of his involvement in Bogotá, Fidel remarks that at that juncture, "I had already some rudiments of Marxism-Leninism, but one cannot say that I was then a Marxist-Leninist," though he was "almost a Communist." He adds that "potentially" he was close to "a Communist political conception" while still being very much under the influence of the struggles of the French Revolution. In his 1985 conversations with the Brazilian friar Betto, Castro claims that after starting out as a "utopian Communist" (a description he often uses), he already "really had contact" with Communist literature, again in his third year at the university. He says that by then he was familiar with "revolutionary theories" and the works of Marx, Engels, and Lenin.

Whatever the true version of Castro ideological progress, there is no question that he never deviated from what he calls his "anti-imperialism" and his deep resentments of the United States and its foreign policy. His old friend Alfredo Guevara says that while Fidel's Bogotá speeches to students were intellectually "impeccable and coherent" in "anti-imperialist terms," it does not follow that they were based on "Marxist analysis." Guevara's opinion is that Castro seriously embarked on "theoretical studies" only after Bogotá.

In 1950, while completing his university studies, Fidel signed the

Stockholm Peace Appeal shortly after the start of the Korean War in June. The Stockholm document has been widely regarded as a pro-Soviet propaganda operation, and while great many non-Communists around the world signed it in good faith, it is unlikely that Castro was too innocent not to realize what he was doing. In Cuba the appeal was sponsored by the Cuban Youth Committee for Peace, and the text with all the signatures was published in the September issue of the periodical *Mella,* a Communist publication at the university. Fidel identified himself as a "student" and member of the national committee of the *Ortodoxo* party (whose founder, Eddy Chibás, abhorred the peace appeal and the Soviet Union). In November, Castro published an article in the Havana daily *Alerta* urging independence for Puerto Rico and stressing that Cuban students were united against "tyrants" in the Americas.

After graduating from law school in September 1950, Fidel Castro decided to go immediately into law practice, but to concentrate on political cases and what his friends would call the "lost causes" of litigation on behalf of the poor. To proceed in this fashion was a conscious political move on Fidel's part, consistent with both his professed beliefs and his political-revolutionary ambitions. It would have been easy for him, particularly through the Díaz-Balart marriage connection, to seek partnership with influential firms and lawyers and rapidly command a lucrative practice, but Castro chose another route.

The partners Fidel selected for his law office were Jorge Aspiazo Nuñez de Villavicencio, a former bus driver, nine years older than Castro, who graduated in the same class with him, and Rafael Resende Viges, another fellow student with a poverty background. Jorge Aspiazo, now retired in Havana, recalls that Fidel met with him and Resende on the *escalinata* of the university one afternoon in September to propose the partnership. Aspiazo, who voted against Fidel for class delegate in their first year in the law school because he suspected him of being a "rich boy," soon became a friend, though to this day he remains apolitical. Resende, the same age as Fidel, was also a friend. Later he would swerve to the right, joining the Batista camp and ultimately emigrating to the United States.

As Aspiazo recalls it, their quarters consisted of a small reception room and a tiny private office on the second floor of a rundown building at 57 Tejadillo Street in Old Havana. This was the capital's banking and business area, in narrow streets and small squares of colonial structures near the harbor, and most law offices were located there. The ancient Rosario building, where Castro and his partners went to rent space, was almost entirely occupied by law firms.

The monthly rent was $60, and the owner, José Alvárez, insisted on a first month's payment and a month's security, for a $120 total. However, the three young lawyers had only eighty dollars between them, and it took persuasion for Alvárez to agree to settle on that. They also talked him into lending them some furniture, including a desk and a single chair, so that they could start working. They bought a typewriter on installment.

Aspiazo says that their first professional arrangement was with a wealthy Spanish immigrant who owned a wholesale lumber business, Madereras Gancedo, that sold materials to local carpenters. A typical Fidel proposition, the deal was that the Gancedo firm would provide the lawyers with free lumber to build their office furniture, and in return they would collect overdue bills from carpenters who had purchased wood from the Spaniard.

This was collection-agency work, but not the way Fidel conducted it. Instead, Aspiazo recounts, Fidel summoned Gancedo's debtor carpenters to the office and asked them for a list of people who owed *them* money for their services. The lawyers then spent their time collecting on behalf of the carpenters, and whenever they succeeded, Fidel would call his "client" to say that they had money for him. Aspiazo says that when the carpenter would ask Castro to pay, for example, twenty dollars to the lumber merchant, Fidel would reply, "No, you need this money now, and our client does not," and he would hand the man the twenty dollars.

On one occasion, Castro and Aspiazo went to collect from a carpenter in the Lawton district of Havana, one of the poorest, but the man was away. Aspiazo says that the carpenter's pregnant wife asked them to wait, stepping into the kitchen to make coffee for them, and Fidel asked his partner to lend him five pesos, placing the banknote under a plate on the table. When the wife served them coffee, Fidel told her that the carpenter should not worry about his debt to Gancedo and to come by their office when he could.

Aspiazo adds that they never paid the lumber company a cent, and that after sending them bills for their furniture materials, "they got tired and never collected from us." But the law firm itself did poorly in collections because Fidel insisted that most of its work be *pro bono publico,* or as a free public service. They represented the stall owners at the municipal market in Havana, peasants in Havana province who were being expelled from farms, students involved in riots, and just about every worker who had a legal problem of some kind and turned to Aspiazo, Castro & Resende. In their three years of partnership (which ended, in effect, when Castro left to lead the attack on the Moncada barracks in 1953), they earned a total of 4,800 pesos, of which 3,000 pesos was for one case and 1,800 pesos for the second one. They also sued the American-owned Cuban telephone company for lower rates on behalf of the subscribers, and the court actually found in their favor, but by that time Castro was already in prison.

Whether concerning his law firm or his own affairs, Fidel had a total disdain for money and, in fact, he never had any. Theoretically, his allowance from Birán was designed to support his family, but, as his friends say, Fidel often gave it away the moment it arrived; it was enough even for a casual acquaintance in his political circle to ask for a loan for Castro to produce the money.

On his graduation from the law school, Fidel received a brand-new Pontiac sedan as a present from his father. Shortly thereafter, a friend borrowed it for a trip out of town, totaling the Pontiac in an accident and suffering serious injuries. At the hospital, where he rushed as soon as he heard about the accident, Fidel met his friend's father, a powerful and rich conservative politician, who told him he would pay for the car. Castro replied: "You are wrong. Your son is dying. How can you be concerned about the car? You don't have to pay anything; you should worry about your son." Three years later, the politician would intervene with his friend Fulgencio Batista to assure that Castro received decent treatment in prison; he may have even saved Fidel from assassination by the wardens.

When his old friend Baudilio Castellanos came from Oriente to visit Havana, Fidel invited him to lunch at home. On Castellanos's arrival, however, Castro suggested they drive first to the municipal market, where he went from stall to stall, picking rice, potatoes, and *malanga* (an edible root, popular among poor people in Cuba), but without paying for anything. Castellanos remarked on it, and Fidel said, "Oh, I never pay here . . . they are my clients, and they pay my fees with food." Thereupon, they drove back to the apartment on Third Street where Fidel cooked the lunch.

Fidel's financial insouciance continually victimized Mirta. One day, when Castro was out of Havana, she telephoned Jorge Aspiazo in tears, begging him to come over immediately. At the apartment, Aspiazo found all the furniture gone, and Mirta weeping on the floor with Fidelito in her arms. The store, she told him, had repossessed all their furnishings, including the baby's crib, because Castro had failed to make the installment payments. Aspiazo somehow obtained enough cash for a down payment on new furniture; when Fidel returned the next day and looked around the apartment, he remarked in surprise, "Christ, this is not *my* furniture. . . ." He took it for granted that Aspiazo resolved the problem.

*   *   *

Very soon, Castro had the first opportunity to act as his own attorney in court, an experience and an idea that would be vital in his revolutionary career before too long. This performance came as a consequence of his arrest on November 12, 1950, in the southern port city of Cienfuegos for participating in an antigovernment student demonstration. After a year-long hiatus, Castro was back in the public view.

Although he was already practicing law, Castro had retained his ties with the university, auditing certain courses that interested him and maintaining his contacts with the FEU. In fact, the Cienfuegos authorities incorrectly identified him after his arrest as president of the social sciences school at Havana University. The reason Fidel became involved in this particular confrontation with the Prío government was that he needed solid political exposure—and his name again in the newspapers.

Castro was not forgotten. Alfredo Guevara recalls that even strangers asked him in Havana, "How is Fidel?" and word of his "public defender" reputation was spreading. Still, he needed what he regarded as a national event, and Cienfuegos seemed to fill the requirement. High-school students there had called a "permanent strike" to protest the ban on their organizations and associations by Education Minister Aureliano Sánchez Arango and Interior Minister Lomberto Díaz, and to Fidel this was a matter of principle worth a good fight.

Castro and a law student from Havana named Enrique Benavides Santos were arrested by soldiers before he could begin speaking to the student rally, and they were taken to jail and hit with rifle butts. For the next four hours, the students and the army and the police fought in the streets, and as an account of the Cienfuegos events published after the revolution put it quite accurately, "The objective that brought Fidel . . . had been reached; the protests of the people against the regime were much more violent than if the meeting had been peaceful."

From Cienfuegos, Castro, and Benavides were taken during the night to the provincial capital of Santa Clara and locked up in the prison there, but Senator Chibás had gone on nationwide radio to denounce the arrests, and a demonstration erupted in front of the prison during the morning. The two men were released "conditionally," but before returning to Havana, Fidel issued a thundering denunciation of "the executioners of the people," printed in full in Cuban newspapers.

In mid-December, Castro and Benavides returned to Santa Clara to be tried for inciting the Cienfuegos disorders, and Fidel stunned the court by informing it that he was a lawyer and intended to conduct his own defense. When it developed that he needed an attorney's black robe and black cap to be allowed to address the judge—as well as five pesos in fees—a collection was taken up at once among the audience, and Castro rose to speak. Always believing that offense is the best defense, he delivered a roaring accusation of the government for "strangling the liberties" of Cuba, claiming that the regime and the army should be judged at the trial—not him and Benavides. The presiding judge heard Fidel out, and pronounced his verdict: acquittal. It was a victory that Castro never forgot and one that he still often and fondly recalls.

Increasingly, Castro was gaining access to public opinion through the press and radio. Ramón Vasconcelos, the editor of the outspoken daily *Alerta* (and a former Prío cabinet minister), had become a friend, opening his newspaper's pages to Fidel's fiery articles. In June 1951, for example, Castro published a lengthy defense of workers' rights, citing the cases of nine hundred employees illegally dismissed by a canning company and of peasants deprived of land. He concluded by stating that "justice for Cuban workers and peasants" must be the nation's principal goal. He also had frequent access to the Voice of the Antilles radio station, constantly hitting the Prío government on issues of corruption and denial of justice in the Cuban society.

In 1951 the Korean War had become a major issue in Cuba because President Prío was believed to be ready to succumb to United States entreaties and send an army unit to fight in Korea. This was not a popular notion, and even as anti-Communist a figure as Senator Chibás went on the air to demand that the country be consulted before such a decision was taken. Actually, an army battalion was being trained for Korean duty at the Managua camp south of Havana, notwithstanding the quiet opposition of many officers there. One of them was Lieutenant "Gallego" Fernández, now Cuba's vice-president under Castro. (Prío eventually let the idea die.)

Fidel had already signed the Stockholm Peace Appeal, and now, in *Saeta,* the campus Communist publication, his younger brother, Raúl (a member of its editorial board since entering the university late in 1950), published an anti-American broadside over the Korean War. Raúl would say many years later that it was Fidel who first acquainted him with Marxist thoughts and texts. This is most likely true, though there is a widespread idea that it was Raúl who pushed Fidel into communism. At this stage the older brother had no compunctions about writing for *Saeta* himself. In Castro's case it was consistent with his own description of his evolution toward Marxism-Leninism (this was now three years after Bogotá), and his attempt to preserve a balancing act between the Chibás *Ortodoxos,* whom he still needed, and the Communists to whom he was ideologically attracted.

Thus Fidel signed a declaration in favor of "Democratic Rights and Freedoms," drafted by a campus committee and published in *Saeta*; it was an accusation against the Prío government over the "repression" of students and the "violation" of the freedom of the press. Although Prío was a freely elected president and Cuba was a democracy in the formal sense (the press certainly was free), his government was immensely corrupt and high-handed, and such disparate groups as the *Ortodoxo* party and the Communists could join in attacking it. This made Fidel's maneuvering easier, and he kept writing in *Saeta* on such subjects as the need for university students to "define themselves on the side of just and revolutionary [causes]" and university reform while at the same time cultivating the *Ortodoxo* electorate. In Cuban politics Chibás and Castro were on the same wavelength.

On the evening of Sunday, August 5, 1951, Senator Eduardo Chibás shot himself in the abdomen during his weekly radio program, attempting to commit public suicide and changing the course of Cuban history when he died eleven days later.

The reasons for Chibás's suicide at the age of forty-three have never been fully understood. He shot himself with a .38 Colt Special revolver, a powerful weapon, at the end of a speech urging Cubans to "awaken" in the name of "economic independence, political liberty, and social justice." En route to the hospital, he whispered, "I am dying for the revolution . . . I am dying for Cuba . . ."

But this does not explain it. The senator was a passionate and popular figure, the leading candidate for the presidency in the 1952 elections, and a man seemingly happy in his personal and family life. His brother, Raúl, who was also his closest political associate, says thirty-four years later that "I still do not find an answer. . . . I think that in part it was disillusion because it is very easy to become disillusioned in politics if things are not going well, if people turn their backs on one. . . . He must have felt a little lonely, and that many people who should have helped him did not help him. They were not cooperating. . . . Possibly he felt he was no longer useful, and that he could be more useful making an example of himself . . ."

Raúl Chibás was unquestionably referring to the awesome dilemma his brother was facing that Sunday when he was unable to produce the promised proof of his longstanding allegations that Education Minister Aureliano Sánchez Arango was guilty of corruption and self-enrichment on a colossal scale. The senator's accusations for more than a month over the radio were Cuba's greatest political sensation, and on the appointed day the nation breathlessly awaited Chibás's documentation. But, surprisingly, he failed to mention the subject, shooting himself as the climax of his appearance. It is generally believed that a group of congressmen who held the documents of proof against the minister, and had promised them to Chibás, betrayed him at the last moment for reasons of politics or subornation. Having spent his political life preaching honesty and truth, Chibás could not face the fact that he would break his word to his people. This assessment is privately shared by Fidel Castro.

In any event, Chibás's death completely altered the Cuban political scene, removing not only a powerful candidate for the presidency but also a man who was the nearest thing to Cuba's conscience. Most important, perhaps, the Chibás suicide paved the way for Batista's coup d'état the following year; Raúl Chibás and many other Cubans are convinced that Batista would not have risked the coup if Eddy were alive because the senator would have instantly become the chief of a powerful opposition movement that might have done away with the dictatorship. Batista, who was another candidate in the 1952 elections, was under constant assault by Chibás, worrying him much more than the attacks by the considerably less important Fidel Castro.

For Fidel the death of Chibás had many meanings. He could not yet foresee the Batista coup in the vacuum that had been created, but he understood better than any other politician in Cuba that the whole equation had changed, that the atmosphere was now filled with uncertainties that could bring together the revolutionary climate Castro so earnestly desired.

In terms of *Ortodoxo* party politics, Fidel was too young to aspire to replace Chibás as the top leader, and this never entered his calculations. He did perceive, however, that without Chibás dictating policy, younger and more independent-minded *Ortodoxos* could be attracted to his revolutionary line. There were no commanding personalities in the party who could replace the senator when it came to influence over the youth. At the same time, Fidel spared no effort to show himself as an absolutely loyal and heartbroken disciple of the dead leader. It would be unjust to portray Castro as simply a cynic and an opportunist, because he must have had some liking and admiration for Chibás even if they disagreed on much in politics, and even if today the Castro regime has erased his memory. Whatever the reason, however, Fidel was ever present after the fatal pistol shot at radio station CMQ.

As Chibás lay dying eleven days, Castro was near the door of suite 321 at Havana's Medical Surgical Center around the clock. When Chibás's body lay in state at the university's *Aula Magna* (Hall of Honor), Fidel stood by the bier as part of the honor guard for twenty-four hours preceding the funeral. A photograph published in Cuban newspapers and magazines shows him standing in the front row, the fourth from the casket, staring at the floor. He wears a gray suit and a tie; most of the other politicians there are in *guayaberas* (this is also the first photograph showing Castro with a pencil-line moustache).

Raúl Chibás and Castro led the victorious fight to place Chibás's body on display at the university instead of at the National Capitol because of his senatorial rank. They argued that his political career had begun at the university, whereas the capitol was seen as the symbol of the corruption the senator had always denounced. When the decision was made to move the body to the campus, Fidel told newsmen that "it is better if we have him there because degenerates cannot go to the university to desecrate the memory of Chibás."

The senator was to be buried the following day, August 17, at Havana's Colón cemetery, at the end of a funeral cortege led from the university by the military. According to his friend Max Lesnick, Fidel conceived the idea of diverting the procession to the presidential palace, where Chibás's body would be placed in the presidential chair and be symbolically proclaimed Cuba's president before the burial. Lesnick says that Castro had learned from a friend, Rosa Rávelo, that her father, who was an army captain with ties to the *Ortodoxo* party, would command the escort of the gun carriage transporting the body. He nearly convinced Captain Rávelo to take the procession to the palace, but in the end reason prevailed, with the officer realizing that he could trigger a mass uprising—which was presumably what Fidel had in mind. He never lacked ideas.

His next one, less than a month after Chibás's death, was to charge two National Police officers formally with the death of a worker, a member of the *Ortodoxo* party, during a riot on Sunday, February 18, when crowds had fought to guarantee that Chibás would reach the radio station. There had been a government-inspired attempt to prevent him from arriving at CMQ, and the riot ensued. In his attorney's capacity, Castro presented the charges against Major Rafael Casals Fernández and Lieutenant Rafael Salas Cañizares before a Havana criminal court. The latter was also involved in violence against students during the Martí statue incident in 1949. The case attracted wide attention, and the government tried futilely to move it to military jurisdiction. In the end, the officers were placed on "conditional liberty" under 5,000-peso bail, but not acquitted (the case was dropped after the Batista coup, and Castro would have to contend again with Salas Cañizares).

A curious political episode involving Fidel Castro and Fulgencio Batista appears to have occurred sometime after Chibás's death. Batista, who ended his elected presidential term in 1944, returned to Cuba from his Florida home in 1950 to reenter politics—hoping to be president again. He founded the Unitary Action Party (PAU), whose principal ideological aim was to restore him to the presidency, and began to gather and buy support. Before dying, Chibás claimed that the Communists were supporting Batista on the theory that the *Ortodoxos* would steal the social-justice thunder from them, and that they could resume the pleasant cooperation they had had with Batista in 1940–1944 when a Communist served in the cabinet.

This is quite plausible, and it was probably because of this situation that Batista supposedly expressed his interest in meeting Castro, about whom he had been hearing a great deal—including Fidel's attacks on him. According to at least three separate and credible versions, the meeting was arranged through Fidel's brother-in-law, Rafael Díaz-Balart, and a mutual friend named Armando Vallibrende who had known Castro through the UIR political gang in the Forties. Díaz-Balart headed Batista's youth organization in PAU. Fidel was taken to Batista's luxurious Kuquine estate, not far from Havana, and received by the general in baronial splendor. In his private office, there was a large painting of Batista as sergeant along with busts of famous historical personages, a solid-gold telephone, the telescope Napoleon used on Saint Helena, and the two pistols the emperor had at Austerlitz.

According to one version, Batista confined the meeting to general conversation, taking measure of Castro and avoiding politics. Another version has it that they did discuss politics, and that Fidel told Batista that he would support him if he ousted Prío in a coup d'état. If this was the case, then Castro was testing the older man, and Batista became afraid that Fidel had somehow learned of his secret plans for a coup the following year and was acting as *agent provocateur*. To Fidel's extremely logical mind, it made perfect sense that Batista would try a coup, particularly with the crumbling Prío regime and Chibás's death, and he was simply seeking a confirmation. Batista, however, terminated the meeting abruptly—never suspecting the grief the young man would cause him in the future.

Castro's line was to step up his attacks on both Prío and Batista as he increased his efforts to run as an *Ortodoxo* for the Chamber of Deputies from Havana province in the June 1952 elections. Roberto Agramonte, a traditional Cuban politician from a famous family, replaced Chibás as the *Ortodoxo* presidential candidate, inheriting the senator's one-hour Sunday radio spot on CMQ. Fidel thought, however, that the party's youth (represented by himself) should share in the CMQ time, and managed to obtain ten minutes' air time from Agramonte for his own pronouncements. Moreno Frajinals says that on CMQ Fidel used a text so as to remain within the time Agramonte had allotted him, but he was very relaxed on the air.

Castro was also given time on Radio Álvarez, another Havana station, and *Alerta* went on publishing his occasional articles.

Toward the end of 1951, Fidel was managing three separate but related political enterprises: He was representing thousands of poor Havana urban dwellers whose homes the Prío government was planning to raze to build a huge civic square in the center of the city, he was investigating Prío's personal misdeeds as president, and he was campaigning intensely to represent a Havana district in the congressional elections. But nobody seems to recollect anything about Castro's family life at that time. He did see his brother Raúl quite often, mainly because he was at the university and increasingly active in politics; Raúl had not yet formally joined the Communist party's Socialist Youth, but he was very close to the party.

The area that the government wanted to level in midtown Havana was a forty-eight-acre district known as La Pelusa, a miserable slum. To protect the residents, Fidel convinced them that their rights would not be ignored, instructed them at street rallies what to say to government inspectors, and went to court to demand that the Public Works Ministry indemnify each homeowner (or shack owner) for the property to be razed. The ministry finally agreed to pay a fifty-peso compensation in each case, which was not unreasonable, but the payments were never made because Batista soon ousted Prío and canceled the agreement. Batista then expelled the area's dwellers to proceed with the civic square. Today it is Castro's Revolution Square.

To investigate Prío in depth, Fidel mobilized not only his law-firm partners but friends from the *Ortodoxo* party's Youth Section as well, notably a young man named Pedro Trigo from Havana province's rural area. Between September 1951 and January 1952, Castro and his investigators came up with impressive evidence against the president. Fidel had learned the Chibás lesson that accusations without proof are worthless and self-defeating. On January 28, 1952, stressing that the date marked the anniversary of José Martí's birth, Castro presented the indictment of Prío to the Court of Accounts (a federal administrative tribunal), summing it up in five specific charges. Each one began with the words "I ACCUSE the President of the Republic." The charges specified that Prío had committed bribery by granting amnesty to a friend serving a prison term for child molestation, and appointing him nominal owner of presidential farms; violated labor laws by forcing workers into twelve-hour labor shifts under military foremen; insulted the armed forces "by turning soldiers into laborers and peons and forcing them into slave labor"; contributed to unemployment "through the substitution of paid workers by obligatory labor from soldiers"; and betrayed national interest by selling farm products at prices below the market.

The enormously detailed accusation, full of names and figures, was published the next day in its entirety in *Alerta* under the headline I ACCUSE (Castro having read Émile Zola) and broadcast over the Voice of Antilles radio station. Prío, Cubans were told, had constructed "ostentatious palaces, swimming pools, airports, and a whole series of luxuries" and acquired "a chain of the best farms and most valuable lands in the vicinity of Havana."

Never quitting when ahead, Castro issued a second indictment of Prío on February 19, 1952, this time charging that the president was paying eighteen thousand pesos monthly to political gangs and *pistoleros* and keeping 2,000 gangsters in public jobs, the price of the 1949 "gangs' pact" negotiated by Prío and denounced by Castro at that time. Fidel also charged that in four years, Prío's land holdings rose from 160 to 1,944 acres. All these revelations stunned Cubans, and Jorge Aspiazo, who was Fidel's principal law partner, remarked later that "friends were assuring us that Fidel would not last more than a week."

Castro succeeded in so shaking up the Cuban political establishment that his own *Ortodoxo* party was scared to have him as a candidate, especially now that the fiery Chibás was dead. Playing it safe, Roberto Agramonte, the party's presidential candidate, simply omitted Castro's name from the *Ortodoxo* electoral list issued in February. However, Agramonte underestimated Fidel's determination to have his own way by any imaginable means (Fidel was forever being underestimated), and such means were instantly devised.

The device was for one or more districts in Havana province to select him as the congressional candidate at their local party assemblies, and this is where Castro's gift for planning ahead paid off. The poverty-stricken Cayo Hueso district of Havana, where Fidel had begun his door-to-door campaign in 1951, was the first one to pick him. Then came the rural district of Santiago de las Vegas, where Castro had been investigating Prío; Jorge Aspiazo says that the dwellers from La Pelusa (where Fidel had been fighting for the rights of the homeowners about to be dispossessed) collected coins in tins in the streets to raise enough money to rent a bus to take them to Santiago de las Vegas so they could attend a Castro rally and support his candidacy. He had a way of exacting loyalty, if not gratitude.

Now a candidate, Castro engaged in what by Cuban standards was a political blitzkrieg, rich in new techniques. Jorge Aspiazo says that in December 1951, Fidel inaugurated a new program on the "Voice of the Air" radio station that within two months attracted fifty thousand listeners, according to the ratings. Max Lesnick, who as chief of the *Ortodoxo* party's Youth Section watched it closely, recalls that Fidel mounted "a fabulous campaign," obtaining mail-franking privileges from five friendly congressmen plus a list of 100,000 names. Then, Lesnick says, Castro had 100,000 envelopes addressed and sent to every *Ortodoxo* party member in Havana province with a personal message in blue ink, signed by him. He had cut a stencil for the message, but the workers and the peasants did not know it. "This had never happened in Cuba before," Lesnick adds, "inasmuch as political leaders simply went to rallies to deliver their speeches." Raúl Chibás, himself a candidate for senator, says that Fidel "had his own group that followed him, his own organization within the party." This was Castro's Orthodox Radical Action (ARO) group inside the party, the young rebels who were the forerunners of his revolutionary movement and whom the older leaders always tried to neutralize. Conchita Fernández, who had been Eddy and Raúl Chibás's secretary, also ran for congress from Havana province, and she recalls how Castro often appeared at the end of her rallies to speak in her support. On the day his first anti-Prío indictment was published, Fidel appeared in the township of San Antonio de Río Blanco, just as the small crowd was dispersing after hearing Conchita. But, she says, he summoned the people back by shouting and waving a copy of *Alerta* with his exposé and the photographs of Prío's *fincas* he had taken from the house of a friend nearby. "And within five or ten minutes," Conchita recalls, "that park was full because he had such magnetism that all people needed was to hear him." Conchita, who had known Fidel since he was a student leader in 1947, and always admired him (she would become *his* secretary after the revolution in 1959), says that crowds applauded Castro "deliriously" because he spoke the truth, "and he didn't care if the next day someone shot him, or did something to him." Thus the Castro legend was being born with the young candidate delivering as many as four hour-long speeches in four localities in one night.

In an interview with Lionel Martin, an American journalist who has lived in Cuba since 1961, Castro said of his campaign that "I addressed myself directly to the masses; I had one hour of radio, and there was the press with all those denunciations. . . . There was a great political vacuum." He told Martin, the author of a book titled *Young Fidel,* about his direct-mail campaign and the use of congressional franchise "to which I had no right, but I had no other way of doing it, either."

"They couldn't brake me," Castro told Martin. "This was a problem I had studied well. They could not brake me in any way. They did not see me with much pleasure, but I was supported by the masses. They couldn't fail to present me [as a candidate]. But this wasn't yet popularity on a general level; it was popularity on party level. I hadn't stepped out yet from this environment, although my writings had repercussion among all the people."

At that stage, the Communists withdrew their support from Batista (they had no presidential candidate of their own for 1952), realizing that the *Ortodoxos* could win even without Chibás; they were also impressed by Castro's possibilities as a congressman. Consequently, they proposed an electoral pact in all the Senate and Chamber of Deputies races, and announced that they would vote for certain *Ortodoxo* candidates anyway even after their offer was turned down. Castro thought the *Ortodoxos* rejected the Communist alliance because of their fear of the United States.

The general assumption early in 1952 was that Castro would be elected to the Chamber of Deputies with the votes of the Havana province urban and rural proletariat. Max Lesnick says that there was not "the slightest doubt" about his victory, adding that "I knew the influence he had among all the young people in the party, the most sincere people in the party, and the working people."

The elections were never held because of the Batista coup, but it is nevertheless useful to ponder how Castro would have acted in the congress, reconciling the representative system with his revolutionary ideas. The next question is whether Castro might have advanced all the way to the presidency through the electoral system, or if he needed the Batista dictatorship to create the revolutionary climate leading to his ultimate victory. In reconstructing Castro's intellectual and political processes, it is relevant to know what he had been planning for his anticipated congressional investiture.

Many Cubans think that without a coup, Castro would have served as a congressman for four years until 1956, then run for the Senate, and made his pitch for the presidency in 1960 or 1964. Given the fact that Cuba was wholly bereft of serious political leadership and given Castro's rising popularity, such a scenario is not implausible. In that case, it would appear that he was fated to govern Cuba—no matter how he arrived at the top job.

Castro himself said in a 1965 interview with an American visitor that even before the Batista coup, "I already had some very definite political ideas about the need for structural changes. . . . I had been thinking of using the parliament as a point of departure from which I might establish a revolutionary platform and motivate the masses in its favor. . . . Already then I believed that I had to do it in a revolutionary way." Speaking of his situation in 1952, Castro volunteered that "in many ways I was still not a Marxist, and I did not consider myself a Communist"; this contradicts his subsequent statements that when the coup came, he had already become a full-fledged Marxist-Leninist, and it is part of his tortuous interpretations of his own ideological growth.

Fidel told the interviewer in 1965 that "once in parliament, I would break party discipline and present a program embracing practically all the measures that . . . since the victory of the Revolution have been transformed into laws," knowing that the program would never be approved but would rally the population around it for subsequent action. And he said, "I already definitely believed in the need for seizing power by revolution." Ten years later, Castro told Lionel Martin that he realized that the Cuban problem could not be resolved through the parliament, and that his plan was to "break institutional legality" at the proper moment and proceed to take power. His parliamentary resources and immunity, he remarked, would help him "to move more freely and to conspire more freely."

In 1985, Castro said in a discussion of his political past that even before the Batista coup he had "a revolutionary concept and even an idea how to carry it out" by passing through a political phase—his anticipated congressional seat—and then attaining the "second phase of 'revolutionary capture of power.'" But he also touched on a fundamental point of his philosophy: That to make a revolution, one must first assume power. This was a concept that most Cubans did not understand until the victory in 1959.

Whatever Fidel Castro had in mind as a candidate, including plans that might never have worked out, it is unquestionable that the Batista coup and the end of the electoral process in 1952 saved him the "political phase" of his secret and seemingly improbable dreams of revolution and conquest of power.

Castro's suspicions that Batista was preparing a coup were reinforced early in February 1952. Raúl Chibás recalls that around that time he ran into Fidel on the stairs of the Havana house of Roberto Agramonte, and a conversation ensued about various perils ahead. He says that suddenly Castro asked him if "I had any news about a conspiracy, that Batista was conspiring and thinking of a coup d'état." Chibás replied that he had not heard anything; in retrospect, he thinks that Fidel "had soldiers, army people, in his campaign, and that was how he learned about the plot." He saw Castro again the following week, and Fidel was even more certain that a coup was in the works.

One version maintains that Castro had been apprised of unusual activity, involving comings and goings by military officers and civilians at Batista's Kuquine estate. Castro hid outside the estate, the story goes, photographing all this traffic a day or so before the coup, but by then it was too late to be of help. According to another account, President Prío had received a letter from a woman in Oriente province, reporting a military conspiracy. But when the Chief of Staff of the army asked the head of his bureau of investigations to look into it, the officer advised his superiors that there was nothing to it; the military intelligence officer was a Batista agent.

At dawn of March 10, 1952, Fulgencio Batista strutted into the army's Camp Columbia in Havana with his officers to be warmly greeted by the troop commanders. This was the coup that, meeting no resistance, ousted Carlos Prío Socarrás from the presidency. It was quick, silent, bloodless, surgically precise, and wholly cynical. The next day, already proclaimed chief of state, Batista moved to the presidential palace that he had left eight years earlier as constitutional president.

III
THE WAR
(1952–1958)

CHAPTER

1

As directed and inspired by Fidel Castro, the Cuban revolution was born directly from the Batista coup d'état, to become both the island's conquest of its full national independence and the implantation of a social, economic, and political order without precedent in the Americas. Whether a revolution so uncompromising and of such magnitude could have occurred without the political conditions suddenly created by the advent of the dictatorship is most unlikely. Batista's rule was so widely hated, it brought more unity to Cubans than any government since the Machado dictatorship twenty years earlier, a unity which Batista failed to understand and which would serve as the main trigger for the revolution. Even with Batista on the scene, the great revolution could be set in motion only through the leadership provided by Fidel Castro, then a twenty-five-year-old Havana lawyer already known for his dedication to rebel causes, his oratorial gifts, and his single-mindedness. The events encompassed between 1952 and the last days of 1958, when the guerrilla army achieved military victory and captured political power, show that while Batista opened the portals to this historical revolution, it was Castro who would inevitably march through them to crown his six-year war.

In Castro's mind, however, the ouster of Batista was only the first tactical objective in his revolutionary enterprise. The strategic objective, which he chose not to reveal until the successful completion of the first phase, was the social revolution that ultimately turned Cuba into a Marxist-Leninist state, today financed by the Soviet Union but in the final analysis still molded by Fidel Castro.

Castro was the only personage who knew exactly where he was going politically at a time when nobody in Cuba, neither the old nor the new generation, had any sense of direction or orientation, let alone historical vision. When one looks back over nearly forty years of Castro's adult life, analyzing his youthful and then the more mature spoken and written views and opinions, listening to his old friends and companions, and, above all, observing the societal transformation of Cuba, it seems evident that he possessed this vision in ways that history bestows upon the chosen few. Fidel often engages in contradictions over causes or the nature of past events or his role in them. If it suits him at a given moment, he embellishes memories and he manipulates them, yet the record shows his absolute consistency in working for a revolution. He talked social revolution and prepared himself for it long before the Batista coup, and his Bogotá experience four years earlier was part of his revolutionary trajectory. Simply to get rid of Batista did not loom as a life goal for Castro.

Castro's opening military challenge to Batista was the assault on the Moncada army barracks in Santiago and a simultaneous attack in Bayamo, the other stronghold in Oriente province, by rebels of the still-unnamed Movement sixteen months after the dictatorial takeover. Fidel personally led the charge on Moncada, barely escaping death, then defied Batista in the courtroom, and vanished for a year and a half in a prison cell where he tranquilly planned the next challenge while devouring hundreds of books on politics, history, philosophy, economics, and literature. Prison was, in effect, his postgraduate school in humanistic studies, and imprisonment also turned out to be a surprising political asset, as a nationwide campaign to grant him and his companions amnesty made Fidel even more famous than the assault itself; he always managed to turn things to his advantage.

In the history of the Cuban revolution, the 1953 attack on Moncada is revered as the equivalent of the launching of the first war of independence in 1868, and of José Martí's uprising against Spain in 1895, both of which likewise failed at the time. But Moncada is the cornerstone of modern Cuban history, and Fidel Castro's courtroom oration in justification of the assault—"History Will Absolve Me"—is regarded as Cuba's real declaration of national independence, *the* great revolutionary manifesto, and something akin to the scriptures, all magnificently fused together. It is also the most cited text in Cuba, constantly analyzed and interpreted, with the dogmatic verities never questioned, but new insights into Fidel Castro's mind and heart devoutly discovered and deepened with the passage of years.

With the dictatorship of "His Majesty King Sugar," as the Cubans called their principal product, unbalancing the economy and pushing unemployment up to frightening levels, social conditions during the Batista years were ripening for an explosion sooner or later. While the luxury Havana Hilton Hotel and casino for rich American tourists and businessmen was built in the capital, the average farmhand had only 108 days of available work at a dollar per day (and no food) in 1952, and 64 days of work in 1955. The rest was "dead time" with no employment of any kind. Social justice had to be a clarion call for a leader like Castro. It would be powerfully sounded in the Moncada oration, and it has remained the unquestioned justification for every revolutionary act after the victory, including the permanent imposition of a politically and intellectually oppressive state apparatus.

Fidel Castro set out to organize his Movement (much later it would formally become the famous 26th of July Movement, complete with black-and-red flag and stirring hymn) and to plan the revolution virtually within minutes of learning that Fulgencio Batista and his officers had occupied the Columbia army camp in Havana at dawn of Monday, March 10, 1952, thereby deposing Carlos Prío Socarrás from the presidency. In Cuba in those days, the real power resided at Columbia, and Batista, who had no chance of being elected to a new presidential term though he had contrived to be chosen senator in the meantime, resolved to seek it there. General elections had been scheduled for June 1, and Castro was a candidate for the Chamber of Deputies from the capital's poorest workers' *barrios,* but the coup put an instant halt to the electoral process. Fidel, who never trusted "liberal bourgeois democracy" anyway, was delighted with this turn of events, though he violently denounced Batista for "a brutal snatching of power." Castro realized that even if elected to congress, his revolutionary aspirations had limited possibilities, as did his political career, in a representative type of government. In 1974, Castro claimed that his notion had been to abandon "institutional legality" and grab power at the "opportune moment." It had been just brave talk to say that he would use his congressional seat as a revolutionary platform. Thus, Batista's coup was a gift from heaven for Castro, putting him seriously and promisingly in the business of revolution. And Fidel did know in principle how to do it, improvising when needed as he went along. Writing about the coup two years later, he noted that its "only positive value" was to set off a "new revolutionary cycle." From the first day of the new regime, he dedicated himself to plots, plans, maneuvers, seductions, feints, and attacks, and to the creation of a conspiratorial revolutionary movement in a brilliant and amazingly systematic manner. He left nothing to chance.

This period of preparation was rich in drama, danger, panache, adventure, and, invariably, classic acts of defiance that added to his public image as Cuba's "pure" leader of the new generation. His personal strategy was to operate on two parallel levels: the invisible level of conspiracy in which secrecy was never pierced, and the more *visible* level of street and courtroom protests against Batista at every imaginable opportunity. These two levels were complementary because, for one thing, Fidel's public performance facilitated recruitment for the conspiracy and the creation of his blindly loyal following.

Melba Hernández, a lawyer who is seven years older than Castro, and one of the two women who participated in the Moncada assault, says of his impact on people that "I think this happened to everybody: From the moment you shake hands with Fidel, you are impressed. His personality is too strong. When I gave my hand to this young man, I felt very secure, I felt I had found the way. When this young man began to talk, all I could do was to listen to him. . . . Fidel spoke in a very low voice, he paced back and forth, then came close to you as if to tell you a secret, and then you suddenly felt you shared the secret . . ."

And Castro did impress enough carefully selected men and women to accept his guidance at any cost to themselves to be able to organize his Movement so rapidly. Pedro Miret Prieto, then an engineering student from Santiago (he had attended the same La Salle boys' school as Fidel, though they had not met there) and the man who personally gave secret military training to each Moncada fighter, says he will "never forget" the day he encountered Castro, six months after the Batista coup. Now a member of the ruling Politburo and one of Castro's closest collaborators, Miret says that he decided to follow him when he realized that "all these other politicians will do nothing" against Batista, but Fidel would act. This story is repeated over and over again as the aging revolutionaries recall meeting Castro in their youth, an unforgettable emotional experience.

Still, it was necessary for Castro to gain enough exposure as the top leader against Batista that potential recruits would know his reputation and accept that first encounter. Certainly the publicity he gathered through the Bogotá adventure, his prominent university leadership, his congressional campaign, and his radio talks made him fairly well known to politically minded young Cubans, but Castro had his eye on an even wider audience. An example of how Castro's method worked for him is the account by Ramiro Valdés Menéndez, once one of Cuba's most powerful men and until 1986 the interior minister, on how he joined the Movement. The son of a poor family in Artemisa, a town in Havana province with unusually strong radical and anarchist traditions, Valdés was a twenty-one-year-old truck driver's helper with virtually no formal education when the Batista coup came. But he was interested in politics, and through a friend in the *Ortodoxo* party youth branch, Valdés arranged a meeting with Castro because he had listened to him on the radio in the past and now wanted to see whether this was a leader to be followed.

They met in Havana in the heat of July at the *Ortodoxo* party headquarters and as Valdés recalls it, "There was Fidel in his proverbial striped winter dark-blue suit, and there we talked . . . and there I enrolled in the Movement." Valdés was entrusted with organizing a secret ten-member cell in Artemisa—each of the ten was to recruit the next ten-member cell and so on, under Castro's compartmentalized conspiracy structure—and he was with Fidel at Moncada, in prison, in Mexico, in the Sierra, and in the revolutionary government. Valdés's experience of seeking out Castro was repeated by scores of others. In time, Fidel's rebels acquired the honor of being called the "Centennial Generation"—the generation launching a revolution in the year of the hundredth anniversary of Martí's birth. This, too, was a marvelous touch of *Fidelista* mythology.

Castro was able to move smoothly into a leadership position among the young *Ortodoxo* party members and sympathizers, chiefly because there was nobody else credible in Cuba to challenge Batista. There had not been the slightest effort to resist Batista's coup on March 10. President Prío fled the country (after refusing to supply arms to students ready to fight in defense of the constitutional government), and traditional political leaders also went into exile or proved to be pathetically ineffectual and scared. It is interesting to note that at first Castro was prepared to support with his tiny band of followers whatever anti-Batista move the very wealthy old-line leaders would launch after promising money, arms, and action from abroad. But Fidel and his friends were to be disenchanted and irritated by the constantly broken promises that arms would be delivered *mañana* and *mañana*. A few idealists in Cuba did attempt to conspire, only to be caught instantly by the secret police. Finally, Castro decided to proceed independently, sick of the "liberal bourgeoisie." As he said later, "When none of these leaders showed that they had either the ability or the realization of the seriousness of purpose or the way to overthrow Batista, it was then that I finally worked out a strategy of my own."

Cuban Communists, the only ones with a professional organizational structure, were just as ineffectual after the Batista coup as all the other traditional political parties. As in the 1940s, they were probably not above cutting a deal with Batista to maintain their key positions in the labor confederation, or at least remain neutral. In the 1950s Moscow-directed

Communist parties were much less confrontational and adventurous than one might suspect. Moreover, Batista may have toyed too with the idea of some *modus vivendi* with the Communists, allowing the Communist daily *Hoy* to be published for a time even after the Popular Socialist Party was declared illegal.

In any event, the last thing Castro wanted at that juncture was an alliance with the Communists. In a 1981 conversation, Fidel put it very plainly: ". . . When I had already conceived a revolutionary plan and I already had a Marxist-Leninist formation, I didn't enroll in the Communist party, but we created our own organization and we acted within that organization." Yet, to organize a revolutionary movement single-handed, starting from zero, is an endeavor so ambitious that it verges on absurdity and arrogance. Fidel, moving from hideout to hideout in and around Havana in his creaking old dark-brown Chevrolet sedan (his latest car), succeeded by the end of the year in both creating the insurrectional nucleus and expanding it into an armed movement. It seems never to have crossed his mind that he might fail, or that, unlike Lenin, Mao Zedong, and Ho Chi Minh, the fact that he had no political party or organization to support him was not a drawback. "It was logical, no?" he says, wholly convinced of it. Lenin, whom Castro regards as his Communist ideological mentor, plotted from the safety of Zürich while the young Cuban lawyer barely kept ahead of the dictatorship's secret police at home.

Ironically, the United States' reaction to the Batista coup revolved around the entirely irrelevant Communist question. On March 24, two weeks after the deed, Secretary of State Dean Acheson wrote President Truman in a secret memorandum that "while Batista when President of Cuba in the early '40s tolerated communist domination of the Cuban Confederation of Workers, the world situation with regard to international communism has changed radically since that time, and we have no reason to believe that Batista will not be strongly anti-communist." This conclusion, plus the fact that most Latin American governments had recognized the Batista regime by April 1 (Dictator Trujillo in the Dominican Republic being the first), led the United States to do likewise. In those days, many Latin American governments were dictatorial, and democracy was not a hemisphere issue in any of the capitals, including Washington. Only the previous December, Truman had expressed his faith that Cubans had attained "maturity" and "affinity" with United States democratic aims. Further, the United States also did not wish to endanger its special economic relationship with Cuba at a time when the island's economy was in dire straits and huge American interests were beginning to suffer.

Finally, Ambassador Willard L. Beaulac, a highly competent career diplomat, informed the State Department that in the absence of any meaningful opposition to Batista, the United States might as well accept him. The conclusion that no opposition leadership existed in Cuba was shared by both the American ambassador and the burgeoning revolutionary chief; this is probably the only time Castro and the State Department agreed on anything. But the State Department had forgotten the role of revolutionary students in triggering the overthrow of the Machado dictatorship two decades earlier, and there is nothing to indicate that the American embassy or the State Department were at all aware of the Generation of 1930 or of Fidel Castro. He may have been big in Artemisa, but not in Washington.

The hard core of Fidel Castro's Movement consisted of no more than eight or ten persons until the middle of 1952, when it began to gain recruits more rapidly. This nucleus was formed by members and sympathizers of the *Ortodoxo* party. Communists from the Popular Socialist Party were automatically excluded because of Castro's early decision not to join the party, creating his own organization instead, and the Communists' own reluctance to submit to his leadership.

Actually, the Communists did attempt to influence his tactics and behavior in their favor immediately after the Batista coup, but Fidel rebuffed them. Alfredo Guevara was the instrument for this approach. In an interview in Havana in 1985, Guevara maintained that on instructions from the Socialist Youth (the Communists' youth organization), he contacted Fidel as soon as he could—Castro had gone into hiding the morning of the coup—to urge him to return as a student "to become the great figure at the university." The Communists' idea was to build anti-Batista opposition on their terms, using Fidel and the university as the leading edge in their strategy. Guevara also acknowledged that the party had lost control of the University Students' Federation and was having a hard time regaining it.

According to Guevara, Castro accepted his proposal at first, "but then he didn't do it, and he disappeared." He said that the reason for Fidel's refusal to go along with the Communists was that while the Popular Socialist Party had in mind "a struggle of the masses," which meant political unity in opposing the dictatorship, "Fidel had the idea of direct action, that is of popular insurrection." Guevara added that the next time he saw him, Fidel had already acquired a clandestine radio station and was engaged in preparing for Moncada.

Fidel's refusal to subordinate himself to the Communists, whose political potential he held in low esteem, did not mean he had no contacts with them or avoided participating in public protests with them and other opposition groups. The only member of Fidel's Movement who was an active Communist was his younger brother, Raúl, who formally joined the Socialist Youth in June 1953. But he was excluded from all secret policy planning and decision-making, not being part of the Movement's "general staff." Fidel and Raúl claim that he learned that Moncada barracks would be the target of the Movement's assault only a few hours before it occurred, and that he was a "plain soldier" in the operation (Luciano González Camejo, a middle-aged worker, was the only other Communist to take part in Moncada, but he joined the Movement late and his ideological persuasion had apparently been overlooked). Raúl devoted most of his energies to university demonstrations organized by the Communist-controlled Tenth of January Committee, often carrying the Cuban flag in the front row of marchers, his short figure almost dwarfed by it.

Castro had resolved to stay away from the university even before Guevara made his proposal and, being informed of the coup around five o'clock in the morning, went into hiding for six days. Contrary to published reports, he was not among the mass of students gathered at the university in mid-morning to demonstrate against Batista (Raúl, however, was there). Fidel thought that he might be arrested, and he also concluded that there were more important things to do politically than to shout "Death to Batista!"

Late the previous year, Castro and his family had moved from the tiny apartment on Third Street to a larger one on the second floor of a building at 1511 Twenty-third Street, also in the residential Vedado section but in a less elegant neighborhood. The rental price was about the same, and Fidel was as broke when the coup came as he had been before. One evening, when Fidel was in hiding (he would vanish when the Batista secret police put on the heat, reappear, hide again, and so on), he and Pedro Trigo, an *Ortodoxo* textile worker and one of the first Movement members, stopped at the Castro apartment building on Twenty-third Street. On the second floor, they encountered disaster. Electricity had been cut off because the utilities bill had not been paid; the apartment was in darkness. Three-year-old Fidelito had a throat infection and a high fever, and the best Castro could do was to arrange for the child to be taken to the Calixto García hospital to be operated on by a friendly physician. He also borrowed five pesos from Pedro Trigo, giving it to Mirta to buy supplies for Fidelito. Actually, Castro was carrying one hundred pesos in his pocket, but this was money he had collected that day for the purchase of weapons, and he felt he could not use it even for family sickness. Subsequently, Fidel's friends took it upon themselves to make sure that his rent and utilities and furniture installments were paid.

Another time Castro could not find his old car in front of the *Ortodoxo* party offices downtown where he had left it. It had been repossessed by the finance company from which Fidel had bought the used car, but failed to make payments. As a friend tells the story, that was the darkest day ever for Castro. Deprived of his car, he walked over to a café where he often stopped for coffee and a cigar. He told the owner that he was hungry because he had missed lunch, but had no money on him. Since he already owed the café five pesos, the owner refused him further credit. Fidel thereupon began walking home, a distance of three miles. Crossing Central Park, he stopped to look at newspaper headlines, but did not have the five cents to buy a newspaper; the vendor shouted at him, "Keep moving, keep moving . . . don't stand here . . ."

At his apartment he collapsed in bed in deep depression, falling asleep. Awakening late in the afternoon, Fidel, as his friend tells the story, had overcome the depression and was again full of fighting spirit. Later, Castro would laugh when recounting these incidents and the way his friends assumed the responsibility for paying his bills. "They even would give me something to take care of my food." He would say he was the "first paid professional member of the Movement."

Nothing is remembered, however, about his wife, Mirta. Until the divorce in 1955, she must have suffered in silence the daily hardships and Fidel's continual absences. And after the coup the situation became still more untenable when her brother, Rafael Díaz-Balart, was named vice-minister of the interior in the Batista government. This ministry was in charge of public order and the secret police, among other political responsibilities. Now the brothers-in-law were in hostile camps.

Fidel had slept at home the night of the coup, but he fled at dawn to the apartment of his sister Lidia, five blocks away. He left Mirta and Fidelito behind as well as his brother Raúl, who was staying with them. Fidel's instincts were correct; secret policemen did appear at the apartment in midmorning looking for the brothers (they also missed Raúl, who had gone to demonstrate at the university).

Normally, Castro appeared on the radio every day in the early afternoon for a fifteen-minute political chat (or harangue), but he knew he would be arrested if he turned up at the studio on the day of the military takeover when all constitutional guarantees were suspended. In any case, Fidel's immediate interest was to be fully informed about events, and several of his friends offered their services. One of them, René Rodríguez, had been told by Mirta that Fidel was at his sister's house, and he reached them there. Castro asked Rodríguez to visit the university numerous times that day to keep him posted on student movements.

From the university, Rodríguez brought the FEU president, Alvaro Barba, to confer with Fidel at Lidia's house. Then, Rodríguez went to the residence of Roberto Agramonte, the *Ortodoxo* party presidential candidate, to check on the mood among the leaders. When Rodríguez reported that Agramonte and his colleagues were thinking of nothing more drastic than passive resistance to Batista—and that they had no message or marching orders for him or the party youth, Fidel exploded in rage, shouting that the *Ortodoxo* leaders were cowards and worse. At night, Castro decided he was no longer safe at Lidia's apartment and moved to the Hotel Andino, a boardinghouse downtown where he had once lived. On the morning of March 11, Rodríguez, who was making the arrangements, went with Castro to the home of Eva Jiménez, an *Ortodoxo* party youth militant, in the middle-class section of Almendares. Eva had bought enough food for several days and given her maid the week off. Castro was wearing dark glasses, which normally he never used (because he was near-sighted, he needed clear prescription spectacles), and the two men took a bus to Eva's place. Fidel had a five-peso bill Lidia had given him, but the driver had no change, and in the end a stranger on the bus paid the sixteen cents for their fare.

Unbeknownst to Fidel that day, another safe house had been provided for him. This was the luxury apartment where Natalia "Naty" Revuelta lived in Vedado with her husband, a leading Havana heart specialist. Naty, an attractive blond woman from a wealthy background, educated in the United States and France, had revolutionary sympathies, friendships in the *Ortodoxo* party, and had heard a lot about Fidel Castro. The day of the Batista coup, she delivered keys to her apartment to several prominent *Ortodoxo* leaders in case they needed a haven from the secret police, with the specific request that one of the keys be given to Castro. But in the confusion of the March events, the key did not reach Fidel until much later. Later, Naty Revuelta became a very important person in Fidel's life, one of an extraordinary contingent of beautiful and/or highly intelligent women who, in effect, dedicated their lives to him and his cause—and without whom he might not have succeeded.

Castro spent two days and nights at Eva Jiménez's apartment, drafting and redrafting in longhand (no typewriter was available) a proclamation against the Batista takeover under the heading "Not a Revolution—A Bang!" The proclamation was written strictly on his own behalf on a kitchenette table, and on March 13, Fidel dispatched René Rodríguez and Eva Jiménez to the newspaper *Alerta,* which had published his articles in the past, to ask the editor to print his broadside. However, Ramón Vasconcelos, the editor, turned it down on the grounds that anti-Batista opposition was "unrealistic and on the moon." Because press censorship had been established, Fidel's envoys made no effort to contact other newspapers.

But Castro was determined to circulate his proclamation, and so he sent Rodríguez to contact a friend who lived above a pharmacy in downtown Havana and had a mimeograph machine in his apartment. Raúl Castro and Antonio López Fernández, a six-foot-six worker whom Fidel had befriended during his electoral campaign and was known as Ñico, mimeographed five hundred copies of the proclamation to be distributed in the streets as soon as possible.

The opportunity to do so came on Sunday, March 16, when Fidel left his refuge and drove to the Colón cemetery to join *Ortodoxo* leaders and students, who gathered at the grave of Eddy Chibás as they did every sixteenth of the month since his death the previous August. The *Ortodoxo* party leaders' speeches were so tepid that Fidel could not restrain himself. He raised his right arm and shouted, "If Batista grabbed power by force, he must be thrown out by force!" As policemen approached menacingly, the tall figure clad in a white *guayabera* was encircled by his friends to protect him (*Alerta,* which had refused to publish his manifesto, reported Castro's cemetery outburst, commenting that his words were "well received by the crowd . . . demonstrating again the sympathy he enjoys among the masses of the party").

Finally distributed at the cemetery, the mimeographed proclamation charged that "there is nothing as bitter in the world as the spectacle of a people that goes to bed free and awakens in slavery," that there is "infinite happiness in fighting against oppression," and that "the fatherland is oppressed, but someday there will be freedom." Castro ended by quoting from the Cuban national anthem: "To live in chains is to live sunk in shame and dishonor. To die for the fatherland is to live!"

As far as Castro was concerned, he was now personally at war with the Batista dictatorship, quixotic as this notion may appear. In his proclamation he urged "courageous Cubans to sacrifice and fight back." Looking back at these events twenty years later, Fidel told Lionel Martin that he had launched his campaign "a bit as a *guerrillero,* because in politics one must also be a *guerrillero. . . .*" Once embarked on his war, Castro would never relent. On March 24, a week after the Colón cemetery gathering, he presented in his capacity as attorney a brief to the Constitutional Court in Havana, accusing General Batista of violating "the constitution [and] the form of government" by his military coup. He then went on to list the punishments for such acts as prescribed by the penal code, concluding therefore that "Fulgencio Batista's crimes have incurred punishment deserving more than one hundred years' imprisonment."

Castro's goal in presenting this brief was not so much to attract new attention to himself (*Alerta* published a report on it the following day), and certainly not to obtain legal satisfaction from the court, but to establish a fundamental revolutionary principle for the future. Specifically, he set out to proclaim the concept of revolutionary legitimacy, making the point that "if, in the face of this series of flagrant crimes and confessions of treachery and sedition, [Batista] not tried and punished, how will this court try later any citizen for sedition or rebelliousness against the illegal regime, the product of unpunished treason?" Knowing perfectly well that the court would never try Batista, Fidel was carefully building the legal base for his planned revolution, justifying it beforehand as a *legal act* against an *illegal regime*. This would grant full legitimacy to the revolutionary government he would install in 1959. To the highly legalistic Latin American mind, such formal legality is paramount.

Fidel gave up his actual law practice, such as it was, in the aftermath of the Batista coup, but his law firm stayed in business and, on his instructions, proceeded to conduct additional legal battles against the dictatorship. Jorge Aspiazo, his former partner, says that Fidel had him sue three Batista cabinet ministers in criminal court for defrauding the State Unemployed Persons' Fund by naming "imaginary persons" to public posts, "firing" them, then collecting the unemployment payments themselves. The case wound up in the Supreme Court, which threw it out.

Castro kept up the barrage. On April 6 the newspaper *La Palabra* (closed by the regime after its first issue) published another ferocious attack on Batista by Fidel along with his poetic warning that "the seed of heroic rebellion is being sown in all the hearts" and that "faced with danger, heroism is strengthened by generously shed blood." Castro understood that revolutions demand romanticism and rhetoric.

On clandestine fronts Castro kept equally busy. Operating from different safe houses and from the *Ortodoxo* party offices on the Prado (strangely, the regime allowed political parties to go on functioning, although the congress was closed and the elections canceled), Fidel held literally hundreds of meetings and interviews with prospective Movement members between the time of the March coup and the start of May when a new phase would begin. According to Jorge Aspiazo, he often drove to the countryside at night to meet in a field with local *Ortodoxo* members from rural sections of the party.

Historically, it is relevant that the Castro Movement was born almost entirely from the rank-and-file of the *Ortodoxo* party, the reformist and radical but essentially establishment organization founded by Senator Chibás in 1947. Great numbers of working-class and middle-class Cubans, including sugar workers and mountain farmers, became powerfully attracted to Chibás and his party as a new and "clean" phenomenon in the country. The now-aging Communist party, on the other hand, seemed unable to penetrate what should have been its natural constituency among workers, and, typically, its strength was chiefly among intellectuals, students, and labor-union leadership. The claim that cold-war propaganda against communism was responsible for this state of affairs is not credible.

When Fidel rejected the Communists' entreaties to join them at the university in opposing Batista, he was being realistic. Young *Ortodoxos* (and even middle-aged ones) from modest backgrounds were much more revolution-minded than the Communists, they had no use for Communist ideological discipline, and they admired Castro. Fidel used the intervening years to develop wide contacts, especially inside the Youth Section and the Orthodox Radical Action (ARO) faction he had organized. Despite the animosity of older party leaders, Castro expanded his contacts and loyalties in the course of his electoral campaign. When the Batista coup came on March 10, 1952, and Fidel made his decision to fight "independently," he already had a potential revolutionary network. Now the problem was how to turn it into a fighting instrument.

The solution emerged unexpectedly from a chance meeting at the Colón cemetery when Castro was introduced to Abel Santamaría, a twenty-four-year-old accountant at a Havana Pontiac dealer's office. (It seems as if cemeteries are a major scene in Cuban politics. This stems from an old tradition of using major and minor anniversaries of dead personages, even of such heroes of the past as Céspedes and Martí, as an emotion-laden pretext for political demonstrations that the police under any regime would hesitate to break up. This tradition no longer exists, however, although Castro used Colón as a venue for political proclamations after 1959.) On this particular occasion, the Batista government had forbidden May Day labor celebrations, and many opposition militants went on May 1 to the grave of Carlos Rodríguez. Rodríguez had been killed by the police during the Prío presidency, and Castro took two key police officers to court on murder charges. Now Castro was at the cemetery as a revolutionary gesture, and he was being presented to Santamaría by Jesús Montané Oropesa, an accountant at the General Motors office. Fidel had met Montané shortly before the coup when he unsuccessfully tried to swap his ailing car for a better one. Since March, Montané, Santamaría, and a few other friends had been looking for ways of fighting Batista.

Castro and Santamaría, a tall, light-haired man from Las Villas province, hit it off immediately, and their May 1 encounter was a turning point for the incipient Movement. For one thing, they shared rural origins: Abel was born at the Constancia sugar mill, where he worked until coming to Havana at the age of nineteen, and Fidel was from the Oriente canefields. That Castro came from a landowning family and Santamaría was the son of workers made no difference. Besides, Abel, along with Montané, was an *Ortodoxo*. Abel lived with his sister Haydée (she was called "Yéyé") in an apartment near Castro's home, and his and Fidel's first and endlessly long conversation took place at the Santamarías'. It produced a deep relationship between them, with Fidel acting as the chief of the Movement and Abel as his deputy. It was the less spectacular but more cool-headed Abel who would be crucial in giving shape to the Movement that Castro had conceived.

In the weeks and months that followed, the Movement grew and improved. Montané joined the nucleus with his friend Boris Luís Santa Coloma, also an accountant (an exiled Cuban scholar wrote in all seriousness that the accountancy background of so many key Cuban revolutionaries deserves to be analyzed); Melba Hernández, the *Ortodoxo* lawyer who was so impressed with Fidel, and her friend Elda Pérez did likewise. Melba brought in Raúl Gómez García, a twenty-three-year-old poet and teacher; Fidel added his friends Pedro Trigo and Ñico López. By midyear, this was the core of the Movement.

From the outset, Castro's authority was undisputed, and he ran his Movement like a military organization. Melba Hernández, who was the closest to Fidel and the Santamarías during the whole preparatory period, recalls it as "militancy twenty-four hours a day," and a life of extraordinary discipline, which also required a complete change in the social relationships of all the Movement members.

Melba, nowadays honored as a Heroine of the Revolution and still very active politically in her seventies, says that Castro demanded a series of "compromises" from the revolutionaries. First, she says, "we had to hate the regime that oppressed us, which was easy, then to repudiate that society that lived from corruption, and to take the decision to fight against it." To be able to fight the corruption, great temptations had to be resisted, she recalls, "and because this was a clandestine movement, it had very rigid discipline, very strict discipline and secrecy, complete discretion, and militant behavior . . . this is how we were being educated, and a militant would be expelled from the Movement for the violation of any of these rules." "As the Movement grew," she recalls, "the groups of young people who already belonged to the Movement met every Sunday. It was like a test. For example, they would be summoned for 5:05 P.M. If someone wasn't there at exactly that time, we would analyze it, and then the person was admonished, or punished, or expelled. Indiscretion, any kind of indiscretion, no matter how small, was cause for expulsion."

Additionally, Melba Hernández says, the Movement's general staff, composed of Fidel Castro, Abel Santamaría, Haydée Santamaría, and Melba, met once a week to discuss the activities and behavior of all the members (some published accounts state erroneously that only Fidel and Abel formed the general staff). Fidel and Abel conferred regularly with the military committee and the political committee. The two headed both committees, each of which had four other members. But only Fidel and Abel had the power to select recruits and make all the strategic and tactical decisions, ruling out any dissent. The Movement was organized into cells ranging in membership from ten to twenty-five, and their orders came from the general staff or from one of the committees, depending on the subject and its importance. It was a totally vertical structure without political bodies or functions, a principal difference between the Castro Movement and Communist or other political revolutionary parties. Fidel had intended it, in effect, as a "caudillo" structure designed to win the war and stay away from politics.

Whatever Castro's inner ideological convictions, he was determined to keep his Movement from ideological identification with Marxism-Leninism. Melba Hernández says that "in our ranks in that period there was never talk about communism, socialism or Marxism-Leninism as an ideology, but we did speak of the day that when the Revolution will come to power, all the estates of the aristocracy must be handed over to the people and must be used by the children for whom we are fighting." Melba makes a point of noting that "the problem of workers' exploitation was not discussed," but "we did talk about the workers' wages, the abuse of the workers, the abuse of the peasant." She says that "in the language that was then accepted, we insinuated what we were doing." And Fidel and Abel insisted on "the importance of the incorporation of the woman in revolutionary struggle."

In a 1977 interview with the Soviet Communist party's theoretical journal *Kommunist,* Castro said that "during all that period I maintained contacts with the Communists [who] had their own objectives in the situation," but "one could not ask them, either, to have confidence in what we were doing." He remarked that it "would have been difficult for a party educated in a classical form, with its plans, its concepts" to have faith in the Movement. Besides, "a Communist party could not contemplate the conquest of power . . . it wasn't possible to attempt the conquest of power in Cuba if one had a Communist label . . . a revolutionary power could be conquered in Cuba, but it could not be done as a Communist party." This analysis is consistent with the position Castro took toward the Communists in 1952, though it can also serve as a justification both for his refusal to deal with the party in preparing the revolution, and for the communists' own unwillingness to help the *Fidelistas* until very late in the war.

Mario Mencía, the only serious historian of the Cuban revolution working in Havana, has written that though Castro had a "Marxist formation" since his student days, "he eluded this [public] identification with iron willpower even in his declarations." Mencía finds that while early on Castro "had a revolutionary project pointing toward socialism," he faithfully followed José Martí's precept that to achieve one's goals "one must conceal [them]" because to "proclaim them as what they are would raise difficulties too great to be able to reach them in the end." The only time that Castro departed from this rule during the early period, Mencía writes, was when he accused Batista, in an underground publication in mid-1952, of being "a faithful dog of imperialism" and an ally of "great Cuban and foreign interests." But even this attack was signed with a pen name.

After Pedro Miret, the engineering student and arms expert, met Castro in September 1952 and agreed to train Movement members for military purposes, internal discipline became even stricter. As Melba Hernández recalls it, constant personal control over the members was the leaders' great preoccupation, and at the weekly meetings Castro and Santamaría analyzed the behavior of each revolutionary over the previous seven days, including their personal lives, and often "criticism was made . . . leading to expulsion." But Fidel and Abel also practiced self-criticism, which is a Marxist concept.

Jesús Montané, the accountant who was in the first revolutionary nucleus, recalls that "in our Movement it was absolutely forbidden to ingest alcoholic drinks," and those who had the drinking habit "could not be militants." Montané says that "the lives of these revolutionaries was guided by the most absolute austerity and morality," and that on one occasion Fidel had suspended an important member of the Movement for drinking, letting him return only when he gave it up completely. Subsequently this revolutionary "offered his precious life to the sacred cause of the Cuban Revolution."

The Movement's principal headquarters was at the Santamarías' apartment at Twenty-fifth and "O" streets in Vedado (it is now a national shrine), but the plotters also used the apartment of Fidel's sister Lidia, three blocks away, the apartment belonging to Melba Hernández's parents (who were fervently pro-revolutionary) on Jovellar Street in downtown Havana, and an office on Consulado Street, just off the seaside MaleCón boulevard, provided by a secret member of the organization who worked for a rich businessman. The office on Consulado, Melba says, offered the best "camouflage."

As usual, propaganda was foremost on Fidel's mind. Before they met him, the Santamarías and Montané published irregularly a mimeographed publication called *Son Los Mismos* (They Are the Same) to attack Batista, the name indicating the present military regime was as bad as the previous governments. But as soon as they joined forces, Fidel proposed that the publication be renamed *El Acusador* (The Accuser), and he began to edit it with Abel and the young poet Raúl Gómez García. *Son Los Mismos* went on being printed simultaneously for some weeks, then it was dropped altogether. At the university the Communists printed their publication, *Mella,* for which Fidel had written in the past and continued to do occasionally later.

Castro signed his articles "Alejandro," his middle name. Raúl Gómez, the poet, signed his articles "The Citizen," another sign of the influence of the French Revolution on the *Fidelistas*. The Movement's publications were produced on an ancient mimeograph machine purchased by Abel Santamaría and Montané for seventy-five pesos; to avoid discovery by the police, the copying machine was constantly moved from place to place by a friendly Spanish taxi driver, and Montané said later it spent most of the time in the trunk of the car parked in front of the Detroit bar on Twenty-fifth Street.

But Castro wanted to be on the air as well. A few days after he met Santamaría, he convinced Abel to come with him and Montané to visit a physician in Colón, some 150 miles from Havana, whose name he had received from a Movement member. The physician was Dr. Mario Muñoz Monroy, and at the age of forty-one, he was an ardent revolutionary, ready to follow Fidel. As it happened, Dr. Muñoz was a light-aircraft pilot and a ham-radio operator, too. What Fidel wanted from him were *two* ham transmitters to announce an antiregime demonstration that was being planned at the university for the following week. He wanted two transmitters in case one failed, and the enthusiastic doctor was miraculously able to provide the first one in time to broadcast the "Free Airwaves of the Resistance and National Liberation Movement" rally on May 20 on the forty-meter band.

The name of the Movement was invented by Fidel for the occasion, and the broadcast was barely heard, but *Son Los Mismos* was able to report the existence of the underground radio, and this was one of the main objectives in having the transmitters built. Castro believed correctly that different types of propaganda feed on each other; Dr. Muñoz's stations were the forerunners of the Rebel Radio in the Sierra six years later. As Montané recalls it, Castro bubbled with new ideas for the Movement all the way to and from Colón in Abel's car.

Anti-Batista opposition also took shape in the formation on May 20, 1952, of the National Revolutionary Movement (MNR) by Rafael García Bárcena, a well-known professor of psychology, sociology, and philosophy at the University of Havana and the National War College. García Bárcena had founded the *Ortodoxo* party with Eddy Chibás in 1947, and now he attracted to his MNR young middle-class opponents of Batista, such as Armando Hart, Faustino Pérez, and Juan Manuel Márquez, all of whom later joined the Castro Movement. In Santiago the MNR recruits were Frank País, who would become a hero of the war, and Vilma Espín, now Raúl Castro's wife.

Fidel took a noncommittal stance toward the MNR—he distrusted middle-class liberals who relied on the conquest of power through military coups—and toward the Liberating Action (AL), organized in July by a politician named Justo Carrillo. He concentrated on his own Movement. At that time, Pedro Miret was busying himself training MNR members in the use of arms at the university; he had not yet met Castro. Justo Carrillo was trying to infiltrate groups of young army officers.

On August 16 a huge rally was held at the Colón cemetery to commemorate the first anniversary of Senator Chibás's death, and Fidel had his associates print ten thousand copies of the third issue of *El Acusador,* an enormous press run for a mimeographed publication, to be distributed there and in Havana streets. As "Alejandro," Castro had written two fiery articles, one criticizing the *Ortodoxo* party for the cowardice of its leadership, and proclaiming that "the movement is revolutionary and not political," and the other accusing Batista of being an "evil tyrant." In his best style of political invective, Fidel informed the dictator that "the dogs that lick your wounds every day will never conceal the awful smells emanating from them," and that "when history is written . . . it will speak of you as it speaks of plagues and epidemics . . ."

However, this was the end of *El Acusador.* Even before the cemetery rally began, the secret police finally found the mimeograph machine at the apartment of Joaquín González, a Movement member. The agents smashed the machine and seized about one half of the copies that were there. As they approached the cemetery with copies of *El Acusador,* Abel Santamaría, Edla Pérez, and Melba Hernández were arrested. Only Fidel and Haydée Santamaría escaped arrest in the ranks of the Movement's directorate. Elda and Melba were released later that day, and were able to find Fidel to tell him about the companions in jail.

The next day, Castro and Melba appeared at the Castillo del Principe prison in their capacity as attorneys to seek freedom for Santamaría and González. On the way, Fidel said, "Let us buy them something," but he had only one peso in his pocket, and all he could get were cigarettes and matches. At the prison, they were stunned to see most of their other companions in detention, too: Montané, the poet Raúl Goméz, and all the others. It was only then that Castro realized that the Movement had been betrayed by an infiltrator, a police informer, and that the arrests had been carried out methodically the previous day and night.

Castro spent the day arguing for the release of his companions, and then devoted days and nights trying to track down the "traitor," who was never discovered. This effort, however, prevented him from going to the hospital where Fidelito again had to undergo emergency surgery. He was only able to see his son back at the apartment days later. Several days after the arrests, the Military Intelligence Service—the murderous SIM—tracked down one of the Movement's two radio transmitters.

These setbacks did nothing to discourage Fidel. All his companions were released within a few days, and during the first week of September, he presided over a clandestine meeting of new Movement members in Old Havana. He told them: "All those who join the Movement will do so as simple soldiers; any merit or post which one might have had in the *Ortodoxo* party will not matter here. The fight will not be easy and the road to be traveled will be long and arduous. We are going to take up arms against the regime."

The next day, Fidel and Abel were arrested in Castro's sedan on a Vedado street by police in a patrol car. At the station, they and the car were searched, but nothing incriminating was found and they were let go. This was Fidel's first arrest as the secret Movement's head, making him even more careful about all his moves.

Then once again a tragic anniversary was the occasion for a revolutionary act. November 27 was the eighty-first anniversary of the execution by the Spaniards of eight nationalist medical students, and Castro and student leaders gathered at the university for an anti-Batista rally. Fidel and his companions brought along the second radio transmitter Dr. Muñoz had built for them to broadcast the demonstration. But the police cut off electric power to the university, preventing the holding of the rally.

It was one more disappointment, and it marked the end of the first phase of the Movement's revolutionary activities. But the evening was not wasted altogether. In the darkness of the campus *escalinata,* Fidel was introduced by his friend Jorge Valls to Naty Revuelta, the rich and beautiful wife of the heart specialist, the woman with revolutionary sympathies who back in March had tried to send him a key to her apartment so he could hide from the police. A romance was born that evening.

CHAPTER

2

The hundredth anniversary of the birth of José Martí fell on January 28, 1953, and for Fidel Castro this was an extraordinary opportunity for spectacular revolutionary gestures. Despite the fact that clandestine military training had acquired momentum in the closing months of 1952, and Castro had practiced extreme caution in his personal visibility, the celebration of Martí's centennial could not be ignored.

The Movement was in its second phase: Planning and preparation for actual revolutionary action had begun. By this time Fidel had cut himself off from any direct identification with the training operations. He never went to the university grounds where Pedro Miret conducted clandestine excercises, and most of the new recruits were unaware that Castro was the man who headed the Movement; he, of course, knew who each individual was and what he was doing.

As a basic rule, Castro believed that he and his fighters should not call unnecessary attention to themselves. They had been involved in university demonstrations the previous November, but now the *Fidelistas* lay low, staying away from university affrays during December and the start of January 1953. On the other hand, Castro was too much of a committed public figure, and he understood that it might be harmful to him politically for his absence to be noted. Fidel had to walk a thin tightrope, a situation aggravated by his penchant for acting on impulse.

On January 13, Castro attended a Havana meeting of the leadership of the *Ortodoxo* party to hear a proposal to form an alliance with other political parties to oppose Batista. The session collapsed in chaos when the most prominent of the *Ortodoxos* walked out in protest against what they suspected would be the end of their party's independence, or what was left of it after the coup, and an ultimate sellout to the dictatorship. Stomping out, too, Fidel shouted, "Let's get out of here. . . . You can't count on these politicians to make a revolution." This was Castro's final contact with traditional politics in Cuba.

At about the same time, Communist and other extreme leftist students formed a committee to erect a statue to Julio Antonio Mella, a university leader and cofounder of the Cuban Communist party, who was assassinated in Mexico in 1928. The idea was to consecrate the memory of Mella and thereby virtually make him a national hero. By placing the statue in the street outside university buildings, they would also expand the territory of campus autonomy which the authorities could not legally breach.

The committee was organized by Alfredo Guevara, and on January 10 the Mella statue was unveiled. Fidel made a point of not attending for tactical reasons, but his brother Raúl was present. There was a certain political ambivalence in the relations between the two brothers (who were personally quite close, with Raúl living for varying periods with Fidel and Mirta), inasmuch as Fidel insisted on keeping him out of the Movement at that time, but kept him partially informed of its progress. That Raúl was involved with the Communists evidently did not trouble Fidel, and he may have thought that it would be useful to have his younger brother become a link between the Movement and the party; Alfredo Guevara could not discreetly play that role.

On the morning of January 15, students discovered that during the night the white marble bust of Mella had been splashed with black paint, and quickly a crowd of angry youths began to gather. By noon, thousands of students were marching down Havana streets, hanging Batista in effigy, and battling the police, who began firing on the demonstrators. It was the biggest riot since Batista's assumption of power, and in midafternoon the youths decided to march on the presidential palace. The police fought back with tear gas and bullets, and a twenty-one-year-old student named Rubén Batista Rubio was fatally wounded.

After dark, groups of students, among them Alfredo Guevara and Raúl Castro, returned to the university to regroup and to await a police attack. Thirty other students, who formed a protective cordon around the Mella statue, were arrested and taken to police stations. At the Third District police station, a student named Quintín Pino suddenly saw a familiar figure entering the precinct around midnight, and he cried to his companions, "Hey, here comes Fidel . . ." Castro, who meticulously avoided the riot in which Raúl had fought all day, now chose to appear as attorney, obtaining the release of the thirty students before dawn. Politically, this had more value than hurling stones at the police.

University tensions and disorders went on until the week of the Martí anniversary celebrations, an occasion on which the Batista regime and its opponents battled over the proprietorship of the memory of the Apostle. The government launched the celebrations with a reception at the presidential palace on the evening of January 25, and the festivities continued with a formal session in front of the National Capitol (where parliament functioned before the coup) on the night of January 27.

Counter-celebrations were simultaneously set in motion by radical opposition groups, ranging from the University Students' Federation (FEU), the *Ortodoxo* party, and the Socialist (Communist) Youth to the Women's Martí Centennial Civic Front, the latter a very new entity. The FEU organized a Martí Congress for the Defense of the Rights of the Youths, and Raúl Castro acted as a key member of the founding commission. The congress produced a permanent committee, with Flavio Bravo, the chairman of the Socialist Youth (and a full-time Communist organizer), as one of its fifteen vice-presidents, and Raúl Castro as one of ten permanent secretaries, personally responsible for propaganda. Fidel Castro had no part in the congress, but inevitably he was caught up in its aftermath.

The evening of the Martí youth congress, January 26, the police raided a house in a Havana suburb where about twenty women belonging to the Civic Front were preparing for distribution the following day a leaflet condemning the regime for collecting additional taxes to pay for the anniversary celebrations. The women were pushed into a waiting police van to be taken to the headquarters of the investigations section, but they shouted and sang inside the vehicle, and their voices were recognized by three occupants of a car that happened to pass them at the entrance to the bridge over the Almendares River. The three men were Fidel Castro, Aramís Taboada, and Alfredo "El Chino" Esquível; all three were lawyers and all three belonged to the secret Movement.

"Let's turn around and follow them," Fidel said, and the three attorneys reached the investigations office immediately after the van. Castro announced that he was the women's legal representative, and, as one of them said later, "He didn't move from there until the last one of us left the police office at dawn." But Fidel Castro's grand revolutionary gesture was yet to come. He and his Movement were to celebrate the Martí centennial by staging a public parade of his clandestine "army" the following evening.

*　　*　　*

The Movement's so-called army had begun to take shape at both the university and in the countryside in the early autumn of the previous year when Castro decided to act "independently" and met Pedro Miret, who then undertook to mold the Movement into a military weapon. At the outset, Miret and others at the university were engaging in sporadic weapons training in the basements of buildings, but because silence had to be maintained, the guns could not be fired except occasionally at the sports stadium. The instruction centered on arms handling, assembling, and disassembling, and crawling with rifles in hand. When Miret began to train the first volunteers (before joining Fidel's Movement, he taught almost anyone who asked), all he had was an ancient Halcón submachine gun, one M-1 rifle, one Springfield rifle, one Spanish-made Mendoza rifle, two Winchester shotguns, and a few pistols.

Miret's operation changed totally after he met Castro on September 10, 1952. Fidel had heard about him, and sent Ñico López, his *Ortodoxo* worker friend, to ask Miret to train "a little group" he had. Miret, then a fourth-year student, was disenchanted by the old-line political organizations for which he was training young men, and after meeting with Fidel, he devoted himself completely to the Movement. When Fidel formed the Movement's military committee shortly thereafter, Abel Santamaría, Miret, José Luís Tasende, who worked in a refrigeration plant, Ernesto Tizol, a chicken farmer, and Renato Guitart, a young man from Santiago, where he worked in his father's small trading firm, were the members.

Most of Fidel's recruits were poor workers from Havana and nearby localities. They spent their Sundays (and their few extra cents for bus fare) at the university, learning the mechanism of a weapon that, as Miret once said, could not even fire, and dragging themselves in the dust under the burning sun. Most of them had never set foot in a primary school, let alone a university, and they had to overcome a sense of social inferiority when they met students.

The recruits came to the university in cells of ten or fifteen men, each cell being assigned beforehand the exact hour and place to report for training. First, the groups stopped at a Havana high school to receive the password for the day. Once, the password was "Alejandro," which was Castro's code name, and a man would say before a locked door at a university building, "We have been sent by 'Alejandro.'" Tasende and Tizol usually checked the identity of the recruits at the entrances before letting them enter.

Security precautions were so great that men from different cells were forbidden to talk to one another, and names were never given or exchanged. Castro himself not only avoided ever being present at training sessions, but he even concealed his relationship with Miret. When messages had to be exchanged or plans coordinated, Ñico López, Tasende, or Tizol would act as contacts between Castro and Miret. By the same token, Miret never went places where he could run into Fidel, Abel, or other senior Movement members. Armando Hart, later one of Castro's most trusted companions and a Politburo member, remembers that after meeting Fidel for the first time at the *Ortodoxo* party headquarters in Havana, he was asked how his military training at the university was coming along. For months Hart could not figure out how Castro knew about it because he had no idea that Miret, who invited him to undertake military instruction, had links with Fidel.

To build his Movement, Castro says he traveled forty thousand kilometers in his car during the fourteen months before mid-1952 and the Moncada attack to contact groups and individuals all over Cuba so that the revolutionaries would be equipped and ready to strike. Given the Movement's lack of funds and access to weapons, this was a miracle of improvisation and secrecy. Fidel said, "We succeeded in having twelve hundred men, and I spoke with them, one by one, I organized each cell, each group, the full twelve hundred . . ." Miret estimates that between 1,400 and 1,500 trainees passed through his hands (not all of them remained in the Movement) between September and December 1952, divided among 150 cells. But because the Movement was woefully short of weapons at the time of Moncada, the attackers were selected from only twenty-five cells in Havana and Pinar del Río provinces. In these cells, members knew only each other, and in one instance in the town of Artemisa, two young workers who were close friends discovered only on the eve of the Moncada assault that they belonged to different cells of the same Movement.

Artemisa, with its working-class and anarchist traditions, was one of Castro's best recruitment pools. It produced 250 volunteers and some of the best sharpshooters of the Movement, among them Ramiro Valdés, the future interior minister. When Pedro Miret decided early in 1953 that field training was required, too, the rebels practiced in rural areas around Artemisa and in Havana province.

Assisting Miret in training the rebels was a U.S. Army veteran from the Korean War named Isaac Santos, known as "Professor Harriman" to the volunteers. It was Castro who discovered Harriman at a friend's house, learning that he had been instructing another anti-Batista faction, and that he was an expert in hand-to-hand combat. Harriman ran a tough commando course, and he taught the *Fidelistas* how to use a compass and orient themselves in any terrain. Eventually, however, the rebels suspected him of being an American intelligence plant, and he was eased out before Moncada. He was seen in a Batista prison in 1953, exhibiting signs of torture, suggesting that the judgment of treachery may have been an unjust one.

When the Martí centennial came in January 1953, Fidel Castro already had a "rebel army" of sorts, and he decided to show it off, a risky gesture, but for him a typical one. To compete with the Batista celebration at the National Capitol on the night of January 27, university students' organizations, the new youths' rights committee, the women's civic front, high-school groups, and young workers planned a huge torchlight parade across Havana, and this is where the *Fidelistas* chose to present themselves to the country.

They did so by fielding a five-hundred-member contingent that marched in the parade in military formation behind Fidel Castro, chanting, "Revolution! . . . Revolution! . . . Revolution! . . ." The flaming torches they carried in their hands were armed with large, pointed iron nails at the top to be used as lethal weapons in the event of an attack by the riot police. The *Fidelistas* had spent all morning constructing the nail torches at Calixto García Hospital and at the university stadium. Late in the afternoon, Abel Santamaría and José Luís Tasende organized the rebels into marching units; when night fell, Castro appeared to assume command of his troops and joined the parade of thousands of torches. Raúl Castro marched behind his brother, as did Melba Hernández and Haydée Santamaría, the two women founders of the Movement, and most of the general staff.

Presumably because of the presence in Havana of distinguished foreign guests, the regime did not interfere with the parade; the police made no move to halt the parade, ignoring the chants of "Revolution" and "Liberty," and there was no violence. The next morning, the actual Martí anniversary day, thousands of youths paraded again through Havana, this time from the university to Martí's monument in Central Park, and again Fidel led his perfectly drilled units, the men marching arm in arm. Melba Hernández recalls that the *Fidelistas'* martial appearance caused a "sensation," but she also heard people in the streets saying, "Here come the Communists." At the same time neither the crowds nor the authorities seemed to realize that the highly disciplined detachments marching behind Castro in the two successive parades were actually under his command, that they belonged to an organized movement, and that they had military training.

There is no rational explanation for Fidel and his associates' decision to participate in the Martí parades as a large group that could so easily be identified by the police and the SIM as a revolutionary organization, thereby inviting its own destruction. It threatened all the painstaking security measures taken over long months to prevent the Movement's public identification with Castro, and it is a mystery how the secret police failed to draw the obvious conclusions from the spectacle of Fidel marching at the head of his columns.

Though it certainly was the bravado on which Castro thrives, he may also have calculated that the Batista regime, unaware of the Movement's existence, would think nothing of the presence of Fidel and his friends among the parading protesters. The temptation to flex muscles in public for the first time may have been too great to resist. Asked about the risks, Pedro Miret says that "we were certain that nobody would know who we were," and that at the time the government saw only the traditional political parties as enemies: "We were nothing to them, we did not exist." Besides, Miret says, the plotters held the authorities' professional competence in low esteem.

Castro remained in public view during most of February, although he and his closest associates were already actively engaged in preparing targets for military revolutionary action. On February 8, Castro published in the mass-circulation weekly *Bohemia* an article denouncing the destruction by the police of the studio of the Havana sculptor Manuel Fidalgo. Among the works destroyed were small busts of Martí with the inscription "For Cuba That Suffers" (a Martí quotation) and death masks of Senator Chibás. Fidalgo sold most of his works to raise funds for Castro's Movement, and Fidel made the sculptor's unexplained disappearance after the police raid sound ominous. He wrote an article that was printed with pictures taken by Fernando Chenard, a professional photographer who came to the Movement from the Communist party. It was the first time since the Batista coup that the Cuban press published an article signed by Castro.

On February 13, Rubén Batista Rubio, the youth shot by the police during the Mella statue disturbances the month before, died at a Havana hospital, and this triggered another protest. Castro, who had been at his bedside almost every day, helped lead some thirty thousand persons in a silent funeral procession from the university to the cemetery on February 14. Riots broke out all over Havana after the burial, automobiles were set ablaze, and the police fired on the students. The next day, formal charges of promoting "public disturbances" were lodged by the secret police against "Doctor Fidel Castro," who was described as an *Ortodoxo* youth leader. The government eventually dropped the case without bringing it to trial.

There were more disturbances in Havana in February and March, including some on the first anniversary of the Batista takeover, but Castro and his Movement concentrated almost entirely on military preparations. Mario Mencía, the historian of the Cuban revolution, believes that after the events of February 14, Castro had made a "tactical change" in the revolutionary movement. During the first year, he writes, the nascent Movement took advantage of every opportunity to defy the Batista regime publicly. In the second year, however, Castro's policy shifted to the avoidance of any situations that would interfere with the Movement's principal goal of direct action. Mencía observes that this "revolutionary agitator" knew how to adjust phases of his Movement to changing circumstances.

For these reasons Castro remained neutral when another anti-Batista faction attempted to overthrow the regime through a civilian-military coup. The MNR, the Nationalist Revolutionary Movement, had been organized the previous year by Rafael García Bárcena, a liberal university professor with links to young army officers. Drawing its following from among middle-class students, intellectuals, and professionals, it had no pronounced ideological tendencies. Its ranks included, however, a number of young lawyers, doctors, and other professionals who later joined Castro in the 26th of July Movement. García Bárcena's concept was to lead about fifty men, armed with pistols and knives and daggers, on the army's Camp Columbia in Havana and to seize it with the support of pro-MNR officers inside the installation. Then, the professor believed, a popular uprising would follow and Batista would fall. Even with a conspiratorial bias, this was a very unconvincing plan, depending as it did on the improbable success of an attack by a small group of civilians on Cuba's largest military base, and on the ability of a handful of young officers to assume command of the troops.

Castro was aware of this enterprise, planned initially for March 8, then postponed to April 5, and considered it pure adventurism. The idea of organizing an army, seizing the barracks, and ousting Batista in twenty-four hours "seemed absurd." When Professor Bárcena outlined his plot and asked for his backing early that spring, Castro conveyed these views.

Subsequently, Castro was accused of fearing García Bárcena as a rival. But discussing it on another occasion, Fidel said he had told the professor that he was willing to "analyze" the plan with him, and that he had enough men and weapons to participate in such an action if it had a chance of success. However, he felt that the MNR leaders must halt immediately their plot discussions with every politician on the island. García Bárcena ignored this advice, and Castro remarked later that the conspiracy had to fail because "it was the most advertised action in the history of Cuba," and consequently the Movement could not risk becoming involved.

There were, of course, deep ideological differences between the two leaders, revolving around such issues as the concept of the class struggle. The professor was against it, and Castro claims that as a Marxist he already believed in it at the time. It is entirely possible that Fidel simply would not subordinate his ideas to a "bourgeois"-minded leader. Moreover, Castro demanded the destruction of the existing army along with Batista's overthrow, while García Bárcena had wanted an alliance with the military establishment. In the end, however, the whole argument became academic because the MNR was betrayed from inside its ranks (as Castro had warned), and on Easter Sunday morning, April 5, the police raided the house in Havana where García Bárcena and his followers were preparing to launch their coup three hours later.

Seventy persons were arrested in the MNR's revolutionary fiasco, and fourteen were tried for subversion in a trial lasting nearly two months, attracting great public attention. García Bárcena, defended by the young lawyer Armando Hart (who was a MNR member but was not arrested), was sentenced to a two-year prison term. There were more disturbances in the streets of Havana and other cities, but General Batista appeared satisfied that the only existing threat to his rule had been removed.

By the same token, Batista experienced no fears over the "Montreal Pact," signed in a Canadian hotel on June 2 by leaders of the internally divided traditional political parties, proposing his removal and the return to constitutional government. The pact failed to issue any call for armed rebellion against Batista, but concerns developed within the Movement's leadership that whatever funds were available to Cubans abroad for purchasing arms would flow to the followers of the political parties on the island rather than to the *Fidelistas*. According to one published version, Castro had agreed to be a regional coordinator for an armed uprising financed by the Montreal group under the overall command of Juan Manuel Márquez, an *Ortodoxo* politician in Havana, but this would have been a wholly implausible role for Fidel. Later, Márquez became one of Castro's closest companions (and one of the first to die after the *Granma* landing in 1956).

These Cuban revolutionaries of distinct allegiances and tendencies at home and abroad were entirely consumed by domestic politics, hardly aware of great world events. Though Cuban students had opposed the United States over the Korean War in 1950 and 1951 (Castro being among the most outspoken critics of American policies), they seemed to have lost interest in it now that the armistice was approaching. On March 5, 1953, Josef Stalin died at the Kremlin, but there is no record of any reaction to it among the "old" Cuban Communists or among the self-proclaimed young Marxists such as Fidel Castro.

And the only interest the United States had in Cuba at that juncture was to announce publicly its support for Batista's proposal to revise the 1943 electoral code (Batista periodically talked about elections) so that, in effect, the Communists under any disguise would be kept out of any future elections. But Washington failed to use this opportunity to insist that Batista actually set an election date; the assumption was that the Havana regime was firmly in control, and nobody had any idea that Fidel Castro and his Movement even existed.

* * *

At the start of Easter Week, Fidel Castro motored to Oriente, stopping briefly in Santiago, Palma Soriano, and several other spots. This was a perfectly normal holiday trip for a Havana lawyer whose family home was in Oriente, and there was no reason for anyone to suspect that Castro was actually surveying the terrain for the armed blow at Batista. In truth, Fidel had selected this particular week for travel because he knew that García Bárcena's conspiracy was planned to erupt on Easter Sunday, and he wanted the whole Movement leadership to be away from Havana. Abel and Haydée Santamaría took Melba Hernández to their native Las Villas province, and Jesús Montané went to see his parents on the Isle of Pines.

The idea of seizing an army base through an attack by several hundred armed men was devised by Fidel late in 1952, when Pedro Miret had completed the first stage of the volunteers' military training. He had settled on the Moncada army barracks in Santiago, Cuba's second largest military installation, sometime between January and March 1953. Driving through the city of Pinar del Río on one of his recruiting trips late in 1952, Castro pointed to the army barracks and asked his companion, José Suárez Blanco, "What do you think of this?" Suárez, who understood the question, but also knew how limited were the Movement's resources, replied, "Nobody can get in there."

The dilemma facing Castro was that he could not start a revolution without arms, and that the Movement had no access to funds to purchase them. Fidel could barely make ends meet to support himself and his family; they were essentially subsidized by friends in the Movement. As there was no real source of money for this secret organization in Cuba or any prospects abroad, Castro concluded easily that if they could not buy arms, they had to grab them. In a pep talk to his companions in the first weeks of January 1953, Fidel summed it up: "But there are places where there are more than fifty M-1s; there are places where there are one thousand oiled rifles, well-kept rifles. . . . There is no need to buy them, there is no need to bring them, there is no need to oil them, there is no need to do anything; the only thing that needs to be done is to capture them . . ."

The risk involved was once again overridden by Castro's self-confidence. The idea of attacking Moncada in Santiago would serve two purposes: as a means of obtaining quantities of modern weapons and as a major military base to become a center around which a national revolution would grow. Castro was able to remind his companions that the Cuban armies in the independence wars started out as guerrillas, equipping themselves with weapons obtained from assaults on Spanish fortresses. During the uprising in Bogotá in 1948, too, crowds had obtained weapons from police stations they attacked.

Castro's studies of military strategy included *Chronicles of the War of Cuba* by General José Miró Argenter, and the *Campaign Diary* by Máximo Gómez, the generalissimo of the pro-independence forces. When three shiploads of arms for the Cubans were confiscated by the United States, Gómez wrote José Martí not to worry because he would take arms away from the Spaniards. Fidel was likewise impressed with Ernest Hemingway's fictional account of Spanish Republican guerrillas fighting the well-equipped Nationalists and grabbing their weapons, going as far as to tell American visitors that he learned guerrilla warfare from *For Whom the Bell Tolls* (strangely, he made no effort to become acquainted with Hemingway, who resided permanently on the outskirts of Havana until shortly before his death). According to Alfredo Guevara, he visited the Communist party's bookstore to buy works on Soviet Army campaigns in World War II (which included much irregular warfare by the Russians). Castro is not known to have sought familiarity with guerrilla writings by Mao Zedong, or Vietnam's Ho Chi Minh or General Giap.

In a 1966 speech, Castro explained the military aspects of his Moncada strategy: "We did not expect to defeat the Batista tyranny, to defeat its armies, with our handful of men. But we did think that this handful of men could seize the first arms to begin arming the people; we knew that a handful of men would be sufficient not to defeat that regime, but to unleash that force, that immense energy of the people that, yes, was capable of defeating that regime." The central political aspect of the Castro strategy, he described, was a rejection of the traditional notion that "revolutions can be made with the army or without the army, but never against the army." Nobody in those days, Castro said, would even think of a revolution against the army, yet his entire philosophy reposed on the belief that the army must be destroyed to clear the ground for a "real" revolution. Castro's master plan, then, called for the capture of arms by his revolutionary "handful" in order to defeat the Batista regime *and* destroy its army, and then proceed to implant the great revolution.

For reasons that Castro describes as "accidental," the Moncada assault failed. He has never accepted intellectually the possibility that his concept or planning might have been faulty. Moreover, it will never be known if the great revolution would indeed have leaped ahead with the capture of the barracks. This is a fundamental question concerning the validity of Castro's initial revolutionary strategy. Castro evidently believed that favorable conditions already existed in Cuba so that Moncada could trigger a nationwide rebellion, but in retrospect this was not necessarily a correct assessment. Besides, it was doubtful that Batista would be deserted by the bulk of the armed forces even if Moncada fell, and therefore Castro would have had to fight for survival against the dictator's superior firepower. Ironically, his defeat combined with his spectacularly defiant stance afterward, helped create a revolutionary climate on the island with the birth of the 26th of July Movement. When he launched the Sierra war three years later, the revolutionary conditions had come into being, making the Castro enterprise much more plausible. Speaking in general terms long after the fact, Castro declared that "it is an error to think that [revolutionary] conscience must come first, then the struggle." On the contrary, he said, "the struggle must come first and, inevitably, behind the struggle will come with rising impetus the revolutionary conscience." This has worked for Castro in an exceptional fashion during and after the Sierra war, and in the context of a changed strategy, but it might have failed if he had found himself sitting inside Moncada with his handful of men with no adequate conscience and support in the country at large. The Castro strategy ran totally counter to Marxist-Leninist theories, which is why the Communists refused to support him for such a long time, and ultimately he triumphed on the strength of an incredible series of gambles and his unparalleled self-confidence.

At his trial after the assault, Castro explained that he had expected to take Moncada by surprise and without any bloodshed, in part because he did not think the government had ever anticipated an attack in Santiago; the coup tradition was to try to seize Camp Columbia in Havana. Asked by the prosecutor what he had proposed to do if he had captured Moncada, Fidel said: "We only counted on our own efforts and the help of all the people of Cuba, which we would have obtained if we could have communicated with it by radio. . . . The people would have responded with firmness if we had succeeded in putting ourselves in contact with it. Our plan consisted of taking Moncada and then broadcasting the last speech of [Senator] Chibás over all the radio stations of the city. We would have read our revolutionary program to the people of Cuba; our declaration of principles touched on all the aspirations of generations of Cubans. At that time, all the opposition leaders would have supported us, joining the Movement throughout the republic. With the entire nation united, we would have overthrown the regime . . ."

Raúl Castro said later that "Fidel had concluded that the small engine would be the capture of the Moncada fortress . . . which, in our hands, would set in motion the big engine, which would be the nation fighting with the weapons that we would have seized for the laws and the measures and the program that we would have proclaimed."

In the meantime, Fidel Castro ordered the acceleration in the training of his forces. The period between August 1952 and January 1953 was mainly devoted to recruitment, organization, and creating the structure of the Movement. Military training, conducted by Pedro Miret, Harriman, Tasende, and Tizol in university basements and atop roofs, was now moved to rural farms and fields to allow actual firing and combat practice; after January 1953, there was no more military activity on the campus.

The *Fidelistas* trained wherever they could. They used the small farm where Pedro Trigo (one of the first Movement members) and his wife lived in the southeastern suburb of Calabazar, and the farm belonging to another friend in Catalina de Güines, not far from Havana. In all, at least fifteen different field locations were used for training, sessions being switched from one place to another for security reasons. The trainees usually came in small groups by bus, often getting off individually at different points, then following instructions to reach the ultimate destination. Sometimes, an unknown person—for example, a tall (or short) man in a sport shirt with blue (or red) squares—had to be spotted to be given final instructions. There was split-second coordination involving different cells who were training.

Oscar Alcalde, who owned a laboratory and worked part time as a Finance Ministry inspector, joined the Cerro Hunting Club, where he regularly brought Movement members as guests (but never the same ones twice) to practice target shooting with shotguns. Alcalde had to pay thirty or forty pesos each time for bullets and additional pesos for tips to avoid problems with club employees. The club was also used for final firing practice with .22-caliber sports rifles for those selected to participate in the Moncada attack (although they were never told where they would be going). There were fake mobilizations and emergency drills, and the military commanders winnowed out one by one those volunteers who were not considered fit for combat.

Castro attended field training sessions very rarely, but when he came, he displayed his meticulous habits. On one occasion during training at a farm near Los Palos in Pinar del Río province, a small metal part was lost from a damaged rifle, and in the rain and falling darkness, Castro searched in tall grass for the tiny spring mechanism until he found it; turning to the volunteers, he said, "Perseverance will give us our victory."

Castro was aware of the vital importance of every weapon in the Movement's small and antiquated arsenal, and he kept track of individual bullets as well as rifles. Modern weapons were not only expensive but difficult to obtain. One day, Pedro Miret and Oscar Alcalde, the treasurer, almost fell into a secret police trap when they went to buy ten old Thompson submachine guns from a man who claimed to be a Spanish Republican refugee, but was actually a military-intelligence agent. The contact had been made through a third party, but the Movement security was so solid that the police had not identified the prospective buyers beforehand; Miret and Alcalde fled when they realized they were being surrounded by plainsclothesmen in identical blue sports shirts at the spot where the transaction was to occur.

Miret says that the military committee finally decided to arm the *Fidelistas* with relatively cheap and easy-to-find .22-caliber sports rifles and hunting shotguns. Still, it was on surprise and not firepower that they relied. In the end, the arsenal included sawed-off .44-caliber Winchester shotguns, automatic Remington shotguns, and several semiautomatic .22-caliber Browning rifles—plus the old M-1, the Springfield, and the Spanish submachine gun that Miret had treasured all along. Ernesto Tizol was in charge of purchasing most of the shotguns because he owned a chicken farm and could make it appear plausible. In other cases, Movement members falsified order forms from businesses where they had friends. Fidel received a Luger pistol purchased for eighty pesos, and he carried it at Moncada. In Santiago, Renato Guitart was able to buy a number of shotguns and rifles, and five thousand bullets.

Raising funds for the purchase of arms, ammunition, food, and everything else needed to equip the little army was a constant preoccupation. More and more Movement members and their families had to be supported because they were too busy with the revolution to be able to work. At his trial, Castro told the court that 16,480 pesos had been donated by individual volunteers, most of whom died at Moncada or were murdered by the army and the police afterward. The best estimate is that overall, the Movement was able to collect up to forty-thousand pesos during its entire pre-Moncada existence, an amount so small that the prosecutors refused to believe it.

There was a tiny trickle from the sale of the works of the sculptor Fidalgo (who reappeared late in May when he failed in his attempt to escape police by sailing to New York as a stowaway) and impressively generous donations from individuals in and around the Movement. Jesús Montané contributed his entire 4,000-peso severance pay when he quit his accountant's job; Oscar Alcalde gave 3,600 pesos from mortgaging his laboratory; Tizol mortgaged his chicken farm; the photographer Chenard gave 1,000 pesos; and Pedro Marrero sold all the furniture from his apartment except for the bedroom set because Castro would not let him do it. Abel Santamaría sold his car. Naty Revuelta turned over her 6,000-peso savings.

Naty also joined a group of Movement women who worked endless hours at Melba Hernández's parents' apartment, sewing uniforms from material bought cheaply from a nearby department store, making military caps, and sewing noncommissioned officer's stripes on the sleeves of a dozen uniforms. Castro had decided the rebels would wear Cuban Army uniforms to conceal their identity until the last moment. His idea was to buy two or three hundred uniforms, and Pedro Trigo located a medical corps corporal named Florentino Fernández Léon who detested Batista and was willing to purchase the uniforms from soldiers he knew in Havana. But he was able to buy only one hundred uniforms by June, and Castro decided to have the Movement sew the additional ones he needed. As usual, nothing was being left to chance, no detail was overlooked.

Fidel Castro arrived in the Oriente town of Palma Soriano on April 3 to confer with a local dentist named Pedro Celestino Aguilera, the chief of one of the Movement's cells in eastern Cuba. Aguilera briefed Castro on the political situation in the region, emphasizing the widespread opposition to the Batista dictatorship. Castro thanked him, told him he would be in touch, and departed without saying what he was planning to accomplish in Oriente.

Having already selected Moncada as the attack target on the basis of information from Movement contacts in Santiago and on the basis of his own conclusions, Castro began dispatching secret emissaries to study the area in detail. Oscar Alcalde went to Santiago in May to observe the change-of-guard pattern and timing at the fortress downtown. Lester Rodríguez, a senior Movement member, followed him to concentrate on every detail of the small area between a nearby courthouse and the barracks, a stretch of street through which some of the attackers would drive. And so on.

During his April tour of Oriente and accompanied by Renato Guitart, Castro instructed Ernesto Tizol to buy a small farm on the road between Santiago and Siboney Beach where Fidel used to go as a high-school student. He, Guitart, and Tizol spent a day walking up and down the road until they agreed on the farm they deemed suitable. Then they went to swim at the beach. Castro calculated that the farm would be the perfect spot to concentrate his forces on the eve of the attack, and as a place to fall back on in case of need. The farm was ten miles from downtown Santiago and Moncada, and seven or eight miles from the Sierra Maestra foothills in the other direction. In the event of a disaster at Moncada, Castro's plans contemplated a retreat to the Gran Piedra Mountains nearby, which he knew well from his schoolboy climbing days. Tizol was instructed to say to the owner that he was buying the farm to raise chickens as well as to have a vacation place not far from the sea. After initial hesitation, the owner sold El Siboney farm to Tizol.

In June, Abel Santamaría departed for the Siboney farm to prepare it to receive clandestine arms shipments. Before Abel left Havana, Castro told him to assume the overall command of the Movement if anything happened to him. Abel did not tell his sister Haydée where he was going; the reasoning was that in case of detention it was better for her not to know her brother's whereabouts. Fidel instructed her to move to Melba Hernández's apartment.

While Abel and Tizol worked at the farm in secrecy, other Movement members in Oriente were instructed by Castro to buy furniture and a refrigerator for the Siboney farm, to rent mattresses for the men who would arrive later, and to rent rooms for others in hotels and boardinghouses in Santiago and Bayamo. Castro was planning a simultaneous attack on Bayamo army barracks on the western approaches to the Sierra Maestra. After his visit to Palma Soriano, he went to the manganese mines of Charco Redondo to establish contact with the miners. Charco Redondo, just outside of Bayamo, was to serve as a support area during the attack on the barracks. The blueprint was rapidly turning into military reality.

On his way back to Havana from Oriente, Fidel stopped overnight at the Birán farm to see his parents and to borrow 140 pesos from his brother Ramón. Fidel had asked him for a larger amount, without saying that it was intended for the Movement, but Ramón refused. This was the last time Fidel saw his father: He would be too busy to see Don Ángel before departing for Mexico two years later, and the old Spaniard would die before Fidel could return in triumph.

CHAPTER

3

Fidel Castro's great adventure, culminating fourteen months of conspiracies and preparations, finally began to turn into reality on the evening of Friday, July 24, 1953, in the suffocating heat and humidity of summertime Havana. Having completed all the last-minute arrangements, delivered updated sets of instructions to those staying behind, and made his farewells to the very few people who mattered to him personally, Castro climbed in a rented 1952 blue Buick sedan with a cream-colored roof for the long journey to Oriente and what he called his encounter with destiny.

Castro, Abel Santamaría, and the rest of the Movement's military committee had picked Sunday, July 26, as the date of the planned dawn attack on Moncada and Bayamo. They were the only six men in the whole organization who knew the places and date of the action. Sometime in May, but just before Fidel left for Santiago, the entire battleplan was reviewed one more time. Castro, Pedro Miret, José Luis Tasende, and Ernesto Tizol, the four members of the military committee still in Havana, went over every operational detail.

As a security precaution, Castro slept at a different home every night in the weeks preceding the departure for Oriente, rarely appearing at his latest apartment on Nicanor del Campo Avenue in Nuevo Vedado where he had moved with Mirta and Fidelito some months earlier. His movements around the city avoided any pattern or routine. On that last Friday he shifted from using a black Dodge to a blue Buick (rented that morning for fifty dollars for a weekend "vacation at Varadero Beach"), with a young black Movement member from Oriente as his driver. The responsibility for driving Castro had been given to Teodulio Mitchel on the eve of the trip to Santiago, although Fidel had only met him a week earlier. Mitchel was an ex-soldier, and for the past year or so he had been a truck driver in his native Palma Soriano, where Dr. Aguilera, the local dentist, recruited him into his Movement cell. The idea was that Castro's driver should be someone from the area who knew the territory and was known there himself, so Aguilera had sent Mitchel to Havana to meet Fidel. They hit it off instantly, and Castro bought Mitchel a meal of steak with fries to mark the start of their friendship. For at least forty-eight hours, Fidel's life would be in Mitchel's hands, based on his instinct that he was someone he could trust.

In the preceding weeks Castro had gradually dispatched key personnel to Oriente, starting with Abel Santamaría, and in the last week he began moving the bulk of his force, traveling in small groups or individually by train, bus, and car. Haydée Santamaría and Melba Hernández, the only two women in the combat ranks of the Movement, went separately by train to Santiago on July 21 and 22, respectively. They each carried weapons in their suitcases; Melba had rifles in a florist's box. Larger shipments of arms and ammunition along with the uniforms had been sent to the Siboney farm during June and July through a variety of means. The last weapons purchase was made in Santiago two days before the attack.

Raúl Castro was away during these preparations. Working very closely with the Communists at the university (though not yet a party member) and notably with the Socialist Youth's chief, Lionel Soto, Raúl left the country in February as a member of the Cuban delegation to the Communist-sponsored Fourth World Youth and Student Festival in Vienna. Then he went on to Bucharest, spending a month in Rumania, and continued his discovery of Communist Eastern Europe with leisurely visits to Budapest and Prague, "visiting factories," which he followed with a nine-day stay in Paris. Raúl sailed home, arriving in Havana on June 6. Part of his European tour was financed by his parents, who gave him five hundred dollars after his sister Juana talked them into it.

As a result of his absence, Raúl played no role whatsoever in the growth and molding of the revolutionary movement. Raúl, who formally joined the Communist Socialist Youth after returning from Europe, said later that when he received word on July 24 to leave Havana immediately for an unknown destination, he failed to inform the Socialist Youth. "In not advising it, I may have committed an error," he said. "I surely committed an error, but I belonged to the party for a month and a half, and I didn't have a very keen sense of obligations [to it]." Still Raúl almost missed the Moncada assault altogether.

On his return to Havana in early June, Raúl had been arrested on charges of carrying subversive propaganda along with two Guatemalans he had befriended during the crossing (another member of the shipboard group was a Soviet citizen who was not allowed to disembark in Havana, and whom Raúl later called "my first Soviet friend"). The Guatemalans were released when Guatemalan diplomats intervened, and they were placed back aboard the liner. Raúl was the only one to remain imprisoned.

Melba Hernández, who went to the Castillo del Principe prison as an attorney to try to free Raúl, says that she found him "very enthusiastic" about his European trip. The police confiscated his diary, and the chief of the secret-police investigations told Melba that Raúl had written in it that "the socialist world is a paradise." She was unable to obtain his release because he had now been charged with "public disorder" and had to stand trial. The next day, Fidel obtained from a judge a "provisional liberty" warrant, but the case never came up: For one thing, Raúl would be in a Santiago prison for participating in the Moncada attack when the date of his court appearance for "public disorder" rolled around.

Raúl was unable to undergo a minimum of military training in the short time remaining before Moncada, but he seemed to doubt the success of any armed action by the Movement because he thought the group was too small. Perhaps he was taking the orthodox Communist view of putsch conspiracies, but when his friend José Luis Tasende asked him whether he would come along if the Movement launched an operation, Raúl replied, "Yes, I shall come. . . . In the Movement there is my brother and my best friends: you, Miret, Juan Almeida . . ." Robert Merle, the French historian who has written widely about Moncada, believes that "at the decisive moment, sentimental fidelity overcame ideological loyalty."

Raúl was sharing a boardinghouse room with Pedro Miret when Tasende came to tell him, "We are taking the train tonight." Suffering from an awesome hangover after attending a late party the night before, Raúl summoned his energy to meet Tasende later that afternoon to pick up a shipment of weapons, and to go to the railroad station to board the train. Sixteen other Fidelistas under Tasende's command were traveling on the same train, but they pretended not to know one another. Raúl received his ticket from Tasende, who sat next to him, and saw that Santiago was the destination. "Moncada?" he asked. "Yes," Tasende replied softly. They arrived in Santiago in the afternoon of the next day.

Between 120 and 130 men traveled from Havana and Pinar del Río provinces to Oriente between July 23 and 25, a feat of Fidel's logistics. In addition to buses and trains, fifteen automobiles were used in this deployment, Fidel going in one, Pedro Miret in another, and so on. Each departure followed a precise timetable that included stopovers and arrival times. The cars were borrowed or rented from travel agencies. The overwhelming majority of the volunteers had no idea where they were going until reaching their destinations, but the discipline was so great that, except in one or two instances, no questions were asked, nothing was questioned.

Fidel Castro had always demonstrated the most acute interest in the personal lives and welfare of his fellow revolutionaries. Melba Hernández recalls, for example, that shortly before the departure for Oriente, Castro suddenly realized that Gildo Fleitas López, one of the most active Movement members, had been engaged for years to a girl whom he had known from childhood and that he might die at Moncada without a chance to marry her. "Therefore," says Melba, "Fidel organized rapidly here in Havana a wedding with all the requirements of the law, a wedding with a veil and with everything that goes through a girl's head, with everything a girl dreams about—the wedding of Gildo with Paquita. Gildo had his honeymoon with Paquita. Gildo fell, he did not return." In Mexico three years later, Castro decided that Arturo Chaumont, one of the rebels selected for the *Granma* expedition, should marry his woman companion, Odilia Pino, so that she did not become a spinster in the event of his death during the landing. As Melba tells the story, neither of them had actually planned marriage, but Fidel kept insisting that "we must marry them." His companions began to complain that Castro was interfering in people's private lives, but Melba says that he conducted the project like a "military campaign" to press Arturo and Odilia to go to the altar. In the end, Fidel prevailed, and money was somehow found not only for a church wedding, but for a honeymoon in Acapulco and a cash wedding gift from the Movement. Three months later, Arturo Chaumont was captured by the Batista troops, never to be seen again. Says Melba Hernández: "Fidel has this preoccupation with giving his friends a bit of happiness . . ."

Even though the specific mission had never been discussed during the long months of preparations, there were vastly more volunteers than equipment to arm them. Finally, Fidel and the military committee determined that 135 attackers would be required for Moncada and 30 for Bayamo, and the chiefs of the individual cells were ordered to select their best men to meet the necessary numbers. The volunteers were told only to bring casual clothes for an extended absence.

The Wednesday before the attack, Castro had supervised the drafting by the poet Raúl Gómez García of the Moncada Manifesto that was to be issued to the nation when the rebels attacked in Santiago and Bayamo. Late that day, Fidel went to the apartment of his friend Naty Revuelta in Vedado. There, he gave her the text of the Manifesto to retype and make a number of copies for him to take along to Santiago and for her to distribute to principal political leaders and editors in Havana the moment he seized Moncada.

Naty, the green-eyed blonde who had been devoting most of her time to Fidel's cause, had already selected and purchased the records that the revolutionaries planned to play in their moment of victory over Santiago radio stations: the Cuban national anthem, independence war hymns, Chopin's triumphant A-Major Polonaise and Beethoven's *Eroica* Symphony. The great sound of patriotism and revolution, the blood-quickening music, the Chibás farewell speech, the appeal for the creation of armed people's militias, and the Manifesto would be the spark for the national uprising against Batista. First, the Manifesto, signed by "The Cuban Revolution," invoked the memory of José Martí, charging that the "true revolution" ignited by him and continued by subsequent generations had been undermined by Batista's "treacherous coup" to bring "the crimes of blood, dishonor, unlimited lust, and theft of the national treasury." It then proclaimed that "before the chaos into which the nation has fallen, the determination of the tyrant, and the godless interests of the men who support him, the youth of Cuba who love freedom and man's dignity stand up in a gesture of immortal rebellion, breaking the insane pact made with past corruption and present deceit."

In the document Castro offered Cuba a nine-point program, in which the first point stated that "The *Revolution* declares itself free from the shackles of the foreign nations." This sense of priorities presumably conveyed his nationalism (if not his anti-Americanism), but it also was the nationalism of his generation, and no Marxist connotations should be read into it. If nothing else, Castro himself insists that at that stage he carefully concealed his Marxism. The Manifesto also called for "definitive social justice" and "respect for workers and students," but such phrases had been employed by Cuban politicians for a half-century. In terms of representative democracy (meaning democratic elections), Castro declared the "Revolution's . . . absolute and reverent respect for the Constitution of 1940," which Batista had suspended with his 1952 coup. In other words, the Moncada Manifesto was hardly a call to the barricades, and in retrospect one might even wonder whether it would have been sufficient to set the country aflame.

In any event, Castro was confident that he and the young poet had produced an inspiring text, and one of his important errands on his last day in Havana on Friday, July 24, was to pick up the retyped copies of the Manifesto from Naty Revuelta. With Teodulio Mitchel driving him, Fidel started the day meeting Movement members at the Santamarías' and Melba Hernández's parents' apartments for quick conferences. Next, he rushed to the southeast suburbs to pick up two platoon leaders, to the vicinity of the airport to locate a cell chief, to Santiago de las Vegas for the same purpose. Back in Havana, Castro stopped again at the Santamarías' home to pick up extra arms, and at his sister Lidia's apartment to hand instructions to a group just arrived from Artemisa. Tauntingly, he also visited the secret-police headquarters to inquire about a "client," on the theory that he would sense if there were suspicions about him.

When night fell, Fidel was racing again along the airport route to contact an aviation radio specialist whose services he required in Santiago. Except for catnaps, he had not slept at all and had eaten virtually nothing in the past forty-eight hours, too busy and too charged up to be able to relax. Because the blue Buick ignored a stop sign, Castro and Mitchel were stopped by a police cruiser and fined on the spot after Fidel calmly informed the policemen that they were en route to the airport to meet family members. Had the highway patrolmen decided to take them to a police station, which was often done in such traffic cases, Fidel's revolution might have been thrown completely off schedule. He always seemed to live on the brink.

Returning to the city, Castro met briefly at a bar with the dentist from Palma Soriano, then drove two blocks to his own apartment. There, he told Mitchel that he was going upstairs "to kiss my son . . . God only knows when I'll be able to do it again." A photograph in official Cuban archives shows Castro with Fidelito, then nearly four years old, pointing at something in the distance with his extended right arm and index finger, his son appearing pleased and amused. Castro wears a suit and a tie, there is a handkerchief in his breast pocket, and, shot from profile, he looks the role of a Latin matinée idol. The caption says that in the picture Fidel is saying good-bye to his son, but it seems too posed, and one suspects that the photograph may have been taken for future historical purposes.

Castro is not known to have said anything to Mitchel about taking leave from his wife, and it is extremely unlikely that he gave Mirta the slightest indication of his plans or of the reason for kissing his son good-bye. Again, she remains the unseen figure in the Castro family drama. At the apartment, Fidel put on a white *guayabera* (presumably having changed from the suit which he wore in the picture), and as one of his friends said later, his only baggage for the trip was an extra *guayabera* and a volume of selected works of Lenin.

Castro's last stop was Naty Revuelta's apartment on Eleventh Street in Vedado to pick up the original and several copies of the Manifesto as well as the records she had bought for the planned revolutionary broadcast from

Santiago. Under the circumstances, Naty was the only person not in the top leadership of the Movement to have advance knowledge of the Moncada attack: Not only had she obtained the records and typed the Manifesto, but now Fidel instructed her to hand-deliver copies to the heads of the *Ortodoxo* party, including Raúl Chibás, the brother of the late Senator Chibás, and the publishers of the weekly *Bohemia* and two main daily newspapers. Naty was to take the Manifesto to their homes at five o'clock in the morning of Sunday, July 26. Then, Fidel said good-bye to Naty.

It was already late in the evening when Castro and Mitchel finally took the central highway that runs east from Havana to Oriente. In Matanzas two hours later, they happened to come across another Movement automobile, and Castro chatted for a few minutes with its occupants. He repeated the warning that he had given earlier to all the drivers to observe the speed limit to avoid being stopped by the police (but that he himself had forgotten), and the blue Buick headed for Colón, where they stopped at the home of Dr. Mario Muñoz. The physician was preparing to leave for Santiago to join the assault teams.

Castro and Mitchel had breakfast early in the morning in El Cobre, then stopped in Santa Clara, the provincial capital of Las Villas, to find an optician. Fidel, who always remembered everything, had left his glasses behind in Havana at Melba Hernández's apartment, and being quite near-sighted, he did not want to lead the attack on Moncada half-blind (in those days, he seldom wore his glasses in public). Although it was Saturday, the optician made a pair for Fidel while he waited.

They reached Bayamo around six o'clock in the evening of Saturday, July 25, and Castro went to the Gran Casino inn to talk with the twenty-five men who were to attack the army barracks the next dawn. They had arrived in groups by train and automobile earlier in the day, and Fidel reviewed with them all the details of the battle plan. After leaving Bayamo at 10:00 P.M., the Buick was stopped at an army checkpoint near Palma Soriano. This was where Teodulio Mitchel proved his real value. When a soldier approached the car to check the passengers' documents and search the trunk, Mitchel recognized him as a friend from their hometown, greeting him by name. "Is that you, Mitchel?" the soldier said. "Go ahead . . ."

At midnight, Castro and Mitchel arrived in Santiago, in the midst of wild celebrations of annual Oriente carnival. In Cuba, carnival is in July, and Santiago was always famous for its music and dancing in the streets, day and night, from Friday until Sunday night. It was for this reason that Fidel had chosen the dawn of Sunday, July 26, for the attack. He assumed correctly that most of the garrison would be away from Moncada on weekend passes, and that the guard would be lax. This strategy, too, may have come from Martí, as he had chosen Sunday, February 24, 1895, to launch the anti-Spain uprising because it was pre-Lent carnival.

The two men had coffee in a shop downtown, and drove on to Siboney. The farmhouse was completely dark, and the blue Buick was challenged by Jesús Montané, who stood guard in the shadows. Inside, Fidel found 118 men of his Rebel Army plus Melba and Haydée. Most of the men had gathered at the farm in the course of the afternoon, coming directly from Havana via Santiago or after an overnight stop there, and they were exhausted from lack of sleep and the heat. Melba and Haydée, who earlier had fixed a chicken-with-rice meal for the men, ironed the 120 army uniforms for the *Fidelistas;* they were in the only room with a light.

Abel Santamaría briefed Castro on the state of readiness, and Fidel insisted on returning to Santiago at two o'clock in the morning for one final mission. Abel went with him as some of the men at the farm tried fitful sleep, or just chatted in low voices and smoked. In the early evening, weapons had been brought out from dry wells on the property where they had been hidden; the automobiles were concealed inside specially built wooden chicken runs. Fidel's foray to Santiago was to find Luis Conte Agüero, a well-known radio commentator and *Ortodoxo* politician who had a house in the capital of Oriente; he wanted to tell him about the impending assault and persuade him to coordinate the revolutionary broadcasts. However, Conte Agüero had stayed in Havana that weekend, and the disappointed Fidel was back at El Siboney at 3:00 A.M. It was his fourth consecutive night without sleep, but he was bursting with energy and anticipation.

Additional men arrived, and the final count was 131 fighters, including Dr. Muñoz, Melba and Haydée, and Fidel himself. Castro ordered the men to change into light-brown army uniforms, suddenly discovering that even the largest uniform available was too small for him. He inspected himself worriedly in a mirror in the half-dark room, concerned that he might not look convincingly soldierlike at the moment of attack—a potential security problem.

Though detailed plans had been worked out beforehand, Castro discovered at El Siboney that Abel Santamaría, whom he had not seen in long weeks, insisted on leading the main assault group instead of letting Fidel do it. The original plan had been for Castro to attack the fortress itself while Lester Rodríquez moved on the courthouse across the street and Abel captured the nearby hospital. Fidel wanted Abel at the hospital because it seemed the safest place; he had also been designated as Castro's successor as the head of the Movement—if Castro were killed. But Abel pleaded with Fidel. "Don't be like José Martí, exposing yourself needlessly." Castro answered, "My place is at the head of the fighters. It can't be anyone else." In the end, Fidel ordered him. "It is decided: You will go to the hospital."

Castro's next argument was with Melba and Haydée. They informed him that they intended to participate in the attack. "No," said Fidel. "You've done enough. You will stay at the farmhouse." It was Dr. Muñoz, wearing a physician's white coat on Castro's instructions, who solved the dispute, proposing that the two women accompany him and Abel in the takeover of the hospital. It was a civilian hospital, not guarded by soldiers, and Melba and Haydée would be useful to him as nurses.

At four o'clock in the morning, Castro assembled the rebels in the darkened house to outline the plan of attack; only the military committee members had known until that moment that it was Moncada they were to assault. There was an instant of terrible fear of discovery when one of the men accidentally fired a shot from his rifle, but there was nobody in the area to hear it. After explaining the operation, which he promised would last no more than ten minutes, Fidel said:

"In a few hours you will be victorious or defeated, but regardless of the outcome . . . this Movement will triumph. If you win tomorrow, the aspirations of Martí will be fulfilled sooner. If the contrary occurs, our action will set an example for the Cuban people, and from the people will arise young men willing to die for Cuba. They will pick up our banner and move forward. . . . You know already the objectives of our plan; it is a dangerous plan, and anyone who leaves with me tonight will have to do so willingly. There is still time to decide. . . . Those who are determined to go should move forward. The watchword is not to kill except as the last resort."

A rebel asked Castro what should be done with prisoners, and Fidel replied, "Treat them humanely; don't insult them. And remember that the life of an unarmed man must be sacred for you." Suddenly, one of the four university students in the group told Castro that they had decided not to go because they thought the armament was not adequate for the mission. With withering indignation, Castro ordered the students placed in the bathroom under guard. Then, the radio expert from the Havana airport announced he would not engage in "illegal" actions; he, too, was taken to the bathroom. Now Fidel Castro had 123 men and two women to launch the revolution.

Castro says that when he embarked on the Moncada enterprise, "there was [only] a small group of those with the greatest responsibility and authority who already had a Marxist formation," and that he had personally acted in this context with this "nucleus of people." Recalling these events in conversations in 1985, Castro emphasizes that the "qualities we did require from the companions were, in the first place, patriotism, revolutionary spirit, seriousness, honorability, disposition to fight . . . and agreement with the goals and risks of . . . armed struggle against Batista."

Castro's definition of the ideological makeup of the Moncada rebels is entirely consistent with the recollections by scores of early Movement members that when the attack was being prepared, political instruction centered on the Martí tradition—never on Marxism-Leninism. It is also borne out by the political backgrounds of the Moncada group. According to Mario Mencía, the historian of the revolution, out of 148 participants in the Moncada and Bayamo actions, there were only two Communist party members: Raúl Castro, who had joined the party six weeks earlier and had no voice in any Movement decisions, and the sugar worker, Luciano González Camejo. Outside the party, only Fidel and one member of the Movement leadership knew of Raúl's affiliation. Castro and Abel Santamaría considered themselves serious students and adherents of Marxism, but the overwhelming number of their followers held essentially moderate social-justice views in terms of the Cuban experience.

As Castro has repeatedly stressed, he recruited Movement members among young working-class followers of the *Ortodoxo* party of Senator Chibás, an organization that stood for social justice, social-welfare legislation for the overwhelmingly poor Cubans, and for anti-Communist nationalism. Castro may have assumed that once exposed to widespread and victorious revolutionary processes, these *Ortodoxo* workers and peasants would be propelled toward socialism, relinquishing their traditional distrust of anything identified with Marxism. He has often charged that this antipathy toward Communism was due to cold war "imperialist propaganda," but the record shows the Cuban Communists' historical inability to attract the masses. On the other hand, when Chibás founded the *Ortodoxo* party, hundreds of thousands of workers who normally ignored political affiliations rushed to join him probably because of his enormously appealing personality.

In any event, the *Fidelista* "army" was of working-class origin. Only the leaders had intellectual backgrounds. Four of them had university degrees: Fidel Castro and Melba Hernández as lawyers, Mario Muñoz as a doctor, and Pedro Celestino Aguilera as a dentist. The poet Raúl Gómez García, Pedro Miret, Raúl Castro, Lester Rodríguez, and Abelardo Crespo were occasional university students. The bulk of the Movement members had only elementary-school education, and many lacked even this background. Though lacking university degrees, five members were public accountants, including Abel Santamaría and Jesús Montané—two top rebel leaders. No more than twenty rebels earned more than two hundred pesos monthly. The largest group was construction workers (carpenters, painters, brick-

layers, and so on), then came farm workers, cooks and waiters, office-work-
ers, drivers, shoemakers, mechanics, bakers, milkmen, ice-delivery men,
street vendors, and self-employed persons (traveling salesmen, for example).
And there were many unemployed young men in the revolutionary ranks.
The oldest fighter was Manuel Rojo Pérez, a fifty-one-year-old peasant.
Castro described his Moncada companions correctly when he said in a 1965
speech that while they could not be called Marxist-Leninists, "we were
capable of understanding some of the essential principles of Marxism—the
reality of a society divided between the exploited and the exploiters. . . .
Our immediate task, our struggle with minuscule resources against that
military power that flattened our country, absorbed the greatest part of our
attention."

In December 1961, when he announced to the world his allegiance to
Marxism-Leninism, Castro noted that he had been asked whether at the
time of Moncada he had thought the same way he did now, and he added,
"I thought very similarly to how I think today. This is the truth." And in
a Moncada anniversary speech in 1965, he made a point of recalling that
"among the books they took away from us [after Moncada] were books by
Martí and books by Lenin."

Speaking in 1975, Fidel volunteered that "the fundamental nucleus of
the leaders of our Movement . . . dedicated part of its time to the study of
Marx, Engels, and Lenin," and specifically to the lecture of Marx and Engels's *Communist Manifesto,* Lenin's *The State and Revolution,* and Franz
Mehring's *Karl Marx: The Story of His Life.* (Castro's critics have seized
upon the remark he made in an interview shortly after the revolution that
he had read only 370 pages of *Das Kapital* before Moncada as proof of his
ideological inadequacies, but this is not a serious argument.) Melba Hernández, who was the closest person to Castro and the Santamarías during
the conspiracy period before Moncada, says that "it was Fidel who formed
Abel in the reading and the study of Marxism-Leninism. . . . It was Fidel
who formed Abel Santamaría ideologically, and this has to be stated
clearly." Melba saw them several times a day, every day, for a year.

Before her suicide in 1980, Heydée Santamaría reminisced about her
brother Abel, assassinated at Moncada, being encouraged by Castro to
study Marxism more intensively when they were organizing their revolutionary movement. But Haydée added that even at the risk of "being im-
politic," she had to say that Abel refused in those days to join the
Communist party because he knew he could not enjoy there the political
freedom he had in the *Ortodoxo* party. After Moncada, at El Siboney farm
the army found a copy of the first volume of a two-part Spanish edition of
Lenin's selected works with Abel's name written flourishingly across the
cover; the Batista regime used it as evidence that the assault was a Commu-

nist plot, and newspapers printed photographs of an army lieutenant at the farm, a rifle in his right hand and the captured Lenin book proudly displayed in his left hand.

There is therefore remarkable consistency in Castro's descriptions between 1961 and 1985 of his own, his friends', and his Movement's ideological evolution. In his own case, the emphasis is on his advance from phase to phase both in the ideological content of his thought and its public disclosure. His principle was to proceed one step at a time, strategically and tactically. At Moncada, Fidel Castro wished to be seen as the defender of Cuban democracy and social decency from the ravages of the Batista rule. Subsequently, he chose to espouse Marxism-Leninism publicly. To his critics, he has practiced deception; in his own eyes, he deserves plaudits for brilliant revolutionary strategy, and he always falls back on Marx's and Martí's thesis of "historical justification"—that history's requirements justify the means. And the record shows that Castro never publicly criticized the Communists during the entire insurrection, even when they denied him help.

Looking back at *Fidelismo*'s trajectory, it is evident that even for the purpose of ousting Batista, Castro had organized a revolutionary movement totally lacking in internal democracy. From the outset, the Movement was dominated by a few individuals (Fidel and Abel Santamaría, then Fidel alone after his friend was killed at Moncada) with the conviction that the military program must be fulfilled before the political program could be unfolded.

In the intervening years, Castro has demonstrated another dimension of consistency in political behavior: Once he imposed Marxism-Leninism as the official Cuban doctrine, he proceeded to bring to power the Communists who were his university friends but had never participated in any of his revolutionary actions, often promoting them at the expense of veteran *Fidelista* fighters. From the university Communist network were Lionel Soto who took Raúl to Eastern Europe, who served for long years as Cuban ambassador to the Soviet Union, and at the Third Congress of the Communist party in 1986 was named to the key post of secretary of the Central Committee; Flavio Bravo Pardo, once secretary-general of Socialist Youth, is a Central Committee member and chairman of the National Assembly; Alfredo Guevara, who was Castro's first Communist friend, commutes between Paris and Havana as a "cultural ambassador," as close as ever to Fidel and Raúl. And from the ranks of the old Communists, the party leaders and guiding lights before and during Batista's time, most have held Politburo and Central Committee posts as long as their advancing age and health permitted. At the 1986 Third Congress, it was the octogenarian

Fábio Grobart, the only surviving cofounder of the Cuban Communist party in 1925, who officially introduced Fidel Castro to the delegates.

When Fidel Castro outlined his plan for the attack on Moncada, he assured his followers gathered in the darkness of El Siboney that with the element of surprise in their favor, the attack itself would be short and swift. He abstained from explaining what would happen afterward. At his trial, too, he said only that he was certain that the people of Cuba would rise to support him and liquidate the dictatorship. While Castro may have deliberately refrained from sharing his strategic concepts with his enemies, he was already thinking of the next revolutionary phase. There is considerable evidence that his plans were extremely ambitious.

Thus Pedro Miret, a member of the Movement's military committee, emphasized in a conversation in Havana in 1985 that the assault on Moncada was specifically designed to "isolate the region of Oriente, which could have been easily isolated." This would have been possible at the time, Miret said, because there were no adequate communications between Havana and regional military commands and facilities; such communications were established after Moncada. Castro, a born strategist despite the lack of formal military studies, reasoned that for army reinforcements to reach Santiago and to try to retake Moncada and the city, it would have been necessary to come down the central highway from Holguín, a city in the north of Oriente province where at least a regiment was stationed. But the mountain highway led through Bayamo, which is south of Holguín and west of Santiago, and this was the reason for the effort to secure the Bayamo barracks simultaneously with the capture of Moncada. Then, the revolutionaries would blow up the bridges over the Cauto River, which flows through Bayamo, there cutting off the Holguín-Santiago highway. Theoretically, at least the southern part of Oriente province would have been sealed off by the rebels, creating in effect a "liberated zone," a concept to which Castro returned in launching the Sierra war at the end of 1956.

As he had said, Castro assumed he would have widespread popular support after taking Moncada in historically rebellious Oriente province, where the Cuban independence wars had been launched, and he had hoped that armed "people's militias" would emerge to bolster his revolutionary power. Moreover, nestling between the Caribbean and the mountains, Santiago would be relatively easy to defend; it had only one major land access route. A study of the Moncada operation by the Historical Section of the Political Direction of the Revolutionary Armed Forces issued in 1973, confirms that Castro's strategy was to make "a rebellion erupt in a region, try to keep it alive, call a general strike, and gain time for a popular mobilization that would raise the struggle to a national level." Santiago being Cuba's second

largest military center, located at the opposite end of the island from
Havana, made the plan even more enticing, the study said, because it
would have been extremely hard to move loyalist troops there. After removing arms from Moncada and Bayamo barracks (Moncada had 4 heavy machine guns, 10 submachine guns, 865 Springfield rifles, and 471 revolvers,
which would have been enough to equip a major rebel force), Castro would
abandon the buildings to avoid being attacked from the air. There were no
antiaircraft guns at either location, and the rebels would have spread
throughout the areas they controlled and would not present a single target
to aircraft in each city.

The Moncada study reported that if the Batista regime did not collapse
immediately, the Castro forces would engage in "irregular war" in the
mountains and fields as the independence fighters had done in the nineteenth century. Closeness to the mountains, it said, would allow a rapid
transition to guerrilla war if the conflict became prolonged. The official
study concluded the Castro blueprint was excellent, but allowed itself the
remark that "the weak aspect of the project was the reliance on the exclusive result of a single action, making the rest of the plan depend on it."
If the attempt to capture the barracks failed, the study said, "the whole
plan would fail." But, it added, this was "the only possibility open" to the
rebels, and "they had to follow it."

This conclusion amounts to the judgment that for Fidel Castro there are
risks, no matter how extraordinary, he is always willing to take because
such are his nature and his instinct. It is reminiscent of the comment made
by an American historian about Prussia's Frederick the Great (whose life
Castro had studied) that the king was a romantic, "and part of Frederick's
romanticism was surely the ability to envision ultimate victory when all
rational calculation indicated it was clearly impossible."

And Castro had instilled his faith in victory in his men. In the darkness
of El Siboney, a few minutes before five o'clock in the morning of Sunday,
July 26, the young Cubans sang the national anthem in whispers. They
were ready to go.

CHAPTER

4

The Moncada barracks is an ugly sprawl of buildings and fields in the shape of an irregular rectangle, occupying a 15-acre area of high ground in downtown Santiago. It was originally built as a fortress by the Spaniards and reconstructed after a fire in 1938. Named for General Guillermo Moncada of the Liberation Army in the independence wars, following the establishment of the Cuban republic, it served an essentially internal security function inasmuch as an external attack on Santiago was wholly unlikely.

At the time of the assault at dawn of July 26, 1953, Moncada was the headquarters of Infantry Regiment 1, known as the Maceo Regiment (General Antonio Maceo was a hero of the last independence war), and its normal complement was 402 men, of whom 288 were privates. A twenty-six-man Rural Guard squadron was also stationed there. Castro's calculation was that on a carnival night, the barracks would contain less than one half of the total number, including those missing on general holiday leave, and those remaining asleep or drunk. For the attack, Fidel would have seventy-nine men with him to try to take Moncada, the rest of the force being assigned to related targets in the fort's immediate vicinity, and a separate detachment being deployed in the town of Bayamo. His plan was designed in such a manner that it had to succeed totally, or fail totally, regardless of whether or not he had additional men.

The Moncada compound was divided into two main parts. The eastern portion of the rectangle was a firing range, the western portion was the fort proper—the objective of the rebel operation. The fort's perimeter was formed by thick-walled, two-story crenellated yellow barracks buildings on its east side, where the official entrance to Moncada headquarters was located, as well as along the south and north sides for a half-block each. The other half-blocks on the northern and southern flanks and the entire length of the western side of the compound were protected by high walls and fences; officers' and enlisted men's clubs were behind in the southwest inside the fort's perimeter, and exercise fields in the northwest.

The fort could be penetrated through the main entrance into the command building with its guardroom, but such a frontal attack would have been foolhardy. There were gateways on the west side of the compound and on the east side; they were too far, however, from the center of the fortress, obliging the attackers to cover much terrain in the open. Castro therefore chose Gate 3 in the southeastern corner of the fort, directly south of the headquarters building entrance. Automobiles could enter the fort through this gate to reach the courtyard, with the rebels then storming the structures from inside the compound and seizing them in a moment of surprise.

For additional insurance, Castro had decided to occupy the Palace of Justice (the Santiago courthouse), a three-story building one block south of the compound's southern perimeter line, to provide covering fire on the courtyard area inside. Small houses and bungalows of regimental officers and noncommissioned officers filled the block between the fort and the courthouse. The military hospital was next to the courthouse, but Castro did not think it was a useful firing position. He did order, however, the occupation of the Saturnino Lora civilian hospital, a two-story building with single-story wards in the back, which looked down on the compound's western wall and could provide additional cover for firing across the street.

Having studied the Moncada layout and security for months, Castro concluded that the fort could be rushed through the number 3 gate, after the commando team in the lead automobile disarmed the three guards in front of two blockhouses with firing slits in the sides, then lifted an iron chain across the roadway to let the other commando cars into the courtyard. Once inside, the rebels were to run up the stairs of the buildings, disarm the sleeping soldiers (killing them only if they resisted with weapons), take over the fort's radio transmitter, and fan out throughout Moncada with freshly captured army weapons. This was the plan and Castro saw no flaws in it; his scouts had even determined with absolute precision the times a two-man army patrol passed Gate 3 during its nocturnal rounds along the perimeter. The assumption was that the patrol would adhere to its split-second schedule that dawn, too.

* * *

The sixteen automobiles with Fidel Castro and his 123 attackers, including Haydée Santamaría and Melba Hernández, began departing from El Siboney at 4:45 A.M. The first car was a 1950 Pontiac driven by Abel Santamaría; the commando team that was to disarm the guards was in the fourth car, a 1950 Mercury, and Fidel drove the fifth car, a brand-new 1953 Buick. The automobiles would take their assigned places in the attack column en route to Santiago because it was easier to organize the caravan on the highway than in the dark confinement of the farmhouse front yard.

All the men wore brown Cuban Army uniforms with black neckties, wide-brimmed campaign hats or visored caps, and knee-high boots or leggings (Haydée and Melba were in slacks and blouses). The only difference between the attackers and the soldiers was in their armament. Whereas the Maceo Regiment's troopers carried modern .30-caliber New-Springfield rifles, the *Fidelistas* were equipped with an assortment of hunting shotguns, .22-caliber semiautomatic sports rifles, a few .44-caliber sawed-off Winchester rifles, a single M-1 rifle, and a single Browning submachine gun. Pathetic or brave, the rebels were superbly trained in the use of their weapons.

To avoid identification in the event of premature detention, the rebels had no documents of any kind on them. Castro and a group of older-looking men (many in the force were barely over twenty years old; Raúl Castro was twenty-two and appeared still younger) wore sergeant's stripes on their sleeves to exact even more respect from Batista's soldiers: Batista himself having led a sergeants' rebellion in the 1930s that handed him power over Cuba, this was a highly respected rank. Fidel had shaved off his moustache a day or so before the assault, possibly so as not to be recognized.

There was no sign, however, that the regime had any advance notice of the uprising. When Castro had reached El Siboney the previous night, he learned to his immense frustration that Colonel Alberto del Río Chaviano, the commander of the Oriente military district to which the Maceo Regiment belonged, had opened Moncada to the public on Friday the twenty-fourth as part of the carnival celebrations. "What a lost opportunity!" he kept muttering angrily, imagining the ease with which he could have taken over the fort had he known of Chaviano's plans in advance and could have gathered his men in time to infiltrate Moncada.

At this peak of the summer, the rest of Cuba was completely tranquil. General Batista was spending the holidays at the luxury Varadero beach resort and had been aboard his yacht, *Marta II,* off the coast for the last few days. Near Batista's hometown of Banes in Oriente (not far from Fidel

Castro's home in Birán), 150 peasants had been arrested by the Rural Guard for putting up squatter homes on land belonging to the Santa Lucia sugar mill, and the story happened to be published in the newspaper *Alerta* on Sunday, July 26, just as the *Fidelistas* converged on Moncada.

But all began to go awry instantly. The third car to leave El Siboney was a black 1948 Chevrolet driven by Mario Dalmau and bringing the five other men, including team commander Raúl Castro. (At the last moment, Fidel made his brother the cheif of one of the three detachments at Moncada, albeit the smallest.) On the way to their target, the Palace of Justice, they made a wrong turn, winding up at a downtown square in the opposite direction, and finally arrived at the courthouse several minutes behind schedule, when the fighting had already begun.

The sixth car, carrying Boris Luis Santa Coloma (Haydée's fiancé and a member of the Movement's civilian committee) and seven other fighters, had a flat tire right after leaving the farmhouse. Boris and three others were able to transfer to other vehicles, but four men had to be left behind for lack of space, and the car was abandoned. Now the force was reduced to 113 combatants and 15 automobiles.

The four university students who had decided hours earlier not to participate in the operation, and were ordered by Fidel to be the last to leave the farm en route home (after their detention in the bathroom), disobeyed him and managed in the darkness to join the middle of the motorcade. As they approached Santiago, they turned left on the central highway in the direction of Havana, but the car behind them with eight rebels assumed that this was the way to Moncada and followed them. The driver realized his error only when far past Santiago. When the eight men returned to the city, it was already too late to fight. Thus the total assault contingent that finally struck the fort area was down to 111 persons, including Fidel Castro, and 14 vehicles. In a 1959 interview with Carlos Franqui, the editor of *Revolución,* Castro said that "the best armed *half* of our troops was delayed at the city gates and so was not present at the vital moment. . . . Our reserve division, which had almost all our heavy weapons—except for those with the advance party—made a wrong turn and completely lost its way in a city that was unfamiliar to them." But this is completely misleading: Castro had not been denied the "best armed half of our troops," but only twelve men from the number that left El Siboney, and his own force, including himself, contained eighty-three fighters when the rebels moved on the barracks. Raúl and his five-man team were a few minutes late, yet in position at the courthouse during the battle. Abel Santamaría with twenty-three rebels, including the two women and Dr. Mario Muñoz, reached the civilian hospital precisely on time.

The whole operation did suffer a delay when the motorcade had to halt at the narrow bridge over San Juan River to let a jeep pass with two innocent hunters going out of town for a day in the countryside. Then Fidel, swearing savagely, had to stop his Buick just before the turnoff to the barracks to let the advance party's car maneuver into position in front of him.

The time for the attack on Moncada had been set for 5:15 A.M., but it was probably a few minutes later when the commando vehicle, a Mercury, braked at Gate 3 to force entry into the courtyard and open the way for Fidel's Buick and the rest of the motorized column. The commando team hit at once. Because the Movement operated on the principle that leaders must always personally lead armed actions, this eight-man commando team was composed of two military committee members, Renato Guitart and José Luis Tasende; a civilian committee member, Jesús Montané; the three best cell leaders, Ramiro Valdés, Pedro Marrero, and José Suárez; and two outstanding marksmen, Carmelo Noa and Flores Betancourt (whose brother was in the Bayamo detachment).

According to plan, Guitart shouted imperiously at the three army sentinels at the gate, "Clear the way, the general is coming!" and the men, seeing his sergeant's stripes, drew up at attention. Montané, Valdés, and Suárez took the Springfield rifles away from the stunned soldiers while Guitart and the four others removed the iron chain over the entrance, following the plan, and rushed up the outside stairs of the barracks building to occupy Moncada's radio communications center and prevent contact with Havana and Holguín.

Fidel, armed with his huge Luger pistol, slowed down the Buick as he drove past the military hospital on his left, about 150 yards before Gate 3. He had slowed down so that he and his six companions could decide what to do about a two-man army patrol, armed with submachine guns, that suddenly materialized in front of them and was now tensely watching the action at the gate. The patrol was either ahead of time in its appointed rounds, or the *Fidelistas* were critically behind schedule. Simultaneously, a real army sergeant also appeared in the street, out of nowhere, examining the situation suspiciously. For all practical purposes, the rebels lost the battle of Moncada at that precise moment because with the first light of dawn they no longer had the element of total surprise in their favor.

"At that moment," Fidel Castro told French historian Robert Merle in 1962, "I had two ideas in my mind. Because each of them had a submachine gun, I feared that the men of the army patrol will start firing on our companions who were busy disarming the sentinels. In the second place, I wanted to avoid that their fire alert the rest of the barracks. I conceived, then, the idea of surprising them and taking them prisoner. This seemed easy because their backs were turned to me . . ."

All these events lasted seconds, perhaps minutes, as Fidel drove slowly toward the patrol, readying his Luger, carefully opening the door on his side, then suddenly accelerating and hurling the Buick at them when they turned and pointed their Thompsons in his direction. The Buick hit the curb at an angle with the left front wheel, and the engine stalled. The army sergeant shifted his body, aiming his revolver at Fidel, but he was brought down by rifle fire from the rebel vehicle behind the Buick. Fidel and Pedro Miret, who rode with him in the Buick's front seat, somehow found themselves on the ground behind their car, with a soldier firing at them from a window of the military hospital on their left. Bullets whizzed past Fidel's face, and he covered his ear with his hand as if he had been deafened. At that moment, alarm bells went off with shrieking fury throughout the barracks.

The entire action was over in less than a half hour. Ramiro Valdés, Jesús Montané, and José Suárez were the only rebels who actually succeeded in entering the fort, and for a few minutes they had as captives some fifty half-dressed soldiers they found on cots along the walls inside the courtyard. But other soldiers, rifles and revolvers in hand, began appearing from all over the barracks, firing at the three rebels. Soon the three were separated, each fighting his way out of the building. Valdés and Montané remember felling several soldiers with their bullets. All three ultimately escaped to the street.

Now soldiers were firing massively at the retreating *Fidelistas* trapped in the twenty-foot-wide street between Gate 3 and Garzon Avenue behind them; it was the short street past the military hospital the assault automobiles had taken to rush the Moncada gate. Fire came from the roofs and windows of the barracks above them, from the military hospital, and, in a diagonal trajectory, from a .30-caliber heavy machine gun mounted atop a tower on the firing range some six-hundred feet away. The rebels were an excellent target for the machine gun, whose emplacement on the range was a painful surprise for them (they were also ignorant that a .50-caliber heavy machine gun was on the roof of the officers' club, neutralizing Raúl Castro's small group at the Palace of Justice).

Renato Guitart, the member of the military committee who had been in charge of most of the arrangements in his native Santiago, was killed in front of Gate 3 along with Pedro Marrero, Carmelo Noa, and Flores Betancourt. They were the first rebel fatalities of Castro's revolution.

Fidel Castro knew that he had irretrievably lost the Moncada battle the instant his Buick stalled outside the fortress and the alarm bells began clanging, but he still tried desperately to regroup his men for a second attack. He stood in the middle of the street, barely visible in the half-light of the dawn and the thickening gunsmoke, shouting commands and encouragements of *"Adelante, muchachos! Adelante! . . ."* (Advance, boys! Advance! . . .) But the men could neither hear him nor understand him, and the spirit of the attack was quickly lost. To set them an example, Fidel climbed back in the Buick to try to start it up, but it was in vain. Furious beyond words, he attempted once more to regroup his troops, waving his Luger in the air. Most of the rebels were hiding behind the low fences of the bungalows across the street, pinned down by withering fire from Moncada.

Yet, Fidel still stood his ground. Spotting two soldiers setting up a heavy machine gun on the roof of the fort, he fired at them with his heavy pistol, then fired again at other soldiers taking up positions above. His companions were yelling, "Fidel, Fidel! . . . Get out of there! . . . Get out! . . . ," but he seemed oblivious to their cries and to the hail of bullets around him.

Pedro Miret, then the Movement's principal military expert (after Fidel), believes that it was impossible to regroup the attackers because "the people faced a situation that had not been anticipated." Reminiscing about the battle, he says that the plan called for all the automobiles to enter the courtyard behind Fidel's car, for Fidel to lead the men up the main staircase to occupy the Moncada command post, and then for others to spread out through the barracks. But when this failed, there was little Fidel could do "as he urgently tried to organize everything in the midst of a firefight." Miret thinks that the "critical moment" was the accident of the sudden encounter with the army patrol. After the rebels were thrown back, he says, Fidel attempted to deploy a defensive line to cover the retreat, and was "again organizing everybody." But Miret adds, "This is easy to say because we had lots of people who never heard real gunfire and, especially, never heard a .50-caliber machine gun, which is very impressive."

In the end, Miret says that Fidel "decided to leave" with as many men as he could, and they became separated. Miret will not admit it, but he and two others seem to have stayed behind deliberately to cover the departure of the Castro group; he wound up inside the military hospital, beaten, left for dead, rescued by a doctor, put in bed, nearly murdered by army intelligence agents in the hospital, and, finally, put on trial. It was in prison that he next saw Fidel.

Meanwhile, Fidel was in an absolute fury at himself, since he concluded that it had been a frightful error to order the capture of the army patrol. The resulting series of clashes had led to the sounding of the alarm in the barracks and the end of his chance at Moncada. With his companions falling dead or wounded, a new attack was out of the question, and there was no alternative but to become resigned to defeat and to pull back. In a letter written in prison on December 31, 1953, Fidel told a friend that "for me the happiest moment in 1953, in all my life, was that moment when I was soaring into combat—just as it was when I had to face the tremendous adversity of defeat with its sequence of infamy, calumny, ingratitude, incomprehension, and envy."

Satisfied that all the surviving rebels had left the vicinity of Moncada, Fidel got into the last vechicle to depart from the scene, a car containing six other men, including one with a serious thigh wound. But as the automobile started, Fidel saw a rebel on foot trying to escape from the gunfire from the fort behind them. He ordered the car to halt, stepped out to make room for the other man, and proceeded to retreat, moving backward and firing at the barracks. As he was turning the corner of Garzon Avenue behind the military hospital and out of the field of direct fire, the automobile belonging to a rebel taxi driver from Artemisa came in reverse from Gate 3 and picked up Castro and three additional companions.

The first idea occurring to Fidel at that point was that the simultaneous attack at Bayamo might have succeeded, and that he should try to join the contingent there. But he also decided that the five of them in the car could start by capturing the small Rural Guard post at El Caney, a few miles north of Santiago, and secure the rearguard for Bayamo. If the action at Bayamo had failed, Fidel reasoned, they would go up the mountains to continue the struggle—as the Movement's plans had anticipated—and El Caney would be the best place to regroup.

The driver of the car, who was unfamiliar with the Santiago region, either misunderstood Castro's instructions or simply had no desire to go to fight in El Caney. Thus, instead of going north, he turned east on the road leading to El Siboney farmhouse and ran into the four rebels who had been left behind before dawn when Boris Luis Santa Coloma's car had a flat and could not continue. Fidel no longer insisted on doubling back to El Caney, and he stopped a passing private car, ordering the two occupants to take the four rebels to El Siboney, the first vehicle following him. (The private car then continued on its way.) Not quite three hours had elapsed since Fidel Castro had left the farmhouse to conquer Moncada—and Cuba.

In Havana, Naty Revuelta had left her Vedado apartment at five o'clock in the morning, as instructed by Fidel Castro, to distribute copies of the Moncada Manifesto to political leaders and newspaper publishers. Several of them were not at home; others were unreceptive. A publisher sent his son-in-law out to meet Naty in the living room to tell her that they just had heard on the radio the news of a failed armed coup in Santiago. Naty was gently but firmly pushed out of the apartment. And she knew that if she were stopped by agents of SIM, the Military Intelligence Service, and copies of the revolutionary document were found in her handbag, she faced imprisonment or worse.

*   *   *

From the high terrace of the Palace of Justice, Raúl Castro and his men witnessed the debacle in front of Gate 3, but they could be of no real help. Strategically, they were in an excellent spot, but their weapons were inadequate to give Fidel covering fire, and the terrace quickly became a target for the Moncada heavy machine guns. Seeing that Fidel's group was retreating and the shooting was dying down, Raúl concluded that their support mission had been fulfilled, and that they had better leave before being surrounded. Downstairs they ran into five armed policemen, but were able to take their revolvers away and to get into the car driven by Mario Dalmau, fleeing in a hail of bullets. Only Lester Rodríguez, whose home was in Santiago, went on foot to his parents' house. The courthouse team suffered no casualties. Raúl ordered Dalmau to drive to the coast.

Occupying the civilian hospital, across the street from the western side of the Moncada compound, Abel Santamaría could not see what was happening at Gate 3, and he was unaware of Fidel's gradual retreat. But soldiers from the barracks were now firing heavily at the hospital, and Abel realized that he was being encircled and trapped. Thinking that by fighting as long as possible he could aid Fidel's group, wherever they were, Abel decided to stay at the hospital until he ran out of ammunition. He told his sister Haydée and Melba Hernández that "we are lost," that they should try to escape, and that they might survive because they were women. Abel added: "Don't take risks. . . . Someone has to live to tell what happened here."

But Heydée and Melba did not even try to flee. The firing stopped around eight o'clock in the morning when Abel and his men had no more bullets, and the soldiers also stopped shooting. But the army waited a full hour before daring to enter the hospital building, and the two women spent the time helping the nurses feed screaming babies in the maternity ward; the babies had been in the midst of a battle for nearly three hours. In the meantime, other nurses put hospital gowns on the rebels and made them appear to be bedridden patients to protect them from the soldiers. Abel's head was bandaged over one eye, and he was put to bed in the ophthalmology section.

It was the rebels' fate that the Moncada civilian press chief, Señor Carabia Carey, happened to be at the hospital during the siege, undergoing some form of treatment, and he betrayed the *Fidelistas* to the army, adding that two rebel women were also hiding in the building. Melba and Haydée saw Abel being yanked out of bed and beaten with rifle butts; his face was a bloody mass. In the late morning, the women were escorted to Moncada, and as they crossed the street, they witnessed Dr. Mario Muñoz being hit on the head by rifle butts, then shot in the back. He died on his forty-fifth birthday, a mild-mannered physician with a small moustache and spectacles. Carabia was arrested after the revolution, and served most of a thirty-year prison term. He was then allowed to leave for the United States.

In a hallway in the Moncada headquarters building, Haydée and Melba saw a young man, his face so smashed by rifle-butt blows as to be beyond recognition, thrown on the floor by soldiers. They made him sit down on a bench, and the youth had the strength to scribble on a piece of paper, "I was taken prisoner—Your son." Melba realized it was meant for his mother, then she looked into his eyes and knew he was Raúl Gómez García, the young poet who had drafted the Moncada Manifesto, written a stirring poem at El Siboney, and chosen to read the Movement's victorious proclamations over the radio. Then a soldier shot him dead.

The soldiers referred to the rebels as "Koreans," a derrogatory allusion to North Korean Communists who had invaded South Korea three years earlier. Haydée and Melba heard them brag about beating to death a very tall and tough rebel who tried to fight them with his fists, who had had to be tied up to be tortured by army interrogators. When a soldier mentioned that the rebel wore black-and-white shoes, Haydée realized it was her fiancé, Boris Luis, as he had been unable to procure boots that fit him. He died without saying a word. Later, Haydée was told by a soldier that Abel was being tortured, and a sergeant brought her her brother's gouged-out eye as proof. Abel died under torture later that Sunday. All the men in Abel's detachment were subsequently murdered by the army except for a young teacher who succeeded in hiding.

Ramiro Valdés, one of the three attackers who actually penetrated Moncada, extricated himself from the battle and climbed into one of the rebels' cars, dragging Gustavo Arcos, a friend who had been shot in the belly. Arcos had traveled in Fidel's car from the farm to Moncada, then became separated from him in the affray. The car had four flat tires, but Valdés drove on the rims, succeeding in getting Arcos to a physician's home blocks away.

Nearly ninety miles northwest of Santiago, the Bayamo rebel detachment was destroyed in less than fifteen minutes. Twenty-seven men had been assigned to the operation there, but only twenty-two actually took part in the attack on the army barracks. Led by Raúl Martínez Arará, an early Movement member, the detachment, divided into three squads, rushed the structure from the back because it was the only side protected by barbed wire instead of a wall.

The rebels, moving at 5:15 A.M. in complete darkness, crossed an open field between the street and the barbed-wire fence. But they discovered to their surprise that the gate in the fence, open during the day, was padlocked at night. Attempting to climb over the fence, the men stumbled upon empty cans covering the ground, alerting the soldiers. A dog barked, horses in the stable kicked the sides of their stalls, and a sentinel shouted, "Halt!" Hiding behind bushes, the *Fidelistas* began firing on the barracks with their sports rifles and shotguns, but the soldiers countered with machine-gun fire—and it was all finished.

Raúl Martínez Arará ordered the retreat, and the men raced to their cars or just down the street on foot. Twelve rebels died at Bayamo (including Mario Martínez Arará, the commander's brother, who was tortured to death). The ten survivors succeeded in escaping. Among them were the commander, Fidel's friend Ñico López, the black poet Agustín Díaz Cartaya who weeks earlier had composed a revolutionary march that would become famous later as "The 26th of July March," and Teodulio Mitchel, the truck driver who had brought Fidel from Havana to El Siboney only six hours earlier (he had rushed to Bayamo to fight as soon as he dropped Castro off).

By noon on Sunday, July 26, twenty exhausted rebels were back at El Siboney farmhouse, three of them wounded. Twenty more appeared by midafternoon, just as Fidel Castro was preparing to leave for the mountains to resume the war, this time as a guerrilla chief.

Within hours of reaching the farm, he announced that he was going to set up guerrilla operations in the Gran Piedra Sierra, a chain running diagonally to the sea northeast of Santiago. The Gran Piedra Mountains rose some ten miles from the city, and the highest peak, also called Gran Piedra, was over three thousand feet. Fidel asked for volunteers, and nineteen agreed to go—though one of them changed his mind within minutes of the march because his new shoes hurt. The next day, this nineteen-year-old boy, named Emilio Hernández, was captured and murdered by the army; the official announcement said he was killed fighting at Moncada.

Fidel's column included Jesús Montané; the Movement's treasurer, Oscar Alcalde; José Suárez and Israel Tápanes, who had fought at Gate 3; Juan Almeida, the black bricklayer's apprentice who began working at the age of eleven and would become one of the top guerrilla commanders; and young Reinaldo Benítez with a festering bullet wound in his leg. An old black woman in a shack above Siboney sent her grandson as a guide with the Castro group, and the next day they reached the village of Sevilla Arriba. Looking down at Santiago Bay at their feet, Fidel raised his arms, and in his best fighting style, proclaimed to his little band, "*Compañeros,* today it was our turn to lose, but we shall return."

Farther uphill, the men came across a hut, but the black peasant refused to sell Fidel the chickens he was raising there; this was one time when Castro's powers of persuasion failed. But the peasant led them to his brother's house a distance away, and the rebels feasted on a pig the man killed for them. Fidel, according to a published account, talked with him about the peasants' oppression by local landowners. Then he gave the man a nickel-plated pistol, saying, "When they come to bother you, open fire with this pistol . . . don't believe in anyone. Defend what is yours."

At another peasant's house, Fidel listened to a radio speech by General Batista giving his version of the events of July 26. The dictator had rushed back to Havana from his yacht and Varadero the afternoon of the Moncada attack, and had set up operational headquarters at Camp Columbia. His Council of Ministers declared a state of emergency in Cuba, and suspended a provision of the prison code under which wardens are held responsible for the lives of the prisoners.

This seemed intended to legalize a decision taken by Batista and General Francisco Tabernilla, the Army Chief of Staff, to murder *Fidelista* prisoners as they were being caught by the military in and around Santiago—even in hospital beds—and as far as Havana, where a handful succeeded in escaping. Over a period of four days starting on July 26, sixty-one men were thus assassinated, including Abel Santamaría and Boris Luis Santa Coloma, who at first were horribly tortured, the poet Raúl Gómez García, José Luis Tasende, who was a member of the military committee, and Dr. Muñoz. In seventy-two hours, Fidel lost some of his best friends. Only eight had died in combat, among them Gildo Fleitas whose wedding Castro had organized in Havana the month before. The final *Fidelista* death toll was sixty-nine, and only five wounded because, as Castro later pointed out, the regime wanted no surviving prisoners.

The army and the police suffered nineteen fatalities in the fighting and twenty-seven wounded. But the regime needed to make the outcome of Moncada loom as a great victory, so in his report to the Santiago Court of Justice, Colonel Chaviano (who had cowered under his desk at the command post during the battle) claimed that between four hundred and five hundred men, "equipped with the most modern instruments of war," had attemped to overthrow the regime. In tune with official propaganda, Chaviano reported that the attackers used knives "to open the abdomens of three patients" in the military hospital, and in firing on the troops they employed double-explosion (dum-dum) bullets in their automatic Remington rifles.

Batista as well as Chaviano charged that the attack was planned, organized, financed, and armed by antiregime groups exiled abroad, notably by former President Carlos Prío, overthrown in the March 1952 coup. Such a propaganda line served to justify the subsequent repression and declaration of martial law. It is also entirely possible that it had never occurred to the dictator that an uprising like Moncada could have been orchestrated so secretly by a young troublemaker from Havana University and a group of penniless followers. Listening to these accusations over the radio, Fidel Castro said, "They wedded us with a lie, and forced us to live with it."

In Santiago the army command sought to prevent newsmen from seeing too much of the barracks the afternoon of the attack, but a photographer succeeded in taking pictures of a number of *Fidelistas* tortured to death and still lying in hallways and passages. Marta Rojas, a young reporter for the weekly magazine *Bohemia,* hid the photographs in her brassiere as she flew to Havana that evening. The gruesome photos were published five days later, shocking the nation and unleashing a wave of sympathy for the rebels.

On July 28, as Fidel and his men were climbing the Gran Piedra Mountains, civic leaders were meeting in Santiago under the chairmanship of Archbishop Pérez Serantes (who twenty years earlier had prevailed on Fidel's father to have the boy baptized) to try to halt the executions. Word that rebel prisoners were being killed had spread in the city, and the group decided that the archbishop should persuade Colonel Chaviano to guarantee the life of any new prisoners. Both the archbishop and the colonel knew that Fidel Castro was at large in the mountains, and their agreement, as far as Monsignor Pérez Serantes was concerned, specifically assured that Castro would be taken and kept alive. Additionally, the archbishop published the next day a declaration, titled "Enough Blood!" outlining the new accord with the army.

As a result of the archbishop's statement, thirty-two *Fidelistas* came out of hiding to surrender to the authorities and were locked up at the Santiago jailhouse, where they were treated without undue violence. Before the intervention of the Roman Catholic Church, Haydée Santamaría and Melba Hernández had been moved from Moncada to the jail at dawn on July 28. During the trip in a jeep, a soldier burned Haydée's bare arm with the red-hot tip of his cigar. But later that week even *Fidelista* leaders who emerged from their hideouts, including Dr. Aguilera and Ernesto Tizol, were given full protection.

However, the rebels who were the most wanted, namely Fidel and Raúl Castro, were not surrendering. They had to be taken, and the archbishop was increasingly concerned about their fate. He did not entirely trust Colonel Chaviano and his officers, and he had already traveled to Manzanillo, a city near Bayamo, to escort personally a *Fidelista* prisoner captured there. On July 29, a Wednesday, Raúl Castro was arrested at a roadblock near the township of San Luis as he walked toward Birán and his parents' *finca.* He had fled by car from the Palace of Justice, but soon realized that he was safer on foot, and in two days and nights he covered the distance from the south to the north of Oriente province. The Rural Guard detained Raúl because he had no identity documents, and at the local police station, he insisted that his name was Ramón González. The lieutenant who interrogated him told Raúl he was lying and locked him up in the tiny jail.

The next morning, a traveling salesman came through San Luis, and the lieutenant summoned him to try to identify his prisoner. In the Oriente countryside, most people know each other, and the man said, "This is Raúl, the son of old Ángel, and the brother of Fidel Castro." Raúl was transferred to Santiago and Moncada the same afternoon, spending the night at the barracks, certain that he would be shot. But Colonel Chaviano was not looking for a confrontation with the Church, and Raúl was sent to the jail, which was rapidly filling with his companions.

For Fidel Castro, Wednesday and Thursday were bad days. His band was in awful shape. Reinaldo Benítez could hardly move because of his leg wound. Jesús Montané, who was flat-footed, dragged himself behind the others. The heat was suffocating, and they had no food. Fidel ordered a halt for the night in a canyon on Thursday, when suddenly a shot rang out: One of the rebels had accidentally fired his rifle and the bullet went into his shoulder. On Friday, July 31, Fidel allowed the group to split up. Five men, including Montané and the two injured rebels, started back to Santiago and were arrested en route. But Castro and thirteen followers were determined to stay in the Sierra. In fact, Fidel developed the notion of returning to the coast and moving east to Sierra Maestra.

Archbishop Pérez Serantes, meanwhile, had become so preoccupied with the fate of Fidel Castro and other missing revolutionaries that he had taken to the road in a car to look for them; an elderly priest in a black cassock, a cross and chain around his neck, stopping every few hundred yards in the wet heat of the equatorial forest road to call out to the rebels must have been a strange sight.

On the evening of Friday, July 31, Fidel and his companions stopped at a hut on a hillside for an overnight rest. They had been moving south slowly toward the coast, but the journey was turning into an impossible undertaking. Five more rebels had abandoned Castro, leaving nine. They were almost entirely out of food. Behind the hut, they spotted a young peasant, crouching over a wood fire over which he was boiling a pot of rice. Without a word, the rebels slid behind the peasant, took his spoon away from him, and began to eat his rice.

"Do you have more food?" one of the men asked. The peasant said no, but he would guide them to the farm of his employer, two hours away. The farm was near the Siboney road, and Fidel recognized the owner as an old acquaintance and learned from him of the archbishop's efforts to locate him. He was also told of Chaviano's guarantees, and this convinced Fidel that only he and the two other Movement leaders—Oscar Alcalde and José Suárez—should attempt to cross over to the Sierra Maestra. The others, he said, should surrender. In the meantime, they returned to the hillside hut to spend the night.

At dawn on Saturday, August 1, a sixteen-man Rural Guard squad led by Lieutenant Pedro Manuel Sarría Tartabull came upon Castro and his sleeping companions and opened submachine gun fire on the hut. The army had been tipped off that Fidel was to be in the area, so Sarría was dispatched to look for him at the vast Mampriyá estate, which included the mountain farm. The gunfire flushed the rebels out, but Sarría cried, "Cease fire! . . . I want them alive!" One of the soldiers called Castro an "assassin" for killing soldiers at Moncada, but Fidel, with automatic rifles fixed on him and his eyes burning with rage, yelled back, "It is you who are assassins . . . it is you who kill unarmed prisoners . . . you are the soldiers of a tyrant!" A corporal shouted to Sarría, "Lieutenant, we'll kill them!" Sarría, a tall, black, fifty-three-year-old professional officer, raised his arm. "No," he roared, "don't kill them! I order you not to kill them! I am in command here. . . . You can't kill ideas. . . . You can't kill ideas!"

Eventually, the soldiers calmed down and proceeded to tie the hands of their prisoners. Fidel did not realize then that he and Sarría had met once at the university in Havana, and he insisted that his name was Rafael González. He was deeply tanned by the Sierra sun and hoped the lieutenant would believe him, especially because he had heard the official radio announce his death.

As Sarría's detachment escorted the *Fidelistas* to the road, Castro whispered to him, "Yes, I am who you think I am . . ." Then he asked, "Why didn't you kill me? You would have received a nice promotion, a promotion to captain . . . ?" Sarría replied: "*Muchacho,* I am not that kind of man . . ." Fidel said, "But if you spare me, they will kill you." And the black lieutenant told him quietly, "Let them kill me. . . . It is one's ethics that must decide what one will do . . ."

Fidel Castro's life was spared as much by the fact that Rural Guard Lieutenant Luis Santiago Gamboa Alarcón was in bed with flu in the Moncada barracks the night of July 31–August 1 when Squad 11 was dispatched to look for the rebels in the hilly countryside above Siboney. Gamboa was a tough officer, wholly devoted to law and order as well as to the regime, and it is certain that he would have killed Castro without the slightest hesitation upon finding him, aware that Colonel Chaviano wanted the rebel chief dead—no matter what he had promised the archbishop. Later, during the Sierra war, Gamboa was promoted to captain after executing six peasants in Oriente whom he suspected of aiding the revolution. In January 1959 he was among the first Batista officers to be shot by a revolu-

tionary firing squad after a quick trial. Gamboa's illness that night gave
Lieutenant Sarría command of the squadron. On the day of the attack,
Sarría had also intervened inside Moncada to prevent soldiers from executing two rebel fighters captured in the street. Many years later, he told an
interviewer in Havana that he had spared Fidel's life not because he sym-
pathized with him, but "because he was a human being . . . I love the
profession of arms, but where I am in command, I believe that no crimes
should be committed."

Sarría's courage included his determination to control his own soldiers at
a psychologically risky moment when Fidel, Oscar Alcalde, and José Suárez
were captured. All the Rural Guard troopers were black, as was Sarría, but
the three prisoners were white. In the complex world of power relationships
in Cuba, and especially in Oriente, black and mulatto soldiers tended to
identify with Batista—himself a mulatto. To them, whites who were
armed were automatically rebels and lawbreakers who should be killed. In
their privileged military position, the soldiers did not experience racial
discrimination, and Batista was their hero. Thus, when they flushed out
the three white rebels, the soldiers' first cries were "They are white . . .
they must be killed," and Lieutenant Sarría had to summon all his authority to halt what would have been, in effect, racial murders.

Moreover, Sarría was a Freemason, a fairly widespread affiliation in
Cuba, and there was a bond among brothers in the supposedly secret
lodges. On a hunch, Oscar Alcalde told the lieutenant he was a Freemason,
wondering aloud whether Sarría might not be one, too. When Sarría nodded, Alcalde said, "From Freemason to Freemason, and because you have
saved our lives, I will tell you where we have hidden our weapons—they
are thirty feet from the hut, under bushes." Because Sarría needed to
strengthen his authority over the troopers, the find of the arms cache—
eight rifles and three pistols—made him look good before the soldiers.
Alcalde had thought this would be the case.

As for Fidel, he was chastising himself for the "terrible error," as he said
later, of having slept inside the hut. He knew that as a rule, military
search parties never fail to go into huts, shacks, and houses, but it had been
"so tempting to sleep under a roof because it had rained and the soil was
wet." They were fatigued after living on fruit six days and nights in the
mountains. Castro said that until his rude awakening "we slept and slept so
deliciously." Thereafter, Castro recalled that during the nearly two years in
Sierra Maestra, "even when rains came down in a deluge, we always slept in
hammocks strung between trees, or rolled in a blanket on the ground."

As Sarría and his squad and the prisoners neared the Siboney road, firing
broke out and the lieutenant ordered them to drop on the ground. Sarría's

point men had spotted five rebels, including the black Juan Almeida, hiding in the tall grass off the road, and began shooting at them.

This time, the situation was saved by Archbishop Pérez Serantes. That Saturday, he had gone out again in his jeep to look for the rebels, leaving Santiago at 6:30 P.M. with a driver and two friends—but no military escort. A peasant they met on the road told them that a group of rebels had surrendered farther ahead, and the archbishop left the jeep to look for the men on foot. He reached the place where Almeida and the four others had been surrounded, just as the soldiers were about to shoot them. At a run, the monsignor lifted his cassock to jump over a fence and interposed himself between the rebels and the troopers, shouting, "Don't kill them . . . I have guarantees from the authorities!"

But the soldiers were furious over his intervention. They ordered him to leave, insulting him, and one trooper chanted, "I'm going to kill me a priest, I'm going to kill me a priest!" Devotion to the Church and its servants is rather superficial among poorer Cubans, and now Sarría had to impose his authority to save the archbishop. The five new prisoners had their hands tied, and Fidel, Alcalde, and Súarez presently joined them. Sarría commandeered a flatbed truck to transport the prisoners to Santiago, placing Castro between the driver and himself in the cab. The archbishop, unaware of the events of the past hours, walked over to the lieutenant to warn him: "These prisoners are under my protection. I have given guarantees to them." The lieutenant replied: "Monsignor, you should tell it to Colonel Chaviano, not me." Fidel, who became concerned that word would spread of his surrender to the archbishop, broke in to say loudly, "I have nothing to do with the monsignor; I was captured, and it was you, Lieutenant, who captured me." Politically, Castro had to avoid the impression of surrender.

The army command in Santiago had been advised by telephone from the farm that Castro and seven companions had been seized, and presently Sarría's truck was stopped by truckloads of soldiers under the command of Major André Pérez Chaumont, the deputy to Colonel Chaviano. Pérez Chaumont, known in Oriente for vying for elegance with Chavian, and nicknamed "Beautiful Eyes," informed Sarría that "I came to take delivery of the prisoners." Sarría looked down at him, saying, "No, these are *my* prisoners; I'm taking them to Moncada." The major then proposed that Sarría keep all the others, if he delivered Castro to him. The two officers argued sharply for several minutes, Sarría being certain that Fidel would be killed if he turned him over to Pérez Chaumont. But when the archbishop pulled up in his jeep, the major gave up his efforts to pull rank.

Instead, he ordered the lieutenant to take the prisoners to the Santiago city jail rather than to Moncada to deprive him of the prestige among the military of having personally brought in Fidel Castro. This was fine with Sarría: At Moncada, Castro would have been in the hands of Chaviano, Pérez Chaumont, and Captain Manuel Lavastida, the Santiago chief of SIM, the Military Intelligence Service. It was SIM and its head, Colonel Manuel Ugalde Carillo, who had ordered and carried out the murders of the captured *Fidelistas*.

At the city jail, however, Colonel Chaviano was already awaiting Fidel Castro, and Sarría had no choice but to surrender his prisoner. However, the lieutenant had saved Castro's life, first by preventing his soldiers from shooting him, then by refusing to deliver him to Pérez Chaumont. Once at the jail, very much in public view, Fidel was protected. As to Sarría, he remained in the army, but was tried by a court martial in 1957 for refusing to fight the Rebel Army in the Sierra Maestra, and was kept under house arrest until the revolution triumphed. Sarría was promoted to captain in the new *Fidelista* army, and regarded as a Hero of the Revolution. When Sarría died in Havana in 1972 at the age of seventy-two, Castro attended the solemn funeral, and Pedro Miret delivered the oration to recall that "we owe Captain Sarría deserved gratitude for having saved the life of Fidel and his companions."

From the moment he arrived there at 8:45 A.M., Fidel Castro was an instant celebrity at the Santiago jail, an ancient two-story structure downtown. Keeping his aplomb, he behaved with humor and defiance, and when Colonel Chaviano, immaculate in a fresh uniform, his waxed moustache bristling, met him in the jail office, the commander had a group photograph taken. Standing under a portrait of José Martí, Fidel, unshaven in dark slacks and a white short-sleeved sport shirt, towered over Chaviano, Pérez Chaumont, and Major Rafael Morales Alvárez. He seemed like the guest of honor.

Sitting on a bench with his fellow rebels as he awaited a meeting with Chaviano later that morning, Castro glanced at the local newspaper, *Ataja,* with a huge front-page headline: DEAD! FIDEL CASTRO! In Chaviano's office in the jail, Fidel was invited to sit across the desk from him, his hands untied, for what became a Castro speech rather than an interrogation. The official army communiqué said that Castro "affirmed that he was responsible for the entire subversive movement . . . revealed all the subversive plans as well as those he expected to unfold later."

Rather than simply "confessing" his deeds, Fidel made a point of narrating in detail how and why he had organized the Movement, emphasizing his certainty that if the Moncada attack had worked, the people of Oriente would have risen in his support. Chaviano allowed a summary of his declarations to be given to the press and let newsmen interview his prisoner. He even held a news conference to discuss his "interrogation" of Fidel. Newspapers and magazines throughout Cuba reported Castro's words widely and amply; the Moncada defeat began to turn into the seeds of a future victory.

Incredibly, Chaviano proposed that Castro repeat his story over the radio so that the whole nation could learn, the colonel reasoned, how dangerous the subversive movement smashed by the army was. Fidel was delighted, and as he told the French historian Robert Merle ten years later, "Imagine the imbecility of these people! They ask me to take the microphone and to defend my viewpoint before them . . . who as a result of their crimes are morally disarmed before me! Obviously, I took the microphone . . ." Castro then said to Merle, laughing: "And at that minute, the second phase of the revolution began." In that broadcast over station CMKR, Fidel proclaimed, "We came to regenerate Cuba."

To be sure, the Movement suffered very serious losses. Three out of six members of the military committee, including Abel Santamaría, were killed. Over one hundred rebels were killed, murdered or arrested. But many key persons had survived, and Fidel found most of them at the Santiago jail: Raúl Castro, Jesús Montané, and Haydée and Melba, looking dejected, with scarves tied around their heads. Up to then, Fidel did not even know whether they were alive. At the same time, he discovered that Raúl had become the chief of the rebels at the jailhouse; a photograph shows him standing at attention in front of some thirty men in an unmistakable leader stance. Now Castro assumed his leadership position, greeting each rebel with words of praise for having attempted with him to "take the sky by surprise."

In the meantime, Archbishop Pérez Serantes telephoned Mirta in Havana to assure her that her husband was well. This was probably the first time in nearly five years of marriage that Mirta was told anything about her husband's activities, and she appeared to be grateful for the call. She had been very worried about Fidel since she heard about the abortive Moncada attack, and she wanted to be in touch with him. Her brother, Rafael Díaz-Balart, now acting interior minister in the Batista cabinet, had gone to Santiago immediately after the events of July 26, but it is unlikely that he would serve as a link between Fidel and Mirta, particularly because Castro would not have it. The Castros in Birán were informed that their sons were safe and well treated. And the word reached Naty Revuelta in Havana, too.

The day after Castro's capture, on Sunday, August 2, President Batista flew to Santiago to tour Moncada, receive reports from Chaviano and other officers, award the Honor Cross to the flag of the Maceo Regiment, and salute the troops as they marched past him inside the fortress. The people of Santiago, however, were in greater attendance at the solemn mass for the dead at which Archbishop Pérez Serantes officiated at the cathedral. Starting at seven o'clock in the morning, thousands of them filled the cathedral to hear the requiem and, on their way out, to kiss the archbishop's ring. He was the other hero of Santiago.

Also that day, Fidel Castro and all the other Moncada prisoners were transferred to Boniato Provincial Prison, some five miles north of Santiago, to await their trial. Melba Hernández and Haydée Santamaría were placed in the backseat of a green Buick; in the front seat, a civilian sat between two officers; all the other prisoners went in police vans. Melba recalls that when she and Haydée entered the car, Fidel was already in the front seat, "and he recognized us without turning his head, talked to us, and asked us how we felt . . . the military let us converse, so we answered Fidel, and asked how he was . . ."

Fidel Castro celebrated his twenty-seventh birthday in the Boniato prison twelve days later, but he was too busy planning his next move in the war against Batista to reflect much on it. Not only had he to prepare his trial defense, but he had decided to issue as rapidly as possible a complete account of the Moncada events, including insisting that the Movement was independent of Cuban political groups. The regime had alleged that the uprising had been directed by the *Ortodoxo* party and the Communists, and that key leaders of both parties had been arrested all over Cuba. By coincidence, top Communist leaders were in Santiago the weekend of Moncada for a birthday celebration, but to Batista this was proof of their involvement in the conspiracy.

New names, soon to be famous, began to be associated with Castro's Movement. Manuel Urrutia Lleó, the chief judge of the Santiago "summer court" and the future president of revolutionary Cuba, on July 27 had undertaken an inquest designed to identify Moncada victims. Humberto Sorí-Marín, representing the National Lawyers' Association, visited Castro in prison to discuss his defense; later Sorí-Marín would be the Rebel Army's judge-advocate, the author of the agrarian reform law, and minister of agriculture under Castro before being executed in 1960 for conspiring *against* the revolution.

Working around the clock (and letting his moustache grow again), Fidel wasted no time organizing the rebel prisoners politically and educationally, demanding books and other materials from the outside. Raúl and Pedro Miret, the latter recovered from his battle wounds, became his principal collaborators in all his endeavors. Letters came to the Boniato cell from all over Cuba, including from Naty Revuelta in Havana. Lina Castro, Fidel and Raúl's mother, and their older sister, Lidia, came to visit them in prison as did Fidel's wife, Mirta.

Fidel Castro was in high spirits: He was fighting again, his head was brimming with plans, conspiracies, and denunciations. The disaster of Moncada already lay behind, if not forgotten; certainly it was not perceived by him as a monumental defeat. It was on that defeat that Castro was now building his next great offensive.

CHAPTER

5

At Boniato Provincial Prison, Fidel Castro was Prisoner No. 4914, Case 37-053, awaiting trial before the Provisional Tribunal of Santiago for his participation in the assault on the Moncada army barracks on July 26, 1953. After the sentence, he would become National Men's Reformatory Prisoner 3859. But he certainly was not an anonymous prisoner without a name, hidden behind a mug-shot number. In fact, this was when Fidel Castro became famous in Cuba for the first time, when he became the center of nationwide sympathy (and focus of General Batista's wrath), and the recognized leader of the antidictator opposition.

Still, the Batista regime failed to comprehend what it had on its hands in Boniato Prison. This prisoner was much more dangerous politically behind bars in the glare of public attention than as the obscure and penniless revolutionary plotter he had been just weeks earlier. And Batista could not even suspect how Fidel Castro was preparing to take advantage of the new situation, his supercharged mind spewing tactics, maneuvers, and strategies to expand his struggle.

In the hilltop prison, Castro was held in solitary confinement in a small cell on the first floor of the building while the other Moncada prisoners shared cells in the adjoining block. The two blocks were separated by a corridor, and Fidel and his companions could not see or talk to each other. Melba Hernández and Haydée Santamaría were in a cell on the same floor. Theoretically, Castro was isolated from his group, but ways for communication among them were rapidly devised in the fifty-one days between his arrest and his trial.

The rebels were charged under Article 148 of the Social Defense Code, which provides for a prison sentence of between five and twenty years for "the leader of an attempt aimed at organizing an uprising of armed persons against Constitutional Powers of the State." Castro's strategy had two main objectives: first, to obtain acquittal for the majority of the Moncada and Bayamo assailants on the grounds that they were not "leaders" in the uprising; and second, to use the trial as a forum to accuse Batista and the military of having implanted a dictatorship in Cuba in the first place, and of having massacred unarmed rebel prisoners after the abortive attack.

Realizing that he would be found guilty and sentenced, Castro concluded that the trial could and should be turned into the most profitable affair possible for the Movement and himself. The initial step was to establish exactly and in the greatest detail what had happened with each of the rebel detainees in Boniato before, during, and immediately after the assaults on the army barracks. Actually, some seventy-five defendants belonged to the Movement; the balance were political leaders, including Communists who had been arrested after the attack. Castro had lost contact with most of his troops when the combat began in Santiago, and he had not seen any of the survivors (except his own companions on Gran Piedra) until entering the prison. To lay the groundwork for his courtroom battle, he had to have maximum available information.

As Pedro Miret recalls, he and Raúl Castro took it upon themselves to interview every single Moncada prisoner in their group, "hearing the version of each one of what he did, and this way we learned a great deal about these events." In this fashion, Miret says, "we were able to collect a mass of information that we passed on to Fidel, and therefore when he delivered his statement [to the court] he had the totality of the most fresh information." During the seven weeks the defendants spent in Boniato, Castro was fed the information piecemeal through oral messages given in quick whispers by fellow prisoners passing past the iron bars of his cell, or even by friendly guards, and through tiny scribbled messages thrown inside the cell by someone passing. With his formidable memory, Fidel built in his mind detailed files on at least seventy-five Movement members. Having become the hero of common criminals serving time in Boniato, he had their support, help, and protection, too. Thieves and murderers were turned into secret revolutionary messengers.

With this data in hand, Castro was in a position to determine how many of the prisoners were the most likely to be acquitted. With Pedro Miret and Raúl acting as his delegates, he prepared all the companions for the trial. As Miret explained it, "We had nothing to gain with everybody remaining imprisoned, and with the information we had gathered, we could tell who in our group simply had to face prison, whether they liked it or not." Fidel's concept was that the hard-core leadership, including himself, had to confess its participation in the attacks, but the majority could safely deny it because they were completely unknown men, and the authorities could probably not prove their guilt. Many of them, for example, were arrested far from Santiago, often simply on suspicion (several persons were murdered by the military on suspicion, although they had not actually been involved in the conspiracy).

It was Castro who decided which of the prisoners should confess, and which of them should not. Miret says the criterion was that "of courtroom probability," and each prisoner was given Castro's recommendation. However, this "was not an obligation, certainly not an order." The prisoners were free to do what they deemed correct, but all of them followed his advice. Castro, who wanted as many men as possible acquitted so as to help rebuild the Movement, provided the final touch to this strategy. In the courtroom he made a point to identify himself who did and did not participate in the uprising. As a practical matter, neither the prosecution nor the judges could challenge Castro's careful selection of the candidates for acquittal, and they went along with him. It was an extraordinary judicial situation in which the chief defendant, in effect, instructed the court who should be sentenced with him.

Miret and Raúl Castro used the rest of their time in Boniato to organize a small library for their fellow prison mates out of books they requested from friends and relatives on the outside. And they also held daily classes for the group in subjects they knew best: history, language, physics, and mathematics. The idea was to keep the men cohesive and disciplined, not to let them lapse into depression or self-pity, and to whip them into the best psychological shape for the trial. Fidel always believed in physical and intellectual fitness and discipline, and in improving the education of the rebels at every opportunity. Because the *Fidelistas* already were a highly disciplined military organization, they gladly participated in this prison "school." Miret and Raúl Castro were the logical leaders in the endeavor as the only university students in this group of Moncada survivors.

And Fidel, too, discovered that prison offered a magnificent chance to engage in his favorite occupation of reading on a massive scale. Until Boniato, he had never had enough time to read as much as he wanted. In his first letter to his wife Mirta, three weeks after his imprisonment, he asked for Julián María's *Philosophy in Its Texts,* the works of Shakespeare,

García Morente's *Preliminary Lessons of Philosophy,* as well as "some novels that you would think would interest me." In September, while awaiting trial, he wrote his older brother, Ramón, that he was "taking advantage" of the time in prison: "I read a lot and I study a lot." And Castro added: "It seems incredible; hours fly by as if they were minutes, and I, with my restless temperament, spend the day reading, barely moving for anything."

Castro also showed family concern, perhaps greater than at any time in his adult past. He exchanged letters with Mirta until he broke with her a year later, expressing preoccupation over Fidelito's welfare and future. But they were not love letters, not even emotional letters; Fidel simply signed them "Kisses for the *niño* and for you." On August 18 he wrote her from Boniato: "I don't know whether you are in Oriente or in Havana. . . . I have heard very little about you; only that you were in Santiago after my detention, and also that you came to the prison to bring me clothes that were delivered to me. . . . I am well; you know that prison bars cannot break my spirit, my determination, or my conscience. . . . I suffer only for those who are outside, and for those who have felt in their flesh in this struggle the pain of sacrifice. . . . Be calm and courageous. Before all else, we must think about Fidelito. I want him to go to the school you have chosen. . . . When you come, bring him along; surely, they will let me see him."

On September 5 Castro wrote Ramón that he had learned from Mirta's letter that Fidelito had spent a week at the family *finca,* remarking that "he likes very much the countryside and animals . . . on the first of the month, he became four years old. . . . Mirta wants to send him to school, I mean a private school; we shall see whether this is possible." Two weeks later, he informed Ramón that Mirta had placed Fidelito "in a kindergarten in English . . . and in a public kindergarten almost across the street from the house . . . she also sent me some photographs, and I can see that he has grown a lot in recent months."

But Fidel wanted his parents to understand his situation as well. "It is necessary for you to make my parents see that prison is not the horrible and shameful thing they taught us it is," he wrote to Ramón. "This is true only when a man goes there for acts that dishonor him. . . . When the motive is high and great, then prison is a very honorable place." In another letter to Ramón, he mentioned that his father had sent him a telegram inquiring whether Fidel and Raúl needed clothes, and that he replied at once that they had all they needed from Mirta. He said he planned to write his parents later that day, adding, "Do they understand that I am imprisoned for doing my duty?" Then he thanked Ramón for the cigars his brother had sent him at his request in an earlier letter, asking for more boxes, "because the last one is almost finished, and it is always necessary to give a cigar to persons who help us . . ."

The trial of Fidel Castro and his companions opened on Monday, September 21, 1953, at the Santiago Palace of Justice, the building the detachment commanded by Raúl Castro had stormed on the day of the uprising, overlooking the Moncada barracks. It was being held before the three-judge Provisional Tribunal of Santiago, one of a network of special political courts whose sentences could not be appealed. The chief judge was Adolfo Nieto Piñeito-Osorio, and the prosecutor was Francisco Mendieta Hechavarría, and both were respected jurists.

It was literally a mass trial. The 122 defendants were represented by 22 lawyers, there were 6 forensic-medicine specialists, dozens of relatives and friends, and about 100 rifle-wielding army soldiers. They were crowded into a courtroom forty-five feet by fifteen feet, and the humid heat of Santiago notwithstanding, the doors and windows were closed as a security measure. Chief Judge Piñeito-Osorio described it as "the most transcendental trial in Republican Cuba," and despite press censorship imposed by the government under the national state of emergency in force since July 26, the nation was well aware that Fidel Castro was being tried. Six journalists, including Marta Rojas from the Havana weekly *Bohemia,* were allowed inside, although they would not be allowed to publish their accounts (their presence, however, would later directly help Fidel Castro as well as assist in the creation of the historical record).

Hours before the trial started, armored units cut off the streets around Moncada and the Palace of Justice. About one thousand soldiers with automatic weapons lined the route from Boniato Prison to the courthouse. Then, the prisoners arrived in buses at the back door of the Palace of Justice. Fidel Castro was transported alone under heavy guard in an army jeep. He wore his old, striped dark-blue wool suit, a white shirt, a red print necktie, and black shoes and socks. His wife, Mirta, had sent him the suit in prison at his request (he wrote her asking that it be dry-cleaned first and sent with a belt), and now Fidel was sweating profusely in the morning sun. His new moustache was well trimmed and his hair neatly combed. The prisoners, all of them handcuffed, were taken to the library of the Palace of Justice where for the first time they had the opportunity to talk, however briefly, with their lawyers.

As the prisoners entered the courtroom, there was a chorus of whispers: "This is Fidel! This is him! . . ." and the chief judge gaveled for silence. Before taking his seat, Castro raised his manacled hands toward the judges and addressed them in a loud, clear voice: "Mr. President . . . I want to call your attention to this incredible fact. . . . What guarantees can there be in this trial? Not even the worst criminals are held this way in a hall that calls itself a hall of justice. . . . You cannot judge people who are handcuffed . . ."

Castro went instantly on the offensive, but the chief judge sided with him. He declared a recess until handcuffs were removed from all the prisoners, and warned the chief of the army escort that he would not allow prisoners to be brought handcuffed into his courtroom again. Fidel had easily won the first round, and suddenly there was a new mood in the tribunal of justice. That Castro and his principal associates would be sentenced was taken for granted from the outset because in the light of the confessions they avidly volunteered, the judges had no choice but to find them guilty under Article 148. What was different, however, was the atmosphere in which the proceedings unfolded. Chief Judge Nieto Piñeito-Osorio and Judge Juan Francisco Mejías Valdivieso were thought to symphathize with the accused, and the third judge, Ricardo Díaz Olivera, was behaving in a similarly benign fashion although he was believed to be pro-Batista. The prosecutor did his job, and no more. In the end, Castro and his companions were given an astounding latitude in making accusations against the regime rather than defending themselves. Captain Manuel Lavastida, the SIM chief in Santiago, later told one of the defense lawyers that politically the government had lost the case: The public was for the defendants, the judges were for the defendants, he said. And in Santiago, Castro had going for him the Oriente tradition of rebellion and freedom wars.

That Castro and his companions enjoyed so much latitude in the Santiago court also stemmed from the fact that Batista evidently failed to understand the political implications of what was occurring with the three judges; there is no indication that he tried to exercise pressure on them to control the rebels' attitude. At the same time, the Cuban judiciary had jealously guarded its independence from interference by the government. The Santiago judges would go on maintaining this stance even during the most climactic moments of the Sierra war, and even then Batista failed to confront them.

The central concept in Castro's courtroom appearances was to establish the principle of legitimacy for the Movement's attack on the army barracks, arguing that the rebels had the *right* to try to overthrow Batista because he was an usurper who had gained power through a military coup and the violation of the constitution. With an absolutely straight face, Castro reminded the court that, acting as attorney, he had brought charges against Batista before a special court in Havana immediately after the coup, and had demanded up to one hundred years in prison for the dictator for violating the whole body of Cuban jurisprudence. Fidel was not attempting to defend himself and his companions from the charges of instigating an uprising; he was confirming the charges and proudly justifying their actions. By the end of the first day before the Provisional Tribunal of Santiago, the accused had become the accuser.

Castro then proceeded to play a dual role before the judges. The first one was that of the indignant defendant delivering political and revolutionary lectures in his replies to the prosecution. When Prosecutor Mendieta Hechavarría asked him whether he had participated in the Moncada and Bayamo attacks "in a physical or intellectual form," Fidel shot back: "Yes, I participated!" Answering the next question concerning the participation of the other defendants, he declared: "These young people, like me, love the freedom of their country. They have committed no crime unless it is considered to be a crime to wish for our country the best there is. . . . Isn't this what they taught us in school?"

Fidel's second role, even more successful, was acting as his own attorney. He demanded it as his right as a lawyer, and the chief judge could not refuse, having granted permission to do so to two political leaders charged in the same trial who had preceded Castro on the stand. Being his own attorney gave Fidel the opportunity to cross-examine his accusers, including the Moncada commander, Colonel Alberto del Río Chaviano, and to heap more insult and scorn on the dictatorship. Castro had to pay five pesos to an attorney for the courtroom license an attorney required to become an officer of the court. He went through the dramatic sequence of donning a black robe when he acted as lawyer (in Cuban courts, judges and lawyers had to wear robes, and that first day a robe large enough for Fidel could not be found, forcing him to perform in one several sizes too small), then taking it off when he returned to his defendant's bench. On his lawyer's table, he placed a copy of the Social Defense Code and a volume of Martí's works.

Castro produced the next dimension of drama when he was asked who was the "intellectual author" of the attacks on army installations on July 26. The question was fundamental in the trial because the law prescribed punishment for leaders of an armed assault against the state, and because in the Cuban legal tradition, the notion of the "intellectual author" of a crime, the person who inspired it, was as important as that of the "leader" physically carrying out an attack. To this question, Castro answered calmly: "The only intellectual author of this revolution is José Martí, the Apostle of Our Independence." Now he was turning the proceedings in the suffocatingly hot Santiago courtroom into a trial of history.

To each defendant the prosecutor confined himself to the single question of whether he (or she) had participated in the assault. The rebels responded with a yes, and then with accounts of atrocities by the Batista forces against their companions. Baudilio Castellanos García, Fidel's childhood friend and law-school colleague, defended forty-six of the rebels, the bulk of the accused (including Raúl Castro), and his statements and questions were perfectly synchronized with the overall Castro strategy. Between them, Castro and Castellanos established for the court who had participated in the attacks and who, in effect, should be acquitted, following up on the prosecutor's same question. Under the secret selection system, each rebel knew beforehand how to answer. Raúl and Miret had done their job well.

And one by one, the accused went on stunning the courtroom with horror tales: the beatings by the soldiers, the tortures, the summary executions of captured rebels, the brutalities, and the insults. The day after the Moncada attack, the Santiago "summer court" (that is, judges on the bench in the criminal tribunal substituting for vacationing regular justices) had ordered the autopsies of thirty-four cadavers found in the vicinity of the barracks—before the authorities had had time to bury thirty-three of them in a common grave at the Santa Ifigenia cemetery. The autopsy reports, describing "unidentified" individuals with smashed skulls and other signs of violent death by beatings or point-blank gunshots, were entered in the trial records of Case 37-053, enormously damaging evidence for the government's political image.

Astoundingly, it was Prosecutor Mendieta Hechavarría who through persistent questioning of Haydée Santamaría brought out the story of what the army had done to her fiancé, Boris Luis Santa Coloma, and her brother Abel before killing them. Haydée testified that after the soldiers had occupied the civilian hospital and captured her and Melba Hernández, a guard approached, saying that Boris was in the next room and "to tell me that they had extirpated his testicles" to make him talk (Jesús Montané had testified earlier that an army officer had approached him in the Moncada detention area, holding in his blood-covered glove a rotting ball of flesh, and said, "You see this? If you don't talk, I'll do the same thing to you I did to Boris. I will castrate you." The officer held a razor in the other hand). About her brother, Haydée said, "They gouged out one of his eyes." An army witness then added that Abel's eyes had been removed with a bayonet.

At the end of the testimony about army brutalities presented by scores of the defendants, and after his own statement on the subject in his capacity of defendant, Castro assumed the role of attorney. In his black robe, he requested the court to separate all the testimony on the army's treatment of the prisoners and the killings from the record of the Moncada trial so that it could form the basis for subsequent trials for "assassination and torture" he proposed to seek against Batista regime officials. To Fidel Castro's own surprise, the court instantly agreed. Thus Castro was now formally both a defendant and an accuser.

It is interesting to note that in the course of the Moncada trial, the Communists chose not to show any support for the *Fidelistas*. Two top leaders of the Popular Socialist Party, Joaquín Ordoqui Mesa and Lázaro Peña, were arrested in Santiago on the day of the attack and charged under Article 148 along with Castro's rebels and a half-dozen well-known opposition politicians. Asked whether they had conspired with others in an insurrection against the government, Ordoqui and Peña denied it, declaring they had come to Santiago, as they did every year, for the birthday celebrations of the party's secretary-general, Blás Roca. The two Communist leaders said they had no contacts with opposition political parties and no advance knowledge of the Moncada assault. They restated that Cuban Communists had opposed the Batista coup. But they volunteered no expressions of sympathy or solidarity with Castro and his revolutionaries. They were acquitted at the end of the trial.

Batista, in fact, must have found solace in the stand of his old Communist allies. Not only did they eschew public support for the rebels when the trial was under way, but the Popular Socialist Party actually denounced the Castro uprising. In a statement issued by the Communist leadership and disseminated by its clandestine publications as well as at cadre meetings, the party repudiated Moncada as "a putschist method peculiar to all bourgeois political factions." It called the attack "adventurist" and "false and sterile," even though the Communists recognized the "heroism" of Castro's men. The Communists thus threw the whole arsenal of Marxist-Leninist invective against the *Fidelistas*, presumably because the party's (and Moscow's) policy was to oppose any initiatives it did not control, especially when they were not in conformity with the dogma.

One does not know what, if anything, Fidel said on this subject to his brother Raúl, a member of the party's youth wing, but he was extremely bitter about the Communist stance. In a letter from prison to his friend Luis Conte Agüero, a journalist and radio commentator, Castro wrote in December that instead of "sterile and inopportune theories about a putsch or revolution," it should have been "the hour of denouncing the monstrous crimes the government had committed, assassinating more Cubans in four days than in the previous eleven years." He did not mention the Communists by name—he always refrained from direct criticism of communism in Cuba—but the context makes it obvious whom he had in mind. Twenty years later, Carlos Rafael Rodríguez, an old-time Communist who after the revolution rose to become one of Castro's most powerful and influential advisers, told an American journalist that at the time the party had judged Moncada by its "external characteristics" and described it as a "putsch" because "Fidel had not made his position explicit" about his political and insurrectional program. But this is a disingenuous effort to justify Communist actions at the time.

* * *

Meanwhile, after five days of the trial, the Batista regime finally concluded that the situation was getting politically out of hand, and that Fidel Castro could no longer be tolerated in the courtroom. His dual-role performance was too damaging and embarrassing to the government, and the Provisional Tribunal was clearly not disposed to curb him. Not wishing to confront the court directly, Fidel then let the army take matters into its own hands; this opened the way for another spectacular episode in the Castro drama.

When the court convened on Saturday, September 26, and the routine roll call of the defendants got under way, the rebels answered, "Present." But when the name of Fidel Castro was called, there was silence—the first time, the second time, the third time. The principal defense lawyer, Baudilio Castellanos, rose to tell the court that Castro was not present; he had not been brought to the Palace of Justice that day. The chief judge said the trial must proceed, but Castellanos jumped to his feet to demand that the court determine first why the principal defendant was absent. When the chief judge summoned the army officer in charge of guarding the prisoners, he was handed a letter from the office of Colonel Chaviano, the Moncada commander, saying that Castro could not be brought to the courtroom because he was ill at the Boniato prison; a medical certificate was appended to the letter. A court official read the communication aloud. Castro, the certificate said, was suffering from a "nervous crisis."

There was commotion, and a woman's voice shouted, "Fidel is not ill!" It was Melba Hernández, who had risen from the defendants' bench and marched toward the judges, removing a scarf from her head. She extracted a folded envelope from her hair and handed it to the chief judge, announcing, "This is a letter from Doctor Fidel Castro." Returning to her seat, Melba whispered to Haydée, "Now they can't kill him anymore . . ." Castro and his friends had again beaten their enemies to the punch.

Sometime earlier, Fidel and his principal associates had reached the conclusion that there was a plot afoot to kill him in prison. His food would be poisoned, he believed, and for some weeks he ate only what was sent the rebels from the outside by families and friends, or purchased through the ever-helpful common prisoners. He would not even smoke prison cigars. Whether or not the Batista regime had actually set out to kill Fidel is impossible to prove. However, Lieutenant Jesús Yanes Pelletier, the prison's military supervisor, was removed from his post shortly after Castro's arrival at Boniato, supposedly for refusing to obey orders to poison him (Yanes later became a captain in the Rebel Army and chief of the Castro bodyguard). Besides, Fidel has a keenly developed sense of dangers facing him; as his former interior minister, Ramiro Valdés, says, "He can smell the dangers and the risks in the air. . . . This Fidel is a sorcerer, a sorcerer . . ."

In any event, Castro's suspicions were confirmed, at least in his mind, when he was informed on the evening of September 25 that he would not be taken to court the next day because he was thought to be ill. The secret network inside the prison was activated at once. While his companions and the common prisoners strolled back and forth in front of his cell, as they did every evening, Castro whispered that he would be writing an urgent letter that had to be delivered to the court the next day. Then, he proceeded to write it on onion-paper sheets held inside a magazine he pretended to be reading. Leonel Gómez Pérez, a prisoner who had been mistakenly charged with participation in the Moncada operation, was assigned the mission of receiving the letter from Fidel. Leonel had a habit of reading a book as he walked up and down the prison corridor, and therefore the guards paid no attention to him that particular night and did not notice that the prisoner was coming closer and closer to the bars of Castro's cell. When the guard in the corridor looked away for a moment, Fidel threw the letter in an envelope through the bars so that it would land inside Leonel's book. Leonel kept walking for a while longer, then calmly returned to his cellblock. He gave the letter to one of the *Fidelistas* who, in turn, had one of the common prisoners deliver it to Melba Hernández. Melba folded the envelope into a tiny square and concealed it in her hair.

In his letter to the court, Castro, describing himself as his own attorney in the proceedings against him, charged that all possible means were being used to prevent him from being present at the trial so that he could not destroy "the fantastic falsehoods" about the events of July 26, and make known "the most terrifying massacre in the history of Cuba." He insisted he was not ill, and that the court was being subjected to unprecedented lies, and he complained that he was kept illegally incommunicado for fifty-seven days, "without being allowed to see the sun, talk with anyone, or see my family." Fidel informed the court that he had learned with certainty that "my physical elimination, under pretext of an escape attempt or through poisoning, is being planned," and that "the two girls" (Melba and Haydée) faced the same danger because "they are unusual witnesses of the massacre of 26 of July."

Castro then demanded an immediate medical examination by the dean of the medical school of Santiago, the assignment of a court officer to the prisoners while being transported between Boniato and the Palace of Justice, and the delivery of copies of his letter to the Supreme Court and the Cuban Bar Association. He named Melba Hernández as his representative in court and concluded: "If I have to sacrifice a particle of my rights or my honor to keep my life, I prefer to lose it a thousand times: A just principle at the bottom of a grave is more powerful than an army!"

The three judges agreed to order an independent medical examination of their famous defendant, and two distinguished physicians undertook it despite the army command's opposition and found Castro in perfect health. On the basis of their report, the court requested his return to the trial. The Boniato wardens refused, however, to let Castro leave the prison, and the judges gave in to the pressure. This was their compromise with the regime over the way they were treating Castro, allowing him the freedom to accuse Batista at will. They ruled that he was to be separated from the main trial, to be tried separately afterward, and that the proceedings against all the other defendants would continue normally. Fidel never again set foot in the Palace of Justice. At Boniato, he was transferred to a cell on the ground floor, far away from his companions. Melba and Haydée were also punished: They were separated from the other *Fidelistas* and placed in a cell from which, in the words of the military warden, "you won't even be able to see the sky."

Castro's transfer to a remote cell revived concern among his companions that he was still an assassination target, and at the next session of the court, on September 28, Raúl Castro rose to announce at the top of his voice that "I fear for the life of my brother! They have mounted a dangerous conspiracy to assassinate Fidel, and I propose that you suspend this trial because our presence at Boniato might protect him . . ." Raúl was ordered to sit down and shut up, but the chief judge also announced that when the court had received Fidel's letter two days earlier, "it took the required measures for the protection of the defendant." This was exactly what Raúl had wanted to hear, but he was on his feet again as the court was adjourning for the day to warn that "if they kill Fidel, they will have to organize a massacre there and do away with all of us."

The trial continued for another week with a parade of witnesses drawn from among army and police officers as well as hospital nurses. But their testimony demolished the charges originally presented by Colonel Chaviano, the Moncada commander, that the rebels used "modern arms," hand grenades, and knives and daggers to behead the soldiers, and that they had murdered patients in their beds in the civilian and military hospitals. The defense lawyers along with the prosecutor forced out the denials or admissions of "I really don't know" from the government's own witnesses, rendering the proceedings another notch more embarrassing for the regime. Finally, the court convened on October 5, to conclude the trial in an aptly bizarre fashion. Prosecutor Mendieta Hechavarría first announced that he was dropping charges against all the political party leaders, including the Communists, who were among the defendants, and against the *Fidelistas* whose links with the insurrection could not be proved; this vindicated Castro's basic strategy of determining who should be acquitted by advising the court of the "innocence" of some men and the "guilt" of others.

Now twenty-nine rebels were left to be judged. In his summation, the prosecutor praised them for "acting with honor, for having been sincere and very courageous, proper in their confessions . . . and noble in their attitudes." He said he "sincerely applauded" their integrity in admitting their guilt. It was an astonishing spectacle as the prosecution virtually apologized for being forced by these men's confessions to ask the court to sentence them to prison terms under the provisions of Article 148. On October 6, the chief judge handed down the sentences: Raúl Castro, Pedro Miret, Ernesto Tizol, and Oscar Alcalde—the top leadership group—were given thirteen years in prison. Twenty other men received ten-year sentences, and three men got three-year sentences. Melba Hernández and Haydée Santamaría were sentenced to seven months detention at the women's prison at Guanajay, west of Havana. Relatively speaking, the sentences were not heavy: The maximum under the law was twenty years imprisonment for the leaders and the minimum was one year.

Still undecided, however, were the fates of Fidel Castro in his cell at Boniato Prison and of Gustavo Arcos, the man who had been in Fidel's car at Moncada and was now recovering from his wounds at the Colonia Española hospital. No date was set for their trials, and the rebels again began to worry that Fidel could be in danger after they had left Boniato for their ultimate destination. Twenty-six prisoners were placed aboard an army aircraft on October 13; to their surprise, they were taken to the Isle of Pines, which lies immediately south of the Cuban coast, to the island prison, while Melba and Haydée continued on to Havana en route to Guanajay. The court had specified that the men would be held at La Cabaña fortress in Havana, but the regime decided otherwise; the Isle of Pines was more remote from civilization.

Fidel Castro's trial lasted exactly four hours on October 16, and he was sentenced to fifteen years in prison. The extraordinary aspect of the trial was that the regime held it in virtual secrecy in a small nurses' lounge, a room measuring twelve feet by twelve feet, in the nurses' school at the Saturnino Lora civilian hospital, captured by Abel Santamaría's detachment on July 26. The other extraordinary fact was that it produced Fidel Castro's defense speech, which immediately became known as "History Will Absolve Me," and remains to this day the fundamental and legendary document of the Cuban revolution, the venerated scripture of the rebel movement.

The decision to hold the trial in the hospital room was designed to perpetuate the official fiction that Castro was too ill to attend the main proceedings in the Palace of Justice, though he had to be transported from Boniato to the hospital. To add credibility, a wounded rebel and a prisoner who had nothing to do with the whole affair were also brought to the nurses' lounge. Fidel could not resist remarking that it was unwise "to render justice in a hospital room, guarded by sentinels with bayonets on their rifles, because the people might conclude that our justice is ill."

The surroundings were, indeed, absurd. A life-size skeleton hung inside a glass case, and a wall was adorned with a picture of Florence Nightingale. Enough desks and chairs had to be put in the hot, airless room to accommodate the three judges, the prosecutor, the secretary of the court, three lawyers, six journalists, and the three defendants (Abelardo Crespo was on a stretcher on the floor). Two army officers and twenty armed soldiers assured security at the trial. The room had a single barred window, and the morning heat was excruciating.

Fidel Castro, handcuffed and wearing his heavy dark-blue wool suit, arrived at exactly nine o'clock in the morning with his military escort, and the proceedings began. There is no official transcript or record of the trial, and all the descriptions are based on the oral or written recollections of those present. Castro delivered his speech extemporaneously, and Marta Rojas and the other journalists took copious notes, but the official text of "History Will Absolve Me" was later reconstructed from memory by Fidel and put on paper.

By that time, Fidel had spent seventy-five days in solitary confinement, and he had lost so much weight that his watchband kept slipping off his wrist, but he was in fine form. Asked the ritual question of whether he had participated in the Moncada attack, Castro replied, "Affirmative." When the prosecutor inquired whether the purpose was to overthrow the government, Castro answered, "It could not have been any other." The prosecutor said he had no other questions. Then, Colonel Chaviano and other military officers testified about the Moncada events, repeating earlier charges. Castro again requested permission to act as his own attorney, and a young mulatto lawyer named Eduardo Sauren lent him a black robe; again it was too small and kept bursting at the armpits whenever Fidel raised his arm to make an oratorial point. Cross-examining Major Pérez Chaumont, the Moncada deputy commander, Castro charged that two additional members of his Movement had been murdered by the Rural Guard; and, as before, he insisted that the army had killed the captured rebels, wanting no prisoners. And this was the end of the proceedings.

Prosecutor Mendieta Hechavarría chose not to deliver a summation, speaking for barely two minutes to ask the court to apply the maximum punishment to Fidel Castro, as the principal leader of the insurrection, under the terms of the Social Defense Code. Castro looked up and said matter-of-factly, "Two minutes seem to me to be a very short moment to demand and justify that a man be locked up for a quarter of a century."

The maximum sentence could be twenty-six years in the case of a top organizer of an insurrection. Then Castro announced that he insisted on delivering his own defense brief, and the chief judge authorized him to proceed. Standing behind a small table, Fidel had a sheaf of notes, a copy of the code, and his book of Martí quotations. Held overnight at Moncada, Fidel had kept his cellmate awake until dawn while he practiced his delivery.

Castro spoke for two hours, one half the time allotted to his trial, and it was a stunning *tour de force* of memory, coherence, erudition, and patriotic and revolutionary emotion. Evidently, he had meditated deeply in his prison cell on this masterful piece of oratory—it is really much more than a speech—and his superb command and love of the Spanish language transformed it into a work of literature as well. Just as he claimed that José Martí was the "intellectual author" of the Moncada attack, Martí was equally clearly the inspirer of this oration. As it happens, Martí's first great feat of political writing was an essay on "The Political Prison in Cuba," published when he was eighteen years old, after serving time at hard labor for conspiring against Spain, and it was the perfect literary model for Fidel. In the copy of the Martí book before him at the trial, Castro had underlined twenty-nine major passages, and he quoted the Apostle fifteen times in his discourse.

Castro opened his oration in low, slow tones. First, he reviewed the "illegalities" surrounding his trial and his thesis that insurrection against a tyrant was legitimate. Then, he traced the history of his Movement, discussed the reasons for the defeat at Moncada, denounced the tortures and the killings of his companions, and stressed governmental corruption in Cuba and the unfair treatment of soldiers in the army (from the beginning, Fidel always sought to attract the "honest" military to his side). But the brunt of his condemnation was aimed at Batista: "Dante divided his hell into nine circles; he put the criminals in the seventh, the thieves in the eighth, and the traitors in the ninth. What a hard dilemma the devil will face when he must choose the circle adequate for the soul of Batista . . ." He called him in Latin, *Monstrum Horrendum!*

Next, Castro turned to the Movement's political program, the awesome social-economic conditions prevailing in Cuba, and the revolutionary laws the rebels would have proclaimed had they won at Moncada: the return to the 1940 constitution, agrarian reform, recovery of stolen resources, educational reform, profit-sharing by the workers, and public-housing policies.

"I bring in my heart the doctrines of Martí and in my mind the noble ideas of all men who have defended the freedom of the people," Castro said. "We have incited a rebellion against a single illegitimate power which has usurped and concentrated in its hands the legislative and executive powers

of the nation. . . . I know that I shall be silenced for many years. I know
they will try to conceal the truth by every possible means. I know that
there will be a conspiracy to force me into oblivion. But my voice will
never be drowned; for it gathers strength within my breast when I feel
most alone, and it will give my heart all the warmth that cowardly souls
deny me."

To Fidel, the struggle for the people meant "the *six hundred thousand*
Cubans who are out of work who want to earn their daily bread honestly
. . . the *five hundred thousand* farm workers who live in miserable huts, who
work four months and go hungry the rest of the year . . . the *four hundred
thousand* industrial workers and laborers whose retirement funds have been
stolen . . . the *one hundred thousand* small farmers who live and die working
land that is not theirs, contemplating it as Moses did the Promised Land,
only to die before owning it . . . the *thirty thousand* self-sacrificing and
devoted teachers and professors who are so badly treated and poorly paid
. . . the *twenty thousand* debt-ridden small merchants, ruined by economic
crises . . . the *ten thousand* young professionals [who] leave the schools with
their degrees, only to find themselves in a dead-end alley . . . To these
people, whose road of anguish is paved with deceit and false promises, we
are going to say, 'Here you are, now fight with all your might so that you
may be free and happy.'"

Reaching into his monumental memory and erudition, Castro built history's case to justify taking up arms against tyrants: He noted that Charles
I and James II of England were dethroned for despotism; he cited the
English revolution of 1688, and the French and American revolutions; the
independence of European colonies in Latin America; and summoned up
the spirit of freedom and defiance of Stephanus Junius Brutus of ancient
Rome, St. Thomas Aquinas (in *Summa Theologica*), John of Salisbury, Martin Luther and John Calvin, the Scottish reformers John Knox and John
Paynet, Montesquieu (in *The Spirit of the Law*), Jean Jacques Rousseau (in
The Social Contract), John Milton, John Locke, Thomas Paine, the seven-
teenth-century German jurist John Althusius, the French revolutionary jurist Léon Duguit, Honoré de Balzac, and, of course, José Martí.

But, Castro said, "It looked as if the Apostle Martí was going to die in
the year of the centennial of his birth. It looked as if his memory would be
extinguished forever, so great was the affront! But he lives. He has not
died. His people are rebellious, his people are worthy, his people are
faithful to his memory. Cubans have fallen defending his doctrines. Young
men, in a magnificent gesture of reparation, have come to give their blood
and to die at the side of his tomb so that he might continue to live in the
hearts of his countrymen. Oh, Cuba, what would have become of you if
you had let the memory of your Apostle die!"

And Fidel Castro concluded: "As for me, I know that jail will be as hard as it has ever been for anyone, filled with threats, with vileness, and cowardly brutality; but I do not fear this, as I do not fear the fury of the miserable tyrant who snuffed out the life of seventy brothers of mine. Condemn me, it does not matter. *History will absolve me!*"

Castro finished, and the three judges and the prosecutor consulted in whispers in the crowded room for a few minutes. The chief judge ordered "the defendant, Doctor Castro Ruz" to be kind enough to stand, then announced: "In accordance with the request of the prosecutor, this court has imposed on you a sentence of fifteen years in prison. . . . The trial has been concluded." Fidel put out his hands to be handcuffed, but when a soldier had trouble doing it, he suggested that the officer of the guard put the manacles on him because of his greater experience. "And careful with my watch," he added. Walking out of the nurses' lounge, Fidel turned to Marta Rojas, the young reporter, and asked, "Did you get it all? Do you have all the notes?" As Castro left the hospital building under heavy guard, people in the street cheered him.

A week later, the last of the rebels, Gustavo Arcos, was sentenced to ten years in prison at a special trial held at the Colonia Española hospital where he lay wounded. At about the same time, General Batista announced incongruously that presidential elections would be held in Cuba on November 1, 1954, more than a year away. As for the Eisenhower administration, it named Arthur Gardner, a financial expert and an admirer of General Batista, as the new United States ambassador in Havana.

In seeking to trace the evolution of Fidel Castro's ideological allegiances, a complex and tortuous enterprise, the character of the "History Will Absolve Me" discourse is an important milestone. The problem, however, is how this pronouncement should be interpreted from hindsight as distinct from the interpretation given it at the time—or, for that matter, from what Castro might have had in mind.

Since the formal implantation of Marxism-Leninism in Cuba, the official version is that "History" was decidedly an "anti-imperialist" document, forged with the tools of Marxist dialectics and Leninist inspiration (apart from the Martí inspiration). Castro himself nowadays accepts this interpretation of his 1953 thoughts, albeit the Communists failed to reach the same conclusion. The widely accepted view is that he had subtly disguised his real objectives in the revolutionary program to avoid antagonizing the Cuban middle class and the Americans. The French historian Robert Merle writes very perspicaciously that Castro was faithful to Martí's strategy of fighting only one enemy at a time. Merle adds that Fidel failed to explicitly attack the United States economic domination over Cuba, even though he hinted at the nationalization of the American-owned telephone and power companies, but, he says, "what was said in this manifesto was less important than what was being insinuated."

The Cuban revolutionary chronicler Mario Mencía observes that when Castro defined the Moncada assault as "the battle for freedom," his growing use of Marxist-Leninist theory made him take that concept beyond "the narrow confines of apparent bourgeois personal and institutional liberties." Theodore Draper, a severe American critic of the Cuban revolution, has written that only "official, orthodox pro-Castro propagandists" insist that "everything Castro has done, in power, is consistent with 'History Will Absolve Me,' that it contained the essential outline of the entire *Fidelista* revolution."

Given that it is impossible to plumb Castro's mind and selective memory, none of the "official" interpretations are necessarily convincing. If Fidel had fashioned his pronouncement so subtly that even Communist intellectuals at the time missed the point and kept suspecting him of "bourgeois" and "putschist" tendencies, the conclusion is inescapable that the whole exercise served no purpose ideologically. If, indeed, he was concealing his true goals in relatively harmless language, there is no way of proving it. Castro's own contribution to this debate was a comment to a visitor in 1966 that "my Moncada speech was the seed of all the things that were done later on" and that "it could be called Marxist if you wish, but probably a true Marxist would have said that it was not." He went on to say that "it would not have been intelligent to bring about . . . an open confrontation. I think that all radical revolutionaries, in certain moments or circumstances, do not announce programs that might unite all of their enemies on a single front. Throughout history, realistic revolutionaries have always proposed only those things that are attainable." In the end, one is left with the impression that Castro was keeping his options open when he dramatically addressed the court in October 1953, and that his ideas evolved in a firm Marxist-Leninist direction much later. It is an empirically and historically flawed notion to assert that "History" pointed Cuba irreversibly toward communism. Perhaps more to the point, the Castro oration very soon overshadowed the Moncada defeat, allowing him to build on the terrible experience of July 26 and become the revolutionary hero of Cuba. It is a good example of words being mightier than arms.

In any event, Fidel Castro was selected by the Havana weekly *Bohemia* as one of the twelve most outstanding world figures of 1953, along with such personalities as the shah of Iran (overthrown that year and restored to the throne by the CIA), the pugilist Kid Gavilan, Costa Rican President José Figueres, England's Queen Elizabeth (crowned in 1953), and Soviet KGB chief Lavrenti Beria (shot after Stalin's death). This was very heady company for an imprisoned twenty-seven-year-old Cuban revolutionary, but Fidel had no doubt that he belonged there. Even before he was flown from Santiago to the prison on the Isle of Pines, he was deep in planning tomorrow's actions against Batista. It had never crossed his mind that he might actually spend fifteen years behind bars.

CHAPTER

6

Fidel Castro's Rebel Army was born in the men's prison on the Isle of Pines where he and twenty-five companions were confined by the Batista regime for one year and seven months. Of this group, fourteen men (including Fidel) were aboard the yacht *Granma* when she sailed from Mexico to Oriente at the end of 1956, to launch the war in the Sierra Maestra; seven were officers in the invading force. In 1986 three of the Isle of Pines prisoners, headed by Raúl Castro, were still among Fidel's closest associates. One of them is *Commandante de la Revolución,* a title bestowed on only three Sierra commanders.

For Castro's Movement, the Moncada and Bayamo attacks were baptisms by fire for a contingent of militarily inexperienced idealists with inadequate weapons and flawed information about the enemy. Politically and ideologically, the 26th of July rebels were immature and essentially vague about their long-range objectives. This assessment includes Fidel and his principal collaborators, even though he spoke at his trial of the Movement's plans for "revolutionary laws" to be implemented by a new government he had hoped would result from a national rebellion triggered by a Moncada victory. There was much naïveté in that first, brave enterprise, and Castro's Marxist stirrings or motivations had little relevance to the cause when the

barracks were stormed. Above all, the Movement had been virtually unknown in Cuba up to that time.

The Moncada trial, with the wide public recognition it gave Castro and
the *Fidelistas* (press censorship mattered little on this island where the word
of mouth disseminates news with lightning speed), and the experience of
the imprisonment represented the great turning point in the history of the
revolution. On one level, the fate of the Isle of Pines prisoners became a
national issue—they were the object of outpourings of sympathy as the
Batista regime, for its part, was held in increasingly low esteem—and
Castro knew how to take advantage of this situation by creating a political
organization from his prison cell. He understood the absolute necessity of
central and unquestioned authority in the leadership of the Movement,
which he reserved for himself, and of the importance of skilled propaganda
to help the organization grow. As he wrote a friend from the Isle of Pines,
"The propaganda and organization apparatus must be so powerful that it
will unmercifully destroy all those who try to create trends, cliques, and
splits or rise up against the movement." Castro remained faithful to this
principle in prison, in exile, in the Sierra war, in the hour of revolutionary
victory, and later, in beating down a challenge by old-line Communists.

But Fidel also knew the value of the moral imperative in politics. His
approach to the creation of the great revolutionary instrument was inspired
by the philosophy of the Prussian military genius, Karl von Clausewitz; in
warfare "the physical seems little more than the wooden hilt, while the
moral factors are the precious metal, the real weapon, the finely-honed
blade." In prison Castro was an attentive student of von Clausewitz's writings as part of his immense program of reading and study in preparation for
future phases in his revolution. Specifically, Fidel's objective was to endow
his Movement with such moral strength, combined with the natural Cuban
penchant for patriotism, history, and nationalism, that in time it would
overcome the physical and economic power at Batista's command.

As a corollary to his moral ambitions, the trial and prison became the
crucible in which Castro could perfect himself and his men. Under him,
they would be the hard core of revolutionary leadership, what he called the
"vanguard" of the Movement. This called for political education and discipline, and, paradoxically, the Batista prison provided ideal conditions for
the honing of the future Rebel Army. And a rebel army was exactly what
Castro had in mind. He had concluded that in order to succeed, a revolution requires its own army, one that yearns for a revolution. By this time,
he had come to reject the idea of a political solution for Cuba that would
seek to change the government with the support of the existing military
establishment. Before the Batista coup, he had been willing to try the
parliamentary route. Fidel now realized that the moment he became allied

with any military faction, no matter how democratic and progressive-minded, he would inevitably be subordinated to it because he would lack control over firepower. In a sense, he shared the Communists' distaste for the "putsch" strategy, albeit for different reasons. Castro's ultimate plan was to destroy the Cuban armed forces and replace them with *his* Rebel Army, and this constituted his foremost concern as he read, wrote, and meditated in his prison cell—and as he directed the preparation of his leadership "vanguard." In retrospect, he agrees that harsh and frustrating as his imprisonment was, it offered him and the Movement an opportunity to create the revolutionary framework. Without the prison experience, the *Fidelista* revolution might never have soared.

The Provisional Tribunal in Santiago sentenced the rebels to imprisonment in Havana, but for reasons that were not made clear, the regime sent them to the Isle of Pines. This was providential for Fidel Castro because it provided him with still more useful symbolism. The Isle of Pines was José Martí's first place of exile from Havana when, at the age of seventeen, he was sentenced to six years in prison for anti-Spanish activities. Martí had first worked in chains in a stone quarry near Havana in 1870, but after a year he was transferred to work on an estate on the Isle of Pines prior to being deported to Spain. Moreover, the Isle of Pines bitterly reminded the Cubans of the worst aspects of American "imperialism." Under the 1901 Platt Amendment, which had defined the terms of Cuban independence after the Spanish-American war, the Isle of Pines was "omitted from the proposed constitutional boundaries of Cuba, the title thereto being left to future adjustment by treaty." Both the United States government and American businessmen therefore treated the Isle of Pines as a colony; only in 1926 (the year of Fidel's birth) did the United States cede sovereignty over it to Cuba under the Hay-Quesada Treaty.

Twenty-four rebel prisoners landed at the Isle of Pines' Nueva Gerona Airport on October 13, a week after their sentencing in Santiago, to begin serving their terms at the model penitentiary. Melba Hernández and Haydée Santamaría continued on the same plane to Havana en route to the women's prison on the mainland. Fidel Castro was still awaiting his trial. Actually, there are published discrepancies concerning the number of rebels brought to the Isle of Pines. Twenty-nine, including Melba and Haydée, were sentenced on October 6, which left twenty-seven men in that group to be transported to prison. But Mario Mencía, the revolution's chronicler, lists only *twenty-six* prisoners on the Isle of Pines, including Fidel Castro and Abelardo Crespo (the wounded prisoner who was tried together with Fidel in the hospital), in his authoritative account on the confinement of the rebels. While Fidel and Crespo arrived in Nueva Gerona four days after the group, three men sentenced in Santiago with the others never appeared on the Isle of Pines for unknown reasons, although their names are included in the warden's report on those arriving on October 13. It appears that twenty-four is therefore the correct number. The only one of them familiar with the Isle of Pines was Jesús Montané, a member of the Movement's civilian committee, who was born there; his parents still lived in Nueva Gerona.

The Isle of Pines is roundish, just over two hundred miles in circumference, and lies some fifty miles south of the coast of Havana province. Farming is its chief economic activity, but the model penitentiary rivaled it in economic activity when it was built by the Machado dictatorship in 1931. With a total capacity of five thousand inmates, it was a Cuban tropical Siberia for political prisoners. The prison consists of four huge five-story circular structures, each designed to hold 930 prisoners, and a half-dozen other large buildings. One of them is a hospital, and the Movement prisoners were installed in its southern ward, known as Building 1. They were all housed in a rectangular hall with metal beds set in two rows; the ward had three showers, two toilets, and a sink for the twenty-six men. A barred door led to a cement-floored inner patio where the prisoners could exercise. They had access to volleyball, Ping-Pong, and chess. The advantage over Boniato was that they were all together in the hospital ward; had they been put in the circular structures, they would have been kept in two-man cells. Thus, their conditions were not the worst possible, most likely because the Batista regime wished a minimum of adverse publicity.

When the rebels reached the model penitentiary, they had no news about Fidel and feared that he had been killed or sent to another prison. Nonetheless, their spirits remained high. Pedro Miret, the weapons expert, became the acting chief of the group, assisted by Israel Tápanes and Raúl Castro, and they wasted no time getting organized. All available books were assembled in a ward library named after Raúl Gómez García, the young poet killed at Moncada; there was a bookcase in their ward. One prisoner was assigned to conduct outside purchases for the group and keep track of individual accounts, and another man was in charge of distributing supplies from the cooperative the men had formed. Then, there were regular meetings chaired by Miret, and the beginning of the rebels' prison school. A ten-article list of regulations set forth the system of the meetings, including these provisions: "Wounding expressions may not be used in the debate, and it is absolutely prohibited to justify mistakes by claiming that the critic would have made the same, a similar or some other kind of mistake"; "The chairman is empowered to take the floor away from a comrade he considers to be obstructing the progress of the assembly"; and "The chairman is also invested with absolute power to conduct the assembly so that it will progress as he deems best."

Recalling the prison days, Pedro Miret says that the prisoners immediately decided to establish a more rigid daily discipline than the penitentiary's schedule. "If we were ordered to get up at six A.M., we would get
up at five-thirty A.M., very well organized," he says. "By being stricter
than the prison regulations, we were able to do there whatever we wanted.
To the authorities, we seemed like very quiet prisoners, so they left us
alone. They were so ignorant that they never realized what we were doing
with education. And they respected us, they located us away from the main
area so that nobody else could see us."

Fidel Castro joined his companions on the Isle of Pines on October 17,
hugging and embracing them in the ward. In the evening, he sat on his
bed in the first row, almost at the entrance to the bathroom, to bring the
men up to date on the final events in Santiago, including his hospital trial
and his "History" discourse. Fidel's prison file—now he was Prisoner
3859—contained photographs showing his well-trimmed moustache, and
noting that he "had education," and that he had a long scar, apparently
from appendicitis surgery, on his abdomen and a scar on his left thigh.

With Fidel back with his companions, he was at once elected chief of the
group, and prison life picked up in activity. In addition to the library,
which grew to over five hundred volumes (including one hundred belonging to Fidel), the men organized the Abel Santamaría Ideological Academy
as a prison "university" to teach philosophy, world history, political economy, mathematics, and languages as well as Spanish-language classics. The
academy functioned in the patio, where the men sat at the wooden tables
where they normally ate. They had a small blackboard. There were nearly
five hours of classes a day—morning, afternoon, and evening—with Fidel
teaching philosophy and world history on alternate days and public speaking twice a week, Pedro Miret lecturing on ancient history (he remarked
later that they were amnestied in the midst of the medieval period), and
Montané instructing in English. Montané wrote later that "from the outset, Fidel told us that our imprisonment should be combative, and we
should acquire rich experience from it, experience that would help in the
continuation of the struggle once we are freed."

Castro also read to the group (everything from Napoleon Bonaparte's
infantry attack on Hugomont to José Martí's pleas to the Spanish Republic
for freedom for Cuba), then stayed up past eleven o'clock at night "when
sleep catches me reading Marx or [Romain] Rolland." Castro and Miret
wrote friends and relatives for books: They asked Havana University Rector
Clemente Inclán y Costa, University Secretary Raúl Roa García, and just
about everyone who could satisfy their literary hunger. Roa helped them
further by publishing in the weekly *Bohemia* a letter from Miret requesting
books; censorship had just been lifted, and this called additional political

attention to the prisoners. Miret believes that their library had four volumes of José Martí, most of the important works on the French Revolution, including all the Girondins' volumes, and a complete collection of Lenin, Marx, and Engels—suggesting that the education of the *Fidelistas* now was taking a pronounced ideological direction. And Fidel wrote to a friend: "What a formidable school this prison is . . . from here, I can finish forging my vision of the world and the sense of my life."

In his first letter from the Isle of Pines to his brother Ramón, Fidel informed him that the prison censors had refused to deliver to him a registered letter from Ramón "because it touched on subjects not permitted by censorship . . . which greatly surprises me." But he urged his brother not to form a negative impression from it because "the persons running this prison are much more decent and prepared than those in Boniato." He wrote that prisoners were not robbed or exploited, and that "men here are much more serious . . . there is discipline, but there is no hypocrisy. . . . I don't want to tell you, brother, that we are in a paradise here; there still are many deserved things to be obtained by us, but it seems there is good will on the part of the authorities, and everything will work." Fidel was being a bit optimistic. The prisoners were allowed to receive visits once a month, and Castro's letter to Ramón indicated that Mirta was planning to fly from Havana to see him at the prison. He urged his brother to come along.

By December, Fidel Castro was again on the offensive. In an immensely long letter to his friend Luis Conte Agüero, the radio commentator, Castro retold the massacres suffered by his rebels at Moncada and asked, "Why have the barbaric and insane mass tortures and murders . . . not been courageously denounced? That is the inescapable duty of the living and to fail to do so is a stain that will never be erased." But Castro also informed Conte Agüero that as a result of the denunciations he made during the Santiago trial, the court there had accepted three lawsuits by him against Batista and three of his top commanders "as the men who ordered the killing of prisoners." He added that the court in Nueva Gerona on the Isle of Pines, the jurisdiction to which he now belonged, had also accepted the lawsuits. In the strange world of Cuba, a rebel chief serving a prison term for insurrection against the regime was able to sue for murder the head of state whom he undertook to overthrow. And just as curiously, various Cuban courts went on hearing depositions in these cases until almost the time of Castro's amnesty.

Fidel then quoted Martí: "When there are many men without honor, there are always some who have within themselves the dignity of many men. Those are the ones who revolt with terrible force against those who steal the people's freedom, which is to steal men's honor." This was Castro's latest challenge to Batista, and the letter went on to say that triumph at Moncada would have meant the transfer of power to the *Ortodoxo* party in the spirit of the "true ideas of Chibás," the first and the only time he had proposed such a course. He repeated from the "History" discourse the revolutionary laws the new government would have promulgated, but now there was not the slightest whiff of Marxist thinking in Castro's letter. Fidel clearly sought *Ortodoxo* support as he prepared his Rebel Army. Moreover, he wanted the letter to be issued by Conte Agüero as "The Manifesto of the Nation," with a subtitle taken from Martí, "Message to a Suffering Cuba." He also asked that this manifesto be handed to his wife, Mirta, for publication in the Havana Univeristy publication *Alma Mater*. Interestingly, Fidel was increasingly seeking Mirta's help in his political endeavors, and she tried to do all she could. But the manifesto was published only much later as a small pamphlet.

The holidays were approaching, but Fidel told Conte Agüero that "needless to say, we shall not celebrate Christmas, we shall not even drink water, to show our mourning . . . make this known as such, because I believe in that way the objective will be more noble and humane." Meanwhile, he continued his dizzying voyage of intellectual discovery, tailoring it to his needs. He became fascinated with Napoleon III (the despised "Little Napoleon"), reading both Victor Hugo and Karl Marx on this topic. He wrote a friend in Havana that much as Hugo's *Les Misérables* had stimulated him, "I grow a little tired of his excessive romanticism, his verbosity, and the sometimes tedious and exaggerated heaviness of his erudition." But he found that "on the same topic of Napoleon III, Karl Marx wrote a wonderful work entitled *The Eighteenth Brumaire of Louis Bonaparte*. . . . Where Hugo sees no more than a lucky adventurer, Marx sees the inevitable result of social contradictions and the conflict of the prevailing interests of the time. For one, history is luck. For the other, it is a process governed by laws." Again, there was relevancy to it in terms of Castro's current interests: Napoleon III had seized power through a coup (or putsch in modern language), and Fidel saw it as an evil act in the context of his own Cuban perceptions.

Castro's reading list at the end of 1953 did seem to confirm that prison was a fantastic university and that his own tastes defied definition: Thackeray's *Vanity Fair*, Ivan Turgenev's *Home of the Gentry*, a biography of the Brazilian Communist leader Luis Carlos Prestes who had led a "long march" of sorts through his vast nation, the Dean of Canterbury's *The Secret of Soviet Strength*, a modern Russian novel by a young revolutionary, A. J. Cronin's *The Citadel*, Marx's *Capital*, Somerset Maugham's *The Razor's Edge*, four volumes of the *Complete Works of Sigmund Freud*, and seven Dostoevski novels, including *Crime and Punishment*. Castro's readings led him to the conclusion that Julius Caesar was a "real revolutionary" in the context of Rome's "intense class struggle." Perusing Fidel's comments on his studies of history, literature, science, and politics, one sees how his superbly methodical mind subordinates all texts to his private interpretation (or prejudice) to confirm what he had already decided was the irreversible and irrefutable course of history. With this type of intellectual assurance about past history, it may be easier to understand Castro's absolute assurance about *future* history and the role he sees foreordained for him to play.

Pending the great swings of history, however, he had to cope with the immediate reality of Cuban politics as represented by Ramón Grau San Martín's announcement in January 1954 that he would run against Batista in the promised presidential elections in November. This annoyed Castro immensely: He felt that no responsible politician should dignify Batista's elections with his candidacy (besides, Batista would surely rig them to win), and he personally despised Grau for his gangster politics during the 1944–1948 term. Most disconcerting of all was the decision of the illegal Communist party to support Grau (who persecuted the Communists during his last presidency) against Batista (whom the Communists supported in 1940, but then were outlawed by him in 1952). It seemed to make no sense for the Communists to play any part whatever in the Batista electoral rodomontade, but the Popular Socialist Party now had a solid track record for entirely irrational and unrealistic policy decisions. Castro refrained from open cricitism of the Communists, who six months earlier had called him an adventurer.

At the prison, Fidel fell afoul of Batista with the most disagreeable consequences. On Saturday, February 12, 1954, the general came to the prison to inaugurate a new power plant, about sixty yards from the hospital ward. As soon as Castro became aware of his archfoe's presence, he gathered the men and proposed that they sing together and as loudly as possible the Movement's revolutionary hymn. Written shortly before the attacks on the army barracks by Agustín Díaz Cartaya, a young black self-taught composer who belonged to a clandestine movement cell in Havana, the piece was first called the "Freedom March." Castro, always on the alert for great propaganda possibilities, had commissioned the hymn, and what Díaz Cartaya wrote turned out to be one of the great battle songs, certainly Cuba's best. Its stanzas urged "Forward, all Cubans, may Cuba ever prize our heroism; we're soldiers united, fighting so our country may be free," damned "cruel and insatiable tyrants," and ended on the triumphant note of "Viva la Revolución!" Over the revolutionary years, the hymn became one of the powerful emotional weapons in the *Fidelista* arsenal, and Castro certainly understood it. After the assault, Díaz Cartaya (who was captured in Havana after escaping from the Bayamo attack) was asked to add a stanza about the death of "our comrades in Oriente," and the song was renamed "The 26th of July March." It is played to this day at great revolutionary rallies, and it is the musical theme of Radio Havana's short-wave world service.

Now, Fidel could not resist the temptation of baiting Batista, and the twenty-six men under the window of the hospital ward burst into song. At first, the general thought the march was a tribute by prison inmates, but as soon as he made out the lyrics, he exploded in fury and left the prison. The next morning, guards removed four *Fidelista* leaders from the ward; each was placed for two weeks in a tiny individual isolation cell, where a man had to stoop, in the mental patients' section of the hospital. One of the men was Ramiro Valdés. In the afternoon, Fidel was taken away from the group and put in solitary in a fifteen- by twelve-foot cell by the door of the hospital, across the corridor from the prison morgue. Then Díaz Cartaya was locked up in an isolation cell, the guards flaying him with ox-dick whips; on February 15 he was beaten so severely that he was left unconscious on the floor of his cell.

Castro and his companions paid a heavy price for their brief act of defiance of Batista. Fidel himself would remain in solitary until he was released from prison under the amnesty fourteen months later (his brother Raúl was allowed to join him six months before they left the Isle of Pines). In their hospital ward they were deprived of newspapers and mail, their radios, and for a time, outside visitors. Though Castro was able to maintain communication with his companions in prison and the outside world through a chain of amazingly inventive clandestine methods—and to go on directing the Movement from his cell—he no longer had direct contact with the group. He could no longer teach and indoctrinate them, and before too long the "ideological academy" quietly folded. The *Fidelistas* were now busy in intricate prison conspiracies, required to support Castro's organization and propaganda efforts.

All things considered, Fidel was not uncomfortable in his cell. It was large enough for him to pace up and down, as was his habit, and it had a toilet and a shower. He had a bookcase and a small hot plate (it took a half hour or more to cook a portion of spaghetti on it, as he tells visitors he now guides through the cell), and the metal hospital bed was equipped with mosquito netting. The big problem was light: In the daytime, a weak light seeped through a window high up on the wall. Fidel had no artificial light for forty days of his imprisonment, and he strained his eyes trying to read at night with a tiny oil lamp, as candles were not allowed. Later, electric light was installed in the cell, and he wrote a friend that enforced darkness and "the humiliation of the shadows" were the "most absurd of all the human barbarianisms I can conceive." After two months in solitary, Fidel told a woman friend in Havana in a letter that "you can't imagine how this solitude devours energy; sometimes I'm exhausted . . . when one is fatigued by everything, there is no refuge from boredom. . . . Days elapse like in lethargy. . . . I always do something, I invent my own worlds, and I think and I think, but this is precisely why I am so exhausted. How did they shrink me as a human being? . . ." Frequent rainstorms flooded his cell, and often he had to hide his beloved books inside suitcases to protect them from the water.

On February 20, 1954, Melba Hernández and Haydée Santamaría were released from the women's prison at Guanajay, after serving five months of their seven-month sentences. For Fidel Castro, this was an event of immense importance: Melba, a skilled lawyer and one of the key personages in organizing the Moncada attack, would now become his trusted agent on the outside to help him revive the Movement. From the original leadership group, Melba was the only one to whom Fidel could now turn—all the others were either dead or imprisoned. Moreover, the prison had made her even more combative. Leaving the prison, she spoke freely to waiting newsmen, and Havana radio stations (temporarily free of censorship because of the Batista electoral campaign) were able to broadcast her words: "We went to Moncada moved by a sacred love for freedom, and we are ready to give our lives for its principles." Clearly, however, the regime was attaching no importance to the opinions of a woman just out of jail.

It did not seem to pay any attention, either, to the increased activities of Mirta on behalf of her imprisoned husband, or those of Lidia, Fidel's revolution-minded older sister, and Naty Revuelta, who not only corresponded with him openly but also served as a clandestine communications channel. In the history of the Cuban revolution and in Castro's success, women have played a role that may well have been decisive, a fact that a great many Cubans do not fully appreciate even today.

Mirta and Fidelito were permitted to visit Castro several times (he wrote to a friend in June 1954 that "I have now spent more than three thousand hours completely alone, except for the briefest moments I have spent with my wife and my son"), and they used these meetings to exchange operational messages for the establishment of a revolutionary network. During June, Mirta attended a tribute at Havana's Theater of Comedy for the *Ortodoxo* radio commentator Luis Conte Agüero, going on stage to read a letter from Fidel, praising his friend and denouncing Batista as a "tyrant" and "despot." But Castro also wrote about his life in solitary: "I only have company when some dead prisoner, who may have been mysteriously hanged or strangely assassinated . . . is laid out in the small mortuary facing my cell."

When Fidel sent his first letter of instructions clandestinely to Melba Hernández on April 17, he told her that "Mirta will give you the means of communication with me every day if you want to." In insisting that "propaganda cannot be abandoned even for a minute because it is the soul of all the struggle," Castro advised Melba that "Mirta will speak to you about a pamphlet of decisive importance because of its ideological content and its tremendous accusations, to which I want you to pay the greatest attention." He urged her to maintain the "most absolute reserve" about the communications channel he had set up with Mirta. He also reported that Mirta had told him in prison "about the great enthusiasm with which [all of] you are fighting . . . I only feel the immense nostalgia about being absent."

Castro's communications system within the prison and with the outside functioned perfectly. Most of his secret messages were written in lemon juice, used as invisible ink, between the lines of open letters he wrote to friends and relatives in Havana, which the prison censor passed. When heat was applied to the white paper, the brown tracing of lemon-juice writing stood out. As part of the system, he was receiving large numbers of lemons along with other foods sent him from the outside; his appetite for lemons was never questioned by the jailers. Fidel also tells of placing tiny scraps of paper covered with nearly microscopic writing inside false bottoms of wooden match boxes. The boxes were switched back and forth between Fidel (and other rebels) and visitors as they kept lighting cigars and cigarettes, and this, too, went unnoticed by the wardens. Pedro Miret reminisced many years later about the means used by the prisoners to keep in touch with Fidel in solitary and with the outside. Their hospital ward was about 150 yards from Castro's cell in the same structure, he says, and one outside patio was used by the whole group of prisoners for exercising while Fidel had a separate patio with a wall and the roof of the hospital building between them.

"In our patio, we often played with rag balls, which we made ourselves," Miret says, "and every once in a while we would deliberately make the ball land on the roof. One of the prisoners would then ask permission [of the guards] to climb to the roof, supposedly to recover the ball, and would throw it to the other side and [Fidel] would pick it up." Written messages were concealed inside the rag balls, and Castro would hurl balls to his companions over the wall with *his* messages. Fidel, according to Miret, sometimes threw "strange objects over to us . . . once he even threw a can." Miret recalls that "we played ball all the time," and that sometimes even the common criminals using the same patio helped them recover the rag balls or threw them over to Castro. Again, the wardens never discovered the system.

At one point, Miret and Raúl Castro found a *Petit Larousse* dictionary in the prison library and discovered the deaf-mute sign-language code. Miret says it took them a month to learn the hand signals, but when, after six months of separation, Raúl was allowed to join his brother in his cell, the code served its purpose. Though the barred doors of Fidel's and Raúl's cell and those of the hospital ward were at a certain distance, Pedro Miret and Raúl managed to signal each other by sticking their hands outside the bars and using their fingers to communicate. "It was very difficult," Miret says, "but it worked," with the two of them straining their eyes. Later, Raúl perfected a system of using fingers for letters of the alphabet to make the "dialogue" easier, Miret recalls, and even today the young Castro playfully addresses him with hand signals.

Food also was a message-delivery device. Miret says that the prisoners sometimes cooked a dish and had a guard take it over to Fidel in his cell. On one occasion, the guard realized what was occurring, "but Fidel spoke with him and convinced him to continue passing the messages." Mashed potatoes were a favorite delivery method, but cigars were the best for sending lengthy pieces of information. Miret and the others learned how to unroll and reroll cigars with great expertise—the cigars were presents the prisoners received from the outside—to insert the messages inside. To ensure that guards did not discover the stratagem, three or four cigars were sent to Fidel at a time via a warden. Fidel then had to unroll all of them to find the slip of paper. To send out messages, tiny pieces of paper were put inside the half a man placed in his mouth. On visiting days, prisoners would go to meet their families holding lit cigars and put them out just in time so that the fire would not reach too far, then pass the cigars to their relatives during an affectionate embrace. In still another method the prisoners carved matchbox holders out of wood and passed them on to visitors with messages hidden inside the holders.

The internal communication was to keep Fidel informed of occurrences in the prison as well as outside, and so he could deliver instructions to his companions and send out materials from the island. The prisoners received such materials from Castro and arranged to have them smuggled out in cigars and matchbox holders.

The entire text of Castro's "History Will Absolve Me" discourse was smuggled out of prison through all these means. Fidel spent several months reconstructing from memory what he had said before the judges in the Santiago hospital lounge, and he proceeded to commit it to paper in small, barely legible writing. He completed it in June. "History" was the pamphlet of "decisive importance" that Mirta mentioned to Melba Hernández on Fidel's instructions, and now they all faced the tremendous task of getting it out, transcribing it, and publishing it. This mass of material, adding up to fifty-four closely printed book pages, was sent out in part as lemon-juice interline writing in letters by Fidel and others, and in part inside cigars. Miret says it took three months to smuggle out the entire text.

Melba and Lidia coordinated this astounding editorial enterprise. The two women and Haydée Santamaría first ironed the letters to bring out the text of the lemon-juice writing. Then the manuscript was typed by five persons working separately, including Melba and her father, Manuel Hernández, at the family apartment. Next, the typewritten pages were taken to Lidia's apartment to be collated. But meeting Fidel's subsequent instructions proved impossible. He had written Melba in mid-June that "at least 100,000 copies should be distributed within four months" throughout the island, with mailings to "all journalists, lawyers and doctors' offices and teachers and other professional groups." Castro had somehow calculated that it would cost only three hundred dollars to print each batch of ten thousand copies, completely underestimating the problems involved, but insisting that it must be done because "it contains our program and our ideology, without which nothing great may be expected." The next day, he wrote that "our immediate task . . . is not to organize revolutionary cells to build our ranks—that would be a grievous error—but our task now is to mobilize public opinion in our favor, to spread out ideas and win the people's backing."

For this reason he was anxious to publish "History" in a mass printing. But lack of money and the need to print the pamphlet clandestinely limited the circulation to only 27,500 copies, and it was not until the end of the year that even this number could be distributed. The public impact was limited.

Publishing "History" was not the only task with which Melba was charged. Castro ordered her to travel to Mexico to establish contacts with Movement members exiled there, notably with his university friend Lester Rodríguez. Fidel was very disturbed over the prospects that other opposition groups, especially the wealthy faction headed by the ousted president, Carlos Prío, might achieve success and attract others to its banner. This would naturally undermine Castro's ambitions to bring Cubans together in support of his proposed revolution. Melba's mission in Mexico was to persuade Castro's friends abroad not to follow Prío. In May Batista had granted amnesty to the ex-president and the other signers of the 1953 Montreal Pact, but this excluded "those who took part in the attack on the Moncada Garrison," and Fidel had every reason to feel threatened politically.

The isolation also began to tell on him. Apart from a few visits in the early part of 1954 by Mirta and Fidelito, and an unexpected one in April from Waldo Medina, a Havana judge whom he had known at the university, Castro saw nobody but prison guards. This monotony was broken on the five occasions when he was taken to the court in Nueva Gerona, the capital of the Isle of Pines, to testify in the criminal cases he had brought against government officials for the killings at Moncada. But the prison director overrode the instructions from Interior Minister Ramón O. Hermida to let the Castro brothers appear before courts in Santiago and Havana, leaving Fidel deeply embittered.

Having completed the "History" text and issued secret instructions to Melba and other Movement members, he turned again to his marathon reading endeavors. He reported in a letter to Havana that he fell asleep finishing Kant's *The Transcendental Aesthetics of Space and Time*, remarking that "of course, space and time disappeared for a good while from my mind." He complained in another missive that "I have nothing at all about Roosevelt's New Deal," and that "I mainly want information on him: in agriculture, his price-raising policies for crops, the protection and conservation of soil fertility, credit facilities, the moratorium on debts and the extension of markets at home and abroad; in the social field, how he provided more jobs, shortened the workday, raised wages and pushed through social assistance to the unemployed, the old and crippled; and, in the field of the general economy, his reorganization of industry, new tax systems, regulation of the trusts and banking and monetary reforms." Five years later, when his revolution triumphed, his policies covered every single point in the New Deal legislation he had studied in prison—before he turned to Marxist formulas. In the meantime, Fidel wrote a friend, "I can't stop thinking about these subjects, because—sincerely—I would revolutionize this country from end to end with joy . . . I am convinced that one could make happy all its inhabitants. I would be disposed to bring upon myself the hatred and the ill-will of a thousand or two men, among whom some parents, half of my friends, two thirds of my colleagues, and four fifths of my former college companions! . . ."

Castro's reading was so all-embracing—from Kant and Einstein to Franklin Roosevelt's legislation to Lenin's *The State and the Revolution*—that it would be imprudent to draw from it firm ideological conclusions concerning his convictions at the time. In 1975, Castro told Lionel Martin that after Moncada "I knew what was the final objective. My program was the antechamber of a socialist revolution. To reach the third floor, one has to start from the first floor." Then he quoted to Martin the phrase of José Martí that "to achieve certain things, they must be kept concealed [because] to proclaim what they are would raise difficulties too great to attain them in the end." But again, these are hindsight pronouncements, fitting an existing political state of affairs, and there is no way of establishing what Castro was thinking in his prison cell.

Two major revolutionary events occurred in the world during the spring of 1954, leaving varying impressions on Fidel. In May France's colonial rule in Indochina crumbled with the victory of the Vietminh Communists and nationalists at Dienbienphu. There is nothing in Castro's writings at the time to suggest an exaggerated interest in this Communist guerrilla triumph, militarily or ideologically.

On June 17 a right-wing military force, organized and financed by the Central Intelligence Agency and the United Fruit Company, invaded Guatemala to overthrow the leftist regime of President Jacobo Arbenz Guzmán. Arbenz had begun nationalizing American-owned agricultural land, imposing advanced social legislation, and giving free vent to anti-American sentiment. The CIA intervention in Guatemala, the first such act in Latin America since the 1930s, had an enormous impact on Castro, confirming all the Martí warnings and corroborating the "historical fatalism" theory that proclaims that nothing may happen in the region without the permission of the United States. In the long run, the Guatemalan intervention may have marked Castro more deeply than Marxist and Leninist theories. Later, Fidel would have firsthand accounts of the Guatemala affair: one from Ñico López, his early Movement companion and Bayamo fighter who had exiled himself to Central America, and the other from the Argentine physician Ernesto Guevara, who had been working under Arbenz. Castro and Guevara were to meet before long. In his prison cell, Fidel was photographed reading a magazine report on Guatemala.

On the evening of Saturday, July 17, 1954, Fidel Castro suffered a devastating emotional blow. Listening to the radio newscast, he heard an announcement that the Interior Ministry had terminated the employment of Mirta Díaz-Balart, his wife. He never knew that she was employed by the ministry, or received sinecure payments from it, even though her brother Rafael (once Fidel's university friend) was vice-minister of the interior in the Batista regime. Publicly as well as clandestinely supportive of her imprisoned husband and his causes, Mirta was believed to be totally loyal to Fidel, notwithstanding her family ties with the government. In fact, Castro at first refused to believe that she had indeed accepted ministry money. The very evening of the broadcast, he wrote her that she should immediately enter a libel and defamation suit against the interior minister, suggesting that perhaps someone else was forging her signature to collect such payments.

Mirta's parents were dead, but she and Fidelito were apparently being supported by the Castro family and by several wealthy friends of Fidel's. It made no sense to him that his wife would, in effect, betray him politically, and in a letter he wrote that same night to Luis Conte Agüero he said that this was "a machination against me, the worst, the most cowardly, the most indecent, the vilest and intolerable." He wrote Conte Agüero that Mirta was too intelligent to allow herself to "be seduced by her family, consenting to be on the government payroll, no matter how hard was her economic situation . . . she has been miserably calumnied." Castro asked Conte Agüero to find out the truth from her brother Rafael, but centered his wrath on Interior Minister Hermida, saying that "only someone as effeminate as Hermida, in the last stages of sexual degeneration," could turn to such behavior of "inconceivable indecency and lack of manhood." This reference, correct or not, was the first recorded expression of his anti-homosexual obsession. Fidel also accused Hermida of having "put in my mouth" declarations about the improvement in his treatment in prison. He told Conte Agüero that "now wrath blinds me and I almost cannot think," but that the radio commentator should take whatever measures he considered convenient, and that he was ready to challenge his brother-in-law to a duel. "The prestige of my wife and my honor as a revolutionary are at stake," he wrote. "Let them see me dead a thousand times rather than having me suffer impotently such an offense!"

But Fidel was wrong. His sister Lidia informed him four days later that Mirta had been on the Interior Ministry's payroll, and that now she demanded a divorce. There is no known explanation for this episode. Mirta, who the following year married an *Ortodoxo* politician named Emilio Nuñez Blanco, has never publicly discussed her personal life, and Fidel himself to this day ignores what happened. Mirta left for the United States with Fidelito immediately after the break with her husband, and presumably they never saw each other again (she has lived in Spain since the revolution, and now she visits Fidelito and his children quietly in Cuba about once a year). Castro, too, insisted on a divorce when he learned the truth. He answered Lidia with a brief note: "Do not be concerned with me; you know I have a steel heart and I shall be dignified until the last day of my life."

On July 26, the first anniversary of Moncada, Interior Minister Hermida and two other cabinet ministers suddenly visited Castro in prison in a display of courtesy and cordiality that bewildered him. But that same day in Havana, the police violently dispersed a university commemoration, organized (on Fidel's orders) by Melba and Haydée. Castro's account of the meeting with Hermida stressed that they both agreed there was nothing personal in their political differences, and that when Fidel protested the incident involving Mirta, the minister blamed her brother Rafael, whom he called "an irresponsible child." The central fact, however, was that Castro was now regarded by the government to be important enough politically to receive personal apologies from three cabinet ministers; Hermida flattered him by saying that no man in Cuba had a clearer political reputation than

Castro, and told him, "Don't be impatient: I, too, was a political prisoner in 1931 and 1932."

Nevertheless, Fidel remained personally shattered. Rafael Díaz-Balart publicly denounced Hermida for the prison visit, and the two had to resign, but this was no consolation for Castro. He wrote Conte Agüero that "I live because I have duties to fulfill. . . . In many terrible moments I had to suffer in one year, I thought how much more pleasant would it be to be dead. I consider the 26th of July [Movement] to be very much above my person, and the moment I know that I can no longer be useful to the cause for which I have suffered so much, I will deprive myself of life without hesitation, especially now that I no longer have a personal cause to serve."

In the divorce battle between their lawyers, Fidel demanded that the first condition be the return of Fidelito to Cuba and his enrollment in a school he would choose for his son, now five years old. He wrote Lidia that "I refuse even to think that my son may sleep a single night under the same roof sheltering my most repulsive enemies and receive on his innocent cheeks the kisses of those miserable Judases. . . . To take this child away from me, they'll have to kill me. . . . I lose my head when I think about these things." Castro continued to insist on having Fidelito's custody when he left prison, telling his lawyers that if a court ruled against him, "it would reaffirm my principles and my determination to fight until death to live in a more decent republic." In April 1955 he issued an ultimatum: Fidelito had to be a boarder in a Havana school by April 1 or he would block the divorce. Fidel and Mirta's divorce decree was granted the next year, when he had already left Cuba, but the struggle over Fidelito would continue for years—until Castro was in a position to win it.

Meanwhile, Castro had become Cuba's most famous political prisoner and, increasingly, a factor in national politics, proving that an isolation cell can be the steppingstone to respectability and leadership. Early in June, *Bohemia* published a lengthy interview with Fidel, illustrated with seven photographs showing him in his cell and in the prison library. This was the first time he had received such massive national exposure, and he minced no words about the crimes of Batista in the interview and his own revolutionary plans. The Hermida visit to him on the Isle of Pines resulted in a major crisis over Castro in the Batista cabinet, and the government now handled him with kid gloves.

In August Raúl was allowed to join Fidel in the cell, ending his total isolation (he was still kept apart from the other inmates) and providing him with a permanent audience. Fidel reported that their cell was considerably enlarged, that they were given a large patio, that prison personnel took over cleaning chores, and that "we don't have to get up until we want to

. . . we have plenty of water, electric light, food, and clean clothes—all free . . . we don't even pay rent." This was his political assessment:

"Our moment is coming. Before, there were a handful of us; now, we must join with the people. Our tactics will be different. Those who view us a group will be sadly mistaken. We will never have a group mentality or group tactics. Now, moreover, I can dedicate myself body and soul to my cause. I will put all my energy and time into it. I will begin a new life. I am determined to overcome all obstacles and fight as many battles as may be necessary. Above all, I see our path and our goal more clearly than ever. I haven't wasted my time in prison, for I've been studying, observing, analyzing, planning, and training the men. I know where the best of Cuba is and how to look for it. When I began, I was alone; now, there are many of us."

Castro was thinking about creating a national movement to replace the pre-Moncada clandestine activities, and he was intrigued by the proposal Conte Agüero sent him in August for establishing "a civic movement that is becoming a pressing need." He replied quickly that he agreed "on the need," but, most pointedly, he warned against a situation in which multiple opinions and interests would have to be accommodated. He believed as always in revolutionary unity under his command, and he told Conte Agüero that the first step would have to be the release of the 26th of July prisoners. "A perfectly disciplined nucleus . . . will be tremendously valuable in terms of training cadres for insurrection or civic organization," he wrote. "It is evident that a great civic and political movement must have the necessary strength to win power, by peaceful or revolutionary means; otherwise it will run the risk of having that power snatched away . . ."

Students of the Cuban revolution have questioned whether at that stage Castro was applying Leninist or "caudillist" principles to the organization of a vertical revolutionary movement; a simpler answer is that he was holding out for absolute leadership, without which he correctly believed nothing could be accomplished in Cuba. He repeated that the "propaganda and organization apparatus must . . . unmercifully destroy all those who try to create trends . . . or rise up against the movement. . . . We must have our feet solidly on the ground, without ever sacrificing the greatest reality of principles."

Fidel Castro had just turned twenty-eight years old, and even while still in prison, he loomed as a major national political leader. When former President Grau, again a candidate for office against Batista, spoke at a rally in Santiago in October, the crowd began chanting Fidel's name; Grau's response was that as soon as he was elected, he would declare a full amnesty, including for the "boys of Moncada." Grau soon realized he could never defeat Batista's machine and he pulled out of the race. But amnesty for the Isle of Pines prisoners was turning into a nationwide campaign, and Batista knew he could not ignore it forever.

Running unopposed, Batista was elected as "constitutional" president on November 1, 1954, an event that plunged Cuba into considerable depression. But the United States, which could not—or would not—appreciate the pitfalls of Cuban politics and the rising revolutionary potential, rushed to embrace the dictator again. After all, he had fully supported the Guatemalan operation. Accordingly, Vice-President Richard Nixon came to Havana in February 1955 to toast Batista at a black-tie palace reception. He was followed soon after by CIA Director Allen W. Dulles, the author of the intervention in Guatemala. Neither of them had ever heard of Fidel, but before too long the young Cuban would cost Dulles his career.

The proamnesty campaign was launched early in 1955 by a committee of the prisoners' mothers who issued a manifesto "to all Cuban mothers" entitled "Cuba, Freedom for Your Sons." It soon became transformed into the Relatives' Amnesty Committee for Political Prisoners; Fidel's sister Lidia was a militant leader; a young architecture student from Santiago, Vilma Espín, became involved in the effort; and Celia Sánchez, the daughter of a revolution-minded doctor at an Oriente sugar mill, organized the deliveries of canned meat, chocolate bars, and other delicacies to the men on the Isle of Pines. All these women would be crucial both to the revolution and to Fidel's life.

Castro inevitably had dramatic ideas to enhance the amnesty effort. On January 1, 1955, he instructed Ñico López and Calixto García, who had escaped from Bayamo and exiled themselves first in Central America and then in Mexico, to present themselves to Cuban emergency courts as "Moncada fighters." This, he wrote, would force the reopening of the basic Moncada trial, "and we would arouse the nation against Batista when he is about to assume power on February 24." Fidel's concept was to make López and García the subject of a huge propaganda campaign, and he sent them for their signature, public statements to be given to the principal radio stations and newspapers on the eve of their return to Cuba. Then, the two men would be met by journalists for interviews, although "you will doubtless be arrested immediately." Castro urged López and García to talk other rebel exiles into surrendering to Cuban courts for trial as well, "but make sure it appears as your own idea; I don't want to put any moral pressure on them. . . . If anybody else decides to follow in your footsteps and return before February 24, the government will go crazy, just when it wants to make a show of political normalcy at all costs, and this might become a decisive factor in forcing it to sign the amnesty." As an afterthought, Fidel added that "if by any chance, they don't want to arrest you when you arrive

. . . present yourselves . . . before the Provisional Court in Santiago, stating that you 'want to suffer together with [your] imprisoned comrades,' and they will be forced to act." The two men, however, failed to return before the amnesty.

Batista was obviously feeling the growing pressure to grant amnesty to Castro's contingent, but he was sufficiently concerned about leftist activities in the country to accept United States' advice and create a special intelligence agency (BRAC) to combat communism. On the day of Batista's inauguration as constitutional president, an impressive group of traditional political leaders, editors, and intellectuals signed a "Public Appeal," demanding "liberty for the political prisoners and guarantees for the return of all those in exile." On March 10, the third anniversary of the Batista coup, amnesty bills were presented to both chambers of the Cuban congress, and the regime made it known it would give them its blessings if the *Fidelistas* promised not to attempt fresh insurrections.

Castro replied with a statement, signed by all his fellow prisoners, rejecting the conditions. He wrote: "The Pharisees once asked Christ whether or not they should pay tribute to Caesar. Any reply he made would be bound to offend either Caesar or the people. The Pharisees of all times have used that trick. Today, they are trying to discredit us in the eyes of the people or find a pretext for keeping us in prison. I'm not at all interested in convincing the regime to grant the amnesty. . . . The regime commits a crime against our people and then holds us hostages. . . . For, today, we are more than political prisoners; we are hostages of the dictatorship. . . . Our personal freedom is an inalienable right as citizens. . . . We can be deprived of these and all other rights by force, but no one can ever make us agree to regain them by unworthy compromise. We won't give up one iota of our honor in return for our freedom."

Fidel had learned the hard way that he must never compromise with any person or government, and now he knew he had a winning hand and that he could wait for an amnesty on his terms—which meant no terms at all. He had also learned José Martí's counsel that a revolutionary must always be patient. In April proamnesty demonstrations occurred in Havana and other cities, and the Cuban press, now free of censorship, was openly in favor of an amnesty, denouncing the regime for keeping the Moncada men in prison. Angered by Castro's defiance, the regime had the penitentiary's administrative council sentence Fidel to thirty days in solitary for illegally sending out his statement, published in *Bohemia* magazine, against a conditional amnesty; the press now seized on this act to attack the government. Batista had to accept Castro's terms to prevent a major crisis.

On May 3 the congress approved the amnesty bill, and Batista signed it on May 6, "in honor of Mother's Day." On Sunday, May 15, 1955, exactly at noon, Fidel Castro and all his companions were freed from the Isle of Pines prison—less than two years after the assault on Moncada. And he came out swinging and fighting. A now famous photograph shows him with his right arm raised in a salute as he walked out of the prison's administration building with Raúl, Juan Almeida, and Armando Mestre, followed by the others. He wore his old gray wool suit (he had sent the dark-blue suit of the trial back home) with an open-collared white shirt; he had written Lidia earlier not to waste money on a new *guayabera* and slacks. From the prison, Fidel was driven to the home of his companion Jesús Montané, and then to the Isle of Pines Hotel in Nueva Gerona to hold his first press conference and sign autographs; as he had so often emphasized, propaganda must not stop even for a moment.

In the evening the ex-prisoners boarded the steamer *El Pinero* for the crossing to the fishing port of Batabanó on the mainland. Their friends on the island shouted farewells as the vessel left, and some three hundred supporters greeted them on arrival at five o'clock in the morning on Monday. The crossing was the first opportunity Fidel had in fifteen months to see his prison companions, and they used these first hours together to plan the launching of the 26th of July Movement as a revolutionary organization of the masses. But Castro also found time to draft the "Manifesto of the People of Cuba from Fidel Castro and the Combatants," declaring that their war was just beginning:

"As we leave the prison . . . we proclaim that we shall struggle for [our] ideas even at the price of our existence. . . . Our freedom shall not be feast or rest, but battle and duty for a nation without despotism or misery. . . . There is a new faith, a new awakening in the national conscience. To try to drown it, will provoke an unprecedented catastrophe. . . . Despots vanish, peoples remain . . ."

The document was published in the Havana mass-circulation daily *La Calle* on the morning the Isle of Pines prisoners arrived at the capital's railroad station from Batabanó. When the train entered the station, a crowd broke into the national anthem, and Castro was carried on the shoulders of his admirers to the street. Now Fidel wore a white *guayabera* and gray slacks, and he carried a large Cuban flag that was handed him on arrival. Speaking to journalists, Castro announced that he would remain in the country, and not become an exile, in order to be active politically in the *Ortodoxo* party. This was part of his short-term strategy: The *Ortodoxos* were well organized and well disposed toward him (the Moncada rebels were mainly recruited from among young *Ortodoxo* militants), and they provided a natural base for the creation of the 26th of July Movement. Nor did Castro wish to antagonize the army as he returned to the political scene; he said that "I am not an enemy of the army, simply an adversary," and went out of his way to praise the army officer in charge of security at the island prison (whom he warmly embraced as he left the penitentiary).

Castro's blueprint for the new Movement included other young anti-Batista activists, and he wasted no time contacting those who were released from Havana prisons under the terms of the general amnesty. Among these men, Castro found two recruits who would play key roles very soon: Armando Hart Dávalos, a lawyer, and Faustino Pérez Hernández, a physician. Naturally, his sister Lidia (with whom he went to stay in Havana), Melba, and Haydée were among the first people Fidel greeted and embraced. So was Naty Revuelta, the only person in Havana who knew beforehand that Fidel was on the verge of assaulting Moncada. Raúl Castro, who rented a room downtown with Pedro Miret, traveled to Birán late in May to see his parents and spend a week with them. Fidel was too busy with conspiracy to go to Oriente.

Though Castro declared on his return from prison that he would devote himself to *Ortodoxo* party politics, he knew that Batista would not tolerate his type of opposition, and that armed struggle would be the only alternative. He said so to his followers, and they immediately began organizing the 26th of July Movement clandestinely (Fidel first proposed this name when they were aboard the steamer from the Isle of Pines). At the same time, Castro embarked on a public campaign that could not fail to make him intolerable to Batista; he believed in self-fulfilling prophecies. He said he and his associates would serve as "guinea pigs" to test Batista's promise of constitutional guarantees for the opposition, adding that "I shall be the first victim of cowardice if there are no such guarantees" and that "I have been informed that acts of aggression are being prepared against me and my companions."

Meanwhile, violence once more exploded in Havana, with assassinations, bomb bursts, fires, and beatings. Students fought the Batista police, and the authorities responded in kind. Castro jumped into the fray with articles in *La Calle* and *Bohemia* denouncing the regime, and with fiery radio broadsides. When Pedro Miret was arrested on vague charges, Fidel issued a statement charging that "amnesty is a bloody hoax," and rushed to court to defend him. The regime banned Castro from the radio, and so he raised the drumbeat of his accusations in the press. At every opportunity, Fidel reminded Cubans of the massacre of his men on July 26, and when Colonel Chaviano, the Moncada commander, published his own version of events ("we did our duty"), Castro lashed back with a savage *Bohemia* article under the headline CHAVIANO, YOU LIE! Then, trying to divide the military, in another article he paid tribute to army officers, whose conduct he described as gallant.

Making his home with Lidia, who fed him and washed his only *guayabera* shirt every day, Fidel was in perpetual motion, in a paroxysm of revolutionary activity, as if to make up for the twenty-two months behind prison bars. He delivered speeches, wrote articles, and held meetings with his followers to give shape to the nascent 26th of July Movement. Armando Hart and Faustino Pérez, who had belonged to the short-lived moderate National Revolutionary Movement (MNR), now became the channel for recruiting MNR activists for the 26th of July Movement. One evening, several friends listened to Fidel's account of the battle of Moncada—and to his conclusion that "now I have enough experience; with new resources, we shall not fail the next time."

Castro was increasingly concerned about his safety as he went on defying Batista. Starting in the first days of June, his brother Raúl, Ñico López (who had just returned from exile in Mexico), and Jesús Montané moved into Lidia's apartment with their weapons to protect him. After a week or so, however, Fidel decided that he should not sleep two nights in a row in the same place, so he kept moving among the homes of various friends. But he never ceased hitting at Batista. After the general delivered a speech on June 4, declaring that "the government wants to be patient" but it is taken by some people as a "weakness," Castro shot back with an article in *La Calle* calling him "dishonest." He titled the article "Murderous Hands!"

The severe beating of Juan Manuel Márquez, an opposition leader, and the killing of Jorge Agostini, a former naval officer who had just returned from exile, marked open warfare between Batista and the Castro-led opposition. Castro accused the government of murdering Agostini, and that same night seven bombs exploded in Havana; Castro said it was the work of Batista's agents. The authorities countered by accusing Raúl Castro of having placed a bomb in a movie theater, and Fidel charged in a brief before the emergency court in Havana that the regime was planning to murder him and his brother. He also called for a strike in support of railroad workers who had suffered a pay cut. On June 15 the government forbade *La Calle* to print any more Castro articles (his last one was titled "Before Terror and Crime"); in effect, Fidel was silenced politically. *La Calle* was closed by the police the next day.

Castro's unpublished article was called "You Cannot Live Here Anymore," and it hinted heavily that he was preparing to go abroad, though it did not give the slightest indication that he planned to prepare an armed insurrection from exile. On June 17, Fidel instructed Raúl to seek asylum at the Mexican embassy in Havana; there were two court warrants out against him, and an assassination was feared. On June 24, Raúl left for Mexico, the first Movement rebel to take the long road to invasion.

Before leaving, however, Raúl attended a secret meeting at an old house on Factoría Street near the Havana harbor where the National Directorate of the 26th of July Movement was organized on the night of June 12. Fidel, realizing he had to leave Cuba soon himself, felt it was imperative to leave behind a well-functioning organization to support the insurrection. He had learned the lesson of Moncada. This was the meeting Armando Hart mentioned thirty years later when he was reminiscing about the foundation of the Movement; every person who had been present, he said, was either still with the revolution or dead. The eleven-member National Directorate was composed of Fidel Castro, Pedro Miret, Jesús Montané, Melba Hernández, Haydée Santamaría, José Suárez Blanco, Pedro Celestino Aguilera, Ñico López, Armando Hart, Faustino Pérez, and Luis Bonito. The Moncada fighters were in the majority, but new blood had been added to the Movement.

On July 6, Fidel made his last preparations to leave Cuba. A Mexican tourist visa had been discreetly obtained for his passport, and Lidia sold her apartment refrigerator so that he could travel with a small amount of cash. Then she packed his suitcases with more books than clothes. On the afternoon of July 7, Castro left the apartment in an automobile with his sisters Lidia and Emma, his son, Fidelito (whom Lidia had brought along from school), and a woman lawyer. At the airport, Fidel embraced and kissed Fidelito, then boarded Flight 566 of the Mexican Aviation Company. He left behind this message, published by *Bohemia* in 250,000 copies:

"I am leaving Cuba because all doors of peaceful struggle have been closed to me. Six weeks after being released from prison I am convinced more than ever of the dictatorship's intention to remain in power for twenty years masked in different ways, ruling as now by the use of terror and crime and ignoring the patience of the Cuban people, which has its limits. As a follower of Martí, I believe the hour has come to take rights and not to beg for them, to fight instead of pleading for them. I will reside somewhere in the Caribbean. From trips such as this, one does not return or else one returns with the tyranny beheaded at one's feet."

CHAPTER

7

Fidel Castro arrived in exile in Mexico with the clear and specific purpose of organizing and training a rebel force that would land in Cuba to engage in guerrilla warfare in the Sierra Maestra. The guerrilla army would then defeat the Cuban armed forces, depose General Batista, and proclaim a revolutionary government on the island. To achieve his aim, he had at his command, as he set foot in Mexico, a few friends, limitless tenacity, and tremendous powers of persuasion.

"Sitting in front of me, Fidel Castro was shouting at me in my own house, gesticulating violently, as if we were in the midst of a great quarrel: 'You are a Cuban, you have the absolute duty to help us!'" This is the recollection of the late Alberto Bayo, the Cuban-born veteran officer and guerrilla specialist in the Republican army in the Spanish civil war, whom Castro tracked down immediately in Mexico City where the old soldier lived in self-exile.

In his book on directing the training of the *Fidelista* expedition, Bayo describes the encounter: "The young man was telling me that he expected to defeat Batista in a future landing that he planned to carry out with men 'when I have them,' and with vessels 'when I have the money to buy them,' because at the moment he was talking to me, he had neither a man nor a dollar. . . . Wasn't it amusing? Wasn't it a child's play? So he was asking me whether 'I would commit myself to teach guerrilla tactics to his future soldiers, when he had recruited them and when he had collected the money to feed, dress, and equip them, and buy ships to transport them to Cuba.' Come now, I thought, this young man wants to move mountains with one hand. But what did it cost me to please him? 'Yes,' I said. 'Yes, Fidel, I promise to instruct these boys the moment it is necessary.' Fidel Castro added, 'Well, I am going to the United States to gather men and money, and when I have them within seven or eight months, at the end of this year, I'll come back to see you and we shall plan what we have to do for our military training.' . . . We shook hands, and all this seemed impossible to me."

Castro must have heard of Bayo's reputation through the Latin American revolutionary grapevine when he reached Mexico City on the morning of July 8, 1955, after a night in Veracruz. He was wearing his old gray wool suit, and on meeting his brother Raúl and several other Cuban refugees the first evening, he admitted that "I almost wept when I took the plane" from Havana. But he instantly designed a plan, as he said later, to "reach influential persons in this country, whose friendship and sympathy could be useful." General Bayo was among the very first such persons whom Fidel went to see.

Alberto Bayo, frequently but falsely described as a Communist agent, was exactly the man Castro needed for his venture if it were to get off the ground. The amateurish military training of the surviving Moncada rebels was entirely inadequate for the invasion he was planning, and their lack of professionalism (including his own) could not be remedied by inspired readings of the memoirs of José Martí and the nineteenth-century guerrilla experiences of Generals Máximo Gómez and Antonio Maceo. What the 26th of July Movement had to have now was expertise in modern guerrilla warfare in tough mountain terrain against the sophisticated Batista arsenal. Bayo, an aging, white-haired man when Castro approached him in 1955, had fought for eleven years with the Spanish Army against Moorish guerrillas in the North African Rif in the 1920s. He campaigned against the fabled Abd-el-Krim, went on to study guerrilla warfare at the Spanish Military Academy at Toledo, then taught his favorite subject at the Salamanca staff school. During the Spanish civil war, he argued for the greater use of guerrilla forces by the Republicans against Franco's better-armed Nationalists. He reminded his superiors that the guerrilla was invented in Spain, and used first to expel the Moors in the fifteenth century, then Napoleon's French legions in 1808. Now General Bayo was telling Fidel Castro that "the man of the guerrilla is invincible when he can rely on the support of the peasants in place."

The Spaniard had spent his years in exile training both leftist and anti-Communist rebels throughout the Caribbean for assaults on dictatorships in Nicaragua and the Dominican Republic. In the mid-1950s, Bayo was teaching French and English at the Latin American University in Mexico, serving as professor at the Military Aviation Mechanics School, and running a furniture factory. Bayo told Castro that he could devote only three hours daily to his rebels after finishing his normal work, but Fidel protested: "No, General Bayo, we want from you the entire day. You must give up all your other occupations, and devote yourself fully to our training. Why would you want a furniture factory if inside a short time you will come with us, and we shall be together victorious in Cuba? . . ." The sixty-five-year-old general wrote that the young Castro "subjugated me, I became intoxicated with his enthusiasm, and he conveyed his optimism to me," and "then and there I promised Fidel to resign from my classes here and to sell my business."

Bayo never collected the approximately six thousand dollars he had been promised for the sale of his factory, and he lost a monthly income of around three hundred dollars. Unbeknownst to his wife, Bayo said, Castro gave him sixty-five dollars monthly "to continue the fiction" that the factory produced an income, but discovering the truth three months later, she took on additional teaching duties to keep the family in food. Bayo refused further payments from the nearly destitute Cubans; he later wrote books to raise money so he could repay Castro the $195 he had given him at the beginning. Having signed up Bayo, Fidel could start moving ahead with the organization of his insurrection.

Despite his lack of material resources, Castro had a fairly precise battle plan in mind even before leaving Havana. Pedro Miret, who worked closely with Fidel throughout the whole period, says that Castro's decision to go to Mexico was based on a decision he had already made to land a rebel force in Oriente. He recalls that "all this had already been thought out, including the landing zone as well as the place where we would go in the Sierra. . . . I knew about it even before Fidel left for Mexico." Miret explains that the concept of going to the Sierra evolved after Moncada (Castro had actually wanted to launch a guerrilla operation in the mountains above Santiago following his escape from the barracks), and that the idea of operating in the mountains was linked to "mass activities" elsewhere in Cuba. He says that "without mass support activities, there would be no possibility of triumph." This was one of the lessons of Moncada. Miret makes the point that after July 26, a new attack on army facilities was out of the question.

According to Miret, he and Fidel had narrowed down the landing area to a zone between Niquero on the west coast of Oriente province that juts out like a huge peninsula into the sea, and the small port of Pilón on the south coast, some forty miles away along the tortuous seashore. They had formulated the idea of landing in a specific spot of Oriente just before Castro departed for Mexico, and he acted rapidly to refine it. Miret had traveled to the Oriente peninsula in September, studying the terrain, the beaches, and the surf along the coast between Niquero and Pilón. He was accompanied by Frank País, the twenty-year-old regional coordinator of the 26th of July Movement in Santiago and one of its most influential leaders, and by Celia Sánchez, the dark-haired thirty-four-year-old daughter of a physician in the town of Media Luna, who would later become Fidel's most intimate companion and associate in the Sierra.

Miret's inclination was to land on the beaches near Pilón, and Celia obtained from the navy office there depth and tide charts for that section of the coast after their little group had inspected it. Celia was one of five daughters of a patriotic and radical-minded physician named Manuel Sánchez. One day when she was an adolescent, her father took her up Pico Turquino in the Sierra Maestra, Cuba's tallest mountain, to place a bust of José Martí at the peak. The Sánchezes were an *Ortodoxo* family, and Dr. Sánchez had seen enough human misery at the sugar mill where he practiced as a company doctor to develop very strong notions about social justice. Celia took after him politically, and she traveled to Havana following the release of Fidel Castro and his companions from the Isle of Pines prison to see if she could be of help to the Movement. Celia visited the *Ortodoxo* party headquarters, apparently hoping to meet Fidel, but she did not find him there. According to friends, Celia wanted to persuade Castro that he should pursue his war in the Sierra Maestra, and she even brought some maps with her. However, she met Pedro Miret, and he remembers discussing the subject with her. In September, Miret and Frank País, who knew Celia very well, contacted her for assistance with the coast survey. Obtaining the charts was the first service she rendered the revolution and Fidel Castro. Miret then flew to Mexico to hand the maps, the charts, and all other relevant information to Castro.

Fidel approved the proposal to land near Pilón (though plans were changed later), but it remained a closely guarded secret. Miret flew back to Cuba to continue organizing the 26th of July Movement inside the country, and to begin the preparations for the landing expected a year or so later. He kept in touch with Celia, and flew again to Mexico at the start of 1957 to bring Castro up to date on all the developments. Miret was able to report that the Movement, now known by its initials, "MR-26-7" (*Movimiento Revolucionario 26 Julio*), was acquiring personnel and importance.

Being clandestine, the Movement at this stage was small by definition, and potential members were carefully screened by the National Directorate Castro had left behind in Cuba. Specific functions, from recruitment and fund-raising to propaganda and preparations to supply a guerrilla war in the mountains, were assigned Movement members, and there were "coordinators" on the national, provincial, and municipal levels. The *Ortodoxo* party, on the other hand, was perceived by Fidel at that juncture as a massive political base for his revolutionary enterprise. He was careful to maintain his identification with the *Ortodoxos* (on his return from prison he had promised to be active in the party), but from Mexico he went over the heads of the leaders to the rank-and-file to urge commitment to armed insurrection.

Eager to appear as the legitimate heir to the great Eddy Chibás, in mid-August Castro sent a message to the Congress of *Ortodoxo* Militants, then being held in Havana, telling them they had a central role to play in "the struggle for national liberation." The message, read to some five hundred delegates by Faustino Pérez, a member of the Movement's National Directorate, urged the party to reject Batista's offer of congressional elections as "a peaceful solution" since it was a sham. Castro told Batista to resign "because the whims of an adventurer cannot be put ahead of the interests of six million Cubans; if you do not resign, and if you keep trying to impose yourself through force, the six million will use force, and we shall sweep you and your clique of infamous murderers from the face of the earth." For the opposition, Fidel said, there were two alternatives: to "cross their arms and cry like Mary Magdalene because they lack the courage to demand anything with honor," or to take the road "called revolution, the right of all people to revolt against oppression!" The delegates, on their feet, broke into chants of "Revolution! . . . Revolution! . . . Revolution!" As far as Fidel was concerned, the *Ortodoxos* had signed up for "armed insurrection," and his Movement was "the revolutionary apparatus of *Chibásismo*." Thus he merged his two organizations politically.

The Communists had no interest in Castro's insurrection, still believing they could control or coordinate militant opposition to Batista in Cuba. Learning that Fidel was about to leave the country, the party dispatched Raúl Valdés Vivó, the secretary general of Socialist Youth at Havana University, to talk him out of it. The Communists preferred that Castro stay in Cuba and work with them to organize a united political front against Batista, again underestimating his intelligence and his ego. Fidel replied that any mass movement had to be built around a direct confrontation with the enemy, and that he was going away to prepare the ground for a revolution. What remains unclear so many years later is what part, if any, Raúl Castro played in the Communist strategy—if indeed there was a long-range strategy. Though a party member since the eve of Moncada, Raúl's loyalty seemed to be first and foremost to his brother, at least until these two loyalties became fused in the aftermath of the revolution. In the meantime, Raúl was in Mexico, a member of Fidel's personal circle of future invasion leaders.

They commemorated the second anniversary of Moncada by laying a wreath at the monument to the Heroic Children of Chapultepec, a Mexican patriotic shrine, and Fidel expanded his efforts to develop contacts with influential Mexicans, radical Latin American exiles living in Mexico, and Cuban refugees. For Castro the new revolutionary activity was around the clock: conferring, scheming and conspiring, reading and writing, looking for money and recruits, keeping an eye on the political life in Cuba, and giving shape to the 26th of July Movement.

Fidel's first home in Mexico City was a tiny room overlooking a court-yard in a cheap downtown hotel. He did all his reading and writing there. For lunch and dinner, he had to walk from wherever he happened to be to the apartment of María Antonia González, a Cuban married to a Mexican wrestler named Avelino Palomo, in the old section of the city on Emparán Street. María Antonia's apartment was the haven, shelter, kitchen, and headquarters for all Cuban political refugees in Mexico, and where Raúl Castro went to live on his arrival. The Castro brothers had known María Antonia back in Cuba, and during the Mexican exile she became the fairy godmother of the *Fidelistas,* one more in the galaxy of Cuban revolutionary women who made the ultimate victory possible. Her generosity kept the brothers alive. Fidel was receiving eighty dollars monthly from Cuba, and Raúl only forty dollars.

A week after arriving in Mexico, Castro wrote Faustino Pérez through a secret channel that he was studying the Mexican revolutionary process under the presidency of General Lázaro Cárdenas. In the 1930s Cárdenas had expropriated foreign oil companies and promulgated drastic land re-form (and he would presently become Fidel's protector). This became part of Castro's draft of a "complete revolutionary program," which he planned to send to Cuba as a pamphlet for mass clandestine distribution. He had to pawn his overcoat to pay for the printing of some copies of his document, remarking in a letter to a friend that "the pawnshops here are run by the state and they charge very low interest . . . if the rest of my clothes were forced to go the same way, I wouldn't hesitate for a second."

Castro caught the grippe, but despite his high fever, he continued to write in longhand his plans for Cuba after Batista's defeat. At dawn of August 2, he scribbled a note to his sister Lidia in Havana that "although it is now already four o'clock and five minutes in the morning, I'm still writing. I have no idea how many pages in all I have written! I have to deliver it to the courier at 8:00 A.M. I have no alarm clock; if I oversleep, I may miss the courier, so I won't go to sleep. . . . I have grippe with a cough, and my whole body aches. I have no Cuban cigars, and I really miss them." In another letter that week, he remarked that his life in exile was "sad, lonely, and hard."

What his solitary writing produced was the Movement's "Manifesto No. 1 to the People of Cuba," signed by Castro and dated August 8, 1955. Based on a fifteen-point program, this document was much more radical that his proposals two years earlier in "History Will Absolve Me." The Manifesto was intended to reach Cubans on August 16, the fourth anniversary of Eddy Chibás's death. Symbolism was crucial in the continued elaboration of Castro's image, so the Manifesto opened with the requisite citations from Martí and General Antonio Maceo. Castro instructed his followers at home to print "at least fifty-thousand" copies of the document, and to start distributing them at Chibás's grave in the Havana cemetery. In Mexico two thousand copies were printed by Alsacio Vanegas Arroyo, a Mexican printer who was a friend of María Antonia's. The original handwritten text of the Manifesto was smuggled by another woman friend, the sister of the pop singer Orquídea Pino, inside the *History of the Incas,* a classic of the Spanish conquest. That week, Fidel had his twenty-ninth birthday.

Castro's Manifesto was essentially designed to transform the Movement ideologically and militarily into a new streamlined structure while maintaining the continuity of the Moncada tradition. Enormously lengthy, as were all Castro's writings and pronouncements, it was "an open call for revolution, and a frontal attack against the clique of criminals who trample the honor of the nation and rule its destiny counter to its destiny and the sovereign will of the people. . . . The bridges have been burned: either we conquer the fatherland at any price so that we can live with dignity and honor, or we shall remain without one." The Movement, Castro wrote, "is formed without hatred for anyone; it is not a political party but a revolutionary movement; its ranks are open to all Cubans who sincerely desire to see political democracy reestablished and social justice introduced in Cuba; its leadership is collective and secret, formed by new men of strong will who are not accomplices of the past. . . . We defended the military when no one defended them, and fought them when they supported the tyranny, but we shall welcome them with open arms when they join the cause of liberty . . ."

The specific points of the revolutionary program provided for: "The outlawing of the *latifundia,* distribution of the land among peasant families. . . . The right of the worker to broad participation in the profits of all the large industrial, commercial, and mining enterprises. . . . Immediate industrialization of the country by means of a vast plan made and promoted by the state. . . . Drastic decrease in all rents, effectively benefiting the 2,200,000 persons who are today spending a third of their income on rent. . . . Construction by the state of decent housing to shelter the 400,000 families crowded into filthy single rooms, huts, shacks, and tenements. . . . Extension of electricity to the 2,800,000 persons in our rural and suburban sectors who have none. . . . Nationalization of public services: telephone, electricity, and gas. . . . Construction of ten children's cities to fully shelter and educate 200,000 children of workers and peasants. . . . Extension of education to the farthest corner of the country. . . . General reform of the tax system. . . . Reorganization of public administration. . . . Establishment of an inviolable military roster safeguarding the members of the armed forces so that they can be removed from their posts only for good reasons. . . . Elimination of the death penalty in the Military Penal Code for crimes committed during peacetime. . . . Generous and decent pay to all public employees. . . . Adequate measures in education and legislation to put an end to every vestige of discrimination for reasons of race or sex which regrettably still exists in our social and economic life. . . . Reorganization of the judicial branch. . . . Confiscation of all the assets of embezzlers acquired under all past governments . . ."

As with the "History" address, the Manifesto has been submitted to endless analysis and interpretation to determine to what extent, if any, it either suggested or brilliantly concealed Castro's real or attributed Marxism-Leninism. Interestingly, those people who accused him of communism before he publicly announced his faith in it, and today's official spokesmen who insist that he *always* was a Marxist-Leninist, both concur in finding the Manifesto to represent Marxist views. Careful scrutiny of the text in the context of the period during which it was written may lead, however, to a different conclusion: that Castro had left his options open so that in the future he could select the interpretation that suited him best politically. Obviously, land reform, profit sharing, rental cuts, state housing, state-managed industrialization, rural electrification, effective education, and antidiscrimination measures were not inherently Marxist notions even in a Latin American country in 1955, but they could be automatically inserted into a Communist or socialist program. Nationalization of public services, even if they belonged to United States interests, need not be a Marxist act (though the Eisenhower administration, for example, might have seen it that way). Mexico under Cárdenas, whose policies Castro studied so closely, had gone much further in nationalization without being accused of communism; though Perón in Argentina was considered a Fascist and a thug, he had recently nationalized British and American public-service interests, and public services of this type had been nationalized in most of Western Europe since World War II.

In this sense, then, it is an idle endeavor to search for hidden ideology in this Castro Manifesto. He knew exactly what he was saying—and how far he could and should go. In 1955, and even four years later, he always retained the freedom to maneuver. It is certainly plausible that Fidel was already a convinced Marxist-Leninist when he sat, pen in hand, in his tiny Mexican hotel room. He told Lionel Martin twenty years later that his early programs were "the antechamber of the socialist revolution," but this is ex post facto management of history.

Beyond question, however, were Castro's "anti-imperialist" or plain anti-Yankee sentiments, which were never disguised. As a student, he had belonged to organizations advocating independence for Puerto Rico, and among the first friends he made in Mexico was the Peruvian-born Laura Meneses, the wife of the imprisoned Puerto Rican *independentista* leader Pedro Albizu Campos. Albizu Campos was serving a long prison term in the United States, and Castro regarded him as a hero. Laura had attended Fidel's commemoration of the Moncada attack on July 26, and for the balance of his stay in Mexico, they remained close friends, spending much time together.

In his speech on Che Guevara's death in 1967, Castro said that "it was a day in the month of July or August 1955, when we met Che." Fidel Castro and Ernesto Guevara met for the first time at the apartment of María Antonia González in Mexico City. Hilda Gadea, a Peruvian with Incan features who was Guevara's first wife, places the meeting at "the beginning of July" in her book on Che (Hilda died of cancer in Havana in 1974, having been divorced from Che in the early 1960s). An account of Castro's Mexican exile published by the Central Political Directorate of the Revolutionary Armed Forces says he and Guevara "established relations around the month of September." Castro's history-keepers seem unperturbed by such imprecisions.

It happens that Castro and Guevara began their active revolutionary careers within two weeks of each other through a coincidence in personal histories. On July 8, 1953, as Fidel Castro was completing the preparations for Moncada, Che Guevara departed Buenos Aires for Bolivia on the first leg of a revolutionary journey. His father, Ernesto Guevara Lynch, recalls that when Che was saying good-bye to him and his mother, he explained that he was undertaking a self-appointed mission "to combat for the liberation of America from United States imperialism."

"Here goes a soldier of the Americas," Che told his parents, adding that he could help in the anti-American crusade he thought Juan Perón, whom he greatly admired as a potential hemisphere leader, had launched from the presidency of Argentina. Che, who was two years younger than Fidel Castro, had graduated from the medical school at Buenos Aires University on

April 11, 1953, but almost instantly decided that he preferred revolution to the practice of medicine.

Considering himself a Marxist, but never formally joining the Argentine Communist party, Che had spent much of the previous year traveling through South America with a fellow medical student, convincing himself of the colonial status of the continent's nation, and acquiring a profound distaste for the United States during a month-long stay in Miami. According to his father, Guevara had run out of money there and had to await an Argentine plane on which he could hitch a free ride home. Living in Miami on a dollar per day, the father says, the asthmatic Che was forced to walk for miles every day; he ate poorly and seemed never to have met any Americans he cared to remember. He was back in Buenos Aires in September 1952, resuming his studies on a crash basis to be able to graduate with his class.

Che was in La Paz, the Bolivian capital, the day Castro attacked Moncada, and he must have read about it in the local press, but there is nothing to suggest that he was especially impressed. He did not rush to Cuba, journeying instead in slow stages across Bolivia, Peru, Ecuador, Colombia, Costa Rica, Nicaragua, Honduras, El Salvador, and finally Guatemala where, in mid-1954, he witnessed the overthrow of the leftist government of President Arbenz by rebels financed and organized by the Central Intelligence Agency. Che reached Mexico in 1955, and it was only then, on "a cold Mexican night," as he put it, that he first met Fidel Castro, about whom he had been hearing reports through the Latin American revolutionary grapevine. From there on, their revolutionary destinies became joined.

The contacts between Guevara and the Cubans had first developed in Guatemala late in 1953, and early in 1954. Ñico López, one of Castro's most trusted lieutenants, had gone to Guatemala after escaping from the Bayamo affray on July 26, and he was soon introduced to Guevara and Hilda, then working for the leftist Arbenz regime. Guevara met other *Fidelista* refugees, learning the background and details of the Moncada and Bayamo uprisings. Dedicated to the notion of great Marxist revolutions in Latin America and viscerally anti-American, he became fascinated by what he was told about Castro. After Arbenz was overthrown by his CIA-directed enemies, Guevara fled the country and reached Mexico on September 21, 1954. There he ran again into Ñico López, who kept assuring him that Castro would get out of prison in Cuba and probably come to Mexico.

By 1955, Guevara worked full time as a physician at the General Hospital of Mexico. His specialty was allergies (he was an asthma sufferer himself), and he lectured for free at the medical school of the National Autonomous University. The pay at the hospital was so low, however, that he was forced to work as a news photographer for the Latina News Agency to make ends meet. Guevara lived in a minuscule apartment on Napoles Street, his life-style was spartan, and his only extravagant gestures were occasional gifts of classical records and bits of silver jewelry for Hilda whom he married in Mexico in 1955. They had first met in Guatemala, and she followed him north after the Arbenz debacle. Clean-shaven and with his hair neatly trimmed, outwardly Ernesto seemed the classic young Latin American professional with an impressive intellectual bent, what in those days was regarded as a "parlor revolutionary." He had the gift of fine irony in conversation, preferring to be quietly in the background. He was superbly well read, he wrote descriptive prose and poetry exceedingly well; his French was excellent, but his English barely passable.

Ideologically, Guevara considered himself a Marxist-Leninist, and he was a serious student of the doctrine. Castro would say many years later that when they first met, "he already was a Marxist in his thoughts" and "a more advanced revolutionary than I was." Not immediately evident then was Ernesto Guevara's profound idealism, an absolute absence of political opportunism, and a passionate dedication to revolutionary causes. Upon being introduced to the freshly exiled secretary general of the Guatemalan Communist party, which had supported the Arbenz government, Guevara chastised the Guatemalan leftists for not having resisted the American-organized attack, arguing that Arbenz should have gone to the countryside "with a group of true revolutionaries" to keep fighting. As Hilda recalled, they parted "coldly." Ernesto was the romantic revolutionary in search of a revolution.

When Raúl Castro arrived in Mexico late in June 1955, he was invited at once by companions from the Movement already there to meet Ernesto Guevara. The exiled Cubans knew Guevara through Ñico López, and the encounter with Raúl was a great success. Hilda recalled that Guevara brought Raúl to their apartment (she and Ernesto had just begun to live together), and that it instantly became a "great friendship." She wrote that Guevara and Raúl Castro met almost every day, and that Raúl introduced Che to other Latin American leftists exiled in Mexico. Of Raúl, she said that "he had Communist ideas, [was] a great admirer of the Soviet Union . . . and believed that the struggle for power was to make a revolution on behalf of the people, and that this struggle was not only for Cuba, but for Latin America and against Yankee imperialism." At the same time, Hilda recalled, "it was stimulating for the spirit to talk with Raúl: He was merry, open, sure of himself, very clear in the exposition of his ideas, with an incredible capacity for analysis and synthesis. This is why he got along so well with Ernesto."

Around the second week of July (by Hilda's account), Raúl arranged for Fidel, in Mexico City since July 8, to meet Guevara at María Antonia's apartment. They hit it off immediately, talking continually for ten hours, from early evening until the morning. Hilda wrote that Ernesto had told her when he came home that Fidel was "a great political leader in a new style, modest, who knew where he was going, master of great tenacity and firmness" and that they "had exchanged views on Latin America and international problems." And Guevara told her: "If anything good has happened in Cuba since Martí, it is Fidel Castro: He will make the revolution. We agreed profoundly. . . . Only a person like him would I be disposed to help in everything."

Afterward, Guevara wrote about meeting Castro: "I met him on one of those cold nights of Mexico, and I remember that our first discussion covered international politics. Within a few hours that night—at dawn—I was already one of the future expeditionaries." In a letter to his father in Buenos Aires the following year, explaining what he was doing in Mexico, Guevara said: "Sometime ago . . . a young Cuban leader invited me to join his movement of armed liberation of his people and I, naturally, accepted."

And this was Fidel's recollection: "An Argentine by birth, he was Latin American in spirit, in his heart. . . . Much is written about all revolutionaries, and this was the case with Che. Some tried to present him as a conspirator, a subversive and shadowy individual, dedicated to devising plots and fomenting revolutions. . . . As a young man, Che had a special curiosity and interest in the things that were going on in Latin America, a special spirit of delving into students and knowledge, and a special yen for going to see all our homelands. . . . He didn't have anything more than his degree [of physician]. . . . But Che wasn't Che then. He was Ernesto Guevara. It was because of the Argentine custom of calling each other Che that the Cubans began to call him Che . . . the name which he made famous later, the name which he turned into a symbol. . . . It was a matter of minutes for Che to join that small group of us Cubans who were working on organizing a new phase of the struggle in our country."

After their first encounter, Fidel and Che met two or three times a week, and late in July Castro went to dinner at the Guevaras'. Che also invited the wife of Albizu Campos and Juan Juarbes, another Puerto Rican exile. Much of the conversation consisted of Castro questioning them about the situation in Puerto Rico, but at one point Hilda asked him, "But why are you here, when your place should be in Cuba?" Fidel replied, "Ah, a very good question, and I shall answer you," and, as Hilda put it, "His reply lasted four hours." She remembered that his main points were that "the Yankee penetration of Cuba was complete, there was no other way than to continue the road of Moncada," and that he had come to Mexico to prepare an invasion of the island, "to launch an open battle against the army of Batista, who was supported by the Yankees," and that "the struggle in

Cuba was part of the continental struggle against the Yankees that Bolívar and Martí had already foreseen." Castro went on to explain how the invasion was being prepared, and how important it was to maintain total security, "always being aware that there may be infiltrators, but knowing that traitors could be detected." The danger of treason was always on Castro's mind.

Hilda's personal impression of Fidel was: "Very white and tall, big without being fat, with very black hair, shiny and curly, with a moustache, and with rapid, agile, and secure gestures. He did not seem to be the leader that he was: He could pass for a well-turned-out bourgeois tourist; but when he spoke, his eyes lit up with passion and faith in the Revolution. . . . He had the charm and the personality of a great leader and, at the same time, truly admirable naturality and simplicity."

After Castro left, Guevara asked Hilda, "What do you think of this folly of the Cubans of wanting to invade a completely armed island?" She answered, "There is no doubt that it is folly, but we must be with it." Guevara embraced her and said, "I think the same . . . I've decided to be one of the future expeditionairies. . . . We shall soon begin our preparations, and I'll go as a doctor." On August 18, Che and Hilda were married in the village of Tepozotlán in the presence of Raúl Castro, Jesús Montané (who had just arrived from Cuba to join the conspiracy), and the Venezuela poetess Lucila Velásquez, the bride's closest friend. Fidel joined them later for an Argentine barbecue prepared by Che; he had originally planned to be the witness but decided against it for security reasons. He thought that Batista agents, the Federal Bureau of Investigation, and the Mexican political police were watching him, and he was not entirely wrong. The Cubans and the Guevaras were now inseparable; on one occasion, Fidel threw a party for his political friends, preparing spaghetti with seafood sauce and cheese.

Even with Che and Hilda, Castro could not resist the temptation to run everybody's lives. When Hilda told him one day that Che had earned extra money for covering the Pan-American Games for his news agency, and that they could not decide whether to buy a car or take a trip, Castro counseled them "to buy something for the house, like a record player" because there was too much red tape in Mexico in owning automobiles. Anyway, he said, they had friends with cars if they ever needed one. The Guevaras did buy a record player, and Fidel was delighted to see it the next time he visited them. That particular evening, Castro met Lucila Velásquez, and they appeared to be interested in each other. He took the poetess out several times, but, as Hilda observed, "he was so busy with his political problems that he put aside all his other interests." Still, Lucila asked Hilda, "Tell me, how did you conquer Ernesto, how did you catch him to marry you?" Everybody laughed, including Fidel.

In September, Guevara and Castro deplored together the army's overthrow of Juan Perón in Argentina. To a great many Argentines his ouster meant the end of a corrupt dictatorship and a gradual return to representative democracy, but to the two young revolutionaries it marked the end of what they perceived as an experiment in social justice. Guevara complained to Hilda that the people did not fight in the streets to defend Perón's regime. Both Che and Fidel had concluded independently that Perón's *Justicialismo,* a vague populism combined with a welfare state, was the beginning of liberation from capitalism and "imperialism," even though Perón was loudly anti-Communist. But he had enormous support among urban workers, and he was anti-American. That was good enough for Guevara and Castro: They saw the military revolution as "reactionary," and Castro always remained pro-Perón, immune to the arbitrariness and corruption of *Perónismo.* It was Perón who had funded Castro's trip to Bogotá in 1948 for the students' congress, and it was Perón who financed the news agency for which Guevara now worked in Mexico. Meanwhile, while planning the revolution with Fidel, Che Guevara still found time in September to go to Veracruz to present a paper on allergies to a scientific congress there.

With the Movement's top command now firmly established in Mexico, Fidel Castro proceeded to set in motion a series of new operations. When Pedro Miret arrived in September with the coastal charts, he instructed him to accelerate the departure of additional Movement members to train in Mexico for the invasion; he sent a stream of detailed orders to the National Directorate on the island; and he prepared for a long fund-raising voyage to the United States. Jesús Montané and Melba Hernández had already arrived from Havana, and with Raúl Castro and Che Guevara, they could coordinate activities in Mexico in Fidel's absence. Juan Manuel Márquez, the forty-year-old *Ortodoxo* party leader from Havana who had been beaten savagely by the Batista police because of his friendship with the *Fidelistas,* also succeeded in reaching Mexico at that time; he too joined the leadership circle. And the rebels also had a friend in Raúl Roa, the university professor who had sent them books when they were imprisoned on the Isle of Pines, and now lived in Mexico as coeditor of the periodical *Humanismo.*

In Cuba the key personality in preparing for the invasion was Pedro Miret. He explained that during that period going to Mexico to brief Fidel was "one part of the plan"; the other part "was for me to send people over there" to join the military force. Miret says that "when Fidel left, I was charged with the confidential, conspiratorial responsibility, the most delicate aspect of it." With plans for an insurrection and, subsequently, a general strike, "the whole country had to be organized, and this took a long time because there were so many people in the *Ortodoxo* party turning toward insurrection, because we had to collect money for arms, make contacts to buy arms, and conceal arms." Then, Miret recalls, "we had to select companions to send to Mexico, and this had to be done systematically." First, the candidates "had to be tested . . . we had to see how they acted. For example, we would order them to paint a '26' sign [for the 26th of the July Movement] on walls or somewhere, which seems easy . . . but people painting the '26' began losing respect for the authorities, and they became reckless . . .

"In many cases," Miret says, "the men had to prove their stability so that we could start selecting those who would be squadron leaders. When they became 'burned,' in the sense that the police were on to them, we undertook to send them abroad. So, we were sending people out at the rate they were getting 'burned.' . . . Sometimes, we could send them out openly, it was incredible what one could get away with. Once, I just changed my name a little bit. . . . But, of course, officials at the airport belonged to the Twenty-sixth of July. . . . Since we had no money, every time a person left, a cell of the Movement in Cuba had to commit itself to send him money abroad. . . . Some people got forty dollars monthly, but that was plenty in Mexico." Thus Fidel Castro's new Rebel Army grew.

And Fidel was bombarding Cuba with his bulletins of instructions. Two weeks after "Manifesto No. 1" was distributed, he issued a letter to the National Directorate containing his ideas for the structure and functioning of the renascent Movement. His emphasis was on propaganda and security. Propaganda, he wrote, "must never fail. . . . I give it a decisive importance because apart from keeping the morale high, materials circulating clandestinely in the country do the work of thousands of activists, converting every enthusiastic citizen into a militant who repeats the arguments and the ideas . . ." On the other hand, Castro demanded "the most rigorous silence" concerning arms, persons handling them, and places where they are stored. "If any *compañero* learns too much because of his activities, he must be removed from the internal front," Fidel ordered, adding that no more than fifteen or twenty persons in Cuba should have knowledge about such matters, and none should know who the others were. He insisted on indoctrination work among the workers.

The new Movement could have "a centralized direction," Castro said, "that will control all the principal links, but a decentralized organization of the masses, acting in specific tasks; these tasks will be given to all the members of the armed forces who sympathize with us." In this fashion, he expected that the best-known leaders of the Movement inside Cuba could be replaced gradually without any break in the work. The main difference in strategy from the past was that whereas Moncada was conceived as a relatively rapid insurrectional war based on the cities, the new phase called for a prolonged war in the rural areas as well as urban centers. The prolonged-war concept was now central in Castro's thinking.

On September 17, Fidel dispatched a lengthy communication to the Women's Martí Centennial Civic Front, insisting on the basis of the Moncada experience that a revolution must be organized in such a way that no unexpected event or accident can derail it. He remarked that "a revolutionary strategy is always more complicated than a war strategy, it cannot be studied in any academy, and the professional military with their rigid mental notions are the least likely to conceive it." Having given it profound thought, Castro had come to the conclusion that a well-directed and imaginative guerrilla operation had a very good chance against a traditional army that operated by the book; this was the real key to his decision to launch an invasion. According to Castro, political groups should play different roles in a revolution, depending on their public record and the "social interests" they represent; this was his first hint that in his revolutionary unity, some groups will be more equal than others (this was the point that was generally missed by a great many early *Fidelista* backers).

Modern sabotage techniques were to be taught to special combat units inside the country for followup action after the invasion, Castro wrote, and 80 percent of collected funds should be spent on buying arms and 20 percent on organization and propaganda. Everybody could participate in the revolution, he said, from the young to the old, men and women, providing "useful collaboration" without necessarily having to use a rifle. But in a letter on October 4, Fidel demanded a commitment from Movement activists in Cuba to provide systematic financial support; this would be his chief test of "loyalty of our militants to revolutionary principles and discipline." Referring to himself as "Alex" (one of his conspiracy names, derived from his middle name, Alejandro), he wrote that "Alex is convinced" that the revolutionary plan will be successful if funds are adequate. For this reason, in October Fidel undertook a speechmaking and money-raising tour of the United States, his first public political exposure on the American scene.

"I can inform you with complete reliability that in 1956, we will be free or we will be martyrs!" Fidel Castro exclaimed before an audience of eight hundred exiled Cubans at the Palm Garden hall at Fifty-second Street and Eighth Avenue in New York on Sunday, October 30, 1955. This was the first time Castro had publicly and formally made the specific commitment, to be repeated over and over in the year to come, to invade Cuba before the end of 1956. He did so for the twin purpose of making his Movement more credible by setting an approximate date for the invasion, creating a deadline for himself and his companions, and keeping the Batista regime off guard. If the Movement was to grow inside Cuba, Castro reasoned, it had to be given something more tangible than vague promises of a revolutionary return—someday. And he believed that a direct challenge to Batista would psychologically weaken him within his own government. He told his enthusiastic New York audience that "the regime is totally disoriented with regard to our revolutionary activities . . . we have developed invincible methods of organization and work. . . . Our apparatus of counterespionage functions much better than their espionage. Whenever their agents abroad inform the government, we immediately get the information here. All information offices and Batista's spies are carefully watched by us."

The New York speech was one of the highlights of Castro's seven-week American tour. Its practical objective was to raise funds for the Movement and to create "Patriotic Clubs" of Cubans living in the United States to support the revolution in a sustained fashion. However, Fidel made sure that the symbolism of following in José Martí's footsteps by mobilizing Cuban exile communities for the "war in liberation" was not lost on anyone at home or abroad. Martí had lived in New York for years before proclaiming the revolution in 1895, and departing on his ill-fated expedition to free the island. Most of the financing for the final war of independence came from Cuban businessmen and workers in the New York area and from cigarmakers in Tampa, Florida. As Fidel moved along the east coast of the United States to deliver his revolutionary exhortations, the spirits of José Martí and Eddy Chibás went with him, their names invoked continuously. (And Chibás's voice was heard on records Fidel had brought with him.)

Castro began his trip to the United States by dispatching a letter on October 8 to the *Ortodoxo* party executive committee in New York, promising "a radical and profound change in national life" as a linchpin in the process of liberating Cuba. He expanded on the theme at the Palm Garden by saying that "the Cuban people want something more than a simple change of command," that "Cuba longs for a radical change in every aspect of its political and social life. . . . The people must be given something more than liberty and democracy in the abstract; decent living must be given to every Cuban." To his audience, these were the themes of Martí and Chibás, and they went over smoothly; nobody read Marxist messages into them until later.

After delivering a fiery speech at the José Martí monument in Mexico City on October 10, Castro left for the United States. His first stop was Philadelphia; then he gave speeches to Cuban groups in Union City, New Jersey, and Bridgeport, Connecticut, before arriving in New York on October 23. Cubans had been emigrating to the United States since the latter part of the nineteenth century, but additional thousands had come in the 1950s because of the economic crisis on the island, and they constituted the principal target of the Castro campaign. Fidel evidently had no problems obtaining a United States tourist visa, presumably because Washington (and the American embassies in Mexico and in Havana) did not consider him sufficiently "subversive," and the Batista regime may not have learned of his planned trip in time to request his visa be denied. Castro wanted to make an impact on Cubans in the United States, but at this stage he was perfectly happy to be ignored by American authorities. Actually, Union City police stopped the car bringing Castro to his speaking engagement, briefly interrogating him and his hosts, but it appeared to be a routine check and he was not bothered again. Another version is that the organizers of the Union City rally had forgotten to obtain a permit, and the police came to find out what was occurring. The New York event was called only on a four-day notice, possibly for security reasons.

It went very well. Imposing in his ancient dark-blue wool suit, Castro instilled enthusiasm with his passionate oratory, and at the end of the meeting, cowboy hats atop the head table filled up with dollar bills from the audience. Juan Manuel Márquez, the *Ortodoxo* leader who had now become one of Castro's favorite advisers, traveled with him during most of the American tour; he had spent some time in Miami in the past, and he was well connected in the exile community. In his New York speech, Castro made the important point of his revolutionary strategy: "We are against violent methods aimed at persons from any opposition organization that disagrees with us. We are also radically opposed to terrorism and personal assault. We do not practice tyrannicide." As he explained many years later in a private conversation, Fidel had taken the view that terrorism, apart from being immoral, is counterproductive because it scares away moderates who otherwise might be potential supporters of the revolution. During the entire anti-Batista war, the *Fidelistas* eschewed terrorism in the sense of political assassination or bomb throwing in public places (Castro says there were incidents of terrorist bombing in Havana at the outset, but he quickly forbade them; sabotage of power plants, for example, was considered highly desirable by him).

On November 20, Castro spoke to one thousand Cubans at the Flagler Theater in Miami; he had spent three weeks since the New York rally in private conversations and negotiations with exiled Cuban leaders, and seeing old friends. His tourist visa had expired, but the Immigration and Naturalization Service in New York extended it automatically. In Miami Fidel was joined by his sister Lidia who brought along Fidelito, and by their fairy godmother, María Antonia, from Mexico. How Fidelito was able to join his father remains unclear. According to one account, Lidia took him out of school in Havana (and from Mirta's control) and traveled with him to Miami in what Castro detractors called a virtual kidnapping. Another account is that Lidia first took Fidelito to Mexico, where their younger sisters, Emma and Agustina, had followed Fidel during the autumn. In any event, six-year-old Fidelito was present at the Miami speech, and when he started playing with the dollar banknotes being placed inside upturned sombreros, his father said, "Don't touch it, Fidelito, because this money belongs to the motherland." In his speech, Castro said, "My son is here; if he were of age, I would take him with me to the battle." Cuban students in Miami handed Fidel a huge Cuban flag. He wound up the tour with speeches in Tampa and Key West, traditional Cuban communities. In Key West, Fidel spent ten days resting and writing at a boardinghouse on Truman Avenue.

The American expedition was a relative success, though Castro never disclosed how much money he had collected; in a letter from Miami to Raúl Castro, he said he would have a $9,000 "surplus" after the printing of ten thousand copies of revolutionary pamphlets. Everywhere he went in the United States, he asked for funds; in the Miami speech, he said that "it does not trouble us to ask for alms for the motherland, because we ask for it with honor . . . nobody will repent of having contributed, but even if aid is insufficient, we shall go to Cuba, with ten thousand rifles or with a single rifle . . ." Castro remarked that there had been surprise over his disclosure of the year he had set for the revolution, and added, "We said which year, but we did not say the month, the day or the hour, nor how or where . . . Martí never denied his revolutionary plans when he was in exile . . ."

Politically, the tour was also worthwhile. He helped organize "Patriotic Clubs" and "26th of July Clubs" in a half-dozen cities, although he would complain later that the Cubans in the United States were not helping more. His idea was that every unemployed Cuban abroad should give a dollar a week, and that the employed ones should contribute one day of their monthly wages. He also thought that members of his clubs should pay two dollars a week. Nothing was ever enough for Fidel, but when he finally launched the Sierra war, Miami-collected funds paid for many arms shipments. In Havana *Bohemia* published very long and sympathetic reports on his New York and Miami speeches, and from Nassau in the Bahamas, Castro issued "Manifesto No. 2 of the 26th of July Movement to the People of Cuba," reporting that "seven weeks of tireless effort dedicated to organizing the Cubans from the Canadian border to the glorious [Florida] Keys have produced the best results." He warned against "imposters" trying to collect money in the name of the Movement, threatening they would be punished. Finally, he said, "no alternative remains for the country but revolution."

Castro returned to Mexico to resume building the invasion army. In the nearly six months since his departure from Cuba, he had created rebel structures in Mexico, Cuba, and the United States, and the year 1956 could be entirely dedicated to military preparations.

However, back in Mexico Castro had to cope with Cuban politics and with Cuban politicians who were finally beginning to realize that he was emerging as the principal opposition leader. An article in *Bohemia,* said that a *"Fidelista* complex" was developing among politicians (it was the first known use of the expression, derived from Castro's first name). Politicians, it went on, "feel dwarfed by the shadow of Fidel Castro that is becoming gigantic," and see in him "too dangerous a rival for some chiefs of [political] opposition." Those politicians should elaborate a coherent stance against "the revolutionary action of *Fidelismo*" and seek a peaceful solution to the Cuban crisis, the article said, but they "already feel displaced by the magnitude of the *Fidelismo* [phenomenon] . . ." Photographs of Castro and Batista illustrated the article, suggesting that they were the only two real power contenders. Two weeks later, Miguel Hernández Bauzá, a political commentator, published an article in *Bohemia* titled "Motherland Does Not Belong to Fidel," a savage attack on Castro. Hernández warned that if Castro should ever seize power in Cuba, he would become "the only dispenser of civic, moral and spiritual grace . . . God and Caesar in one piece of flesh and bones," and that "all those who are not partial to Fidel, would be executed as immoral." Yet, the author conceded that "nobody can claim that Fidel has profited from public funds" as had most Cuban politicians.

Hernández Bauzá must have touched a raw nerve in Castro by depicting him (rather prophetically) as something of an egomaniac because Fidel replied in *Bohemia* with a violent diatribe in nine columns, noting that "four years ago, nobody occupied himself with my person . . . I went unnoticed among the all-powerful masters who discussed the destinies of the country. . . . Today, strangely, everybody rises against me." But this is not, he wrote, because he had abandoned his ideals, but because "they know my rebelliousness cannot be bought with any money or position." Fidel was furious, but he savored every word of the attacks on him by the Havana Establishment: Finally he was recognized, finally everybody was paying attention to him, finally he was feared. And this was exactly the climate he needed to proceed.

Inside Cuba, tensions and violence were growing during the latter part of 1955, heating up the revolutionary mood. At Havana University a group of students led by José Antonio Echeverría, who before long would die for the revolution and enter the pantheon of great Cuban heroes, formed a secret Revolutionary Directorate (DR) to launch armed urban struggle against Batista. The DR, however, had no links with the 26th of July Movement, and it would remain a fairly independent revolutionary group until after Castro's triumph. In the meantime, the DR almost became trapped in a plot to capture the presidential palace and kill Batista. The plan was apparently designed by former President Prío, who was preparing to return to Cuba in mid-August and had arranged to have large caches of arms and ammunition concealed in downtown Havana, supposedly for an assault on the palace. The DR students were persuaded to carry out the plot, but the secret police seized the caches on August 4 and 5, thereby liquidating the conspiracy. The students were arrested nevertheless, and remained detained for over a month. On his return, Prío announced he was abandoning his commitment to insurrection and would combat Batista politically; this earned him a contemptuous broadside from Fidel Castro in Mexico.

Castro also denounced the efforts by a group calling itself the Society of Friends of the Republic (SAR), headed by an octogenarian politician named Cosme de la Torriente, to negotiate new elections with Batista. To Fidel no negotiations were acceptable, and he announced that the 26th of July Movement would consider elections only if Batista resigned and left before they were held. Confrontation was the only strategy Castro regarded as plausible, particularly as his Movement grew. In Havana students battled the police late in November and early in December, and José Antonio Echeverría was seriously injured in a street affray. The DR responded by firing on the police, wounding a dozen policemen. On December 5, hundred of women belonging to the Women's Martí Centennial Civic Front, an organization allied with the 26th of July Movement, fought the police in midtown as they tried to march to a rally organized by the SAR. Scores of women were beaten, and many arrested. On December 7, the police fired on a crowd of students and workers protesting against Batista; one of the wounded that day was Camilo Cienfuegos Gorriarán, a worker who would soon become another hero for the great revolution.

On December 23, nearly a quarter-million sugar workers in the mills and plantations went on strike over a wage dispute. They were instantly supported by the 26th of July Movement, the students' DR organization, and even the illegal Communist party. As the strike continued over Christmas, strikers and students fought with the police in a dozen towns; at the same time, tens of thousands of leaflets with the *Fidelista* slogan "In 1956, We Will Return Or We Will Be Martyrs" were disseminated across the island, and the sign "MR-26-7" appeared in red or black paint on walls everywhere.

In Mexico Fidel interrupted his conferences with Che Guevara and Raúl Castro long enough to celebrate Christmas Eve with his friends and fellow rebels. He prepared the traditional Cuban festive meal of rice and black beans, roast pork, nougat, apples, and grapes. But no sooner had the dinner ended than Fidel launched into a nightlong monologue about economic and development projects he planned for Cuba when the revolution triumphed. Hilda Guevara recalled that "Fidel spoke with such naturalness, such certainty, that we had the impression we were already in Cuba in full constructive labor." Then Fidel and Che spoke of the need to nationalize natural resources and the principal sources of wealth. Yes, Fidel told them, "in 1956, we will return . . ."

CHAPTER

8

The training of Fidel Castro's invasion army started in earnest at the start of 1956, when the first "serious" money began reaching the conspirators in Mexico. Pedro Miret had brought about $1,000 from the first collections by the 26th of July Movement when he came to Mexico on a second visit to Fidel in December 1955, and Faustino Pérez, a member of the National Directorate, arrived with $8,250 early in February. The Rev. Cecilio Arrastía delivered $10,000. This hardly represented a revolutionary bonanza, but Castro had been complaining that in the first two months he had been in Mexico he had received only $85 from Cuba, and that "each of us lives on less money than the army spends on any of its horses." Now the Movement was able to pay the very modest living expenses of the rebels concealed in safe houses throughout the city, but the personal weekly allowance per man was eighty *cents;* the calculation was that it cost eight cents a day to feed each *Fidelista*. Anyone receiving under $20 monthly from home had to turn over half of it to the Movement's treasurer, and 60 percent was taken from remittances over $20. When Max Lesnick, the chief of the *Ortodoxo* youth branch and an old personal friend, visited Mexico on December 30, he found an unshaven and hungry Fidel awaiting him at the Regis Hotel; they lunched on steak Milanese, a Fidel favorite.

By mid-January, over forty handpicked men arrived from Cuba and the United States to join the rebel force in Mexico; counting Fidel, Raúl, Che, and those already in place since the previous year, the total stood at over sixty fighters—plus a group of deeply involved Movement supporters, mainly women and several Mexican friends. To house the future *guerrilleros,* six small dwellings were rented, each just large enough for about ten men; more houses were obtained later. Fidel Castro stayed with Melba Hernández and Jesús Montané (who were engaged to be married), and two bodyguards. Fidelito had gone to Mexico from Miami with his father's sister Lidia, and they lived with a rich Mexican-Cuban couple in a villa with a swimming pool.

Fidel was determined to keep his son with him once Lidia succeeded in flying him out of Havana (though he was not certain where he expected Fidelito to be after he left on the invasion), and he saw him as often as he could. In the Havana revolutionary archives, there is a photograph showing a very serious Fidelito in shirt and tie and an oversize military overseas cap, standing in a safe-house garden with his father, Raúl, Lidia, María Antonia, and several friends. María Antonia's apartment remained the message center and point of coordination, and all the new arrivals were processed there.

For reasons of security and discipline, life in the safe houses was one of monastic rigor and secrecy. It was forbidden for any individual to reveal his personal activities or those of his house group, to any other Movement member, and each house had a commander who was responsible for discipline as well as household and management problems. Members who lived in different houses were not allowed to disclose their addresses when they met at exercises, and they could not visit with each other. Rebels were not allowed to establish outside acquaintanceships, they could not go out except with at least one companion, they could not date women alone (double-dating was usually permitted), they had to be home by midnight, they could not make telephone calls, and alcoholic beverages were strictly banned. Mealtimes were on a rigid schedule, the men took turns cooking and cleaning, and no excuses were accepted. Free time was used for study and lectures, especially on "military and revolutionary themes," according to Castro's instructions. Each house commander was responsible for high morale among his men and for friendly relationships, on the theory that those who could not get along well would not be able to fight together well; there were channels for complaints and suggestions. So severe were the rules that any indiscretion could be regarded as treason; Fidel Castro assumed correctly that Mexico was full of Batista spies and agents seeking to destroy the Movement, and he was accordingly obsessive about security. Betrayal or indiscipline could—and did—result in death sentences in Castro's secret army.

When Fidel collected enough money and gathered enough men in mid-January, he informed General Bayo, the Spanish guerrilla specialist, that he was ready to proceed with the training. At first, Bayo conducted classes and drill exercises in the safe houses; he says in his memoirs that "I was the only one to know the location of all the houses, apart from Fidel, because I had to go to all of them to teach." He pretended to be an English teacher whenever someone in the neighborhood asked him what he was doing there.

In the training, Castro and Bayo's emphasis was on the physical fitness of the fighters. They had to be ready for day and night marches over the worst terrain in the most adverse weather, to sleep on the ground, to go for days with little or no food and water, and to be immensely resistant to fatigue. Nocturnal training was particularly important. Che Guevara decided to improve his physical condition by losing weight although he was already slim, and Hilda recalls that he gave up his Argentine habit of eating a steak for breakfast, confining himself to a sandwich for lunch at the hospital, and a light dinner of meat, salad, and fruit.

First, the rebels were made to walk for long periods because walking was the guerrilla's principal transportation mode; they marched endlessly along Mexico's streets, especially the long Insurgentes Avenue. Every morning, groups of *Fidelistas* rented rowboats on Chapultepec Lake to row for hours; not only was it good exercise and cheap, but Castro thought experience on the water might come in handy during the crossing to Cuba. Alsacio Vanegas, the Mexican printer who was also a wrestler (like Marí Antonia's husband), was drafted by Fidel to teach the men hand-to-hand combat at the Bucarely Street gymnasium. They also played basketball and soccer in the suburbs to improve their agility. Then, Castro and Bayo ordered mountain climbing, beginning with Sacatenco and the higher Chiquihuite just outside the capital. The men from different safe houses converged in buses at their meeting point in front of the Linda Vista movie theater in the northern section of Mexico City to start the climb; they moved in small groups so as not to attract attention. Gradually, Bayo made the men carry heavier and heavier backpacks. Some of the men, including Che Guevara, went far out of town on weekends to climb the 17,343-foot Iztaccíhuatl and the 17,887-foot Popocatépetl (Che considered it a point of revolutionary discipline never to let his asthma attacks interfere with these physical efforts).

In February Castro arranged for his group to use Los Gamitos firing range, also near the city. There were mountains around the range, and Bayo and his instructors used this terrain for guerrilla training. Armed with .30-'06-caliber rifles with telescopic sights—these were the first weapons the Movement could acquire—the rebels practiced firing and ballistics in the most professional detail: deviation rate, trajectory tension, line of fire, plan of fire, angle of fire, trajectory origin, projection line, sighting angle, range, initial velocity, correction in firing on aircraft, and on and on, hour after hour. Bayo was convinced that in a guerrilla war, individual proficiency makes up for enemy numbers and firepower. Then he taught them compass reading and map reading, and how to transfer terrain measurements from a 1:300000 to a 1:100000 map, how to dig shallow trenches, and how to string communication lines. "You men need to learn military culture if you're going to be guerrillas," the one-eyed Spaniard kept telling the men.

Bayo's two principal assistants were Miguel A. Sánchez, a Cuban who had fought with the United States Army in the Korean War and was known as "El Coreano," and José Smith, another U.S. Army veteran. Sánchez was brought from Miami by Castro and lived with Fidel as one of his two bodyguards, but he was later accused of betraying the Movement. Both Sánchez and Smith were top specialists. Fidel rarely participated in the exercises because overall organization, political contacts, and fund raising took up most of his time, but he made a point of checking periodically on progress in the training, observing firing exercises through a theodolite. Melba Hernández recalls Fidel once spent a whole day, from dawn until late afternoon, observing the training, then decided to calibrate the rifle sights. "We were terribly tired," Melba says, "having practiced all day, and having had only a half orange each for food, most of us just relaxed. I, for example, stretched on the ground, chatting with someone. The only one to go on working with Fidel was Che. When they finished, Fidel gathered us, telling us with infinite sadness that the struggle ahead was very long, that if we became exhausted so easily, we wouldn't be able to keep up; and that he was very upset that Che, an Argentine, a foreigner, hadn't gotten tired, that he had gone on training, while all of us, the Cubans, just did not. . . . He spoke with such sadness that afterward it never occurred to us to get tired—we had no right to be tired . . ."

Castro, as seen by his closest associates, was absolutely single-minded about the revolution, but he tried very hard to be sociable on a variety of occasions, and he may even have fallen in love with a young woman in the months immediately preceding the invasion. He had a blend of patience and impatience—patience in terms of knowing how to wait for the correct historical moment to act, and impatience in hating to see even one revolutionary moment wasted. His friends soon learned how to cope with his temperament and his moods. One morning in January 1956, Fidel burst into Melba Hernández, and Jesús Montané's bedroom at 5:00 A.M. shouting, "But this is not possible! . . . We came here to make a revolution, and not to sleep until nine o'clock in the morning!" He had evidently concluded that an extra hour or two could be gained for revolutionary practice (none of them actually ever slept until 9:00 A.M.), and he wanted instant action. From then on, they started marching down Insurgentes at dawn for some eight or nine miles to a street corner, where they took a bus to the Gamitos firing range.

But a few days later, Fidel made up for his martinet behavior. It was February 14, St. Valentine's Day (Cubans call it Lovers' Day), and he started out by wishing Melba a happy holiday, making breakfast (he had taught her earlier how to fry eggs "the best way"), and proposing an evening on the town. As with his military operations, this was planned with equally precise mathematical calculation. "Well," he said to Melba and Montané, "we have sixteen Mexican pesos [roughly $1.20]. Now, a woman, a *compañera* named Lucy, has just arrived from Havana. So, we could invite her somewhere for refreshments, or go to the movies, which would be four pesos per person. But this is your day, so you decide." Melba chose the movies, and the picture they decided to see was a Hollywood production called *Four Feathers,* concerning Indian wars. After fetching the *compañera,* whom Melba thought was very pretty and "in whom Fidel later became interested," they ran into Jesús Reyes, a Movement member often assigned to protect Castro. Reyes insisted on going with them because he had no money at all, and they could not get rid of him, being four pesos short for movie tickets for five. Walking past a pastry shop, they saw tamales and realized how hungry they were. But as Melba remarked, "This was forbidden fruit; we were broke." Finally, the young woman said, "Come on, tell the truth: you have no money, and you feel like eating the tamales. Let's eat them; I'll invite you." Fidel refused flatly, but finally allowed himself to be persuaded to accept a loan. They ate tamales, took Reyes along to see the picture, "and really it was a very improved evening." And as soon as new money arrived, Castro rushed to repay Lucy. He saw her again, but *she* was not the woman to whom he would propose a few months later.

The problem with Fidel Castro and women was that he insisted on their being as passionately interested in politics and revolution as he was. One evening, for example, Melba, Jesús Montané, and Raúl Castro persuaded Fidel to go out with two young Mexican women who were being very helpful in the invasion preparations. Fidel agreed, and Raúl said, "But Fidel, we are not going to talk about politics; we are going to pay attention to the girls, we're going to make it a fiesta evening for them, because otherwise your date will get bored—so let's make a deal not to discuss politics." Fidel nodded pleasantly, promised to keep his date amused, to pay attention to her. The three couples, Melba and Montané, Raúl Castro with a girl named Piedad, and Fidel with his date whose name was Alfonsina González, went to a Mexico City nightclub. Presently, Montané and Raúl and their women companions rose to dance, but Fidel did not move from the table. Melba recalls that when they returned to their seats, Fidel was lecturing Alfonsina on politics. They kicked him under the table, and he briefly changed the subject, soon to return to politics.

Melba says that Fidel lost interest in a girl "as soon as he realized that she was not interested in his goals; then the relationship would grow cold." But, she recalls, "on other occasions, I saw him very interested in a girl if she understood the Cuban situation, and especially if she was involved in working for the Revolution . . . but for various reasons, he never wanted to link his life with such a person." Melba believes that Fidel is "stimulated" by the presence of women, that he always "needs a woman" because "he has a great trust in women," but with him "the intellect is above everything else" in a woman.

Teresa Casuso, a Cuban novelist who became Castro's friend in Mexico and later a diplomat in his service (until she broke away from the regime), thinks that he fell in love with an "extraordinarily beautiful" eighteen-year-old girl from Havana who was her house guest. Casuso identifies the girl only as "Lilia" (other friends say her last name was Amor), adding that she had "an exceedingly polished and liberal education . . . disconcerting frankness, and an ingenuousness that enabled her to talk on every subject on earth." Castro often visited Teresa Casuso in the latter part of 1956—he had convinced her to let him store weapons and ammunition in locked closets in her villa—and usually tarried, waiting for Lilia to come home. Casuso remembers that "as the days and weeks went by, the romance between Lilia and Fidel flowered." Casuso says that Castro "sought her out with a youthful effusiveness and impetuosity that both startled and amused her . . . the amusement gave her an appearance of imperturbability which, combined with her great beauty, enchanted him, although it may also have been the quality which finally exasperated him, for Fidel cannot suffer people to remain unconquerable before him." According to Casuso, "Fidel had made a proposal of marriage to Lilia, and she had accepted," and he proceeded to obtain her parents' consent. He bought her "a pretty bathing suit to replace her French bikini, which infuriated him." But the engagement lasted no more than a month because Fidel had virtually no time to see Lilia as the invasion preparations quickened; in the end, Lilia decided to marry her former fiancé instead. When she informed him of her decision, Casuso says, Fidel "with that terrible pride of his, told her to marry the other man, that he was 'better suited' to her." Later, as he assembled a machine gun in Casuso's house, he told her that the revolution was his real "beautiful fiancée."

Whenever possible, for example, Fidel made a point of visiting the spot where Julio Antonio Mella, the student leader and young cofounder of the Cuban Communist party, had been assassinated by Machado agents in Mexico City in 1929. He evidently feared the same fate at the hands of Batista's agents, and held forth before his friends about Mella's life and death, but, Melba Hernández recalls, "as a man, not as a Communist." Castro was hungry for books, but his shortage of funds prevented him from buying enough volumes for himself and the rebels' safe houses until the day he persuaded a bookseller named Saplana to grant him credit. After their first conversation, Fidel loaded his car with books as if it were a truck, then went on getting them until the invasion. Thus, he acquired volumes about the history of Mexico and the Second World War, about Simón Bolívar, about economics and political science, and everything he could find about Marxism and Leninism. Afterward, the bookseller refused to accept money in repayment.

Fidel's obsession with his personal independence extended to his health. After agreeing to see a physician for a checkup, on the insistence of his companions, he was given a sedative to slow him down a bit. Melba Hernández tried to make him take a pill every morning at breakfast at their house, but Fidel refused. Finally, he told her: "Look, I can't develop a habit and depend on a little pill. I have to depend on myself because this is going to be a long struggle, and when we are in the midst of the war, I won't be able to have any pills. If I'm going to link my activities to pills, then I'm very bad off. I won't be a prisoner of pills." Once, while talking in his bedroom to Melba and Jesús Montané, Castro became depressed and the mood affected them, too. Suddenly, he jumped up and began pacing up and down the room, saying that "this is a bad example, a chief must never do this sort of thing, and, besides, this is a very transitory state of mind because I have full confidence in the Revolution." Melba says Castro apologized profusely, "promising never again to commit this error that he considered very grave."

Most of Castro's free hours in Mexico were spent with Melba and Jesús Montané, the Guevaras, his brother Raúl, his sisters Lidia, Emma, and Angelita, Fidelito, and Eva and Graciela Jiménez, Havana sisters who were old friends. On February 18, three days after the Guevaras' daughter, Hildita, was born, Fidel visited them. He was the baby's first visitor and told them, "This little girl will be educated in Cuba." On another occasion, Che Guevara offered Fidel bitter Argentine *mate* tea in a metal mug from which one drinks with a thin tube, passing the mug from person to person. Fidel at first refused because it was not "hygienic," but then joined the *mate* drinkers. Once, Guevara produced at dinner a bottle of *mezcal* firewater with a worm inside in the Mexican manner, and defiantly proceeded to swallow the worm. With infinite repugnance, Fidel swallowed a worm, too.

Early in 1956 the Military Intelligence Service (SIM) in Havana announced the discovery of a "subversive plot to overthrow the government, which is directed from abroad by Fidel Castro." Numerous Movement members were arrested, and Colonel Orlando Piedra, chief of the investigations bureau of the National Police, was dispatched to Mexico to try to uncover the conspiracy there. Castro heard from friends in Havana close to the regime that Batista had ordered his assassination. Twenty thousand dollars had been offered to two hit men to disguise themselves in Mexican police uniforms, "arrest" Castro in the street, and take him somewhere out of town to kill him and make his body vanish. A letter would be sent to María Antonia González over Castro's forged signature informing her that he had had to leave Mexico suddenly, and that there should be no concern about him. According to Castro's information, this operation was conducted by the Cuban naval attaché in Mexico City, but it collapsed when the rebels were somehow warned of the planned murder (there were sympathizers within the Batista secret police), and announced it publicly. The regime's next move against Castro would be through the Mexican authorities.

In the meantime, Castro finally decided to break his ties with the *Ortodoxo* party, proclaiming the 26th of July Movement as the only real opposition to the Batista rule, "the revolutionary organization of the humble, by the humble, and for the humble." The pretext for the rupture, announced in a long declaration on March 19 and published in *Bohemia,* was the *Ortodoxo* position that Castro's insurrectional line had not been authorized by the party's directorate council. The policy of insurrection and armed struggle had been approved in August of 1955 by the *Ortodoxo* activists' congress, and Castro now denounced the leadership for its "infamy" in rejecting it. He accused it of "cowardice" and submission to the regime.

In truth, however, the break was most convenient for him. The link with the party, which belonged to the traditional political establishment in Cuba, was no longer necessary to legitimize the growing 26th of July Movement, which was ready to emerge as an independent organization controlled by Castro. He felt free to denounce the wealthy landowner leaders of the party and charge them with betraying the Chibás heritage, and to proclaim that his Movement was "the hope of the Cuban working class . . . the hope of peasants for land, living like pariahs in the motherland their ancestors liberated . . . the hope of return for the refugees who had to leave their country because they could not work or live there . . . the hope of bread for the hungry and justice for the forgotten." The Movement, Castro emphasized, is "a warm invitation . . . extended with open arms to all the revolutionaries of Cuba without picayune partisan differences . . ." Significantly, Castro was telling the nation that the removal of Batista no longer constituted the foremost or the only objective of the revolution, and that what was at stake was the nation's whole structure.

In Cuba organized opposition to Batista was rising rapidly. On April 4 the secret police uncovered plans for an uprising by a group of liberal army officers, hours before it was to occur. This was known as the "Conspiracy of the Pure," and its leaders were part of a network established in the Superior War School with contacts in all the Havana commands. The rebellion was aborted because of a spy infiltrated among the officers, and a court-marshal sentenced thirteen of the leaders to six years in prison. Among them were José Ramón Fernández, then a lieutenant (and later a vice-president of Cuba under Castro), and Colonel Ramón Barquín, who was one of the most respected army officers and subsequently helped to build the Rebel Army. On April 20 a Students' Revolutionary Directorate (DR) commando seized the studios of Havana's Channel 4 television station, but a student was killed in the ensuing shootout, and the revolutionaries were unable to get on the air. On April 29 a group of militant members of former President Prío's political organization assaulted the Goicuría army barracks in the port city of Matanzas, sixty miles east of Havana, to force Prío to abandon his conciliatory attitude toward Batista and employ his huge financial resources in armed actions. Fourteen of the attackers were mowed down by machine-gun fire, and Prío was forced into exile in the United States. To Fidel Castro, Goicuría was reminiscent of Moncada, and he charged the government of deliberately staging a "massacre" inasmuch as it knew beforehand that the assault was being prepared. Hundreds of persons were arrested throughout Cuba in the aftermath of this incident in order to destroy the Prío organization. A leader of the DR was murdered by the police on May 15. At the same time, the first issue of the 26th of July Movement's clandestine mimeographed publication, *Aldabonazo* (The Blow), was put out in Havana by Carlos Franqui, a former Communist editor who had joined the Castro forces.

Propaganda and political contacts abroad were also of major importance to the Movement because it needed the support of international public opinion in the struggle against Batista and in tandem with invasion preparations. Mexico was a strategic point for such activities. Fidel himself was in charge of political contacts, which ranged from the so-called Democratic Left, Latin American groups dedicated to representative democracy and social justice, to the extreme left, including Mexican and other Marxists. In those days, one of the pillars of the Democratic Left was ORIT, the Mexico-based Inter-American Regional Labor Organization, which fought the

Communist-backed International Federation of Trade Unions in Latin America, and interestingly Castro made a point of cultivating its leaders. He frequently visited ORIT headquarters on Vallarta Street to meet with Secretary General Luis Alberto Monge of Costa Rica or the assistant secretary general, Arturo Járegui, of Peru, who were among the best-informed sources on Latin American politics (Monge would become his country's president in the early 1980s). ORIT was also a haven for democratic political exiles from all over the area, including Venezuela's Rómulo Betancourt and Peru's Víctor Raúl Haya de la Torre. Betancourt fought the Venezuelan military dictator, Marcos Pérez Jiménez, and Haya (the founder of the leftist-nationalist but anti-Communist APRA party) was the enemy of the Peruvian president, General Manuel Odría. ORIT's Járegui was an APRA member, as was Hilda Gadea, Che Guevara's wife. The antidictatorial tide was beginning to rise in Latin America, and the Democratic Left was a natural (if temporary) ally for Castro. All these men were close to Mexico's president, Adolfo Ruíz Cortines, and his labor minister (and successor), Adolfo López Mateos.

Ben S. Stephansky, then labor attaché at the American embassy in Mexico, remembers running into Fidel Castro at least twice at ORIT offices, once with Monge and once with Járegui. Castro seemed to know both men quite well, and had no reluctance in discussing politics with them in front of Stephansky. The American diplomat recalls that Labor Minister López Mateos had mentioned having met Castro, furthering his impression that the Cuban rebel was extremely well connected in the Mexican political world. Talking with Járegui, according to Stephansky, Castro was curious about ORIT's relations with the United States and Mexican governments, and its role in Latin America. At the meeting with Monge, his interest was in ORIT's ties with former Mexican President Cárdenas, and in the Mexican labor movement. Stephansky, who was probably the first United States official ever to meet Castro (he later became U.S. ambassador to Bolivia), was struck by what seemed to him to be the Cuban's "arrogance" and the speechmaking quality of his conversation. Fidel, he thought, was "professorial and cold." Later, Monge told Stephansky, "This Castro is a strange one." Stephansky also remembers that his colleagues at the American embassy, including the CIA station chief, showed absolutely no interest in his encounters with Fidel Castro; they did not even seem to know who he was.

By mid-spring of 1956, Castro decided that the Los Gamitos firing range was no longer adequate, so he instructed General Bayo to find and lease a ranch better suited for the organization of his small army. Weeks went by as Bayo drove around the Mexico City area with Ciro Redondo, one of Castro's most trusted aides, until he came upon a perfect spot near the town of Chalco, some twenty-five miles from the capital. The Santa Rosa ranch covered ninety-six square miles, including fields and mountains. The ranch house had 6,600 square feet, large enough to accommodate more than fifty men comfortably, and was surrounded by a nine-foot stone wall. It had four towers for protection from bandits, and Bayo remembered that it looked like "an ancient castle."

Santa Rosa belonged to Erasmo Rivera, a rich landowner in his seventies who once fought against Americans with Pancho Villa's guerrillas and was left for dead after a firefight; Bayo thought the significance was striking. Castro had authorized Bayo to spend $240 monthly to rent the ranch, but Rivera wanted to sell it for $240,000. After long days of negotiation, Bayo convinced the owner he was fronting for a millionaire Central American colonel who would certainly buy Santa Rosa, but the ranch house would have to be repaired and painted first. He offered to bring some fifty "Salvadorans" in this colonel's employ who would work on the house for the two or three months it would take to get it in shape—this was the period for which Castro needed the ranch—and finally Rivera agreed to the proposal and to a "nominal" rent of $8 monthly during the repairs. Bayo warned him that the arrangement had to be kept secret because if the press in El Salvador learned about it, the deal would be off. It was necessary, he said, to keep the "Salvadorans" away from the villagers, especially from women, to avoid problems. And when Bayo complained that the area was very arid, Rivera told him where water was to be found. Bayo was very proud: He had saved Fidel $232 a month.

The first rebels arrived at the ranch within a day or two, and Castro named Che Guevara chief of personnel under Bayo at Santa Rosa. Che wrote later that the *Fidelistas* "learned plenty" with Bayo, and that after he attended the Spaniard's first class, "my impression was that there existed a possibility of triumph that I had considered very doubtful when I signed up with the rebel commander [i.e., Fidel Castro], which whom I had from the beginning ties of romantic adventurous sympathy and of shared belief that it was worth to die on a foreign beach for such a pure ideal." Bayo worked his men day and night: They rose at 5:00 A.M. and trained until dark, always sleeping on the floor. As the training advanced, Bayo led nocturnal marches that started at 8:00 P.M. and ended at sunrise, with only a compass as a guide. To create conditions similar to the Cuban Sierra, two camps were set up in the mountains where the fighters would spend several days in simulated combat, forced marches, and guard duty. Water and food were brought by donkeys. Weapons were obtained in Mexico City through a friendly arms merchant, or in the United States by several Castro emissaries. Now that the Movement's treasury was richer, the revolutionaries had twenty Johnson automatic rifles, several Thompson submachine guns, twenty hunting rifles with telescopic sights, two .50-caliber antitank rifles, a Mauser light machine gun, and numerous smaller weapons. Uniforms and sleeping bags were sewn at the warehouse of the Mexican arms dealer. Fidel was delighted with the progress of his guerrillas. He took time out to serve as godfather at the Basilica of the Virgin of Guadalupe in Mexico to the daughter of Bayo's son, Alberto, a pilot who subsequently became an officer in the Cuban revolutionary air force. This was, Bayo said, the first of hundreds of Castro's godchildren. Bayo selected Che Guevara, his chess partner at the ranch, as the top *guerrillero* at Las Rosas. The general also wrote epic romantic poems in tribute to Fidel and several of his favorite fighters.

Toward the end of the training, the rebel leadership was faced with an extremely serious problem. A fighter named Calixto Morales Hernández, a rural schoolteacher, one day broke the rigid discipline by refusing to continue marching, sitting down, lighting a cigarette, and ignoring the officer who ordered him to rejoin the column. The men were then ordered back to the camp, and Fidel was instantly informed by telephone of Calixto Morales's defiance. This was the kind of disciplinary breach Fidel would not tolerate, and he arrived from the city with Raúl Castro and Gustavo Arcos. Fidel convoked a court-martial at once (there were such courts at Santa Rosa as well as at every safe house in Mexico City); he presided over the court, the other members being Bayo and Arcos. Raúl Castro was the prosecutor. The jury was composed of all the men at the camp. When ordered by Fidel to explain his behavior, Calixto Morales insisted that thirteen-hour marches were unnecessary and that he refused to take part in them.

Fidel said that he would not allow a major breach of discipline to go unpunished because without discipline the revolution could not triumph, and that it was therefore imperative that Calixto's punishment serve as an example. Bayo recounts that Castro "was eloquent, passionate, convincing. . . . He sweated indignation through all the pores of his skin, and he demanded in shouts that his *compañeros* halt this pus because otherwise it would spread gangrene among all of them." In effect, he was demanding the death penalty for Calixto. Then Bayo argued that it would be an error to execute Calixto on Mexican soil because no matter how great the secrecy, there was always a risk the body would be discovered and then Mexican police would enter the case. If that happened, Bayo added, it would be "*adiós* to the expedition . . . to the freedom of our adored Cuba . . . to the dream of entering Havana triumphantly . . ." Calixto spoke again, repeating his arguments calmly. Bayo thought that "his defense made no sense."

At that point, Bayo recalled, "Raúl, the prosecutor, leaped like an infuriated lion, interrupting the proceedings. . . . 'What disillusion have I felt this afternoon with the words of General Bayo,' he shouted. 'What a tremendous disillusion with what I have just heard! . . . You spend your life speaking of military discipline . . . of the dignity of the revolutionary uniform, but when a saboteur of our ideals, a bad soldier, appears among us to destroy our revolutionary mystique . . . you, General Bayo, our professor of military ethics, of military morals, you throw in the towel and attempt to save the life of this individual. . . . No, a hundred times, no! We cannot start our history with a hoggishness, we cannot stain our history by dirtying our hands with the pus flowing from this undisciplined individual. . . . I come to you, begging that you be implacable with the companion who broke our laws. . . . Calixto's attitude is incomprehensible, leaving us all filled with perplexity and bewilderment. How was it possible? Has he gone mad? His attempt to dismantle our forces has failed . . . I must ask you to be inflexible with our companion, that you apply to him the code of your conscience since we haven't yet drafted a war code, that you hurl upon him the full weight of your ire . . .'"

Reflecting on this intervention, Bayo later wrote: "I didn't know this Raúl. I knew a youth, almost beardless, but how did he grow before my eyes with his oratory? . . . He grew before me like a giant. . . . We have in Raúl a colossus in the defense of revolutionary principles. . . . If the mad assassins some day sever the life of our idol, our Fidel . . . thinking that in sacrificing him they will extinguish the light of the Revolution, they do not know, they do not have the slightest idea of the man who would pick up the torch, because Raúl is Fidel multiplied by two in energy, in inflexibility, in fiber. . . . Fidel is a little more flexible, Raúl is tempered steel. Fidel is more reachable, Raúl is a calculating machine: You press the lever, and what comes out, comes out. Fidel is a goal attained through talent, persuasion, and personality; Raúl is the ray aimed at a target. . . . Let us halt those who think of eliminating Fidel, asking more softness in his revolutionary laws. His substitute in his own right, Raúl, whom we the lovers of Revolution will follow blindly, will be more implacable with them."

Afterward, according to Bayo, the sentence on Calixto was pronounced by Fidel. He was to be expelled from the Movement and kept prisoner under armed guard until the rebels left Mexico. Fidel did not explain his reasons for saving the man's life; Calixto listened in silence. As Bayo tells the rest of the story, Calixto volunteered to cut wood, clean the kitchen, set the table, and do every chore in the camp. So that his two guards would not be deprived of continued training, he obtained permission to participate with them in the long marches, fourteen or fifteen hours at a time, without uttering a word. In the end, Fidel ordered that Calixto be permitted to march without a guard. One evening, Bayo asked Calixto why he had rebelled in the first place, but the ex-teacher only asked for a chance to fight and die in Cuba. He was given the opportunity to sail on the *Granma*, fight in the Sierra Maestra, and to redeem himself in the eyes of his companions. Bayo ran into him in Havana after victory over Batista, and Calixto told him he could now reveal the reason for his insubordination in training camp. He removed his trousers, and Bayo saw that his lower back was wrapped in tight bandages. "You see," Calixto said, "I have a bone deviation that prevents me from walking a long time because when I get tired, the pain I feel prevents me from going on. If I had told the truth, my companions in camp would have eliminated me from the invasion, which I feared the most, and I knew that in the end the death penalty would not have been applied to me . . . I would have then revealed my physical impairment . . ." After victory, Castro named him military governor of Las Villas province.

Universo Sánchez, who belonged to Castro's inner circle and was Che Guevara's deputy at Las Rosas, recalls that Calixto had actually been sentenced to death for indiscipline, and that he was charged with carrying out the execution. Fidel, however, reversed the decision. A former Communist who joined the Movement before Moncada (but missed the attack), Universo Sánchez was so highly regarded that Faustino Pérez and Armando Hart had personally arranged for him to sail to Mexico aboard an Italian ship late in 1955, to join Castro. In Mexico he became Fidel's shadow and the commander of a safe house. He was entrusted with the most delicate missions, of which counterespionage was the most important. Universo recalls that an infiltrated agent could and would be executed for treason or spying, and added: "We have shot people there." He says that in one instance, he was able to prove that a man who had never been suspected of anything, actually turned out to be a Batista spy. The man, whose identity is unknown, was sentenced by a safe house court-martial and executed on Universo's instructions by one of the rebels. "He was shot and buried there in a field," he says. Every day, Universo recalls, Movement leaders were told of suspected infiltrators, and Fidel always assigned someone to check on the reports, often leading to confrontations. As Universo says, "We always went armed with one or two pistols."

Fidel Castro was armed when Mexican policemen surrounded and arrested him in the street on the evening of June 20, 1956. He pulled out his pistol, but the policemen used Universo Sánchez and Ramiro Valdés, whom they had grabbed seconds earlier, as a screen to prevent him from firing. Fidel was disarmed at gunpoint, and forced into a police cruiser together with Sánchez and Valdés. The three of them had left a safe house on foot after being informed by a Movement member that the police were inspecting cars parked in front of the house. Fidel and his two companions were taken to the Interior Ministry jail on Miguel Schultz Street, and in the course of the night the police rounded up twelve more rebels and their friends, including María Antonia González and Alberto Bayo Cosgaya, General Bayo's aviator son. The general went into hiding for weeks, escaping arrest. Seizing documents and arms, Federal Police agents learned about the existence of the Santa Rosa ranch and prepared to raid it. Fidel, however, insisted that he be allowed to go along with the police to avoid a bloody clash. Arriving in Santa Rosa on the afternoon of June 24 with heavy police contingents, Fidel urged his companions to surrender peacefully. Thirteen men were arrested at the ranch, Che Guevara among them. But most of the weapons and ammunition had been removed to Mexico City the day before. Fidel and twenty-seven of his followers were in jail; of the Movement's leadership, only Raúl Castro was at large. It seemed as if the Castro revolution had come to an abrupt end that June week, and General Batista had succeeded in having the Mexican authorities smash the 26th of July organization. Now, Havana demanded the prisoners' extradition.

Castro had known that the plot on his life had failed, but had not realized to what extent the Mexican Federal Police were willing to act on Batista's behalf; nor did he understand adequately the labyrinthine ways of Mexican politics in which various official factions and agencies often acted independently both of each other and of the country's president. In fact, Castro had felt sufficiently relaxed about the situation that he flew to San José, Costa Rica, on June 10 to meet with Cuban refugees and Costa Rican politicians, returning to Mexico just in time to be arrested.

The assumption within the *Fidelista* camp was that they had been betrayed from within by an agent or a traitor, because the Mexican police knew everything about the location of the safe houses and addresses of their Mexican friends. But the urgency at hand was to obtain the freedom of Castro and his colleagues before they were deported to Cuba. Juan Manuel Márquez, one of Fidel's top associates, rushed back from the United States, where he was buying arms and collecting money, and he and Raúl Castro were able to retain two influential Mexican lawyers to defend the prisoners. On July 2 a judge ruled that the Cubans should be released, but the Interior Ministry refused to do so; the judge's order, however, prevented their deportation. In the meantime, Castro found it necessary to defend himself from charges that he had Communist ties. He had been accused of planning Batista's assassination in Havana, and the Mexican police charged that Castro had come to Mexico with the help of Lázaro Peña, the Cuban Communist labor leader, and Vicente Lombardo Toledano, the Mexican Communist labor leader. Moreover, the police described him as one of "seven Communists" detained in the June 20–21 roundup.

Clearly, it did not suit Castro to be identified as a Communist; so on

June 22, he wasted no time in issuing a denial from prison, a denial couched in extremely careful language. Published in *Bohemia* in Havana the following week, Fidel's statement read: "No one in Cuba is unaware of my position toward communism, for I was a founder of the *Partido del Pueblo Cubano* [the *Ortodoxos*] along with Eduardo Chibás, who never made a pact nor accepted any type of collaboration with the Communists." He added that his Movement had no contacts, either, with former President Prío. In a long article in *Bohemia* on July 15, Castro went back to the Communist issue in considerable detail:

"Naturally the accusation of my being a Communist was absurd in the eyes of all who knew my public path in Cuba, without any kind of ties with the Communist party. But that propaganda is elaborated for the consumption of Mexican public opinion and international news agencies and for the purpose of adding the pressure of the American embassy to that which they have been applying to the Mexican authorities. . . . Captain Gutiérrez Barros himself read me the report forwarded to the president of Mexico after a week of minute investigation; among its observations it was categorically affirmed that we had no ties whatsoever with Communist organizations. . . . What moral authority, on the other hand, does Mr. Batista have to speak of communism when he was the Communist party presidential candidate in the elections of 1940, if his electoral posters took shelter under the hammer and sickle, if his pictures beside Blás Roca [then the Communist party's secretary general] and Lázaro Peña are still around, if half a dozen of his present ministers and trusted collaborators were well-known members of the Communist party?"

This was the first time that Castro had publicly connected the Communists with Batista, and the question arises whether he was doing it because he was truly angry at the party, which was ignoring him, or as a tactical maneuver to save the Movement. Blás Roca told an American journalist in 1974 that "it was a tactic," but at that stage there was nothing else he could have presumably said; Castro himself is not known to have returned to the subject in public. But the fact remains that at the time of his imprisonment in Mexico, Castro and the Communists were very much at odds over revolutionary strategy.

Fidel wrote in the *Bohemia* article that "it seems a common thing in my public life to have to wage the most difficult battle in favor of truth from a cell," and he devoted his Mexican detention to writing and speaking without pause on behalf of his revolution. Behaving with considerable aplomb, he received visitors in the patio of the jail (the Mexicans were lenient about his social life in detention), always wearing a suit and tie, making new friendships, and sending out instructions to his underground Movement. Teresa Casuso first met him in the jail yard when she paid a call on him.

Describing the jail scene, she wrote: "More than fifty Cubans were gathered in the large central courtyard. . . . In the middle, tall and clean-shaven and with close-cropped chestnut hair, dressed soberly and correctly in a brown suit, standing out from the rest by his look and his bearing, was their chief, Fidel Castro. He gave one the impression of being noble, sure, deliberate—like a big Newfoundland dog. . . . He looked eminently serene, and inspired confidence and a sense of security." For years, Casuso has belonged to the sisterhood of highly intelligent and impressive women supporters of Fidel.

Universo Sánchez remembers that Castro had borrowed the "brown suit" which Casuso saw Castro wearing in jail because "Fidel had very bad clothes, and since he was put on television at lot, we wanted Fidel to look elegant." The suit was borrowed from Armando Bayo, the Spanish general's other son, and Universo remarks that "for Fidel to represent the group, he needed a little *cachet* . . ." It was also Universo who, with Castro's permission, tried to bribe a senior Mexican official with $25,000 to release the group (although, as Fidel admitted later, the Movement only had twenty dollars in its treasury at that point), triggering a shocked reaction. But the shock stemmed from the size of the proposed bribe, and Universo says he subsequently learned that "they thought they had grabbed something very important, much more important than what it really was, because I had offered so much money."

The bribe did not work, but the Mexican authorities began releasing the Cubans anyway. Twenty-one rebels were sprung by July 9, including Universo Sánchez, Ramiro Valdés, and Juan Almeida, and four more later that week. That left only Fidel, Che Guevara, and Calixto García still in jail on charges of residing in Mexico with expired permits. In desperation, the Movement lawyers succeeded in contacting ex-President Lázaro Cárdenas, the old revolutionary, and he agreed to intercede directly with President Ruíz Cortines. Because nobody in Mexico, president or not, could ever refuse any request from the legendary Cárdenas, the government released Fidel on July 24, more than a month after his arrest. Che and Calixto García were let go on July 31. Fidel made a point of calling on Cárdenas to thank him.

The revolution was back in business as Castro ordered the Movement reorganized, and most of the fighters and arms moved to Mérida in Yucatán (where the police seized, then returned a weapons cache) and to Veracruz and Jalapa on the southeast coast of the Gulf of Mexico. Fidel remained in Mexico City with a contingent. Che Guevara wrote later that the Mexican police, paid by Batista, had "committed the absurd error of not killing [Fidel] after taking him prisoner." Guevara also noted that when he and Calixto García were the only ones left behind in jail, he had urged Castro not to let the revolution wait for him—he did not know how much longer he would be imprisoned—but Fidel replied simply, "I won't abandon you." To Che, this was a gesture that subordinated his revolutionary attitude to personal friendship, and he remarked that "these personal attitudes that Fidel has with people whom he appreciates are the key to the fanaticism that he creates around him . . . and to an adhesion to principle is added a personal adhesion, making this Rebel Army an indivisible block." Che also composed a "Canto to Fidel," an epic revolutionary poem, ending on the note "If iron halts us on our way/We ask a kerchief of Cuban tears/To cover the *guerrillero* bones/In transit to American history/Nothing more . . ." Guevara told his wife he would give the poem to Castro "on the high seas, en route to Cuba."

Fidel Castro turned thirty on August 13, deep in revolutionary politics and invasion planning, making up for the month in prison and the loss of Santa Rosa. Forty new carefully selected recruits arrived from Cuba and the United States, ten of them Moncada veterans. The new contingent included Camilo Cienfuegos Gorriarán, a young Havana worker who had been wounded in an affray with the police, then had emigrated to California where he married an American woman. His older brother, Osmany, was already in Mexico, but not as a member of the Castro Movement; he belonged to a clandestine group of Cuban Communists that maintained liaison between Havana and Mexico City. During August Frank País, the twenty-one-year-old Oriente coordinator of the 26th of July Movement and the son of a Protestant minister, traveled secretly to Mexico to confer with Fidel about the support to be given the rebels the moment they landed in Cuba. Castro, who remained determined to keep his word and to land in Cuba before the end of the year despite his summer reverses, proposed that the arrival of the invasion force be accompanied by armed uprisings throughout Oriente, and that the political climate be prepared for a general strike. The idea was to force the Batista army to become distracted by multiple armed actions, making it easier for the rebels to reach the Sierra Maestra. In the meantime, Castro and País agreed not to send any more men to Mexico, but to start building up the Movement militarily at home in anticipation of the invasion. Simultaneously, Fidel instructed the Movement in Oriente to retain for itself the collected funds that were normally sent to Mexico, or 80 percent of the total (the balance went as usual to the Movement's National Directorate).

But Castro had also concluded that the time had come for unity among various revolutionary movements in and out of Cuba. In Havana, where Batista was attempting to persuade the opposition to participate in congressional elections in November 1957, Castro's call for unity became the central subject of political conversations. In an interview printed in Cuban newspapers, he acknowledged that the unity he now advocated represented a change in his "tactical line," saying that it was necessary to absorb the lessons of reality. "We can discuss later," he said, "but now only the struggle is honorable." Castro had broken away in March from the *Ortodoxo* party, yet the summer setbacks required him to seek outside support and alliances. Privately he believed he would always remain in control.

The first step was the pact he worked out with José Antonio Echeverría, the twenty-four-year-old president of the University Students' Federation (FEU) and secretary general of the Students' Revolutionary Directorate (DR). Echeverría arrived in Mexico on August 29, meeting with Fidel for forty-eight hours, virtually without rest, at the apartment on Pachuca Street where Castro now lived with Melba Hernández, Jesús Montané, and Cándido González. On August 30 they signed the "Mexico Letter," declaring that their respective organizations had decided "to unite solidly their efforts in order to oust the tyranny and carry out the Cuban revolution." Castro and Echeverría added that the social and political conditions in Cuba were "propitious" and the revolutionary preparations sufficiently advanced to offer liberation to the nation in 1956, and that "insurrection backed by a general strike in the whole country will be invincible." In practice, the 26th of July Movement and the DR committed themselves to organizing armed actions throughout Cuba ahead of the invasion to create a revolutionary climate and after the invasion to coordinate with Castro's rebels. However, no specific plans were outlined, and Castro and Echeverría made it clear that their organizations would remain independent; no joint command was contemplated by either side. Besides, the immediate problem was that neither group had enough weapons to launch serious operations.

In the meantime, the Batista regime sought to undermine Castro by accusing him of accepting money and arms from the Dominican Republic's rightist dictator, Rafael Leónidas Trujillo, and to plan with him a joint invasion of Cuba. This was patently absurd, given Castro's involvement in the 1947 conspiracy against Trujillo and his general opposition to dictatorships. He fired off a four-column letter to *Bohemia,* declaring that "the barrage of calumny hurled against us by the dictatorship exceeds all limits. . . . It had [just] been imputed to me to be a member of the Mexican-Soviet Institute and a Communist party militant." Castro wrote that he continued despising Trujillo, and that he believed that a "revolution of principles is worth more than cannon. . . . We shall never change our principles for the arms that all the dictators may have together. . . . Batista, on the contrary, will not renounce the tanks, the cannon, and the aircraft the United States sends him, not to defend democracy, but to massacre our defenseless people." Now Castro was the favorite theme of discussions and polemics in the Cuban press—to his immense satisfaction. It was propaganda at work for his cause.

The cause, however, urgently needed funds to succeed. Many of his friends at the time claimed that Castro worked out a secret arrangement with Carlos Prío, the millionaire ex-president of Cuba he had so often denounced, to secure money for the 26th of July Movement. Castro never confirmed it, but in an article published in *Bohemia* in August, he went out of his way to say that "Batista has been merciless with Prío beyond all limits, by insults, taunts, and humiliation. . . . When we were arrested in Mexico, and people spoke insistently about our deportation, Prío—a man I have fought several times—was very much a gentleman. . . . He wrote in his capacity as former president of Cuba an open letter to the president of Mexico asking him not to deport us."

According to most accounts, Castro's search for funds came under the heading of revolutionary unity among anti-Batista groups. After his release from jail, he met in Yucatán with Justo Carrillo, the president of the Cuban Agricultural and Industrial Development Bank before the 1952 coup and who was close to the Montecristi military anti-Batista faction whose rebellion had been recently aborted. Reportedly, Carillo gave Castro five thousand dollars. But the crucial source of funds was Prío. Teresa Casuso—Fidel's new friend and the widow of the famous poet Pablo de la Torriente Brau killed in the Spanish civil war—was Prío's friend also. Casuso says that on Castro's request, she flew to Miami where she spent five days with Prío, who "was eager to talk to Fidel." As a result, Castro entered the United States illegally sometime in September, according to friends' accounts, to meet Prío at the Casa de Palmas Hotel in McAllen, Texas. Many published accounts claim that Castro immediately received perhaps as much as fifty thousand dollars from the former president, whom he had once investigated for corruption. Casuso says that "Prío helped to sustain the expense of two years of costly expeditions . . . and clandestine shipments of arms and men."

Late in the spring, Castro saw in an arms catalog a photograph of a PT (patrol torpedo) boat for sale in Dover, Delaware, on the Delaware River. Equipped with torpedoes and 40-mm. cannon, this boat was known for both its speed and maneuverability. Antonio del Conde Pontones, a Mexican arms supplier known as "El Cuate" and Castro's friend, was dispatched to Dover with Jesús Reyes to look the PT boat over. The craft was in good condition, so the two emissaries agreed to pay twenty thousand dollars for her, making a down payment of ten thousand dollars in mid-June; they were to take possession of the PT boat and make the final payment on their return within a few weeks. However, Castro was then arrested and El Cuate was picked up by the police on his return to Mexico. In August El Cuate,

Onelio Pino, and Rafael del Pino, the student who had gone to Bogotá with Castro in 1948 and now belonged to the Movement, were sent back to Delaware with another ten thousand dollars to sail the PT boat to a Mexican port. Because everything seemed in order, El Cuate was ordered to Miami to meet with Prío while the two Cubans were completing arrangements for the craft.

El Cuate says that in Miami he was introduced to Prío by Juan Manuel Márquez, who was then representing the Movement in the United States, and the ex-president handed him twenty thousand dollars, "which I delivered to Fidel in Mexico." The fresh money was providential because at this juncture it was impossible to obtain an export permit for the PT boat from the United States government, and for obscure reasons the owner refused to refund the money. Normally, export licenses were granted easily, but now the State Department was reluctant to do so in view of unrest in the Caribbean, and it was too risky to sail the boat illegally. The revolution was now short a precious twenty thousand dollars, and it had no vessel to take it to Cuba in time to fulfill Castro's pledge. Briefly, Fidel considered buying a Catalina PBY flying boat to land him off the Cuban coast, but he dropped the idea along with Universo Sánchez's proposal that they start the war by bombing La Cabaña fortress in Havana from a plane.

Late in September, Castro and El Cuate went to the hills above the port of Tuxpán on the Gulf Coast between Tampico and Veracruz so as to test .30-'06-caliber Remington automatic rifles in topographic conditions resembling Cuban sierras. El Cuate told Castro that he wanted to go down to the Tuxpán River to look at a yacht he was buying for himself, but when Fidel saw her, he said, "In this boat, I'm going to Cuba . . ." The Mexican argued that the white craft was a luxury boat, too small for an expedition. But Castro said, "If you can get it for me, I'll go to Cuba aboard this one," and El Cuate agreed. As he said later, "You just can't say no to Fidel . . ."

The yacht was the *Granma,* and she belonged to Robert B. Erickson, an American who lived permanently in Mexico City. The wooden thirty-eight-foot yacht had been built in 1943, and she could carry up to twenty-five persons safely. *Granma* was propelled by two diesel engines, carrying fuel tanks holding less than two thousand gallons. But she had sunk during a 1953 hurricane, remaining for a time under water, and much work was required to make her seaworthy again. Erickson was willing to sell her for twenty thousand dollars, providing that he could get another twenty thousand dollars for a modern house he owned on the Tuxpán River. Castro decided to go ahead with the purchase because, he reasoned, they did need a house for the men who would be working on *Granma* and for those ready to leave. A seventeen-thousand-dollar down payment was made to Erickson, and Castro ordered that work on the yacht start instantly. Two rebels were assigned to live in the house, and Onelio Pino was named *Granma*'s captain.

Fidel was supremely optimistic now that he had acquired *Granma,* but he kept an eye on revolutionary politics as well. Pedro Miret, Faustino Pérez, and Ñico López had arrived in Mexico to join him on the expedition; thus the entire general staff of the Movement was with their chief. The FEU student leader, José Antonio Echeverría, returned to Mexico City in mid-October for another conference with Castro; the differences in their personalities and approaches to the revolution were etched even deeper than at their first meeting. Apart from being natural political rivals, Echeverría and Castro saw tactics in different ways, the former insisting on continued violent action, and the latter preferring more coordinated operations; clearly at stake was the ultimate leadership. One week after Echeverría returned to Cuba, however, Students' Revolutionary Directorate commandos ambushed and killed Colonel Manuel Blanco Rico, the chief of the Military Intelligence Service (SIM), as he was leaving the Montmartre nightclub. The attack, carried out by Juan Pedro Carbó Servía and Roland Cubela (who in the 1960s tried to assassinate Castro on behalf of the CIA), had originally been intended for Interior Minister Santiago Rey, but when he did not turn up as expected, the colonel was shot instead. Castro took a dim view of this event, saying in a newspaper interview: "I do not condemn [assassination] attempts as a revolutionary weapon if the circumstances require it. But such attempts cannot be indiscriminately perpetrated. I do not know who carried out the assault on Blanco Rico, but I do believe that, from a political and revolutionary standpoint, assassination was not justified because Blanco Rico was not an executioner." In Havana the police raided the Haitian embassy looking for Carbó, who was not there, but killed ten other youths in political asylum.

On October 24 Frank País brought more bad news from Santiago. País came to persuade Castro to postpone the rebel landing until sometime the following year because, as he had said in an earlier letter sent through secret channels, he had doubts about the efficiency of his armed groups in Oriente "because they were unprotected, unprepared, and uncoordinated." País and his friend Pepito Tey had worked hard to organize a clandestine network in the province, collecting arms and readying the 26th of July units to rise in the cities while others covered the rebels' landing along Oriente's western coast between Manzanillo and Pilón. In charge of the Movement groups in the landing zone was Celia Sánchez, one of Frank País's closest collaborators. Still, País felt an immediate invasion was inadvisable. Fidel and he spent five days arguing, and the young man finally accepted that the landing would come within the next two months. Castro was simply adamant, stressing that his credibility would be destroyed if he broke his promise to return to Cuba in 1956, and that after the June imprisonments it was too risky to stay longer in Mexico.

Then Castro faced a problem with the Communists. In mid-October Osvaldo Sánchez Cabrera, a leader of the illegal Popular Socialist Party (PSP) who spent most of his time in Mexico, had attempted to talk Fidel out of invading Cuba. Speaking in the name of the party, Sánchez Cabrera proposed that the expedition be postponed until late January, when the sugar harvest would begin, so that a sugar workers' strike could be launched to support the landing. Again, Fidel explained that it was vital to keep his word. In mid-November the PSP leadership met secretly in Havana to dispatch another emissary to Castro; this time, it was Flavio Bravo, formerly the secretary general of Socialist Youth, a university friend of Fidel's, and the ideological mentor of Raúl. Lázaro Peña, who also resided in Mexico, was to bring them together, but Flavio ran into Melba Hernández and Jesús Montané in the street, and the contact with Fidel was made at once.

The Communists' message was that in the party's opinion the internal situation in Cuba was "unfavorable to military action prior to December 31," that an invasion did not take realities into account, and that it could result in a failure. Flavio Bravo reminded Fidel that at one stage in the independence war, José Martí himself recognized the need to "postpone military action and create more favorable material and subjective conditions." But Fidel thought he understood "subjective conditions" better than the PSP (not to mention Martí), and he said so to Flavio Bravo. The envoy continued to insist that anti-Batista opposition was "very disunited," and that the Communist party's line of "united front and mass struggle" was not progressing well. The PSP, he said, wished to bring all the young revolutionary groups together before embarking on an insurrection because otherwise the masses would not follow. Specifically, the Communists urged Castro to accept this concept of unity—which presumably they would strive to direct—and postpone the invasion. Then, Fidel would issue a short document denouncing Batista, calling for unity of the opposition, demanding general elections with guarantees for all the political parties. The Communists suggested that it would best be couched as an "open letter" by Castro to workers, students, peasants, the youth, and all the civic institutions. It would be, Flavio Bravo said, the "final call" for a peaceful solution, and the anticipated refusal by Batista would justify before public opinion the turn to armed action against the dictatorship.

Castro patiently explained to Flavio Bravo that he had no alternative but to move soon. Not only had he a promise to keep, but the Mexican police were on the offensive again. A cache of arms was confiscated, and both Pedro Miret and Teresa Casuso had been arrested. If he did not sail as soon as possible, Castro said, he could lose all his men and all his arms. He hoped that his arrival would be met by uprisings, and though there was very little time left, he was asking the Communist party for its cooperation. Then Castro advised Flavio Bravo to fly directly to Havana instead of traveling circuitously, so that the Communist party at home would be apprised immediately why the Movement's tactics could not be altered. Subsequent official accounts make this exchange appear as a friendly review of the situation, but for Fidel it was a tough political fight; the Communists still had no faith in him, and still intended to take over his revolution.

On that same day, Fidel Castro learned that his father, Don Ángel, had died in Birán on October 21. Nothing is known of Fidel's reaction; there are no known letters about it.

On November 19 General Francisco Tabernilla, the Army Chief of Staff, told the press in Havana that "there is no possibility of a landing as announced by Fidel Castro" because "from a technical viewpoint, a landing by a group of exalted and undisciplined persons without military experience and without means for combat must result in a failure." At the same time, however, Cuban warships and aircraft were patroling the coasts from Pinar del Río in the west to Oriente in the east, and army and Rural Guard garrisons were on the alert.

Fidel Castro realized that the decision to sail had to be taken immediately, but he was delaying it mainly because *Granma* was not quite ready; much of the work on the engines and the general seaworthiness was being done too quickly and carelessly because of time pressures. Then, on November 21, when two rebels defected from the Abasolo training camp near the United States border south of Matamoros, Castro knew that he could not wait any longer. On the night of November 23 he traveled from Mexico City to the house on the river in Tuxpán to supervise the loading of arms, ammunition, and supplies aboard the *Granma;* simultaneously, rebel commanders at all the training camps from Abasolo and Veracruz to Mexico City were ordered to move their men to Tuxpán. They began arriving by bus and car under torrential rain.

Before departing Mexico City, Castro sent coded messages to Frank País that he would land on November 30, and that the chosen spot would be Playa las Coloradas below the town of Bélic, south of Niquero on the west coast of Oriente. Then on November 24, while driving to Tuxpán, he wrote his will, to be sent to the friends with whom Fidelito had been staying for more than a year. He wrote: "I leave my son in the custody of Engineer Alfonso Gutiérrez and his wife, Orquídea Pino. I am making this decision because I do not want, in my absence, to see my son Fidelito in

the hands of those who have been my most ferocious enemies and detractors, those who, in a base act without limits, and using my family ties,
attacked my home and sacrificed it to the interest of a bloody tyranny that
they continue to serve. Because my wife has demonstrated herself to be
incapable of breaking away from the influence of her family, my son could
be educated with the detestable ideas that I now fight. I am adopting this
measure not out of resentment, but thinking solely of my son's future. I
leave him to those who can educate him best, a good and generous family,
my best friends in exile, in whose house the Cuban revolutionaries found a
true home. I leave my son to them and to Mexico, so that he can grow and
learn in this friendly and free country where children have turned into
heroes. He should not return to Cuba until it is free or he can fight for its
freedom. I hope that this just and natural desire on my part with regard to
my son, the only one I have, will be fulfilled."

(On December 15, days after the landing of the *Granma*, Fidel's sister
Emma reported to the Mexican Federal Police that "three unknown persons, armed with pistols, intercepted the automobile in which we were
traveling at the corner of Revolución and Martí avenues, seizing my
nephew, Fidel Castro Díaz, seven years old . . ." In Havana Foreign Minister Gonzálo Güell announced that "the child is with his mother, which
excludes the possibility that it could be considered a kidnapping . . .")

Wearing a black cape over his dark wool suit, Castro stood in the rain on
the dock, watching his eighty-one men going aboard the small white
yacht. Tuxpán Harbor was closed because of stormy weather, but El Cuate
convinced his friend the harbormaster to let him sail "because I was planning a little party aboard." Universo Sánchez asked, "When do we get to
the real ship? Where is the mother ship?" At 1:30 A.M. on November 25,
Granma started her engines, slid out of her slip, and sailed without lights
along the river, west toward the sea and Cuba. El Cuate followed the yacht
in his car along the river road until she reached the open sea and vanished
in the darkness.

CHAPTER

9

The crossing of the *Granma* from Tuxpán on the Mexican coast of the Gulf to the coast of Oriente province in Cuba was pure nightmare, lasting seven days and four hours instead of the five days and nights that Fidel Castro had planned. Terrifying weather whipped up by a powerful *El Norte* wind, mechanical breakdowns, the yacht's staggering burden of eighty-two heavily armed men in lieu of the twenty-five passengers she was built to transport (Castro had left an additional fifty rebels behind simply because there was not even a square inch of space to spare aboard), and the presence of only three professional sailors contributed to the misery and the delay. The delay, in turn, provoked an unnecessary tragedy ashore.

Castro's plan was risky because of the number of men aboard the *Granma,* but it was not unreasonable. In relatively good weather, it should have worked to the extent that they would arrive at the landing point roughly on time, and the chosen route—adding up to 1,235 miles—made strategic sense. The yacht was to sail a virtually straight west-east track: from Tuxpán to the exit from the Gulf of Mexico at the tip of the Yucatán peninsula; then crossing the passage between the peninsula's tip and the westernmost point of Cuba (potentially the most dangerous area because Cuban Navy and Air Force patrols could spot the *Granma* the most easily there); dipping south at a safe distance from Cuba's southern coasts; and hitting the western shores of Oriente province below Niquero. The risk of being caught in the Gulf of Mexico was minimal because operationally it was too far for Batista's air and sea forces, and once the vessel had entered the Caribbean, Castro capitalized on the deception factor of sailing east far from the Cuban coast until Oriente coast was sighted, then making a run for it. Batista's ships and planes were not patroling that far south; Castro was practically in British waters off the Cayman Islands.

The extremely foul weather rendered the crossing exceedingly difficult, with the *Granma* riding the waves, much too heavy, mechanically unfit, and poorly steered. Castro could not delay the departure because he feared the Mexican authorities would pounce on the expedition at any moment, liquidating the whole enterprise. His decision was to sail immediately, betting that luck would stay with him. He dispatched a telegram from Mexico City to Frank País in Santiago on November 27 reading: ORDERED BOOK IS OUT-OF-PRINT, signed by *Divulgación* publishers, which was the code for announcing the departure and the anticipated landing early on November 30. Two other coded wires went to Havana for the Students' Revolutionary Directorate and to Santa Clara for the local underground.

Bad weather, however, worked against Batista, too. The regime began to believe privately that Castro was mad enough to make the dash to Cuba, and its intelligence services came up with a list of vessels in Mexican ports that the rebels might be planning to use in an invasion; the list included the name of *Granma*. Starting on November 5, and in total secrecy, the air force began flying constant patrols along the north and south coasts of Oriente province using one or more B-25 light bombers or C-47 transports (Cuban history had persuaded Batista, as it had Castro, that the landing *had* to be in Oriente). But along the southern coast, the aircraft patroled only 20 miles south of the coastline, which was what Castro had figured they would do, while he sailed 170 miles south of it. The generally bad weather in the last days of November also seriously curtailed all the Batista air and sea operations. The regime did begin deploying ground forces to Oriente. Artillery units were flown from Havana to Holguín the day after the *Granma* sailed, and the Santiago garrison was reinforced and placed on alert.

For the expeditionaries the horror began the instant they entered the Gulf of Mexico, just before daybreak on November 25. They hailed the open water by singing the Cuban national anthem and the 26th of July March, and shouting, *"Viva la Revolución!"* and "Down with Batista dictatorship!"—then the sea attacked them. Immediately, most of the men became violently seasick (and as Universo Sánchez recalled, "shitting in their pants"). They were not a fighting force, just a band of very sick men.

Che Guevara wrote that "the entire boat had a ridiculously tragic aspect: men with anguish reflected in their faces, grabbing their stomachs; some with their heads inside buckets, and others fallen in the strangest positions, motionless, their clothes filthy from vomit . . . except for the two or three sailors and four or five others, the rest of the contingent were seasick." Che, Fidel, and Faustino Pérez were among those who did not succumb to illness; Guevara frantically searched the ship for antihistamines for the men, but there were none. Then, the *Granma* began to take on water, the pump turned out to be broken, and they had to bail water with two buckets until the leak was located and fixed.

On the third day the weather improved, and Castro ordered rifles calibrated again and some firing exercises began. But it also developed that the yacht was running at 7.2 knots instead of the 10 knots Captain Onelio Pino and Castro had calculated. Zigzagging to cope with the weather forced further delays, and one of the engines began to fail. The expedition fell badly behind schedule. When the men began feeling better, they became hungry, and Castro had to order rationing as he realized that the *Granma* would never reach land within the planned five days. In the rush of departure from Tuxpán, they could take aboard only 2,000 oranges, 48 condensed milk tins, 4 baked hams, 2 sliced hams, a box of eggs, 100 chocolate bars, and 10 pounds of bread; this could not keep 82 men adequately in food for over a week. During the last two days, there simply was neither food nor potable water. On November 29 the *Granma* came within sight of two fishing boats, but as Castro readied their two antitank guns for possible combat, the fishermen disappeared.

At dawn on Friday, November 30, the expedition was cruising toward Great Cayman Island, only three fourths of the way to the landing zone. But that was the moment when Frank País and the 26th of July armed groups in Santiago thought Castro would be coming ashore as planned, and País ordered an uprising to coincide with the landing, according to the battle plan he had elaborated with Castro. There was no way he could know that the *Granma* was running forty-eight hours late, and therefore at 7:00 A.M. his pathetically small detachment of twenty-eight men attacked the National Police and Maritime Police headquarters as scheduled, hoping to be able to follow this action up with an assault on Moncada. Wearing olive-green uniforms with 26th of July red-and-black armbands, the rebels set fire to the National Police barracks, but lost Pepito Tey, one of the top Santiago leaders, to an enemy machine-gun barrage. At the Maritime Police they captured weapons, but were unable to move on Moncada; the army had four hundred highly trained antiguerrilla troops in the city. Though some street fighting went on for another day, the rebellion had collapsed. The Movement lost three key fighters (Frank País managed to escape), and nine others were killed by the police. Not only had the uprising failed, but now the regime knew that Castro was about to land—somewhere. In Havana and elsewhere on the island, the Movement and the Students' Revolutionary Directorate lacked sufficient means to undertake any kind of armed action. Fidel listened to the Cuban radio's reports of the Santiago tragedy, gritting his teeth in impotent rage. He said to Faustino Pérez: "I wish I could fly . . ."

On the beaches between Niquero and Pilón, 26th of July Movement members also awaited Castro in vain that dawn. Celia Sánchez Manduley, the Movement's coordinator in Manzanillo, had assembled five trucks, gasoline drums, and several dozen men near the town of Bélic and the Colorada beach. The plan was to transport the *Granma* expeditionaries from the beach to Niquero and Media Luna, to capture arms from the local Rural Guard garrisons, then move with 26th of July adherents to the Sierra Maestra, where Castro was to launch his guerrilla war. Inland, the Movement's peasant supporters had prepared their mountain homes to receive and feed the *Fidelistas.* But when Castro did not land and word of the Santiago fiasco reached Celia, she ordered the reception parties pulled back in the evening of December 1. The *Granma* force would be on its own—if and when it reached land.

Late at night on Saturday, December 1, the white yacht was wallowing in high seas as she approached the Oriente coast in total darkness: no moon and no visible coastal lights. Castro ordered the rebels to change into their olive-green uniforms and he distributed the weapons. Crewmen kept climbing to the roof of the cabin to spot the Cabo Cruz lighthouse for a navigational fix—Cabo Cruz was on the southwestern tip of the Oriente coast—when Roberto Roque, the navigator, slipped and fell overboard. Castro ordered the *Granma* to undertake a search for Roque despite the darkness. After sailing in circles for an hour, the rebels heard a weak voice respond to their calls, and incredibly they found him, using only a lantern shining over the waves. Che Guevara and Faustino Pérez, both physicians, revived the nearly drowned Roque, and Castro proclaimed that now they were on their way to victory.

Resuming its careful progress toward the coast, *Granma* entered the Niquero channel, but as he noted the buoys, Captain Pino realized that his charts were wrong, and he did not know his way. The dawn of Sunday, December 2, was just breaking when the yacht suddenly hit mud at low tide and came to a dead halt at 4:20 A.M. The spot was Los Cayuelos, more than a mile south of the beach where Castro had wanted to land (and just below the ironically named Purgatorio Point). The men were ordered to jump into the water, carrying only their personal weapons. All the heavy equipment and stores were left behind. René Rodríguez, slight of build, was the first to go, and the bottom held him; the much heavier Castro followed, sinking up to his hips in the mud. Che Guevara remarked later, "This wasn't a landing, it was a shipwreck." The yacht was stuck some hundred yards from what appeared to be the coast, and Fidel and his men managed to wade to it. Che Guevara and Raúl Castro were the last to leave the *Granma,* trying to salvage some equipment.

Reaching the shore, the rebels realized they were in a huge mangrove swamp with water up to their knees or even their necks. Gnarled tree roots rose like an awesome obstacle course, vines and razorlike leaves slashed and beat their faces, and vast clouds of mosquitoes tried to devour them alive. The men's brand-new heavy boots slowed their advance; some boots and uniforms were so soaked and cut they began coming apart; rifles and ammunition became wet; equipment was lost. The general staff, consisting of Fidel Castro, Juan Manuel Márquez, and Faustino Pérez (the former with the title of Commander in Chief, and the two others with the rank of captain), led the way, the men constantly tripping over submerged tree trunks, falling down, picking each other up, leaning on one another, and somehow succeeding in moving ahead. One must attempt to cross this mangrove swamp oneself even to begin to understand the lung-bursting effort it represented.

At one stage Castro developed the paralyzing fear that they had landed on a coastal key (there are close to two thousand such keys off the Cuban coasts) and not on the mainland, and that they were trapped without means of escape by water. But soon one of the men, Luis Crespo, was able to climb a tree and, in the first light of the winter morning, discern land, palm trees, huts, and mountains in the distance. It took Castro's guerrilla army over two hours to reach firm ground across the mangrove and a lagoon in the center of it, a distance of less than a mile in a straight line; it was a frightening and exhausting experience for them after a week at sea in the overcrowded little yacht. When they finally reached firm ground, they collapsed, panting, to rest. But Juan Manuel Márquez and seven men were missing; they seemed to have been swallowed by the swamp, and their companions were immensely concerned about them.

Still, Fidel Castro had fulfilled his promise: He had returned to Cuba before the end of 1956, and now he was ready to open his war on Batista. Like José Martí, who had landed at Playitas in the dark with a handful of companions sixty years earlier, Fidel Castro stood on the coast of Oriente on this December 2, anxious to liberate Cuba from her domestic enemies.

It was the beginning of a twenty-five-month war, and it required astonishing optimism and faith on Fidel Castro's part at that moment to think that he and his expeditionary force could survive, let alone be ultimately victorious. On the afternoon of December 2, Castro's army had eighty-one men (Juan Manuel Márquez and the seven others emerged from the mangrove slightly to the north, rejoining the main group), minimal armament, no food, and no contact with the Movement ashore. The Batista regime had a standing army, navy, and air force of over forty thousand men—plus the militarized Rural Guard and the National Police. It had Sherman tanks and artillery, and in mid-November, just before Castro sailed from Tuxpán, the American ambassador in Havana, Arthur Gardner, had turned over a squadron of T-33 jet trainers (usable in combat) to the Cuban Army air force. The United States continued its support for the Batista government, oblivious that opposition to it within Cuba was rising daily, and obviously ignorant that a revolutionary band was on the verge of landing on the island.

Batista, however, was ready for Castro. A forty-five-day suspension of constitutional guarantees was decreed when Frank País rose in Santiago on November 30, and less than two hours after the landing, the military authorities knew that Castro was back in Cuba. The sand-carrying barge *Jibarita* and a coastal craft had observed the *Granma* stuck in the mud off Los Cayuelos, and immediately informed the navy. The news of the landing by unknown men on the southwestern coast reached the commander of the Rural Guard at Manzanillo after seven o'clock in the morning, just as the rebels were reaching firm ground. A patrol was sent out at once, but it returned late in the day without finding the *Fidelistas*. The commanding officer reported, however, that local peasants spoke of "some 200 men, well armed, and directed by Dr. Fidel Castro."

It was Castro who had identified himself to the peasants. On the high ground above the mangrove, the rebels came upon the shack of Ángel Pérez Rosabal, a charcoal burner and the first person they saw ashore. Castro said, "Have no fear, I am Fidel Castro, and we came to liberate the Cuban people." This was exactly the sort of thing that Castro would say, but official accounts that Rosabal, a destitute and illiterate peasant living in a poor and thinly populated area, had heard of Castro beforehand are less than credible. In any event, Rosabal invited Castro and several of his men inside his hut and shared food with them. At that moment, hearing powerful explosions along the coast, Fidel ordered a forced march into nearby hills, with Rosabal guiding the column. The explosions came from a bombardment of the mangrove by a coast-guard vessel and army aircraft after the discovery of the *Granma,* and Castro feared that air attacks would be made on coastal huts as well. Most likely, a peasant in the area had heard Castro introducing himself to Rosabal, and had passed the word on to Rural Guard soldiers. Rosabal himself returned to his home by midafternoon. In the hills the hungry rebels came upon two other peasants, who showed them a well and gave them water; they also found a beehive and helped themselves to honey.

While the Castro contingent halted for the night on a wooded hill, their first night in Cuba, the regime claimed total victory. Rural Guard units and an artillery battalion converged on Niquero, where a rebel attack was expected, and additional reinforcements were ordered to the region. Although government forces failed to locate the rebels that Sunday, General Pedro Rodríguez Ávila, the army inspector general in command of the operations in Oriente, informed the press that military aircraft had "strafed and bombed tonight the expeditionary force, annihilating forty members of the supreme command of the revolutionary 26th of July movement . . . among them its chief, Fidel Castro, thirty years old." The general said further that the army had collected the bodies of the rebels, and that, besides Fidel, the cadavers of Raúl Castro and Juan Manuel Márquez were identified by documents in their pockets. The rebels, the general reported, were "literally pulverized" by the air attacks, and official sources indicated that the bodies would be brought to Havana by navy ships after having been "temporarily" buried in shallow ground. This army report was the origin of the news, disseminated worldwide by Francis McCarthy, the United Press bureau chief in Havana, that Castro had been killed, and that his identity was confirmed by the passport he carried in his pocket. At first, the story was believed in Cuba and abroad, but before long the Batista regime paid a dear price in credibility for false reporting (Castro never forgave McCarthy for prematurely announcing his death; the UP newsman had to leave Cuba when the *Fidelistas* won).

The Sierra Maestra is a massif rising along the south coast of Oriente province, from the western foothills right past Cabo Cruz all the way to Santiago in the east. It runs roughly eighty miles on a west-east line, and some thirty miles at its broadest north-south stretch. The spine, called *el firme* in Spanish, averages 4,500 in altitude, and its highest point (also Cuba's highest) is Pico Turquino, slightly over 6,000 feet. The Sierra Maestra's terrain is forbidding—mountain peaks and valleys, forest and boulders, rivers and creeks—even today the region is poor and sparsely inhabited. At the time of Fidel Castro's sojourn there, it was almost wholly isolated from the rest of the country, with no major paved highways and dirt roads often impassable because of drenching rains that turned them into ribbons of deep, red mud. Tough as it was to move through these mountains, the Sierra Maestra was ideal guerrilla territory. From the moment his expedition came ashore at Los Cayuelos, and the plan to take Niquero and Media Luna was aborted, Castro had to break out of the western foothills, where the rebels were so vulnerable to air and ground attacks, to the impenetrable safety of the Sierra Maestra. As they marched east, the men's orders were always to keep the low-flying sugarcane fields on their left and the mountains on their right in order to reach the Sierra haven.

Although Castro always planned to launch the war from the Sierra Maestra and all the preparations ashore by his local supporters were geared to it, doubt remains as to the precise strategy he had in mind when he landed in Cuba. Faustino Pérez, who was one of Fidel's two Chiefs of Staff, says that while they were sailing on the *Granma*, "none of us was convinced that the struggle would develop fundamentally through [the creation of] an army in the mountains." He explains that "the vision we had was that of a nationally organized Movement, thinking about a general strike and also about a *guerrilla* [force] focus that would have a very great symbolic importance, but not in the sense that it would signify at a given moment the possibility of defeating the army of the tyranny." But Faustino adds, "What happened was that the companions who stayed in the mountains began acquiring confidence, the *guerrilla* was growing, blows were being dealt to the army of the tyranny, and it became conceivable that in this fashion a revolutionary army capable of defeating the forces of the tyranny could be built." At the same time, he says, urban groups of the 26th of July Movement believed that much could be achieved through struggle in the cities.

Fidel Castro, on the other hand, had a different appreciation of the basic rebel strategy—or, at least, he has it now. In an interview twenty years after the start of the guerrilla war, he said that "we did not arrive there [in the Sierra] with the purpose of creating a center of disturbance throughout the island and the problem would be solved by a military coup, we always fought the idea of a coup." This difference between Pérez and Castro in interpreting the initial strategy of the guerrilla war is extremely important in understanding the whole Cuban revolutionary process, because the argument over whether the overall leadership should be centered in the Sierra or shared with the urban underground soon turned into the war's central political issue. Subsequently, it led to the disappearance of the 26th of July Movement and the emergence of "unity" under the sway of Castro's new Communist party. In all fairness, however, the record shows that on numerous occasions *before* the December 1956 landing, Fidel had forcefully argued that a revolution could be accomplished with the existing army and that Faustino himself recognized that the rebels must create their own armed forces. Faustino himself now recognizes that once Castro was established in the Sierra, a true Rebel Army became possible. Meanwhile, Castro's immediate priority after spending that uncomfortable first night on Cuban soil was to reach the Sierra Maestra.

The story of how Castro was able to recover from a terrible initial defeat, regroup, fight, start winning against Batista units, and form an ultimately victorious Rebel Army is the story of the extraordinary support he received from Sierra Maestra peasants. Without this support, first from individuals and then from whole networks of peasants, Fidel would never have survived the initial weeks in the mountains nor would he have been able to organize the guerrillas. In the beginning peasants and their families hid and protected the little, poorly armed, and famished rebel band; they served as principal channels for obtaining food, arms and ammunition, and all other kinds of supplies from what could be found in the Sierra or brought up from the urban underground groups, and, finally, they served as the source of manpower. It was not a peasant revolution that gave Fidel Castro power, but there would have been no revolution without the peasants. Moreover, it was Castro who knew how to inspire this astounding display of solidarity and sacrifice despite the danger it presented to the peasants' lives.

The story of Fidel Castro's rebels and the Sierra Maestra peasants begins with the debacle at Alegría de Pío on Wednesday, December 5, when the Rural Guard ambushed, dispersed, and nearly annihilated the expeditionary force on its fourth day in Cuba. The men had spent their second night, from December 3 to December 4, in a clearing on another wooded hill called La Trocha. They had marched east all day over a boulder-strewn path, guided by Tato Vega, the son of a peasant at whose shack they had stopped at noon. The expeditionaries dined on rice and black beans and had a good rest. Tato Vega left them, saying he was going home, and it never occurred to the still inexperienced guerrilla leaders that he would look for Rural Guard units to report the rebels' presence in the area. This was one of the rare acts of betrayal of the rebels in the Sierra, and it resulted in a catastrophe. On Tuesday, December 4, the Castro column resumed the march east, coming to a tiny charcoal burners' village called Agua Fina where a Spanish storeowner gave them some canned sausage and crackers. Because the terrain ahead of them now turned into canefields, where they could be spotted by aircraft, Fidel decided to march all night after a brief dinner pause. They arrived at Alegría de Pío on the morning of December 5, in a state of absolute exhaustion; it had taken them three days and two nights over rocks and boulders to cover the twenty-two miles from the landing point to the low hillside where Castro now ordered them to set up camp.

In this he committed two major mistakes. The first was the choice of a low unprotected hill jutting into the canefield instead of a higher and wooded hill nearby. But the men were so fatigued that he hesitated to ask them to advance another several hundred yards; the nocturnal uphill march over rocks that Cubans call "dog fangs," with expeditionaries continuously stumbling and falling and even fainting (several of them smashing their eyeglasses), had been too much. The second mistake was to deploy sentries too near the camp, thereby cutting down the warning time when the attack came. Castro had paid no attention to the trail of sugarcane debris the men were leaving behind as they sucked pieces of cane during the march. And, finally, there was the treason of Tato Vega, the guide. Shortly after 4:00 P.M. on December 5, the men awoke and were each given a piece of sausage, a cracker, and a mouthful of condensed milk. Quite a few rebels had taken off their boots to wrap bandages around their bleeding feet. At 4:30 P.M., in the words of Raúl Castro, "The hecatomb began . . . we were ambushed by the army." The hundred-man Rural Guard company, firing machine guns and rifles at the rebels, turned the hillside into what Raúl described as an "inferno." The revolutionary contingent simply came apart.

Fidel kept firing his rifle while roaring out commands for an orderly retreat, hoping his men could hide in the canefields and regroup. But as at Moncada, it was too late, and it had to be each man for himself. Che Guevara was hit by a bullet in the lower shoulder, and Faustino Pérez, who was next to him, thought the Argentine had been killed. Guevara wrote in his diary that "I thought of myself as dead, and I told Faustino from the ground, 'They fucked me'. . . . Immediately, I began to think about the best way of dying in that minute when all seemed lost. I remembered an old story by Jack London in which the protagonist, leaning against a tree trunk, prepares to end his life with dignity, knowing he was condemned to freeze to death in the frozen plains of Alaska. This is the only image I remember." Actually, Che's wound was superficial and he escaped the trap, moving east with four companions, including Juan Almeida and Ramiro Valdés. Fidel, Faustino, and Universo Sánchez found themselves together later that afternoon and started their odyssey under the sugarcane straw. Several men had to physically drag Castro away from the battlefield.

The Rebel Army was destroyed. Three expeditionaries are known to have been killed in battle that afternoon. Many of them inside the burning canefields, the remaining seventy-nine men broke up into twenty-six separate groups, like Fidel's and Che's, though many of the "groups" consisted of only a single individual. One of them was Juan Manuel Márquez, the second Chief of Staff, who was quickly captured by the army and brutally murdered. Jesús Montané was captured and taken to prison in Havana. Twenty-one others were also known to have been executed within a day or two, twenty-two were caught and imprisoned, and nineteen more simply vanished; some made their way out of the Sierra to return home and hide, or surrender, some were never seen again. Of the eighty-two men who landed from the Granma on December 2, only sixteen survived Alegría de Pío and the dispersal to pursue the war. However, the entire leadership, except for Juan Manuel Márquez and Ñico López (also captured and executed), stayed alive to fight another day. Apart from Fidel and Che, among the sixteen survivors were Raúl, Faustino Pérez, Juan Almeida, Ramiro Valdés, and Camilo Cienfuegos, though Fidel would not know about them for a number of days. Still, as he lay under the sugarcane *paja* with his two companions, Fidel whispered day and night how and when they would regroup and go back into battle. As Faustino Pérez remembers, "This was the great lesson of faith and optimism—as well as of realism— that Fidel taught us in those days." But none of it would have helped Castro, had it not been for the peasants of Sierra Maestra.

In the aftermath of Alegría de Pío, General Batista and his government were absolutely convinced that Fidel Castro was dead and his expeditionary force completely smashed. With so many prisoners captured by the army, Batista was justified in assuming that the danger was over; even if Castro had not yet been physically located, it was obviously a question of time before he was. Consequently, on December 13 the army high command withdrew most of the combat units from the Sierra Maestra region, leaving behind normal Rural Guard garrisons in towns and villages; aerial surveillance was likewise canceled. An official communiqué said the command considered "the insurrectional movement" to have ended. And Fidel's relatives, friends, and followers were just as certain that the great revolutionary adventure had collapsed and that Fidel and Raúl were most likely dead. Their mother, Lina Ruz de Castro, told a Holguín newspaper that "if they let me go up the Niquero mountains, I shall make them come down with me. . . . I suffer as the mother of soldiers and revolutionaries, but if Fidel and Raúl decide to die, I wish they would die with dignity. . . . I weep for my sons, and I would embrace the same way the mothers of the companions of my sons as the mothers of the soldiers who have died in this painful war."

Still in mourning after her husband's death in October, Lina Ruz de Castro then traveled to Santiago with her oldest son, Ramón, to settle inheritance problems—including Fidel's and Raúl's inheritances—and hoping for news of them. Ramón Castro said they had learned nothing, and that "there is nothing concrete to demonstrate that they killed [Fidel] or that he is alive. . . . I believe that if my brother had died in combat, the government would have already announced it officially." He was right on that score because three weeks after Alegría de Pío, there was absolutely no information about Fidel's fate. Marta Rojas, the reporter who had covered Castro's "History Will Absolve Me" speech in the Santiago hospital after Moncada, wrote in *Bohemia* magazine that Fidelito, now back in Havana with his mother, Mirta (whose new husband was the son of Cuba's chief delegate to the United Nations), kept asking, "Has Papa written? Where is Papa?" Mirta also insisted that she had not kidnapped Fidelito in Mexico, and that his paternal aunts, Lidia and Emma, had voluntarily surrendered him to her.

Throughout December the Cuban press maintained the sense of uncertainty about Castro (censorship had been lifted again), with growing hints that he might well be alive and preparing to reappear. Statements by two captured rebels provided a fairly accurate account of the *Granma*'s voyage and the expedition's first days ashore, offering Cubans some idea of what had happened. Che Guevara was still unknown in Cuba, and published reports spoke of "an Argentine physician named Guevara" as part of the Castro force. Late in December the Holguín newspaper reported that Fidel, Raúl, and forty men had been camping since the eighteenth of the month south of Turquino peak in the Sierra Maestra; their guide was identified as a peasant named Crescencio Pérez who was said to have been arrested.

This was a fascinating piece of confusion for Fidel's friends and foes alike. Actually, it was on December 16 that Fidel, Faustino, and Universo had reached the Cinco Palmas farmhouse of Ramón "Mongo" Pérez in Purial de Vicana on the Vicana River, some thirty-five tortuous miles northeast of Alegría de Pío. Mongo Pérez was the brother of Crescencio Pérez, one of the two key members of the 26th of July Movement's peasant underground who under Celia Sánchez's command had vainly awaited the *Granma*'s arrival on the last day of November. The other member of the peasant underground was Guillermo García Frías. The original plan did provide for the expeditionaries to make Mongo Pérez's farm their first operational headquarters in the Sierra Maestra, but the farm was at least seventy-five miles in a straight line *west* of the Turquino area where Fidel's presence had been reported. Crescencio Pérez ran the Movement's peasant network, but he was not Fidel's guide. It was not Crescencio who was arrested; it was his son Sergio, who was accompanying a fourteen-man group of rebels fleeing south to the coast from Alegría de Pío. Six of these expeditionaries were captured, tortured, and executed. Fidel was naturally unaware of newspaper reports, but they were most helpful in that they served to confirm that he was alive, thereby preventing the collapse of the entire Movement in the country, while totally misleading the authorities as to where he was really hiding.

It took Fidel and his two companions six days to reach Mongo's farm, with crucial help from Sierra peasants. They left their canefield hideout after dark on December 10, five days after the disastrous battle, walking slowly and carefully in single line, with Universo usually the point man. Guided by the stars and their instinct, they covered two and a half miles that night, moving northeast. Spending all day in another canefield, the trio resumed the march on the evening of December 11 and reached a forest-canopied mountain called La Conveniencia after midnight. The canefields were behind them, the terrain was less rocky, and they could advance more rapidly. The silhouette of the Sierra Maestra massif, now discernible in the moonlight, served as a reference point. Below La Conveniencia, the mountain dipped sharply to the Toro River, with the Sierra Maestra itself beginning on the far side. Fidel spotted a thatched-roof peasant house down the hillside, but he chose to observe it very carefully from about three hundred yards in the forest before risking a contact. In a downpour that seemed never to end, the three men watched the house the rest of the night and much of the next day; they had no food, no water, they were drenched, and Universo who had lost his boots at Alegría de Pío, was suffering from having to walk barefoot. When Fidel was convinced that the peasant family was engaged in normal activities—and that there were no soldiers around—he told Faustino to go down to the house. It was 4:00 P.M. of December 12, over sixteen hours after they had arrived at La Conveniencia.

The peasants were mountain coffee growers, Daniel Hidalgo and his wife, Cota Coello. On Fidel's instructions, Faustino requested food for twenty or twenty-five men to give the impression they were a large force. Then, Fidel and Universo joined Faustino at the house. Their hosts slaughtered a piglet, and the three rebels feasted on the meat and vegetables; it was also the first time in seven days that they had drunk water. Daniel Hidalgo had heard about a landing by armed men on the coast, and he told Fidel he had heard about him, too. They spent the rest of the afternoon discussing the best ways of penetrating the Sierra Maestra, and at night Fidel decided to move out again. Led by one of Hidalgo's sons, the three revolutionaries went down a narrow canyon, crossed the Toro River, climbed Copal heights, and continued several miles to Yerba heights. Now they were inside the Sierra Maestra.

On the morning of December 13, after walking all night and covering eight miles through the Sierra forest, Castro and his companions arrived at the house of Rubén and Walterio Tejeda, two brothers who belonged to the peasant network of the 26th of July Movement. Finally, they had made contact with the organization, and they no longer had to depend on luck and their own and virtually nonexistent resources. Following a three-hour rest and a meal of *malanga* roots and milk, the trio continued through the mountains to a farmhouse near the village of El Plátano. This was the mountain fief of the García family, really a tribe, and it was also the home of Guillermo García Frías, a childhood friend of Celia Sánchez's and for nearly two years an active member of the Movement—recruited by Celia—who had been among those awaiting the rebel landing on November 30. But the first García whom Fidel met that noon in a field was Guillermo's father, Adrían García, who was carrying a bucket filled with rice, bread, coffee, and milk. He had heard that there were rebels in the area, and he was looking for them with food literally in his hands. Castro had not yet realized how efficient the Sierra communications network was, so he introduced himself to the old man as "Alejandro," his code name, only to be greeted as "Fidel." Later in the day, around twenty young peasants came to the farmhouse to offer their services to the Rebel Army, and Castro promised to take them as soon as he could organize this army. However, his immediate concern was to get to the Mongo Pérez farmhouse, still farther to the northeast, because he hoped to find enough of the other expeditionaries there to rebuild his force. But first he had to break through the isolation cordon the army had left around the Sierra Maestra's main massif; this meant crossing the heavily guarded Pilón-Niquero highway that separated Castro from Mongo's house. To get across the highway, he needed Guillermo García to guide him. Guillermo was somewhere in the mountains searching for lost rebels (he had watched the Alegría de Pío battle from a mountaintop) and abandoned arms. Fidel decided to wait for him at El Plátano, where he felt reasonably safe. Word went out through the Sierra for Guillermo to rush home.

Guillermo García arrived at his *finca* at one o'clock in the morning of Friday, December 14, and he and Fidel immediately launched into an all-night conversation. It was a memorable occasion in the history of the Cuban revolution because Guillermo not only became the first peasant to join the Rebel Army, but he would become one of its top commanders and, later, a key member of the *Fidelista* regime. If there was a single individual in the Sierra who can be credited with helping Castro survive and win, it certainly was this tough, squat twenty-seven-year-old peasant.

Guillermo is also one of the most interesting personages of the Cuban revolution. His background and allegiances go far to explain why from the outset Fidel Castro was able to command the tremendous support of the mountain peasantry. He was one of eleven children in a family that barely subsisted on what it could grow in the rocky soil and what it could earn from landowners of the large properties in the area. One peso (a dollar) a day was normal pay in the 1950s. Guillermo recalls that the nearest doctor was in Niquero, a day's horseback journey from the *finca* (two of his brothers had died of gastroenteritis as babies), and he charged two pesos for a visit. "My physician was my mother with her medicinal herbs from the fields," he says. The nearest school, a multigrade elementary school, was three miles away in the mountains, and Guillermo quit after the fourth grade at the age of ten to go to work. He helped the family in the fields, tended the landowners' cattle (becoming something of a cattleman himself), and while still a child, accompanied an uncle selling produce in the region.

He met Celia Sánchez for the first time when he was twelve years old. He and his uncle went every week to deliver vegetables to her family's house in Media Luna, which was twenty-seven miles, or twenty-four hours, away on horseback. Celia, who was eight years older than Guillermo, was always politically active, he says, and she enlisted him in the 26th of July Movement in 1955, two years after Moncada. Guillermo García recalls that he was captivated by her explanations of how the Castro ideas represented the aspirations of all young Cubans, particularly given the culture of poverty in which families of his social class had to live. Cuba being a small island, historical and political traditions are important, and they are passed on from generation to generation, even among the poorest of the population. However, peasants and slaves had fought colonizers in the nineteenth century, and Guillermo emphasizes that his grandfather, Bautista Frías Figueredo, was a veteran of the 1895 and 1898 independence wars. During the Spanish-American War over Cuba, Guillermo's ancestors fled to the mountains from the plains of Oriente "to constitute something of a tribe, and then we emerged as a new generation, the third peasant generation." In this sense, Guillermo thinks it was only logical for him to join the 26th of July Movement. Three of his brothers also joined the Rebel Army. One was killed, one ascended to the rank of general, as Guillermo did, and the third one returned to the mountains.

After the Alegría de Pío battle, Guillermo was all over the place. Knowing the area like the back of his hand, he coordinated rescues of lost rebels and picked up a half-dozen survivors personally. Among them was Calixto Moráles, the soldier almost executed for indiscipline in the Mexican training camp when he refused to take part in forced marches because of a back condition to which he would not admit. Moráles was found by another peasant who then turned him over to Guillermo, and soon he rejoined Fidel to fight the rest of the war. García says the rebels were passed "very safely, from peasant to peasant because there, in that zone, all the peasants were really organized. . . . I had perfect security because I knew the political affiliation of every peasant as well as their morality. Those who had bad morality, I sought to isolate totally." But some of the men saved by Guillermo did not choose to go on looking for Castro.

Fidel and his two companions were in a terrible physical state when he met them that night, Guillermo says, and Faustino Pérez was a "human rag" from exhaustion, hunger, and deep cuts from mountain vegetation. Castro, however, "was incredible" He started interrogating me about our organization, about the program he had, how we were going to organize the peasants, how we were going to collect arms for the *sierra*, how many shotguns we would need, and so forth. . . . It seemed like he already had an army with him. . . . So I decided to stick with him." Without having slept for forty-eight hours, Castro talked all night. He wanted to know everything about army movements, the sentiments of the population in the region, and which individuals were reliable and which were not.

Late on Friday, December 14, Guillermo and two of his friends escorted Fidel, Faustino, and Universo down the mountain to a canefield near the village of La Manteca, stopping twice en route for meals at peasants' homes. The canefield was just off the highway they had to cross to enter the heart of the Sierra Maestra, and Guillermo wanted to wait for a safe moment when there were no army patrols around. They waited for over twenty-four hours until Guillermo decided that the Saturday night sound of music from the jukebox in a bar at the edge of the highway, singing and shouting, and noise from a nearby power plant would drown whatever commotion their movements might cause. Flat on the ground, the six men dragged themselves to the mouth of a drainage culvert under the highway, traversing it to the other side. It took three minutes to get through the mud and rotting, stinking matter inside the pipe; Guillermo had proposed this as the safest way to cross, and Fidel approved the idea. Afterward, the men marched for eleven hours up and down mountainsides, covering twenty-five miles and stopping only once to rest, and finally arrived at Mongo Pérez's farmhouse at seven o'clock in the morning on Sunday, December 16. It was exactly two weeks after landing in Cuba that Fidel Castro attained safety—and the real possibility of waging a war.

At Mongo Pérez's farm, Fidel set up camp in the middle of a canefield; now the three of them could eat, drink, rest, and sleep. Castro's plan was to wait there a number of days for some of the other expeditionaries to join them, then start moving again. Through the peasant information network he learned that *Granma* companions were in the area the day he arrived, so he sent out the tireless Guillermo to look for them. On Tuesday, December 18, Raúl Castro and four others reached a farmhouse less than a mile from his brother's camp; for a week now they, too, had been guided and fed by Sierra peasants. At every stop, Raúl made a point of leaving a handwritten note, signed by him as "Captain," to be displayed after the revolution as proof of the help the peasants had given the rebels. Informed that Fidel was in the nearby canefield, Raúl sent a peasant over with his Mexican driver's license so his brother would know he was approaching with his men. Fidel, always careful, sent the peasant back with test questions for Raúl, to confirm his identity further. Just before midnight, the brothers embraced in the canefield. Fidel asked Raúl: "How many rifles did you bring?" and Raúl replied, "Five . . ." Fidel shouted: "And with the two I have, this makes seven! Now, yes, we have won the war!" The next day, Calixto

Moráles arrived, unarmed. The Rebel Army consisted now of nine men and seven rifles.

There was no time to waste. On December 20, Fidel sent Mongo Pérez to Manzanillo on the northwest coast of Oriente province, and then on to Santiago to inform Movement leaders there that the *guerrilleros* were alive and well. He also gave Mongo an enormous list of instructions concerning his needs for food, weapons, supplies, and men. Raúl noted in his diary that day that the peasants "have a fairly good organization, and we are perfecting it, especially in liaison and espionage . . . any movement by anyone in these surroundings is immediately communicated to us." In the evening Fidel moved their camp to a coffee field nearby, close to a creek where they could bathe and swim.

The group of eight rebels, including Che Guevara, Juan Almeida, Ramiro Valdés, and Camilo Cienfuegos—all future revolutionary chiefs— met their first peasants on December 13 after a week of wandering aimlessly between the south coast and the foothills of the Sierra Maestra. At a farmhouse sheltered by trees, the eight, so exhausted they could not take another step, spent all night in what Che described as "an uninterrupted festival of food." Then since their stomachs were unaccustomed to copious eating, all eight became violently ill. The next day, the peasants gave them fresh clothes to replace their torn uniforms, and the rebels split up into two groups. Except for Che and Almeida, who held on to their submachine guns, the rebels left their weapons behind at the peasants' farmhouse as they resumed the march to the northeast; they already knew from their hosts that Fidel was alive and awaiting them. Hearing that the army had picked up their scent, Che, Almeida, and two other rebels hid at a house in El Mamey, a few miles away, that belonged to a Seventh Day Adventist lay preacher named Argelio Rosabal, another extraordinary figure from the Sierra. Not far from them, at another farmhouse Camilo Cienfuegos was concealed inside a dry well.

Rosabal, who was a sugarcane field worker during the week and "with the church on Saturdays," remembers meeting a group of four rebels three days after Alegría de Pío near his house in the mountains just west of Pilón. He knew there had been a battle involving armed men from a ship, and he gave coffee and clean clothes to the four men who went on, never to be seen again. They told Rosabal that they had come to "liberate Cuba." The lay preacher thereupon went to his church, gathered fifteen or twenty fellow Adventists, and told them that "the men who came with the mission, they say, to do away with a little of our misery . . . must be saved." He said that "all of you must take an interest in their lives, and when you learn that there is one or more of them around, take them in . . . if you have no courage to do it, advise me . . ." A few days later, Rosabal was informed that a group of eight men were hiding in the Sierra house of one of his friends (this was the band that included Che and Almeida) and at that point the lay preacher moved the four to his house. The next step, he says, was to get them marching again. Rosabal recalls: "As I am a man of God, I say, 'The situation is not easy, so let us pray . . .' We all knelt, and I begged God to help me in this situation." Che Guevara, who affectionately referred to Rosabal as "the Pastor," knelt probably for the first time in his life.

Led by Rosabal, Che, still suffering from his shoulder wound, and the three other men marched all night until they reached the house of the Adventist's sister-in-law, where a chicken was killed for a meal. Che threw up twice before his stomach could hold the chicken broth, then Rosabal removed the men's boots and stood guard over them as they slept. The next morning, Che asked Rosabal if he could send "anonymous" telegrams to their families; the Pastor said he would try. Placing a sheet of paper with family addresses in a basket of red beans, he walked down the mountain to Pilón and to the house of Celia Sánchez's father, whom he knew well. Since he was unaware that the Sánchez family belonged to the clandestine Movement, he produced the list of names Che had given him only when he satisfied himself that the doctor would protect the rebels.

Returning to the farmhouse, Rosabal discovered that the army had captured one of the rebels from the original group of eight—he had been left behind because of a high fever—and had seized the weapons left in the house. Now it was urgent to send Che and his three companions far into the Sierra Maestra, and on December 16, Guillermo García (who happened to be Rosabal's brother-in-law) arrived from Mongo's farm to be their guide. He reassembled the men hiding in the area into a group of six, and at dawn on December 21 they arrived at Mongo's homestead to rejoin the Castro brothers (Raúl wrote in his diary that among the survivors was "my inseparable friend, Ramiro Valdés"). Che reached the *finca* in the midst of an asthma attack, which he overcame. And Guillermo told Che and his companions: "You will never know how much this man Rosabal did for you . . ."

Indeed, there seemed no limit to what the peasants were willing to do for the Castro guerrillas. Argeo González, a storekeeper and itinerant merchant in the Sierra when the rebels arrived, explains that "the reason all the peasants helped them was that they learned the truth about the struggle against tyranny. . . . The landowners didn't let anybody else work the land, it was all theirs. . . . Peasants had no way out without a revolution." Argeo was among the first regular volunteer suppliers of the incipient Rebel Army, running food and arms up the Sierra Maestra from lowland towns, and he says that although the peasants did not know Castro at the outset, "he earned their confidence, helping them, and not mistreating them." When a Rural Guard trooper visited a mountain house, Argeo recalls, "he would receive bread, eat a chicken if there was one there, take away a daughter if there was one there—but the rebels were different; they respected everything, and this was the basis of the confidence that they gained." When a rebel would get out of line, he says, Castro would instantly punish him, sometimes have him executed. Peasant women, according to Argeo, were soon "the first ones to want to join Castro, to help him." At the same time, the rebels told the peasants that after the revolution the land would belong to them, to those who worked the land. Universo Sánchez remembers that Fidel insisted on paying ten pesos for a chicken even if it was worth only five.

Mario Sariol, another Sierra merchant turned secret rebel quartermaster, remembers meeting Castro in a coffee field early in 1957 and being embraced by him after offering to prepare food for the rebels. He recalls that Fidel's beard was just beginning to grow, and Raúl had "a few hairs, nothing more, and they seemed like youths." (Actually, Castro had decided in the beginning that the rebels should not even try to shave while in the mountains; in their Mexican training camp, General Bayo had made them throw out their razors and toothbrushes "because you won't have them where you're going . . .") As Sariol tells the story, the peasants developed a protective feeling about the *Fidelistas*. When Sariol ran out of funds to purchase food for them in the town of Las Mercedes, the local merchant told him, "Mario, don't let these people go hungry even for one day; come here and take what you need." He never accepted payment, and Sariol says that later Castro ordered him to keep track of all food caches, making certain there was enough for the peasants as well as the rebels. Sergio Casanova, an enthusiastic peasant volunteer, was turned down by Castro when it turned out he had six children and no income other than occasional day's work. Fidel told him: "You can't go with us. . . . Who would look after your family?" Remembering those days, Casanova says, "To me, Fidel was a god."

At Mongo Pérez's farm in Purial de Vicana, Fidel Castro gave his expeditionaries the rest they needed while at the same time he kept them in maximal readiness. The day after Che and his group arrived, the entire contingent was relaxing on a hillside when Castro suddenly shouted, "We are surrounded by soldiers! Take your battle positions!" The rebels responded at once, hitting the ground or hiding behind trees, their weapons at the ready, but nothing happened. Nobody moved. Then Castro, smiling, informed them that this was a false alarm, a training exercise.

At this juncture, the Rebel Army had twenty men, including Fidel.

Sixteen of them were *Granma* expeditionaries, and four were peasants who had formally enlisted: Guillermo García, Crescencio Pérez, his son Ignacio Pérez, and Manuel Fajardo. Subsequent propaganda emphasized that Castro resumed the war with twelve men, but this was a symbolic apostolic touch. As Guillermo García recalls that period, Castro had taken the decision "not to fight" until his force could be adequately reorganized. While carefully selected peasants were being allowed to sign up with the rebels, "it was not convenient for too many people to gather there because there would be no possibility of mobility in the case of an enemy attack." Besides, volunteers had to be recommended by peasants well known to the guerrilla leadership to prevent infiltration by Batista agents.

Castro's greatest problem was arms. He had twelve weapons for twenty men, and he had been furious at Che and Almeida's group for having left its arms behind in a Sierra farmhouse. "To leave behind rifles in such circumstances is to pay with one's life . . . for such crime and stupidity," he berated them. But Mongo's trip to Manzanillo on December 20 paid off very quickly: Peasants arrived at the *finca* on the morning of December 22 with a Thompson submachine gun and eight rifles. On December 23 two men and two women (one of them Mongo's daughter) arrived from Manzanillo, sent by the 26th of July Movement in response to the word brought by Mongo that Fidel was alive. Eugenia Verdecia, the other woman, carried three hundred submachine-gun bullets and nine dynamite cartridges under her skirt.

More relaxed about arms, Fidel now turned to the politics of the revolution. His first major decision was to send Faustino Pérez to Manzanillo, Santiago, and Havana for the twin purposes of informing the 26th of July Movement's National Directorate of the rebel situation in the Sierra Maestra and bringing newsmen—foreign newsmen if possible—to the mountains to convince the world that Castro was well and fighting. In Fidel's mind armed struggle and propaganda were always linked.

Faustino says that he was chosen because he was a member of both the National Directorate and the military general staff, and thus in an excellent position to organize support for the Rebel Army in the mountains and provide credible confirmation of Castro's fighting presence. The immediate need was for a small group of armed fighters from the lowlands to strengthen the army, and for newspapermen to write about Fidel. On Sunday, December 23, three weeks after the *Granma* had brought the *Fidelistas* to Cuba, Faustino got into the jeep that had carried the four Movement members up the mountain from Manzanillo earlier that day. As the army had canceled its search-and-destroy operations against Castro the week before, it was not difficult to get through to his destination. Faustino was dressed like a *guajiro* with a straw hat, and Eugenia Verdecia, the girl who had concealed submachine-gun bullets and dynamite under her skirt on the way up, pretended to be his fiancée. Reaching Manzanillo in the evening, Faustino saw Celia Sánchez immediately on arrival, and now contact was formally established between the Rebel Army and the Movement. Faustino and Celia talked all night, though at first "they gave me a meal I can never forget because I was suffering from organic hunger, and they served me that marvelous cream of asparagus soup . . ."

The following day Faustino drove to Santiago for meetings with key members of the Movement's leadership: Frank País, the provincial coordinator who had thought Fidel was too precipitate in the invasion, Vilma Espín, his local associate, and, from Havana, Armando Hart, Haydée Santamaría, and María Antonia Figurea. It was Christmas Day when Faustino Pérez slid quietly into Havana. It would be almost a year and a half before he would return to the Sierra Maestra.

At Mongo's *finca* the expeditionaries and their peasant friends spent Christmas Eve in a coffee field, eating a roast piglet and washing it down with wine. On Christmas Day Fidel decided it was no longer safe to remain in the same place, and that the time had come to move deeper into the Sierra Maestra. Before departing just before midnight, fifteen rebels signed a letter of thanks to Mongo Pérez, drafted by Fidel, declaring that "the help that we have received from him and many others like him in the most critical days of the Revolution encourages us to continue to struggle with more faith than ever, convinced that a people such as ours deserves every sacrifice . . ." This was Castro's first document of the Sierra war, and he could not wait to engage in military activities. Thus after leaving Mongo's farm the rebels spent the entire night in exercises that ranged from an assault on a mud hut to crossing and recrossing the Vicana River eighteen times in the dark. Then Fidel turned southeast, moving toward the Caribbean coast through the Sierra's high mountains. On December 28 the rebel column was augmented by three expeditionaries who had been believed lost and by three peasant volunteers. They brought a rifle and, for Fidel, magazines and newspapers—the first he had seen in Cuba since the *Granma* landing. Reading them, he learned that José Miró Cardona, the president of the Cuban Bar Association, and Elena Mederos, a liberal member of the Society of Friends of the Republic, had met with Batista's prime minister, Jorge García Montes, to demand decent and humanitarian treatment for the *Fidelista* rebels captured after Alegría de Pío. Castro filed the names away in his mind: Both Cardona and Mederos would be invited to join the revolutionary government.

On December 29, Eugenia Vardecia, concealing sixteen explosive charges, four submachine-gun clips, three dynamite cartridges, and eight hand grenades, caught up with Castro's column in the hills. Again, a

Cuban woman was playing a crucial revolutionary role. Her companion brought volumes on the geography and history of Cuba to be used in teaching the peasants who were joining the Rebel Army. Calixto Moráles, the ex-schoolteacher, was put in charge of education and indoctrination, an important function in building the new army. Che Guevara received an algebra text he had requested. Then, another all-night march in cold rain with only a two-hour halt at a peasant home where a hot meal awaited the rebels. The Rebel Army, now composed of twenty-nine men (additional peasants had joined it), spent New Year's Eve asleep guarded by sentries in a large shed without walls on a wood-covered hillside.

The year 1957 opened with a downpour of freezing rain that prevented the Castro guerrillas from advancing for two days. They had nine small nylon covers for the rifles, but nothing to protect themselves. One night Raúl Castro slept inside a sack of corn flour. The march resumed on January 3, and two days later the rebels stood atop Tatequieto heights on the spine of the Sierra Maestra. In the distance, five miles away, Fidel could see the triple peaks of the Caracas Mountains to the east. "If we can get there," he said, "neither Batista nor anybody else can defeat us in this war."

Then they were on the march again, still moving southeast because Castro had concluded that he could reach the center of the Sierra Maestra more easily from the south, up the natural ridges and canyons, rather than going straight east across them. It was longer but less punishing. Besides, Fidel had developed the notion that he could seize small coastal military garrisons to acquire more weapons. The march from Tatequieto to the coast, with the column advancing sometimes at night and sometimes in the daytime, took eleven days, until January 16. More peasants joined up during the first weeks of January, and now the Rebel Army had thirty-three men. On January 8 the rebels halted for two days at the farmhouse of Eutimio Guerra, a trusted peasant, in El Mulato, a village directly south of the Caracas peaks. The men ate, drank brandy with honey, then suddenly learned that their presence had been somehow reported to Batista's army, and Castro ordered instant departure in the middle of the night. The mountains were so steep that they had to hold on to the vegetation and sometimes moved on all fours. Ramiro Valdés fell, chipping a knee bone. After Che treated him, Valdés dragged himself along the best he could. On January 11 five peasant guerrillas decided to return home, and Castro let them go. He had already resolved to attack the garrison at La Plata on the coast, and he wanted to be only with men he could fully trust. Moreover, Batista authorities suspected that an attack could come from the mountains, and on January 13 the army arrested eleven local peasants; all were murdered.

On January 14 the rebels came to the banks of the Magdalena River, just west of La Plata, leaving the injured Ramiro Valdés and another ailing rebel at a mountain farmhouse. Crossing the Magdalena, they ran into two beekeepers. They paid ten pesos for sixty pounds of honey, but decided to keep one of the beekeepers as a hostage to protect themselves; the other beekeeper was let go after taking an oath of silence. The hostage was paid five pesos a day during captivity, and Castro let him sleep in his hammock. The next day, guided by the beekeeper, the *Fidelistas* reached the heights overlooking the estuary of the La Plata River. They could see uniformed soldiers below, around the post's four structures, the military barracks, and the house of the foreman of the company owning the land in the area.

In Havana General Batista announced on January 15 that the United States had sold his regime sixteen brand-new B-26 bombers. Batista was still skeptical about a Castro revolution, but now he knew that Fidel was alive somewhere in Oriente, and he thought it prudent to modernize and beef up his armed forces. The Eisenhower administration was glad to oblige. Above La Plata that same day, Fidel Castro was preparing his first attack on the Batista forces since Moncada, three and a half years ago. The little garrison on the beach was made up of five Rural Guard soldiers and five sailors under the command of an army sergeant (a coast-guard cutter sat offshore), and Castro was determined to win this battle.

He had twenty-six men with him and, for once, numerical superiority. With a few rebels, Fidel moved on the night of January 16 to a point about three hundred yards from the barracks, awaiting passers-by who could tell him exactly what the soldiers below were doing. Four peasants were caught there, by the *guerrillas* and Castro learned that Chicho Osorio, the land company overseer, feared and detested by the local peasants, would be coming by on his way home. Presently the fifty-year-old Osorio, mounted on a yellow mule, a brandy bottle in his hand, and completely drunk, appeared on the trail where Castro captured him. Osorio's .45-caliber pistol was taken away. Fidel introduced himself as an army colonel, and the overseer, putting in his false teeth, told him that "the order is to kill Fidel Castro. . . . If I find him, I'll kill him like a dog. . . . You see this .45 you took away from me? I'll kill him with this gun if I catch him . . ." Then, Osorio proceeded to give Fidel the names of the peasants in the region who cooperated with the army and those suspected of helping the rebels. He added: "You see the boots I'm wearing? They belonged to one of those who came with Fidel Castro, and whom we killed around here . . ." Che Guevara wrote that at that moment Chicho Osorio had signed his own death warrant.

Osorio's hands were tied behind his back, and at dawn the next day, January 17, Castro asked him to guide his force to the military barracks, pretending that as a colonel he wanted to surprise the slothful soldiers there. Still drunk, the overseer happily agreed. Just then, a government soldier on horseback rode by, dragging behind him five peasants tied up with thongs, and Osorio said proudly he was his friend. The rebels were divided into four squads: Fidel, Che, and four other fighters deployed to the right of the target area, the other squads closed the circle. At 2:30 A.M. the *Fidelistas* started firing on the garrison. Simultaneously, Chicho Osorio was executed on Fidel's orders by the rebels guarding him. Raúl Castro wrote in his diary that "Chicho's fate was sealed a long time ago, just like the fate of any land company overseer who falls into our hands, and the punishment is summary execution, the only way to deal with those thugs . . ." The rebels were totally unforgiving when it came to traitors and "exploiters" of the people, men said to have killed and mistreated peasant families.

The combat was brief. Two soldiers were killed, five were wounded (three of them died later), one escaped, and three were taken prisoner. The barracks and other structures were set on fire, and weapons and ammunition were collected. The booty was nine Springfield rifles and a Thompson submachine gun, plus plenty of munitions and other supplies. This was the first time since Alegría de Pío that Fidel had more weapons than men. It was his first victory, and with no losses, and he and his companions showed magnanimity. Che treated the wounded, and Fidel told the prisoners: "I congratulate you. You behaved like men. You are free. Look after your wounded, and leave whenever you want." The rebels left medicine for the enemy wounded before vanishing back in the Sierra Maestra. At La Plata Castro set the policy toward his enemies for the rest of the war: Prisoners were always sent back alive, traitors and "exploiters" were mercilessly executed.

La Plata was a psychological and military milestone in the guerrilla war. Pedro Álvarez Tabío, the official historian of the Sierra Maestra era, says that this battle "demonstrated for the first time the axiom that Fidel would apply throughout the whole war: that a *guerrilla* army must live on weapons and supplies captured from the enemy [and] except for a few shipments of arms received from outside the *sierra,* this would be the state of affairs during the entire war." At 4:30 A.M., January 17, 1957, exactly two hours after the start of the battle, Fidel Castro ordered his triumphant men to march again, back into the heart of the Sierra Maestra. He set the peaks of Palma Mocha as the next objective. As he observed later, "We had our first successful battle when no one believed we were still alive."

CHAPTER

10

Two weeks after the victory at La Plata, Fidel Castro's guerrillas barely escaped alive from an extraordinarily precise surprise aerial attack on their camp in the heart of the Sierra Maestra by B-26 bombers and P-47 fighter planes of the Batista army. It was sheer luck that the rebels lost no men in this raid (a bomb exploded on top of the big kitchen stove on which breakfast was being cooked that morning), but they were once more dispersed and disorganized. Only on the third day after the furious bombing and strafing of the *guerrilleros* high up on the side of the Caracas peaks was Castro able to reassemble his force; it had been divided into three groups led, respectively, by him, Raúl, and Che.

The attack also served to make it virtually official that Raúl and Che had become the principal rebel leaders under Fidel. In the two months since the landing of the *Granma,* they had easily eclipsed their surviving fellow officers and such prominent fighters as Ramiro Valdés, Camilo Cienfuegos, and Universo Sánchez. To be sure, Raúl and Che had a degree of education and sophistication their present companions lacked, and they were much more politically mature than the others. Ideologically, Raúl had been a member of the Cuban Communist party for nearly four years, and Che, three years older, was a serious student of Marxism-Leninism, never concealing his allegiance.

With respect to their relationship with Fidel, there is no question that Raúl was the closest, personally and politically. He was a practical politician and the natural number two figure in the guerrilla band—and after. Intellectually, it was Che who had the greatest kinship with Fidel, with his erudition, fine irony, and quick mind. Both were superb chess players and masters of the mental rapier. Though Che could never quite overcome his complex about being a foreigner among Cubans, he was the conscience of the Cuban revolution, or at least he tried to be. He was not a practical politician, his revolutionary principles were above compromise. He did not hesitate to tangle with Fidel over matters of ideology during the war in the Sierra or to take on the Soviet Union many years later when he thought revolutionary ideals were at stake. Che may have been naïve, but he was the purest and the most honest idealist of the revolution. Despite appearances, Che's relations with Raúl were not as warm as they were with Fidel. Still, they were friendly and close, after a fashion. Che taught Raúl to speak French during the long months in the Sierra, but he never sought to compete with Raúl in his relationship with Fidel.

When men are thrown together to live and fight a war, their relationships, character traits, strengths, and weaknesses are defined more sharply and quickly than under other circumstances. This was especially true in the Cuban guerrilla war with its hardships and constantly shared dangers. After two months of this war and companionship, it was already very clear what Fidel, Raúl, and Che represented then, and would in the future. In their actions, beliefs, personal behavior, conversations, official and private letters, and, in the case of Raúl and Che, the campaign diaries they kept, they were acutely aware of their historical roles. The campaign diaries were poetic, romantic, and downright lyrical in acknowledging this. Fidel's literary output during the two years in the Sierra was of an epistolary, order-of-the-day, and political manifesto nature. Knowing Fidel, Raúl, and Che in the Sierra Maestra was to know them afterward, when they wielded power. They never really changed, though the nature of the revolution that sustained them would.

After two months in Cuba, the bombing on January 30, 1957, was the third time the rebels came close to annihilation, the first being the *Granma* shipwreck and the second, Alegría de Pío. Responsible for the bombing was a traitor named Eutimio Guerra at whose house they rested and prepared for the assault on La Plata and who guided them up and down the mountain for weeks. That Fidel with his sixth sense about danger and betrayal, and Raúl with his obsession about espionage and counterespionage, were not able to see through this man represented one of their great failures in the guerrilla war. A week earlier, the Batista army had mounted its own attempt to surround and destroy the rebels, but Fidel was able to ambush the soldiers, killing five of them. This was his first encounter with Lieutenant Ángel Sánchez Mosquera, probably the best field commander in the Batista army, who would remain Fidel's nemesis until the end of the war.

Eutimio, slim, thirtyish, and ever-smiling, was so highly regarded as a guide and peasant supporter of the revolution that on January 20, Fidel granted him permission to make a quick visit to his home in El Mulato, northwest of where the rebels were resting that day. In fact, Fidel gave him money. On Eutimio's way back, however, he was detained by the army, on the morrow of the defeat of Sánchez Mosquera's elite unit at Llanos del Infierno above La Plata. He was given the choice of being executed for collaborating with the rebels or betraying them. Specifically, Eutimio was offered ten thousand pesos, a major's rank in the army, and a farm of his choice if he succeeded in assassinating Castro or locating the guerrilla army so that the Batista troops could destroy it. Apparently he agreed because he was given an army safe conduct and sent off on January 25. At that point, Fidel's force was moving west toward the Caracas Mountains, en route to the southwestern rim of the Sierra for a planned conference with 26th of July Movement leaders from the cities. Two days later Eutimio encountered the rebels in a coffee field at La Olla, near El Mulato. He was warmly greeted by the unsuspecting guerrillas, and told them a long story about crossing the Llanos de Infierno battlefield, finding burned-out houses, and hurrying to warn Fidel that the army was in the vicinity. He also brought them candy.

On the strength of these reports, Castro decided to move during the night to a high saddle in the Caracas Mountains and to remain there. Because the night was cold, Fidel shared his blanket with Eutimio as they lay down to sleep on the ground. His Colt pistol and two hand grenades under the blanket, the peasant proceeded to ask Fidel questions about the locations of sentries around them. Instinctively, Fidel gave him evasive answers. Eutimio evidently hadn't the courage to shoot Castro then and there, preferring to let the army do the job. The next morning, January 28, Eutimio again left the rebels, this time ostensibly to look for food and locate several *guerrilleros* who had become separated from the main force. But he went straight to the army forward command near El Mulato to report on the rebels' deployment. Meanwhile, Fidel suddenly decided to move his men some three hundred yards higher on the mountain from the canyon where the big kitchen stove had just been installed; his instinct again saved his companions' lives.

In Castro's judgment their Caracas-peak camp was safe from an encirclement by the army; it was too steep and hard a climb. He dispatched four men to Manzanillo to deliver additional instructions concerning his approaching meeting with Movement leaders. The Rebel Army on the mountain now had twenty-five fighters, seventeen of them *Granma* expeditionaries (additional Alegría de Pío survivors had been reaching Castro for weeks), and more volunteers were sent up from Manzanillo. The little army's numbers fluctuated daily with arrivals and departures. Ramiro Valdés and another rebel were still convalescing in a farmhouse not far away, but hid in the forest when an army unit was spotted approaching. The previous evening Fidel's group had been resting up in their mountain camp. Che wrote in his diary that "Fidel delivered a speech to the troops to warn them about the risks of indiscipline and loss of morale. . . . Three crimes would be punished with death: insubordination, desertion and despotism . . ." At that same time, Eutimio Guerra arrived in Macho, south of the Caracas Mountains, to confer with army commanders on how best to destroy Castro. Given the terrain, the commanders decided that air strikes would be the most effective method, and Eutimio was taken by jeep to the port of Pilón, where the next morning he would be flown in a spotter plane to pinpoint the guerrilla camp.

On January 30, just after seven o'clock in the morning, Batista aviation struck the *Fidelistas,* and the Rebel Army broke up into three groups and fled the area. Fidel's group of thirteen men crossed the spine of the Caracas range to the southeast gradient where the aircraft could not see them through the foliage. Castro was enraged that his force allowed itself to be dispersed for the second time, just when he thought he had consolidated the guerrillas. But at noon of the next day, he was joined by Raúl and his four-man group, and things looked better again. Che, Guillermo García, and three other men got lost in the forest, and it was two more days before they caught up with Fidel. On February 1, Castro learned that three army columns were advancing on the Caracas Mountains, he ordered the rebels to resume their westward march. Guerrilla war in mountain forests is something like blindman's buff: Neither band can see or hear the other, unless they suddenly collide in the dark, and usually the advantage is with those having the best scouts and the best knowledge of the terrain. Castro decidedly had this advantage and thus he again evaded the army.

Hunger and thirst now became their real enemy. They had virtually nothing to eat for two days. On February 3, while crossing a forest, a rebel fighter collapsed, unable to move from thirst. Castro gave him a dry lemon he had in his pocket, and the man sucked it and swallowed it. Another rebel drank putrid water from a beer bottle he found on the trail. Che Guevara suffered an attack of malaria, made it to the spot where the column would spend the night, and collapsed. He had to stay behind the next day with two companions to look after him. Trying to walk, Che fainted several times and, as he wrote later, "I had diarrhea ten times." One night, a downpour drenched him in the camp. On February 5, Che and his companions again got lost; then they were found by Raúl and a patrol, which brought them hot chicken broth.

Eutimio Guerra reappeared late that day, when the men were resting at a farmhouse, wearing new white trousers, a cream-colored *guayabera,* a new hat, and carrying fifty tins of condensed milk. Again, he was greeted with joy by the expeditionaries. Castro decided to divide his army into two teams to make it easier to cross the Sierra to the west, and the first team, including Ramiro Valdés and eight others, left at night. Fidel and twenty men, including Eutimio, stayed behind for another day. Eutimio then asked Castro to meet him alone in a coffee field, but Universo Sánchez accompanied them, visibly unnerving the peasant guide. Finally, Fidel became suspicious of Eutimio's frequent trips, his ability to obtain food, and his constant questions. Unaware that both in Manzanillo and Santiago several Movement members had overheard Eutimio's name in conversations among army officers, Castro wanted more evidence before acting.

On February 7 Batista aircraft bombed and strafed the Caracas Mountains again, and Eutimio recommended that Castro set up camp around an abandoned shed at the bottom of a deep canyon. Fidel agreed, deploying sentries around the canyon. But the planes were back the next day, and bombs again fell on the guerrillas. Eutimio joked nervously, "I didn't tell them to bomb here," and Castro was almost convinced that he had a traitor on his hands. Raúl wrote: "Aviation fills us with plenty of fear despite so many raids." Heavy rain came on Friday, February 8, and Raúl bet Fidel and Che that it was raining harder inside the shed in the canyon than outside. Eutimio returned from still another mysterious expedition, volunteering to stand guard at the entrance to the canyon. Apparently, the plan was for him to let the army units into the canyon, but the rain was so hard that it was abandoned. Now the troops were close to the guerrillas. On Saturday, February 9, Raúl noted in his diary that "Eutimio's behavior preoccupies Fidel." Then Universo Sánchez came running into the house, shouting that a large army column was approaching. Eutimio had gone out again that morning, allegedly to buy food.

During the day a local peasant, detained by a rebel sentry, told Fidel that 140 soldiers were deployed above the canyon. Fidel climbed a rock to watch the enemy positions through his telescopic rifle sight. He heard the peasant say that he had seen Eutimio Guerra "down there" that morning. Now Castro informed his men that he was convinced that Eutimio was a traitor. He led the rebels out of the canyon and up the Espinosa peak, which overlooked the area. Suddenly, several rebels spotted Eutimio running behind a clump of bushes, after which army soldiers concealed by nearby ridges opened fire on the *Fidelistas.* Júlio Zenon Acosta, one of the first peasants to join the Rebel Army, was killed instantly; he had been standing a few steps from Fidel. Che Guevara wrote that Acosta was his first literacy pupil: "We were just beginning to tell the A from the O, and the E from the I. . . . This illiterate *guajiro* who was able to understand the enormous tasks the Revolution would face, and who was preparing himself for it by learning the first letters, could not complete his labor . . ."

The firefight went on for hours, and the rebels again divided into three groups to withdraw to the next mountain. Fidel and Raúl led a group of five under intense fire, dragging themselves through high grass to avoid detection. Brambles tore the flesh on their hands. Che, Juan Almeida, and ten others worked their way up the mountain from the other side, pursued by soldiers firing at them with automatic weapons. Guevara lost all his medicines, his books, and his rifle: "I was ashamed," he noted in his diary. Again, the Rebel Army was dispersed, but it did not allow itself to be surrounded. It lost a fighter, but the army's sustained effort in the mountains was a fiasco. By prearrangement, all the groups were attempting to reach a nearby mountain called Lomón, and the army, in a final effort to destroy the rebels, sent out planes on February 12 to strafe the forest there. By nightfall the *Fidelistas* were back together again, decreased in number to only eighteen men, since some of the *guerrilleros* were away on missions for Fidel, and some peasants had had enough and returned to their homes. The rebels found a farmhouse in a clearing where the peasants served them roast pork for a late dinner. Four days later, Fidel Castro reached the farm at the southwestern end of the Sierra Maestra where a new chapter in the political history of the revolution would be soon written. The march was easy: Only a few harmless mortar attacks by the army. Still, Eutimio Guerra would not be forgotten.

The eighteen men who arrived at the farm of Epifanio Díaz and his wife, María Moreno, at four o'clock in the morning on Saturday, February 16, to meet with the Movement's National Directorate and for a carefully prepared interview with a *New York Times* editorial writer, formed a hardship-toughened and extremely cohesive fighting group. Castro was certain they would make a very positive impression on the visitors. For him the meetings were important to assert his undisputed leadership of the revolution— so much so that he had risked coming to the outer edge of the Sierra Maestra, only twenty-five miles as the crow flies from the city of Manzanillo. Despite the Batista regime's new determination to assassinate him, Fidel felt confident now that he was in control of the situation.

"We identified so completely with the natural surroundings of the mountains," he recalls, "we adapted so well that we felt in our natural habitat. It was not easy, but I think we identified with the forest as much as the wild animals that live there [actually there are no wild animals in the Sierra Maestra.] We were constantly on the move. We always slept in the forest. At first, we slept on the ground. We had nothing with which to cover ourselves. Later, we had hammocks, and nylon . . . and we used plastic for covers to protect ourselves from the rain. We organized kitchen duty by teams. Each team would carry the cooking equipment and the food up the hill. In the beginning, we had to stop at houses to eat, but later we freed ourselves from it. And we did not know the region well. We had practically no political connections in that region. We established the relations with the population. We studied the terrain as we fought. . . . Batista was carrying on a fierce repressive campaign, and there were many burned houses, and many murdered peasants. We dealt with the peasants in a very different manner from the Batista soldiers, and we slowly gained the support of the rural population—until that support became absolute. Our soldiers came from that rural population."

Che Guevara, always more critical than the others, was skeptical in his assessments. He wrote that "the peasantry was not prepared to become part of the struggle, and communication with [the Movement's] bases in the cities was practically non-existent." To him, the first moments after their invasion represented a "subjectivist mentality," a phase when there was "blind confidence in a rapid popular explosion, enthusiasm and faith in the power to liquidate Batista's might by a swift armed uprising combined with spontaneous revolutionary strikes and the subsequent fall of the dictator." In a sense, Guevara was right in thinking that his faith was exaggerated, but the key to Fidel Castro's whole approach to the revolution *was* "subjectivism," a belief that inspiring leadership (his) combined with skilled propaganda could rally the masses behind great national causes. Interestingly, Fidel's own fight with the Cuban Communists was precisely over the issue of his "subjectivism" versus the party's classic Marxist view that "objective conditions" must be created before an insurrection is launched.

In the third month ashore, Castro had run into every imaginable disaster: the shipwreck at the edge of a mangrove swamp, a major military defeat, three dispersals of his rebel force, awesome betrayal, and now almost continuing attacks on his tiny army by the entire Batista air force and at least three thousand elite troops (the figure is Castro's). Yet none of it could break his determination to fight on or his supporters' readiness to help the insurrection. What Castro understood, probably better than Guevara, was that revolutions are not strictly rational affairs. To be sure, the setbacks had taught Fidel the accuracy of Che's comment that "characteristic of the small group of survivors, embued with a spirit of struggle, was the understanding that to imagine spontaneous outbursts throughout the island was an illusion." But in the next phase, Castro relied on his magnetism and imagination to keep the morale of his men high and to expand the guerrilla war. This brand of "subjectivism" was very much alive, and it worked in the mountains.

As far as the peasants in general were concerned, their fears of the rebels disappeared once Castro was able to demonstrate the *guerrilleros'* kinship with them. Eutimio Guerra was an exception, not a rule in terms of the *guajiro* response. Che Guevara himself recognized that the guerrillas "were the only force which could resist and punish the abuses [of the army], and thus to take refuge amidst the guerrillas, where their lives would be protected, was a good solution" for the peasants. Castro's application of "revolutionary justice" as the rebels roamed the Sierra Maestra was applauded by the peasants; whenever a Rural Guard torturer or land company overseer was executed, word spread instantly throughout the mountains. Fidel's rebels endeared themselves even more to the *guajiros* when they set their weapons aside to help with the coffee harvest in May 1957; without their assistance, much of the crop would have been lost by the individual families who depended on it for survival, since many of the peasants had been taken away by the army during punitive expeditions. Naturally, Castro delivered a revolutionary speech to the coffee growers in the middle of the Sierra.

To protect morale among the rebels, and especially the peasant and lowland volunteers, Fidel enforced rigid discipline (he had decreed that insubordination and desertion were capital offenses) as well as displaying a warm personal relationship with his rebels. Guillermo García remembers that Fidel's relations with his subordinates was "a constant theme of his life—talking to people, explaining things to them, listening to them, asking everybody's opinions about everything." At the end of every march, no matter how long, Castro "would analyze for us all that had happened that day, our problems and the enemy's problems, then he would offer an evaluation of the territory where we were." In this fashion, García says, the men had "a complete knowledge of everything . . . that was happening," and they responded with "extraordinary respect for Fidel." He remembers that "if anyone was ill, if anyone felt bad, it was a tremendous preoccupation for him, and this makes one feel respect . . . he always asked people, 'How do you feel?,' 'Did you sleep well?,' 'How was the food last night?' . . . He cared for each soldier." According to García, Fidel also knew how to listen to the peasants and how to talk to them: "He arrived with a program to discuss it with the peasants, hearing their opinions, asking what they thought of different social and political problems . . . for Fidel it was fundamental that each peasant, each child, each youth, each adult would understand the reasons for this revolutionary struggle . . ." On the other hand, the peasants would not listen to government officials or the army, even if they tried to address them, García says, because "they had

404 / F I D E L

nothing to tell them . . . they could not keep up a conversation for five minutes with a humble peasant, a poor worker, because they would be asked, 'What do you bring me?' 'Whom do you defend?'"

And Fidel and his men were at the same time a rough-looking and rough-sounding band. By the end of February, most of them were *barbudos,* sporting beards of varying sizes and colors, long hair, and filthy, torn clothes. Headgear ranged from captured army helmets to straw hats and green caps like Castro's. They easily scared anyone they encountered, and their smell was not faint. Raúl noted in February that he had just taken his third bath (in a creek) since leaving Mexico at the end of November. And their sound, too, was frightful. Carlos Rafael Rodríguez, the Communist leader who spent several months around Castro in the mountains in 1958, remarked that "in the Sierra Maestra, he developed a brutally filthy language, if I can put it that way . . . he was then living the exaltation of the battle, and his every third word or his every fourth word was . . . well, you know . . ." But, Rodríguez says, Castro always watched his language in front of women.

Shortly after five o'clock in the morning on Saturday, February 16, Fidel Castro met the woman who was to become the most important person in his life. Celia Sánchez Manduley was thirty-six years old, unmarried, extremely intelligent and efficient, dark-haired, attractive without being beautiful, and wholly dedicated to the ideals of the 26th of July Movement as defined by Castro. One of five daughters of Dr. Manuel Sánchez Silveira, Celia lived in Manzanillo and Pilón in the southwest of the province. She was acquainted from childhood with just about everybody in that tight little world, from politicians to Sierra peasants, and she was deeply involved in politics. After the 1952 Batista coup, Celia visited Havana to meet *Ortodoxo* party leaders, but never came across Fidel Castro.

When Castro exiled himself in Mexico in mid-1955, after his release from prison, and began planning the invasion, Celia already had solid contacts with the nascent Movement. She obtained navigation charts for the southwestern Oriente coast for Pedro Miret, during his reconnaissance of the area with Frank País. Armando Hart, a Movement founder, recalls that Celia came to Havana afterward to ask to be included in Castro's projected invasion. However, he says, Frank País wanted her to stay in Manzanillo to organize the support for the landing expeditionaries.

In Oriente, working with Frank País, she put together the clandestine peasant network of the Movement that would meet the *Granma* and transport the rebels to the Sierra; she waited over forty-eight hours for Castro to land in her area. In the aftermath of Alegría de Pío, as a new member of the Movement's National Directorate, Celia organized in Manzanillo the first urban support system for the guerrillas (Frank País's Santiago group had been shattered by the abortive November 30 uprising), and dispatched arms, ammunition, food, supplies, and volunteers to the mountains through her peasant network. It was not much help at the start, but for Castro even a single bullet counted then. It was Celia who received Faustino Pérez when he came down from the Sierra at Christmas to tell the world that Castro was alive and fighting (she had sent a jeep up to the Sierra to fetch him after the first messenger brought the word), and she helped him continue to Santiago and Havana. Then, Celia went to work on preparing Castro's forthcoming meeting with the National Directorate, and arranging the interview with the *New York Times* writer. She turned Manzanillo into the rebels' logistics center, right under the noses of the Batista police, the Military Intelligence Service, and the army garrison. Celia was the latest in the providential women who always appeared at crucial moments in his life.

Celia, whose code names were "Norma" and "Aly," left Manzanillo with Frank País on the evening of Friday, February 16, in a car driven by Felipe Guerra Matos, a Movement member. Reaching the point of penetration into the Sierra around midnight, they walked all night with a rebel guide. Right after dawn they ran into Luis Crespo, one of Castro's top fighters, and then into Fidel and his group, who also had been marching all night to the Epifanio Díaz farmhouse. Now at 5:00 A.M. Fidel and Celia met face to face in the middle of a pasture several hundred yards from the house. Castro had met Frank País on his two visits to Mexico, but Celia had been an elusive figure. Neither of them ever described this first meeting, but the mutual impression must have been formidable. The pasture meeting marked the birth of a twenty-three-year association, lasting until her death.

Raúl joined his brother in the field to meet Celia and Frank, and the four talked until high noon, when for security reasons they decided to move to a canefield a half mile away to lunch on delicacies the visitors had brought from Manzanillo and to continue their talk. Fidel gave Celia and Frank a detailed account of what had happened to the rebels from the time they left Mexico. Celia and Frank reported on the failed Santiago uprising, the progress in their efforts to expand the Movement, and the rumors they had heard that Eutimio Guerra was in the pay of the army. Fidel insisted on the urgent need for recruits and arms and ammunition from the cities. Together they outlined a plan for making Manzanillo a staging point en route to the Sierra Maestra under Celia's direction, and for using the Epifanio Díaz farm as the channel into the mountains. The farm, called Los Chorros, had been chosen for this meeting because the Díaz family was fully trusted politically—two of the sons belonged to the Movement—and Castro thought that the farm would make an excellent gateway to the guerrilla world. By midafternoon, it began to look like rain, and Fidel asked Guillermo García and two other rebels to put up a shed with a roof in the field; the four went on talking until nightfall. Meanwhile, Castro had decided to stay away from the farm until the arrival of the other Movement leaders and the American journalist and spent the night outside.

Faustino Pérez, Armando Hart, Haydée Santamaría, and Vílma Espín arrived early in the evening, and were taken to the field to meet the Castro brothers and Celia and Frank. Fidel had not seen Armando and Haydée (who had meanwhile become engaged) since he left Havana in mid-1955, and this was the first time he met Vílma. The daughter of a Santiago physician and American-educated, Vílma was not a member of the National Directorate, but she was so active in the Movement in Oriente that Frank País made a point of inviting her to come along. Vílma also spoke English and could interpret for the newsman. She presently met Raúl, whom she would marry right after the revolution. There was quite a bit of intermarriage in the top rebel leadership—Melba Hernández and Jesús Montané were already married—but divorces came later in most cases. Raúl and Vílma would be an exception.

Frank País and Haydée had come to the Sierra with the idea of persuading Castro to leave Cuba for a Latin American country where he could reorganize the Movement in greater safety. As Haydée later recalled the conversation, Frank had told her, "I haven't talked to him yet, but let us see how we should say it to him. . . . They could kill him [here] and we cannot afford this luxury." But before they could broach the subject that evening in the Sierra field, Fidel said to them, "Look how the soldiers just fire from below and don't dare to come up here! If you can bring me so many bullets and so many rifles, I promise you that within two months I'll really be in combat. . . . We only need some thousands of bullets and a reinforcement of twenty armed men, and we'll win the war with Batista." Frank and Haydée had underestimated Castro's optimism and self-confidence before meeting him that evening, and Haydée recalled later that "we couldn't say anything to him because he spoke with such conviction, and he wasn't asking much . . ." Celia commented afterward that "Frank left as convinced as I was that Fidel saw things correctly . . ." When night came, a peasant from the farmhouse brought a big pot of chicken and rice and *malanga* root. After dinner Luis Crespo told them there was an abandoned shack nearby where the three women could sleep. Fidel, Raúl, and the other men volunteered to accompany them, but nobody could find the shack. Celia recounted later that "we walked around so much that we couldn't find the [dinner] camp, and we slept in the open field." Fidel had decided it was not safe to wander blindly at night, so he picked a spot in a pasture among a stand of palm trees. It was 2:30 A.M. when they finally went to sleep, and as Castro said, "The night was fresh, the mosquitos abundant, and the soldiers close."

At dawn, the *New York Times* emissary had arrived at a camp at the farm, and Universo Sánchez located Castro to tell him about it. Fidel instructed him to tell the journalist that he was at a meeting with the general staff at one of the other rebel camps in the area, and that he would come over as soon as possible. Castro was determined to prevent the journalist from finding out that the Rebel Army consisted of only eighteen men. He had not seen the camp where the meeting would be held because of having spent the previous twenty-four hours in the fields talking with Celia, Frank, and the others, but he had instructed his men to make the site appear like a busy guerrilla command post. Castro was as much a master of detail as he was a master of the grand sweep.

The emissary from *The New York Times* to the Sierra Maestra was Herbert L. Matthews. He was a highly respected member of the paper's editorial board, and he specialized in Latin American affairs. As a reporter for the *Times* he had covered the Italian invasion of Ethiopia, the Spanish civil war, and World War II, and in the 1950s he became the editorial voice on the Western Hemisphere. When Faustino Pérez came to Havana from the Sierra at Christmas to arrange a visit to the Rebel Army by an American journalist, the 26th of July Movement immediately contacted the local *Times* bureau. The contact was made by Javier Pazos, an economics student whose father had been president of the Cuban National Bank before Batista and was himself very close to the 26th of July Movement. He learned that Matthews was planning a trip to Cuba anyway, so Ruby Hart Phillips, the *Times* local correspondent, cabled New York suggesting that Matthews come soon. Matthews had had an interest in the Cuban story when most of the American press was barely aware that a story—or a Fidel Castro—even existed, so he agreed instantly, asking no questions. The *Times* had no idea either what awaited Matthews.

To the newspaperman, the assignment had great personal as well as professional significance. Scholarly and reserved, he was a romantic at heart. The defeat of the Republic in Spain by the fascistic Nationalists that he had witnessed was an emotional jolt from which he never fully recovered. He wrote later: "A bell tolled in the jungle of the Sierra Maestra." A champion of Latin American democracy on the *Times* editorial page, Matthews saw the emergence of the Castro rebellion as the latest worthwhile cause in the Americas; in Fidel Castro and his Movement he sensed a vindication of the Spanish tragedy. At the age of fifty-seven, Matthews felt almost paternal about this movement of young people in Cuba.

Although Castro had never heard of Matthews, the visit had a calculated and immediate political importance for him. Once more, he was taking a page from José Martí's book. As it happened, Martí had arranged to have an American journalist cover his guerrilla war against the Spaniards shortly after landing in Oriente on April 11, 1895. George E. Bryson of the *New York Herald* interviewed Martí on May 2 as the apostle marched through the mountains northwest of Santiago, and then Martí and General Gómez, his military commander, wrote a long letter to the *Herald,* outlining the program of the Cuban revolutionary movement. Bryson returned to New York with the letter, but Martí was killed in combat on May 19. In his diary Martí remarked that he worked with "the correspondent of the *Herald* George Eugene Bryson until three o'clock in the morning," and the whole of the next day. "I work the entire day on the manifesto for the *Herald,* and more for Bryson." He notes that Bryson left on May 4.

Now, sixty-two years later, the Bryson scenario was being replayed with Matthews in another Cuban revolution. Matthews was not being caught unawares, and in his own personal account of his involvement with Cuba, he observed that Castro "was a myth, a legend, a hope, but not a reality . . . and like General Gómez, he must have been saying to himself, 'without a press we shall get nowhere.'" But Matthews also remarked that "with a press [Gómez] got American intervention." In any event, Matthews and his English-born wife, Nancie, flew from New York to Havana on February 9. On the evening of Friday, February 15, the Matthewses left by car for an unspecified destination with Javier Pazos, Faustino Pérez, and Liliam Mesa; all Herbert Matthews knew was that he would meet Castro in the Sierra Maestra at midnight the following night. He did not know Faustino, who used the code name of "Luis" and was posing as the husband of "Marta," the name used by Liliam Mesa. Matthews described Marta as "young, attractive . . . from a well-to-do, upper class Havana family" and as "a fanatical member of the 26th of July Movement, typical of the young women who risked—and sometimes lost—their lives in the insurrection." Matthews then observed most accurately that "the extent to which the women of Cuba were caught up in the passion of the rebellion was extraordinary, for like all Latin women they were brought up to lead sheltered, non-public and non-political lives."

With Marta driving, they arrived in Manzanillo sixteen hours later, on the afternoon of February 16 (Castro had reached their rendezvous spot that dawn). Matthews departed for the Sierra early in the evening, leaving Nancie behind at a Cuban family's home. The trip to the edge of the Sierra was made in Felipe Guerra Matos's jeep with Javier Pazos and two other youths; it was Felipe's third trip through army-patrolled territory that day, having first taken up Celia and Frank, then Faustino and his companions. At midnight Matthews and the others left the jeep to begin climbing up the mountain. They got lost, had to wait for two hours "in a heavy clump of trees and bushes, dripping from the rain . . . crouched in the mud . . . trying to snatch a little sleep with our heads on our knees." Then a rebel scout appeared, identified himself with two low, soft, toneless whistles, which was the guerrillas' signal, and led Matthews and his escort to the camp at Los Chorros where Castro was to meet him. At that point, Fidel's staging of the event began. In his first dispatch to the *Times,* Matthews wrote that "Señor Castro was encamped some distance away and a soldier went to announce our arrival and ask whether he would join us or we should join him. Later he came back with the grateful news that we were to wait and Fidel would come along with the dawn." Castro had succeeded in giving Matthews the impression that he had many camps and, as the *Times* article put it, "had mastery of the Sierra Maestra."

It was theater, literally guerrilla theater, that Castro put on for Matthews. An official account of the Sierra war, published in 1979 in the Communist party newspaper *Granma,* says that "before entering the camp [to meet Matthews], Fidel had given instructions to his companions to adopt martial airs." But, they add, "for some it took a lot of labor to reconcile the martial air required by Fidel with the condition of their clothes and their general appearance. . . . Manuel Fajardo, for example, had no back on his shirt, torn to shreds by his knapsack harness. During the time the journalist remained in the camp, Fajardo was obliged to walk sideways." At one point, Raúl Castro brought sweat-covered Luis Crespo over to where Castro and Matthews were talking, to say, "Comandante, the liaison from Column number two has arrived," and Fidel replied airily, "Wait until I'm finished."

As the *Granma* article explains, the purpose was "to impress Matthews about the total numbers of the *guerrilla* army, without openly telling a lie" and "finally, the journalist believed he had counted some forty fighters where there were no more than twenty and he left convinced that the group he has seen is part of a much larger force." In his *Times* story, Matthews quoted Castro as saying that Batista troops work in columns of two hundred, and "we in groups of ten to forty, and we are winning." Elsewhere, Matthews wrote that "the reports reaching Havana that frequent clashes were taking place and that the Government troops were losing heavily proved true." He observed that Castro had "kept the Government troops at bay while youths came in from other parts of Oriente . . . got arms and supplies and then began the series of raids and counter-attacks of guerrilla warfare," and that therefore "one got a feeling that he is now invincible."

Matthews had no way of knowing that up to then the Rebel Army had engaged in only two very minor clashes with the army, that Castro had just barely made it across the Sierra to meet him, and that he controlled only the ground where they sat. He would never have believed that the Rebel Army consisted of only eighteen men, all of whom he saw time after time during his three hours with Castro. Matthews, however, must never be accused of being a dupe or naïve: He was in an environment totally controlled by Fidel, the Cuban leader was eminently credible, and above all he was very much in existence when Batista was still claiming him dead. In the end, Matthews was correct in concluding: "From the look of things, General Batista cannot possibly hope to suppress the Castro revolt. His only hope is that an Army column will come upon the young rebel leader and his staff and wipe them out. This is hardly likely to happen . . ."

As soon as Matthews departed the camp to be taken back to Manzanillo, Castro resumed his discussions with the National Directorate. For four hours they covered the question of recruiting an armed contingent in Oriente cities to join the Rebel Army in the hills, with Fidel insisting on the Movement's support for the guerrilla war being its first priority. In so doing, Castro was addressing for the first time the differences of priority between the Sierra and the *llano* (the lowlands), differences that soon escalated to a power struggle over leadership roles within the Movement, and that led after 1959 to the liquidation of the 26th of July Movement and consolidation of the rule in Cuba by the new Communists, headed by Fidel Castro. In this sense, the political battle over the future fate of the revolution began two and a half months after the *Fidelistas* landed on the island and while their strength was still limited to eighteen men.

During this conference Faustino Pérez, identified with both the Sierra and the urban insurrectional groups, proposed that a "second front" of the guerrilla war be opened in the Escambray Sierra in the central province of Las Villas in order to lessen the pressure on the Castro force. Faustino had in mind a 26th of July Movement guerrilla army, compatible with the Rebel Army in Oriente, and he argued that weapons more easily available in Havana could be effectively used in the Escambray. According to the version printed in the historical account in *Granma,* this proposal was accepted, "although Fidel was not convinced of the convenience of this idea because he considered that at this moment the important thing was to concentrate available resources in the already existing *guerrillero* nucleus." The version circulating privately was that Castro simply vetoed Faustino's suggestion because he saw it as a threat to his overall leadership; the Escambray was much nearer Havana than the Sierra Maestra, and it could have merged politically with the urban groups. In any event, an Escambray "second front" was established more than a year later by the Students' Revolutionary Directorate, a rival of *Fidelismo.* Che Guevara did not attend the conference because he was not in the National Directorate (and he was a foreigner), but he wrote that the Movement in the Sierra and in the cities

"were practically two separate groups, with different tactics and strategy."
He added that "there was no sign yet of the grave differences that would
endanger the unity of the Movement several months later, but it was already clear that we had different concepts."

Castro concluded the meeting by emphasizing the importance of women
in the revolutionary struggle and by announcing that he would draft a
manifesto to the people of Cuba for the Directorate members to take down
the mountain with them. As usual, Fidel was not wasting a moment in his
around-the-clock military, political, and propaganda revolutionary enterprise. He even found time to demonstrate the functioning of his beloved
telescopic-sight automatic rifle to Celia Sánchez; magically, he had encountered a woman who not only shared and understood his political and philosophical concepts, but was also an expert with weapons. They were clearly
made for each other.

In the afternoon Castro ordered the execution of the traitor, Eutimio
Guerra. Incredibly, he had turned up near the farm where the *Fidelistas*
were holding their meetings that day, evidently in the hopes that he could
finally spring a lethal trap for Fidel. He did not know that he was already
under suspicion. Exploring the farm's vicinity, Eutimio ran into a relative,
who was in the Rebel Army, and a companion. The relative, familiar with
the suspicions about Eutimio, rushed off to find Fidel, leaving Eutimio
with the other man. Castro was not surprised: He had predicted Eutimio's
return, and now he dispatched a squad under Juan Almeida to capture the
traitor. Army safe-conduct passes were found on Eutimio, and he was manacled and taken to Fidel's camp. Raúl wrote that the rebels' first idea was to
turn Eutimio into a "triple agent" working against the army, but he refused. Then Fidel interrogated him at length (Raúl commented later that
"he might have given us more information if we tortured him, but we did
not apply such methods even to such miserable people"). A tremendous
storm exploded overhead with thunder and lightning, and at 7:00 P.M.
Eutimio Guerra was shot to death.

Castro spent the next three days at the farmhouse drafting his Manifesto,
the "Appeal to the People of Cuba" to be distributed throughout the island
about the time the Matthews articles appeared in *The New York Times*. The
approach was reminiscent of Martí's dealings with Bryson: first the interview, then the carefully drafted document for publication in the *Herald*. In
this instance, Fidel calculated that the Matthews reportage would establish
the fact that he was, indeed, alive—and that this in itself would be a
sensation—and then the Manifesto would be circulated with enormously
enhanced credibility. Also, it would be the first formal document of the
26th of July Movement to be issued from the Sierra Maestra, making it

absolutely clear that Fidel Castro alone was the Commander in Chief of the revolution.

The visiting Movement leaders had brought newspapers and magazines with them, and Fidel (who normally depended on his battery-powered radio for news) was able to catch up with affairs elsewhere. Among the events reported in the Havana press was the ceremony at which United States Ambassador Arthur Gardner presented seven Sherman tanks to General Batista, and referred to Castro as a "rabble-rouser." The same week, the aircraft carrier *Leyte* and four destroyers paid an official visit to Havana. The American policy remained one of unquestioned support for the regime, it having never occurred to the administration that a wholly new situation was developing on the island. Even without reading about these latest acts showing United States support for Batista, Castro had complained to Matthews that the regime was using American weapons not only against him, but "against all the Cuban people. . . . They have bazookas, mortars, machine guns, planes and bombs." Answering a follow-up question, Castro said: "You can be sure we have no animosity toward the United States and the American people."

It is understandable, then, that Castro reacted with rage upon learning that Colonel Carlos M. Tabernilla y Palmero, the head of the air force that had been bombing and strafing the rebels and Sierra peasants, was awarded the United States Legion of Merit by Air Force Major General Truman Landon. The American general flew to Havana to present the decoration to Tabernilla (whose father was the Cuban Army's Chief of Staff) for "the furtherance of amicable relations between the Cuban Air Force and the United States Air Force from May, 1955, to February, 1957." At the same time, the U.S. Congress was informed that between 1955 and 1957, the United States had delivered to Cuba 7 tanks, a battery of light mountain howitzer artillery, 4,000 rockets, 40 heavy machine guns, 3,000 M-1-caliber semiautomatic rifles, 15,000 hand grenades, 5,000 mortar grenades, and 100,000 .50-caliber armor-piercing cartridges for machine guns.

At the farm, Fidel and his men had an opportunity to read the Movement's clandestine newspaper, *Revolución,* which Carlos Franqui, a former Communist and an early *Fidelista,* had been publishing in Havana on an irregular basis since mid-1956. The issue, printed late in January with the headline FIDEL IN THE SIERRA, had been sent to Santiago to be reprinted, and the total circulation was said to have been twenty thousand copies. Frank País and Vilma Espín, took some copies to Castro and the guerrillas feasted their eyes on it while they waited for Herbert Matthews's bombshell to burst.

This happened a week later on Sunday, February 24. The most vital part as far as the guerrillas were concerned was in the opening sentences: "Fidel Castro, the rebel leader of Cuba's youth, is alive and fighting hard and successfully in the rugged, almost impenetrable vastness of the Sierra Maestra. . . . Batista has the cream of his Army around the area, but the Army men are fighting a thus-far losing battle to destroy the most dangerous enemy General Batista has yet faced in a long and adventurous career as a Cuban leader and dictator." Elsewhere, Matthews wrote: "The personality of the man is overpowering. It was easy to see that his men adored him and also to see why he has caught the imagination of the youth of Cuba all over the island. Here was an educated, dedicated fanatic, a man of ideals, of courage and of remarkable qualities of leadership."

Predictably, the impact of the Matthews articles (the *Times* printed three of them on successive days) was immense. Because censorship in Cuba had been lifted that week, Matthews's stories were reprinted in the national press, instantly elevating Castro to hero status. The Batista regime made matters even worse for itself when Defense Minister Santiago Verdeja issued a statement on the day after the last article, claiming that Matthews had written "a chapter in a fantastic novel," that he had not interviewed "the pro-Communist insurgent, Fidel Castro," and that even if Castro was alive, he commanded "no supporting forces." This was compounded by the remark that if the interview had taken place, there would have been a corroborating photograph of Matthews and Castro. The *Times,* of course, did have such a photograph and published it the following day, but even then Batista disbelieved it. In his memoirs written in exile, Batista admitted that "I, myself, influenced by the statements of the High Command, doubted [the interview's] authenticity. . . . Castro was beginning to be a legendary personage and would end by being a monster of terror." The Matthews visit was a major turning point in Castro's career, and subsequently a magazine in New York published a cartoon of Fidel, underneath which was the newspaper's advertising slogan, "I Got My Job Through The New York Times." Today, Herbert Matthews is a forgotten man in Cuba: only oldtimers like Faustino Pérez remember him fondly, and Castro no longer mentions him.

Dated February 20, Castro's "Appeal to the People of Cuba" was a call for violent action throughout the island in support of the revolution for which, "if necessary, we shall fight in the Sierra Maestra for ten years." This was a moment of great weakness for the eighteen-man army, but Castro, having already mounted his guerrilla theater for Herbert Matthews, knew how to appear powerful and victorious. In his mastery of the use of propaganda to mobilize resources, he had learned well from Martí and Lenin, his favorite authors.

The six-point revolutionary program Castro signed on behalf of the

Movement urged "intensification in the burning of sugarcane . . . to deprive the tyranny of the revenue with which it pays the soldiers it sends to their death and buys the planes and bombs with which it assassinates scores of Sierra Maestra families." He asked, "What does a little hunger matter today to conquer bread and freedom tomorrow?" After the cane is burned, he wrote, "we shall burn sugar in the warehouses . . ." The next point proposed "general sabotage of all the public services." Then Castro called for the "summary and direct execution of the thugs who torture and assassinate revolutionaries . . . and all those who pose an obstacle to the Revolutionary Movement." He demanded the creation of "civic resistance" in all the Cuban cities and a "general revolutionary strike as the final and culminating point in the struggle." The document was retyped, and the Movement leaders made their way home to Manzanillo, Santiago, and Havana with copies of the "Appeal" concealed on them.

At this juncture, Castro was not interested in taking ideological positions, and his "Appeal" steered clear of them, much more so than had "History Will Absolve Me." He was concerned with the physical aspects of the struggle against Batista, with the inadequacy of his guerrilla band, and with the Movement's inability to bring other revolutionary groups into its fold (and under its control). The question of unity—and unified leadership—was increasingly on Fidel's mind, and he was careful not to upset the volatile revolutionary politics with unnecessary statements and proclamations about his future plans. Proclamations could await a more favorable political climate and greater power. This is why, for example, he chose to disassociate himself personally from a detailed program drafted by Mario Llerena, the Movement's director of public relations in exile (named by Frank País, not by Castro), but not brought up to the mountains for his approval. Instead, he put all the pressure behind the organization of the Civic Resistance Movement along the lines of his own "Appeal," and this became extremely successful. Returning from the Sierra conference, Armando Hart and Faustino Pérez made this project a priority, and the clandestine Civic Resistance turned into a key adjunct of the 26th of July Movement. Concentrating on propaganda, fund-raising, and general support activities, it was meant to attract those who were ready and able to help but not to fight.

For Castro's Rebel Army, the next three months were a period of expansion, preparation, and immense hardship. For another group of revolutionaries in Havana, it was a time of pure tragedy and grief. In terms of the overall contest between Batista and all his opponents, it was also a time of stalemate. Army troops could not destroy the guerrillas, nor could the police in the cities smash the clandestine organizations involved in sabotage, propaganda, and support for the Sierra Maestra fighters. Castro, on the other hand, was too weak to venture out of his constantly changing mountain hideouts. For Batista not to be winning, however, was for the regime to be losing; each day the rebels continued to operate, the greater the danger to the dictatorship. As Fidel told his men in mid-March, "On three occasions we have almost perished . . . the enemy threatens us everywhere, while denying our presence here. . . . We have only twelve rifles and forty rounds of ammunition for each of us, but we have fulfilled our promise to the Cuban people. . . . We are here! . . ."

While Castro was compulsively optimistic, for Che Guevara these were "the bitter days . . . the most painful stage in the war." A series of asthma attacks had virtually paralyzed him, hitting him on the third day of the march when he had run out of his adrenaline-based medicine. Although they were walking slowly, Guevara could not keep up. On a day when the army mortars attacked the rebels, the retreat was easy, but it had to be fast, and Guevara wrote, "My asthma attack was such that it was actually difficult for me to take a step." Luis Crespo, their best Sierra man, dragged Che and both their rifles and equipment for hours on end, muttering affectionately, "Walk, walk, you fucking Argentine, or I'll give you the butt of my rifle . . ." Later, at a shack in a place called Purgatorio, Castro introduced himself to the owner as "Major González of the Cuban Army" in the presence of another peasant. The peasant criticized "that rebel Fidel Castro" for a spell, and departed. Castro then told the owner his true identity. The old man embraced him warmly and offered to go to Manzanillo to buy medicine for Guevara. Castro moved on with his group, leaving Che behind with a young rebel and the outfit's best rifle. The old peasant returned with the medicine, Che spent ten days walking back to the farmhouse, leaning on every tree trunk and on his rifle, while the young soldier "had a heart attack each time my asthma made me cough . . ."

When Frank País left the Sierra after his meetings with Castro, he promised to have a group of volunteers at the same farm on March 5, two weeks later. Fifty-eight recruits from Santiago and Manzanillo, led by Captain Jorge Sotús, reached the farmhouse on March 25, three weeks behind schedule, and only thirty of them had weapons. Among the recruits were three young Americans, sons of servicemen from the Guantánamo naval base: Charles Ryan, Victor Buehlman, and Michael Garney. Now, Castro observed, the Rebel army was almost back to its original strength of eighty-two when the *Granma* landed. This latest contingent was organized and sent up the Sierra by Celia Sánchez from the secret staging area she had set up at a small farm known as La Rosalia, a block from the Manzanillo city jail. She had the men hide inside clumps of bushes (there were no trees on the farm) for several days. Then they were loaded in small groups aboard trucks belonging to a rice farmer and part-time teacher named Huber Matos, an active Movement member, to be taken to the edge of the Sierra. From there, the inexperienced recruits climbed up to the rebel outposts.

While Fidel and his men waited at the farm for Sotús's column, they learned from radio news broadcasts on March 13 that an attack on the presidential palace in Havana by the Students' Revolutionary Directorate (DR) had failed. At least thirty-five DR members were killed at the palace, the DR's president, José Antonio Echeverría, was shot dead in an affray near the university, while scores of others were captured, tortured, and murdered. Pelayo Cuervo Navarro, a very well-known *Ortodoxo* party leader, was assassinated in his luxury Havana neighborhood. In a manifesto written on the eve of the attack, Echeverría stressed that his intentions stemmed from "our commitment to the people of Cuba [which] was established by the Mexico City Pact, which joined our youth in conduct and action." This was the accord he and Fidel ironed out and signed on behalf of the DR and the 26th of July Movement, respectively, when Echeverría went to Mexico in September 1956 to see Castro. No specific actions were assigned to either side, but Castro and Echeverría were so clearly rivals for national revolutionary leadership that any major move by one instantly triggered psychological and political pressure on the other to match it. At the same time, the two young leaders represented different societal groups with all the implications this carried in Cuba: Castro had chosen to lead a working-class constituency with the aid of a handful of intellectuals, while Echeverría spoke for the young Cuban middle class—and a sprinking of middle-class, intellectual veterans of the Republican army in the Spanish civil war.

In the context of their rivalries, Castro had made the first spectacular move when he invaded Cuba in December and was able to survive and grow in the months that followed. The DR, meanwhile, was unable to mobilize an uprising in Havana to support the landing. Castro's standing was immensely enhanced by the Matthews articles in the *Times*. It was therefore logical that in his manifesto Echeverría said that "the circumstances necessary for our youth to carry out its assigned role were not forthcoming at the right time, forcing us to postpone the fulfillment of our obligation." He added: "We think that the moment has now come. We are confident that the purity of our motives will bring us God's favor so that we may achieve the rule of justice in our country." Many years after the revolution, at a commemoration of the palace attack, Castro grabbed the microphone to protest furiously when the chairman of the event omitted Echeverría's mention of God in reading his manifesto. Fidel had had fundamental differences with the student leader about almost everything, but he felt it was insulting to the memory of Echeverría to excise the exhortation to God.

Among the many mysteries of that day in March is why the DR and Echeverría chose this particular moment and this particular type of action to assert their commitment to the revolutionary bargain and to their political *machismo*. It is not absolutely certain that the idea of storming the palace and killing Batista actually came from this student of architecture, Echeverría; he did not personally lead the assault on the palace, being instead in command of a simultaneous strike on a radio station. It remains unclear exactly how the weapons were obtained for the palace coup and by whom they were financed; it may well have been done by wealthy Cuban political exiles, and Echeverría and his companions may have been persuaded (or dared) to kill the dictator. The DR had already engaged in selective political assassination, and secret sponsors of the action against Batista may have surmised that the students would apply it to the usurper in the palace. The day's attacks were superbly planned and their executors were supremely heroic—it all failed because Batista was barricaded on the third floor of the palace when the students took the second floor, to be dislodged later by powerful army reinforcements. Had the attack succeeded, it would have left Fidel Castro in his mountains as a suddenly irrelevant factor in the revolutionary equation.

According to one version of these events, it was claimed that the DR had amassed arms in Havana not only to kill Batista, but also to block Castro's attempts to seize power in the capital once the dictatorship was overthrown in one fashion or another. This is reasonably credible because weapons for the DR's second Escambray front established the following year came from caches in Havana, and because armed DR units did capture the presidential palace after Batista fled the country on the last day of 1958, to prevent Castro from taking it.

In any case, Castro and the leadership of the 26th of July Movement had little use for the DR and its tactics. Faustino Pérez, who had just returned from the Sierra Maestra, is said to have declined a DR invitation for the Movement to participate in the attack on the palace; the decision, however, would have been Pérez's own because Castro was unaware of these plans and there would have been no time to communicate with him. Fidel himself did not hide his absolute disapproval of the DR action, making it clear that he thought it was part of the overall leadership struggle. In a radio interview with an American reporter a month after the DR assault, he said it was "a useless spilling of blood. The life of the dictator does not matter . . . I am against terrorism. I condemn these procedures. Nothing is solved by them. Here in the Sierra Maestra is where they should come to fight." But Castro was consistent in his views against political assassination as a revolutionary instrument. Twenty years later, discussing the assassination of John F. Kennedy, he returned to the theme, saying that "we have never believed in the assassination of leaders . . . we fought a war against Batista for twenty-five months, but we were not trying to kill Batista. It would have been easier to kill Batista than to assault the Moncada, but we did not believe that the system is abolished, liquidated, by liquidating the leaders when it was the system that we opposed. We were fighting against reactionary ideas, not against men."

Castro's disparaging comments nevertheless emphasized the leadership issue. When he said the students should have come to the Sierra Maestra to fight, he obviously meant for them to be under his command. Ironically, Cuban Communists also came out against the DR attack on the palace, putting it under the same heading as Fidel Castro's guerrilla operations. Four days after the assault, Juan Marinello, chairman of the Communist Popular Socialist Party and a minister in the Batista government in the 1940s, wrote Herbert Matthews that "our position is very clear; we are against these methods." Marinello wrote that there was no need for "a popular insurrection," and what Cuba needed was "democratic elections," and "a government of a Democratic Front of National Liberation" which the Communists would have attempted to dominate. Therefore, Marinello informed Matthews that "we think that [the 26th of July Movement] has noble aims but that, in general, it is following mistaken tactics. For that reason we do not approve of its actions, but we call on all parties and popular sectors to defend it against the blows of tyranny . . ." The Communists' position had not changed from November to March despite Castro's ability to survive, and, like Batista, they were not taking him seriously. This is worth noting in the light of subsequent official propaganda claims that the Communists were supposedly helping the guerrillas in 1957.

Another unanswered question, this time involving the Communists, was the "Crime at 7 Humboldt Street." This was the address of a safe-house apartment in Havana where the surviving leaders of the students' attack on the presidential palace had hidden during Easter Week. The leaders were Fructuoso Rodríguez, who was elected president of the University Students' Federation (FEU) after Echeverría's death, and Joe Westbrook (who was with Echeverría at the radio station), José Machado, and Juan Pedro Carbó Serví. On April 20, Easter Sunday, the four were killed by the secret police after the location of the apartment was given them by an inside informant. After the triumph of the revolution, it developed that the traitor was a student named Marcos Armando "Marquito" Rodríguez. Marquito had extremely close personal ties to the top leaders of the "old" Communist party, although he was not believed to be a party member at the time. In 1964, Marquito Rodríguez, who in the meantime had been favored with a university scholarship in Prague, was arrested and tried. At the trial, Fidel Castro acted in effect as the prosecutor, and Rodríguez was found guilty of the Humboldt Street treason and ultimately executed. The Communists were never specifically linked with Rodríguez's actions, but the names of all his powerful friends in the "old" Communist party surfaced at the trial, Rodríguez having confessed his crime to them in exile in Mexico. As it happened, most were purged by Castro in 1962 for leading a "sectarian clique" against him inside the new Communist party he was then forming. The "old" Communists were a classic Moscow-oriented party, and it was never clearly explained why the leaders had not informed Castro earlier of the Rodríguez treason "confession," if that was the case, and why it took the revolutionary regime five years to discover the truth about one of the most painful and dramatic episodes of the struggle against Batista. But Castro has always had his ways of sending messages.

In the meantime, Fidel was shaping his strengthened guerrilla army into an effective fighting force, teaching the new recruits the secrets of irregular warfare, getting them accustomed to difficult marches and other hardships. He rejected Che Guevara's recommendation for engaging into immediate combat as he did not think the men were ready for a major encounter; he waited two months for the right moment. For the time being, it was a nomadic life, moving camp almost every day, marching at night, and sometimes going without sleep for days. Food was short, and, as Che told the story, one day the rebels had to eat their first horse. It was an "exquisite" meal for some of them, and "a test for the biased stomachs of the peasants who thought they were committing an act of cannibalism while chewing the old friend of man."

Late in April, Castro was joined in the mountains by Robert Taber and Wendell Hoffman of the Columbia Broadcasting System. Castro, who had asked for more American journalists after the success of the Matthews trip, did not know who Taber and Hoffman were until they arrived (he only had a brief message that American journalists were on their way up the Sierra), or how complicated it was to deliver them and their bulky television equipment to the Oriente hideouts. Armando Hart and Haydée Santamarí were in charge of the Havana arrangements, but Hart was arrested, and Haydée drove the Americans to Bayamo with Marcelo Fernández, the Havana coordinator of the 26th of July Movement (Haydée had managed to hold on to several thousand pesos in collected funds Hart was carrying). In Bayamo they met Celia Sánchez and Carlos Iglesias, a Movement leader from Santiago, and the six traveled together to Manzanillo and up to the Sierra.

The American television team and Celia and Haydée remained with the guerrillas for nearly two months while Castro concentrated on building a logistics and supply system in the region. Food depots were set up in peasant houses to receive shipments from the lowlands, permanent camps were built for the *guerrilleros* for use during their constant moves, and peasant liaison teams were organized to serve as couriers and a rudimentary intelligence service. Locations were selected where Castro and the others could contact the local population. At this stage, the rebels were beginning to control an ever-expanding area in the Sierra that they called the "Free Territory."

Then the entire column climbed Pico Turquino, Cuba's highest mountain, for Castro to hold a CBS television interview in front of the bust of José Martí that Celia Sánchez and her father had installed there many years earlier. He said: "We have struck the spark of the Cuban Revolution." Che Guevara noted that Castro, who had never been there before, checked his pocket altimeter atop Turquino to assure himself that it was exactly as high as shown on the maps; he never trusted anybody or anything. By reaching Turquino, Castro had shifted his operations back to the central part of the Sierra Maestra where he felt more protected and where he could develop a better infrastructure for what he now thought would be a protracted war. Less is known of the strategic planning by the Batista high command—and their battle projections at the time—because in the euphoria of revolutionary victory the crowds destroyed key military archives in Havana and at Bayamo.

Because Castro's presence in Cuba was becoming both an accomplished fact and a major element in all the political calculations, other rivals sought to challenge his sway. Former President Carlos Prío financed an expedition of twenty-seven men, led by a United States Army veteran named Calixto Sánchez, to establish a separate anti-Batista guerrilla front in the mountains in the north of Oriente. Cuban politics being what they were (and Prío being as wealthy as he was), there was no real contradiction in financing Castro in Mexico, and then a competitive operation later; Prío wanted his fingers in every pie. The group sailed from Miami aboard the yacht *Corintia* on May 19, landed in Cuba on May 24, and were betrayed by a peasant to the army on May 28. Twenty-four men, including several DR members, were caught and shot.

The war continued therefore to be Castro's show. On the same day the *Corintia* expedition was liquidated, Fidel led his men into combat for the first time since January. This was the battle of Uvero on May 28, when the Rebel Army moved the farthest east to take a government garrison on the Caribbean coast. Castro's principal reason for his eastward shift was a message from Santiago that Movement couriers would place an important shipment of modern arms at a specific point in the Sierra Maestra, east of Turquino, for his forces to recover. But as usual there were delays, and the cache was not found until May 20. In the meantime, Che Guevara got lost for a day (thereby discovering that a compass is not enough in the Sierra, and that a knowledge of the terrain is essential), the rebels executed a Batista spy they had caught, and another American journalist, Andrew St. George, joined the Castro column. Additional peasants also signed up with the rebels (in each case, Fidel interviewed volunteers at length about their backgrounds and motivations before accepting them), and toward the end of May, the *Fidelistas* were up to 120 men. But it was the new weapons that gladdened Castro's heart: three tripod-mounted machine guns, three submachine guns, and nineteen automatic rifles, including American M-1s. Che Guevara was handed a submachine gun, becoming for the first time a full-time fighter; until then had spent most of his time as a physician for the troops and the peasants in the villages they crossed, fighting only when necessary. Despite this, the impression Guevara conveys of himself in his war diary is one of toughness and mercilessness. He relates, for example, an incident just before the Uvero battle when the rebels caught a Batista army corporal with an unsavory record. "Some of us," Guevara wrote, "proposed his execution, but Fidel refused to do anything to him." Castro was always willing to shoot traitors, spies, deserters, and rapists, but not war prisoners. Che also, according to his own recollections, ordered the shooting of a peasant *guerrillero* who turned out to be a thief, then began passing himself off as Che Guevara, pretending he was a physician and demanding, "Bring me women. I'm going to examine them all . . ."

The Uvero battle was extremely tough and costly: Of the eighty rebels, six were killed and nine wounded (including Juan Almeida, a general-staff officer); of the fifty-three army personnel, fourteen were killed, nineteen wounded, and fourteen taken prisoner. It was the bloodiest encounter since Alegría de Pío, starting at dawn and lasting nearly three hours, but this time the *Fidelistas* won. They took two machine guns and forty-six rifles, and Castro declared that "thus began a new phase in the Sierra Maestra." Che looked after the enemy wounded and, according to Fidel, left them "in the care of their own doctor, so that the army might pick them up and move them to their own hospitals, thanks to which none of them died." Pedro Álvarez Tabío, the military historian, says that Uvero "had a very important strategic significance because it demonstrated for the first time to the rebel fighters . . . that the Rebel Army could defeat the army of the tyranny, and that the seizure of power through a military defeat of the army was possible."

Immediately after Uvero, Celia Sánchez left the Sierra with the CBS team, continuing on to Santiago, where Herbert Matthews was visiting on his return to Cuba. Castro wanted her to bring Matthews up-to-date on rebel activities. She had, however, taken part in the Uvero battle, the first woman to fight with the Rebel Army. Four months would elapse before she would return, as she was needed in Manzanillo to coordinate the flow of men, arms, and supplies to the Sierra. Addressing her by her underground name, "Norma," Fidel wrote Celia shortly after she left the mountain: "We have such pleasant memories of your presence here that one feels your absence has left a real vacuum. Even when a woman goes around the mountains with a rifle in hand, she always makes our men tidier, more decent, gentlemanly—and even braver. And after all, they are really decent and gentlemanly all the time. But what would your poor father say . . ." After hearing a false report that Celia had been detained, Castro had Raúl, Che Guevara, Camilo Cienfuegos, and other top fighters sign a letter, addressed to Norma, declaring that "you and David [Frank País] are our basic pillars. If you and he are well, all goes well and we are tranquil . . ." A formal statement by the rebel commanders proclaimed that "concerning the *sierra,* when the history of this revolutionary epoch is written, two names must be printed on the cover: David and Norma." Che Guevara wrote that Celia "constituted our only known and safe contact . . . her detention would mean isolation for us." Without her, Cuban history might have been different.

Frank País was killed by the Batista police in Santiago on July 30. Castro wrote Celia Sánchez the following day that "for the moment, you'll have to assume for us, a good portion of Frank's work, especially as you know more about it than anybody else." País was not only the Movement's principal leader in Oriente from the standpoint of providing logistics support for the Rebel Army in the mountains, but, increasingly, a political thinker on a national scale—possibly the best revolutionary mind, after Fidel Castro, in Cuba (Che Guevara was an Argentine and at that stage still tended to stay away from overall Cuban strategies). País's death was a particularly telling blow to the Movement because it now found itself in the midst of a fundamental political crisis.

The growth of the *Fidelista* army and the expansion of the "Free Territory" pitted Castro in the mountains against the urban wings of the Movement in a struggle for revolutionary leadership and policy-making. By July, the guerrilla army had grown to two hundred men whom the Batista armed forces simply could not dislodge, and the earlier Sierra-lowlands frictions turned into an acute if muted dispute. Castro took the view that the first and foremost priority of the Movement in the lowlands was to support the guerrilla army. This implied, of course, the Movement's acceptance of his national leadership, and Fidel was as ready to fight for this as he was to fight Batista. In letters to Celia in July and August, Castro insisted: "The proper order should now be: *All guns, all bullets, and all supplies to the Sierra.*" In the cities, however, the feeling was growing that

Castro should share decision-making with the National Directorate, ostensibly because he was too isolated in the mountains to be sufficiently informed about events in Cuba, but really because many Movement leaders thought they, too, were entitled to a voice in the country's future. Not only were they sending arms, money, and supplies to Castro, they argued, but the 26th of July Movement and the Civic Resistance were also assisting him through sabotage efforts against the economy—blowing up utility plants, factories, and government offices, and torching sugarcane fields—and scores of these clandestine fighters were being killed, arrested, and tortured in the cities.

Having barely consolidated his military position in the Sierra Maestra after the Uvero battle, Fidel Castro now had to turn his attention to the emerging political struggle within revolutionary ranks. Although Castro knew what was coming, it was Frank País who finally forced him to face the political problem. Constituting, in effect, the principal link between the Sierra and the *llano,* País wrote Castro a long letter early in July informing him that because of the vast chaos and confusion reigning in the Movement throughout Cuba, he and Armando Hart had "decided on an audacious move to revamp the Movement in its entirety."

The significant aspects of this decision were that Castro had not been consulted beforehand, which in itself was a challenge to his authority, and that País was now telling him that "the leadership would be centralized for the first time in the hands of a few, the distinct responsibilities and tasks of the Movement would be clearly assigned." Though País spoke of poor coordination within the Movement as the reason for the changes, it was clear that his concept was to divide power between the mountain and the cities; for example, that the National Directorate would include the six provincial coordinators. The Rebel Army would be represented by one delegate. Moreover, he stressed that one of the "defects" of the Movement was the "lack of a clear and precisely outlined program, which is, at the same time, serious, revolutionary, and within the range of achievement." País then informed Castro that he had already ordered the drafting of such a program by a group of intellectuals. Finally, he urged the creation of armed Movement militias nationwide. Again, Castro saw prerogatives taken away from him by the twenty-two-year-old Santiago leader. To be sure, País respectfully told Fidel that "you will all decide on this, but I ask that your opinion be communicated to this Directorate as rapidly as possible."

Whether or not Castro suspected in Frank País a potential or actual rival for leadership—or at least a very independent voice (official Cuban accounts insist that País was a devoted Castro disciple)—he did understand the developing situation. Even before País's communication, he summoned to his side two respected moderates of an older generation, Raúl Chibás and Felipe Pazos, and used their prestige to enhance his own political standing in the country. A great tactician, Castro chose for the moment to form a highly visible alliance with the political center (the ever-outspoken Che Guevara described it as "the makings of treason" in his diary, bad-mouthing Chibás and Pazos as natural betrayers of the Revolution). However, Castro used both men only until they became expendable after 1959. The Communists still believed that the Movement was "putschist, adventurist, and petit bourgeois," and in any event public collaboration with them did not yet suit Castro.

Chibás, an unassuming, quiet, but courageous man, recalls the circumstances of his arrival in the Sierra on July 4, after meeting with Frank País in Santiago and Celia Sánchez in Manzanillo. País was disturbed that "Fidel was making so many decisions . . . it was a *caudillo* problem, that Fidel was making decisions without considering an existing Directorate." Chibás was guided to a house in the mountains to await Castro, who then arrived with his column of some 160 men, followed by Che Guevara with a contingent of wounded rebels. The scene, he says, "was like something out of the movies, watching them coming, taking positions all around, and all in complete silence because there everything was in whispers. I spent a month speaking in whispers: It was their discipline, the difference between the Rebel Army and the Batista army. The Batista army always arrived shouting, and it was easy to surprise them because it was known they were there. But the rebels had their own ways. For example, if it was an open field, only one person would cross it at a time, so if a plane came overhead, it looked like a peasant walking over a field. Then, the next man would cross."

In his long talks with Castro, whom he had known for nearly ten years, Chibás insisted that elections be called in Cuba within a year of the victory over Batista. He told him that with control over the congress, "all the necessary revolutionary laws can be passed." Chibás emphasizes that Castro agreed with the electoral concept and that "nobody forced him" to accept it. The Sierra Maestra Manifesto, issued on July 12, was a result of Castro's meetings with Chibás and Pazos, and was signed by all three. It declared that within a year "we want elections, but with one condition: truly free, democratic and impartial elections." It represented Castro's firm and formal commitment to "free, democratic elections" as the central point of the Manifesto that he intended to be the Movement's program, thus superseding Frank País's manifesto in Santiago. Two years later, Castro would break that commitment in the euphoric climate of revolutionary victory on the island when he offered successive rationalizations why no elections were needed in Cuba.

According to Chibás, Castro drafted the Manifesto "totally by himself," without being influenced by him and Pazos, as charged by Che Guevara. But Chibás says Pazos pushed Castro into proposing in the Manifesto the creation of a "civic-revolutionary front with a common strategy of struggle" and the immediate designation of "a person to preside over the provisional government, whose election will be left to the civic institutions." Castro liked the idea of the "civic-revolutionary front" because it would transcend the 26th of July Movement and, since he proposed it, could be controlled by him from the Sierra. The notion of a provisional government was likewise appealing to Fidel because it would be a prestigious "front" for his own revolutionary activities, and because *he* expected to be able to choose its president from the Sierra even though the Manifesto stated the selection would be made by "civic institutions" such as the bar association and the medical associations, all of which now opposed Batista. Both Chibás and Pazos were also given the impression that they would be considered for the chairmanship of the provisional government, though all along Castro planned to name a figurehead.

A product of Castro's imagination and manipulative talents, the Manifesto had the desired effect of at least temporarily resolving the political crisis in the Movement by invoking "national unity" under his leadership. Patronizingly, the document said that "it is not necessary to come to the mountains to discuss this; we can be represented in Havana, in Mexico, or wherever may be necessary," and it gave the *Fidelistas* an appearance of moderation and reason. As for the endless debate on whether the tone of the Manifesto shows that Castro was not a Marxist-Leninist at the time or that the document was designed to conceal his true purpose, probably the best explanation was provided by Che Guevara in writing subsequently about these events. To him, the Manifesto was a "compromise," even though Castro failed to include more explicit proposals for an agrarian reform. Guevara concluded that "we were not satisfied with the compromise, but it was necessary; it was progressivist at that moment. It could not last beyond the moment at which it would have signified a halt in the revolutionary development. . . . We knew it was a minimal program, a program that limited our effort, but we also knew that it was not possible to impose our will from the Sierra Maestra and that we had to count for a long period on a whole series of 'friends' who tried to utilize our military force and the great confidence the people already felt for Fidel Castro for their macabre maneuvers . . . to maintain the domination of imperialism in Cuba . . ." Castro never disputed the substance of Guevara's interpretation of the Manifesto, and later that year he would again take the offensive to assure his absolute leadership of the revolution. The Movement was given a new National Directorate, but it remains unclear who named the members.

Fidel also had personal problems to make him miserable. During the summer he wrote Celia: "When are you going to send me the dentist? If I don't receive weapons from Santiago, Havana, Miami or Mexico, at least send me a dentist so my teeth will let me think in peace. It's the limit; now that we have food, I can't eat; later, when my teeth are all right, there won't be any food . . . I really feel that I'm just not lucky when I see so many people have arrived here, and not one dentist." But Celia had sent him a new uniform, and he wrote her that "I'm going to begin the fourth campaign in it. . . . And you, why don't you make a short trip here? Think about it, and do so in the next few days. . . . A big hug."

During the balance of 1957, the Rebel Army went on consolidating its control of the Sierra Maestra, expanding the "Free Territory" farther east. Che Guevara, promoted by Castro to *comandante* (major, which was the highest rank in the guerrilla force), moved southeast with his own column, taking the small Batista garrisons at Bueycito and El Hombrito, and capturing weapons. In the El Hombrito area, Guevara set up an armory, a bakery, and *El Cubano Libre,* a newspaper for his men. But the anti-Batista opposition suffered reverses as well. An uprising by the navy against Batista in the port of Cienfuegos on September 5 was smashed by loyalist armored units and aviation, and thirty-two officers and sailors were killed. The Cienfuegos rebellion was part of a larger military conspiracy extending to Havana, Santiago, and the port of Mariel, but at the last moment the plans for an organized revolt were canceled; because of a breakdown in communications, the navy at Cienfuegos was not informed of this and so moved to capture the city without parallel uprisings occurring elsewhere.

The timing of the military conspiracy came as a surprise to Fidel Castro, who had not been taken fully into confidence by the plotters, and it strengthened his suspicions that schemes were being developed to bypass the Rebel Army and remove Batista through other means. What he feared the most was the ouster of the dictator by a military coup d'état, thus shifting the power to a military junta and making him irrelevant.

For all these reasons, Castro reacted with fury when he was belatedly informed that in November seven opposition groups had signed a pact in Miami to create a Cuban Liberation Junta and, among other steps taken after Batista's fall, to incorporate his revolutionary forces into the regular army. Not only had he not been consulted beforehand, but the Miami Pact was signed on behalf of the 26th of July Movement by Felipe Pazos and two other Movement leaders without specific authorization from the National Directorate. In a scathing communication on December 14, Castro announced that the Movement "did not designate or authorize any delegation to discuss such negotiations," that "the 26th of July Movement claims for itself the function of maintaining public order and reorganizing the armed forces of the Republic," and that "while the leaders of the other organizations who endorsed the pact are abroad fighting an imaginary revolution, the leaders of the 26th of July Movement are in Cuba, making a real revolution." He said the Movement "will never relinquish orientation and leadership of the people . . . and we alone shall know how to overcome or to die. . . . To die with dignity, one has no need of company."

Castro's rejection of the Miami Pact meant the death of the Liberation Junta even before it could be officially born, because it was evident that without the Rebel Army it would be a wholly meaningless organization. At the same time, Castro acquired new enemies in the opposition, notably in the Students' Revolutionary Directorate (DR) whose new leadership had signed the Miami document. The DR's political leader, Faure Chomón, wrote that "no organization can, or should, as Dr. Castro has done in a sectarian fashion, claim for itself the representation of a revolution being made by all of Cuba." Castro would have new problems with the DR and the 26th of July Movement the following year, but for now he was again in full control. He also regained the confidence of Che Guevara, who, for some strange reason, had assumed at first that the Miami Pact was signed with Fidel's authorization. Upon learning that Castro rejected it, Guevara wrote him that now he was filled "with peace and happiness." He said that it was clear "who is pulling the wires behind the scenes," adding that "we unfortunately have to face Uncle Sam before the time is ripe."

Uncle Sam, however, was engaged in a number of actions in Cuba that were both contradictory and mysterious. On one hand, the United States continued to supply the Batista regime with weapons to fight the rebels, while on the other hand it secretly channeled funds to the 26th of July Movement through the Central Intelligence Agency.

The story of CIA financial support for the Castro rebellion, a selective form of support, is a surprising one, though it is unclear whether this operation was formally authorized by the Eisenhower administration or undertaken by the Agency entirely on its own. It is not even certain that Castro himself knew that some of the money reaching him or his Movement came from the CIA. A new reconstruction of this United States involvement with Castro shows that between October or November of 1957 and the middle of 1958, the CIA delivered no less than fifty thousand dollars to a half-dozen or more key members of the 26th of July Movement in Santiago. The amount was quite large, relative to what the Movement itself was able to collect in Cuba. The entire clandestine operation remains classified as top secret by the United States government; therefore, the reasons for the financing of the Movement cannot be adequately explained. It is a sound assumption, however, that the CIA wished to hedge its bets in

Cuba and purchase goodwill among some members of the Movement, if not Castro's goodwill, for future contingencies. This would have been consistent with CIA policy elsewhere in the world whenever local conflicts affected United States interests.

These funds were handled by Robert D. Wiecha, a CIA case officer attached to the United States consulate general under the cover of vice-consul, who served in Santiago from September 1957 to June 1959. The late Park Fields Wollam, who as consul general was Wiecha's superior in Santiago, had told State Department colleagues at that time of the CIA role in dealing with the Castro organization. Wartime correspondence from the Sierra era and now in Cuban official archives shows furthermore that Wiecha had tried hard to arrange a meeting with Fidel himself from the moment he arrived in Santiago. The CIA officer's contact was with the Frank País group, and sometime early in July, País wrote Castro that a United States diplomat wished to meet him (Wiecha's CIA identity was unknown to País). Castro replied that "I don't see why we should raise the slightest objection to the U.S. diplomat's visit. We can receive any U.S. diplomat here, just as we would any Mexican diplomat or a diplomat from any country."

Castro went on: "It is a recognition that a state of belligerence exists, and therefore one more victory against the tyranny. We should not fear this visit if we are certain that no matter what the circumstances may be, we will keep the banner of dignity and national sovereignty flying. And if they make demands? We'll reject them. And if they want to know our opinions? We'll explain them without any fear. If they wish to have closer ties of friendship with the triumphant democracy of Cuba? Magnificent! This is a sign that they acknowledge the final outcome of this battle. If they propose friendly mediation? We'll tell them no honorable mediation—no mediation is possible in this battle."

On July 11, Frank País wrote Castro that "María A. told me very urgently at noon today that the American vice-consul wanted to talk with you, in the presence of some other man, but she didn't know who." This was the first direct reference to Wiecha, and País added, "I told her I would consult with you, but that we would first have to find out who the other man is and where they wanted to go and what they wanted to talk about." País's letter also showed that secret contacts also existed with the United States embassy in Havana: "I'm sick and tired of so much backing and forthing and conversations from the Embassy, and I think it would be to our advantage to close ranks a bit more, without losing contact with them, but not giving them as much importance as we now do; I see that they are maneuvering but I can't see clearly what their real goals are."

At that time, a new American ambassador had arrived in Cuba to replace

Arthur Gardner. Earl E. T. Smith, a political appointee, had gone to Santiago on August 1, the day after Frank País was buried. Two hundred women clad in black gathered there to request United States intervention against the terror practiced by the regime. When the women started to chant *"Libertad! . . . Libertad! . . ."* the riot police charged them with truncheons and water cannon. At the news conference that followed, Smith delivered his protest against the "excessive use of force." Raúl Chibás, who was in the Sierra with Castro that week, recalls that the news of Smith's statement caused "a moment of happiness among the troops . . . people were saying, 'You see, the ambassador has already changed, it looks like he isn't supporting Batista.'" Chibás states that even Castro said that American policy could change, "and people there talked about it as a favorable thing that would facilitate the struggle. . . . Anti-Americanism did not exist in that Rebel Army . . ."

On October 16, Armando Hart wrote Castro that "I have been in contact with people close to the embassy. These contacts have told me that people who are on our side—but who do not appear to be—have had conversations with the ambassador himself. I think this is the best policy, since we are kept up-to-date about everything happening there and of all the possible U.S. plans, and at the same time the Movement does not officially commit itself." In all probability, Hart's contacts were officers in the CIA station at the American embassy; this would be consistent with Robert Wiecha's efforts in Santiago to meet Castro and his handling of Agency funds for the Movement. Robert Taber, the CBS reporter who had visited Castro in the mountains, wrote in his book about the Cuban revolution that "Wiecha rendered invaluable and humanitarian service to the Batista opposition." Taber, who was unaware that Wiecha worked for the CIA, identified him as the vice-consul who made inquiries about Armando Hart, Javier Pazos, and Antonio Buch who had been arrested in Santiago, "with the result that General Chaviano was compelled to produce the prisoners, unharmed, to prove that they had not been tortured or killed." Hart's unwitting contact with the CIA began with this intervention by Wiecha, and the Agency's secret relationship with the rebels may have started then. An unconfirmed but credible version of this relationship is that the CIA had also arranged to fly arms for Raúl in the "Second Front."

The meeting between Castro and Wiecha never took place, however, for reasons Castro ignores. It may have been blocked by senior United States officials, particularly when Ambassador Smith himself expressed the hope that Batista would proceed with elections acceptable to Cubans. Curiously, American support for Batista seemed to grow just as Castro's power was increasing, and, typically, it would be withdrawn when it no longer mattered politically. The Wiecha story suggests that the United States may have missed an extraordinary opportunity to establish a dialogue with Castro when he was still in the hills and, theoretically at least, open to a positive future relationship. But it is idle to second-guess history.

The Americans did not come to Fidel Castro's mountain in 1957, but the Communists did—even though the meeting was arranged in great secrecy. The visitor in October was Ursinio Rojas, a sugar worker and a member of the Communist party's Central Committee. He was the first senior Communist leader to see Castro since Flavio Bravo's visit to Mexico a year earlier to dissuade Fidel from the invasion. (Gottwald Fleitas, a party leader from Bayamo, went up the Sierra in the spring, supposedly to tell Castro that Communist activists among the peasants had been instructed to "cooperate" with the guerrillas.) Rojas had been in prison in Havana early in 1957 with Armando Hart, Faustino Pérez, and Carlos Franqui, and in long chats there with them, he maintained the party's official line that Castro was running a "putschist" operation in the Sierra. But by the end of the year, the Communists were beginning to rethink their policy, and Rojas told Castro that certain party members would be authorized to join the Rebel Army as individuals. One year after Castro had established himself in the Sierra Maestra, this was as far as the Communists were prepared to go. *El Campesino,* the party's underground newspaper, declared that "there exists a great difference between the level of the struggle in the Sierra Maestra . . . and the rest of Cuba, that is almost all of Cuba." Fidel Castro seemed unable to impress either the United States or the Communists that he should be taken seriously: Both did no more than hedge their bets.

And before the year ended, Castro received a permanent visitor: On the verge of being caught by the police, Celia Sánchez abandoned her operations in Manzanillo to go up the Sierra. She would stay there with Fidel through the year of victory.

CHAPTER

11

It was not a happy time for military dictators in Latin America. On January 23, 1958, after nearly ten years in power, General Marcos Pérez Jiménez in Venezuela was overthrown by the armed forces as the climax of a revolt by the civilian population led by students, intellectuals, and businessmen. The previous year, General Gustavo Rojas Pinilla had been ousted by the military in Colombia under roughly similar circumstances—with the Roman Catholic Church playing a major role in the rebellion—after a three-year rule. General Juan Perón in Argentina and General Manuel Odría in Peru departed in 1955 and 1956, respectively (Odría stepping down voluntarily, an unprecedented fashion in Latin American history). It was indeed a twilight of the tyrants, yet in Cuba General Batista seemed impervious to the idea that a hemispheric trend was under way—as was the Eisenhower administration in the north.

What was understood even less in Havana and Washington, as 1958 opened, was the character of the antidictatorial struggle conducted in the Sierra Maestra by Fidel Castro and his steadily growing Rebel Army. Whereas the liquidation of dictatorships in the four South American nations led mainly to the restoration of representative democracy in a modern, liberal mode, and even though Castro continued to insist publicly that

Batista's removal would be followed by "free, democratic elections" within a year, he was quietly preparing a fundamental social revolution. Soon he would inform Cubans that free elections were incompatible with a social revolution—and, strictly speaking, his assessment was politically and ideologically accurate—but in the meantime Castro strove to convey an impression of dedication to democracy along with social justice.

An article published under Fidel Castro's by-line (and most likely written by him) in the February issue of *Coronet* magazine in New York went to astonishing and exaggerated lengths to portray him and his "armed campaign on Cuban soil" as the way to pure liberal democracy under which free enterprise would flourish. Even if he applied Martí's and Marx's concept of "historical justification" to conceal his real purpose until the right moment in the interest of a higher cause, and even if this were justifiable, the *Coronet* article borders on intellectual dishonesty. There was even no need for it: The bulk of American public opinion backed him anyway, while, as he surely knew, the Eisenhower administration would never be convinced by his claims.

Castro proclaimed that "first of all and most of all, we are fighting to do away with dictatorship in Cuba and to establish the foundations of genuine representative government." He felt a "personal reluctance" to seek Cuba's presidency after Batista, apart from the fact that under the constitutional age-requirement clause, "I am, at 31, far too young to be eligible for the presidency, and will remain so for another ten years" (actually, Castro would assume the presidential title *seventeen years* later, in 1976, but in terms of real power this was a technicality). He said truthfully that "we want to wipe out corruption in Cuban public life" and "we want to sponsor an intensive campaign against illiteracy." Misleadingly, he declared that "we will support no land reform bill . . . which does not provide for the just compensation of expropriated owners" and that "we have no plans for the expropriation or nationalization of foreign investments here." He acknowledged that government ownership of American-owned utilities "was a point of our earliest programs, but we have currently suspended all planning on this matter." He urged industrialization, saying that "a million unemployed in a nation of six million bespeaks a terrible economic sickness which must be cured without delay, lest it fester and become a breeding ground for communism." He explained that he had to take the "terrible decision" to burn Cuba's entire sugarcane crop in order to paralyze the regime and force Batista to "capitulate," adding that "my family has sizable cane holdings here in Oriente, and my instructions to our clandestine action groups state clearly that our crop must be the first one to burn, as an example to the rest of the nation." He repeated the same points in a February interview with *Look* magazine, violently denying his Movement was Communist inspired.

In any event, in the opening months of 1958, Castro still believed that he was facing a protracted war against Batista. He acted accordingly in the military field by establishing three new battlefronts in the Sierra Maestra. Politically, he made it clear that he welcomed a longer war because it would give him more time to prepare the Rebel Army and the country for the wide revolution he had in mind. (In a letter to Celia in July 1957, Castro wrote that "I feel that the fall of the regime in a week's time would be far less fruitful than four months from now. . . . Here, as a joke, I usually assure the comrades that we don't want to give birth to a *seven-month* revolution.")

The Rebel Army, in particular, was being turned into the cutting edge of the revolutionary process, to become later the ideological and operational center in implementing the great changes. Whereas the Chinese and Vietnamese revolutions were directed by pre-existing Communist parties, Castro vested this role in the new Rebel Army, thus assuring his control of the entire Cuban political picture then and in the future.

For the Rebel Army to play such a political role successfully, a myth had to be built around it, and to Fidel Castro this was a high priority. This myth, preserved thirty years later in the Revolutionary Armed Forces, was based on its function as a social revolutionary force as much as on its military prowess. Because rebel soldiers increasingly concerned themselves with awarding land to peasants who worked it in the "Free Territory" of the Sierra—it was a "mini land reform" carried out in place—as well as helping farms (as in the 1957 coffee harvest), protecting peasant families from the landowners' overseers and the Rural Guard, applying "revolutionary justice" to rapists and exploiters, and opening a few schools and clinics, they loomed as friends of the population. This was the real myth. Guevara, who ran his own zone fairly independently since autumn of 1957, remarked that "we became revolutionaries in the revolution . . . we came to overthrow a tyrant, but we discovered that this immense peasant zone, where our struggle is being prolonged, is the area of Cuba that needs liberation the most." To a greater extent than Castro, the Commander in Chief always on the move, Guevara soon engaged in the political education of his own troops as well as of the peasants; he was not yet preaching Marxism but land reform and other structural changes required by Cuba. His newspaper, *El Cubano Libre,* printed the comment that Communists "are all those who take up arms because they are fed up with poverty . . ."

As self-appointed social revolutionaries, the Rebel Army evoked a peasant response far beyond the initial support that made it possible for Castro to survive during his first months in Cuba. At first, the peasants helped him from instinct, then out of admiration for the rebels' attitudes. Again, this support was crucial as the guerrilla war entered a new phase in 1958. The *guajiros* brought food for the rebels, helped them obtain weapons, provided free labor for the construction of armories and warehouses (they would not accept payment), served as an early-warning system when Batista army forces penetrated the mountains, and acted as couriers among rebel groups and between Castro and the *llano* Movement leaders; some of the best couriers were women. And the peasants remained the principal source of recruitment for the Rebel Army; they knew their way around, they were tough, they were believers in *Fidelismo,* and it was easier to integrate them in the guerrillas than the urban volunteers. And now Castro needed the best fighters available to expand the war.

On March 10, 1958, the sixth anniversary of the Batista coup, Raúl Castro left with sixty-five men from Fidel's camp to establish a new "front," or rebel war theater, in the Sierra Cristal along the north coast of Oriente province. It was directly east of the Castro brothers' birthplace at Birán, and northeast of Fidel's principal operational area. The idea was to create a second "liberated zone" in Oriente that would bring the war to the north and increase the strain on the Batista forces. Always myth-conscious, Fidel named this "Second Front" after Frank País. Raúl's force was designated as "Column 6," to create the impression that the Rebel Army was composed of many units. About the same time, Juan Almeida was dispatched with another column to set up the "Third Front" in the eastern range of the Sierra Maestra, immediately northwest of Santiago. In April, Camilo Cienfuegos moved north toward Bayamo to harass the Batista army there. Guevara and his "Column 4" had been operating in the central Sierra Maestra around El Hombrito since August 1957. By mid-spring 1958, therefore, the Rebel Army occupied or controlled most of the mountain regions in Oriente. Fidel's strategy became one of breaking out of the Sierra strongholds among which the rebels had been moving for a year, and denying the enemy more and more territory. However, Fidel also believed in prudence; in a note sent Guevara in February, he recommended that a planned attack be canceled if support was not available from other rebel units. "I do not believe anything suicidal should be done as we shall be risking too many casualties and fail to reach the objective," Castro wrote. "I strongly recommend that you be careful. As a final order, you should not fight. Take charge of leading the men well . . ."

Gradually, the guerrilla army was being transformed into a more conventional and better-equipped force. In April Fidel acquired a brand-new Toyota jeep, and the rebels held enough territory in the area for him to be able to drive rather than walk within a radius of ten to fifteen miles; most of the mountain paths were passable for the jeep, and even a few roads were now under Castro control. Colonel Arturo Aguilera, who served as Fidel's first driver and *aide-de-camp,* recounts that they moved from place to place every two or three days for security reasons; Castro still exercised enormous caution. Because of possible detection by aircraft, they drove at night without lights, "with Celia or Fidel carrying a lantern along for use when necessary." Aguilera says that they always traveled alone without an escort because they were in rebel-held areas. In late April and early May, Castro spent time inspecting his forces in preparation for what he was certain would soon be a large-scale enemy offensive. He ordered fortifications to be built at the approaches to the spine of the Sierra Maestra, believing that he would lose part of his territory, but that he would triumph in the end so long as he continued to dominate the main crest of the mountain chain.

Morale was high everywhere. A hospital for the seriously wounded was established in the mountains in Guevara's territory, well concealed by vegetation from aerial observation, but hard on the patients because of extreme humidity. Another hospital was set up on the western slopes of the Sierra Maestra, and small medical installations were strung out throughout the rebel territory. Medicine arrived regularly from the lowlands, if not always in desired volume or type. A low-power radio transmitter was installed in Alto de Conrado in Guevara's territory in February, but it could be heard only by a few patrols and peasant families in the immediate vicinity. Guevara also began manufacturing explosive devices known as M-26s; he also set up a slaughterhouse for captured cattle and a small cigar-making plant.

The high morale stemmed in part from the rigid discipline imposed by the commanders, but also from their personal example. Fidel, Raúl, Che, and the others were always in the vanguard, never asking the men to take risks they would not accept themselves. When a peasant officer accidentally shot dead a soldier who was being subject to disciplinary action, the troops demanded that the officer be executed. After Fidel and Che argued for hours that he did not deserve the death penalty, the men agreed grudgingly—but mainly out of personal respect for the commanders. Guevara never sat down at a meal until he was satisfied that all his soldiers had the same quantity and quality of food. Universo Sánchez remembers sharing a chicken-and-rice meal with Fidel on the same metal plate; after each had consumed exactly one half, a single piece of meat was left, but neither of them would eat it. As Guevara recalls, the only thing almost never in shortage was coffee: One could always get a cup at a peasant's hut. Little escaped Che's sharp eye; he was the most complete, if sharp-penned, chronicler of the Sierra war—as one can appreciate from his campaign diary and subsequent writings. Much of these writings touch bitterly and violently on the political dissensions tearing asunder the 26th of July Movement.

*   *   *

It was the question of a general strike that brought the Movement's political crisis to a head, forcing a fundamental confrontation among its leaders. This, in turn, had a critical impact on the course the Cuban revolution would ultimately take. A general strike in support of a national rebellion against Batista certainly was not a new idea; Castro had originally planned one to be timed with his landing in Cuba. In July 1957, Frank País informed Fidel that a National Workers' Front (FON) had been established throughout the country, to be followed by the creation of strike committees so that a general strike could succeed. In his December communication to the Cuban Liberation Junta rejecting the Miami Pact, Castro singled out a general strike among "concrete acts . . . useful in the overthrow of the tyranny" to be carried out through "the effective coordination of the efforts of civic organizations in conjunction with the 26th of July Movement."

Nevertheless, there are reasons to believe that in Castro's mind support for his Sierra war was far more important at this point than anything the Movement could and would do in the *llano*—the lowlands and their cities. His stream of complaints about not receiving enough arms, ammunition, supplies, and money from the *llano* finally led Armando Hart, a cofounder of the Movement, to write him from Santiago in October that "all our comrades here have always considered the Movement there and here as one single entity . . . supplying you up there is so vital for us that we consider it our foremost and fundamental revolutionary obligation." By early December Hart wrote in desperation to Celia in the Sierra Maestra that if the Movement's work in the cities was regarded as unnecessary, "then that raises the question of whether we should consider ourselves members of the present Directorate solely as the instruments for supplying the Sierra. . . . We think we have a duty to organize the workers, to strengthen the civilian resistance, to build provincial and municipal cadres with real revolutionaries, who, together with the revolutionary army of the Sierra Maestra, will guarantee the accomplishment of our program. We must also help the militia, which, outside the Sierra Maestra, without resources or arms (all that we had were sent to you) have heroically succeeded in extending the Revolution beyond the frontiers of the Sierra Maestra and have created an organization that you, as much as ourselves, are duty-bound to protect." René Ramos Latour, who replaced Frank País as coordinator in Oriente (and whose underground name was "Daniel"), wrote Castro to express surprise "by the note signed by you, which essentially was an expression of mistrust toward us, tacitly accusing us of responsibility for the state of neglect in which our forces in the Sierra found themselves, as well as of holding back for the cities the supposed newly arrived weapons destined for you. . . . No, we have never belittled the Sierra. We think the battle ought not to be limited solely and exclusively to the mountains; we must fight the regime on all fronts."

Against this background of rising tensions between Fidel Castro in the mountains and the Movement's Directorate below, the problem of the general strike returned to the fore in March. Neither side would admit that a political power contest was developing, but it was evident that Castro did not care for rivals. When the Students' Revolutionary Directorate established a guerrilla front in the Escambray Mountains in central Cuba late in 1957, Castro acknowledged it begrudgingly and belatedly in February 1958, remarking in a message that "regardless of the revolutionary militancy of the group, we have given instructions to the Movement to give you all possible help." As for the DR, they had even less use for the Rebel Army, and soon the two groups would be embroiled in bitter disputes. Then, a "Second Front of the Escambray" was set up by a rival university group and political opportunists from Havana. And Castro had to face the issue of a general strike in the context of his overall relationship with the *llano* revolutionaries. It was inevitable that other anti-Batista groups would engage in operations of their own. The creation of the two guerrilla organizations in the Escambray Mountains was one example; another was the murder in Holguín, a garrison city in northern Oriente, of Colonel Fermín Cowley Gallegos, one of the most hated and brutal Batista army chiefs, by a Civic Resistance action commando at the end of 1957.

For Castro the general strike thus became—even if grudgingly—one way of imposing revolutionary unity. In his capacity as delegate from the Movement's National Directorate in Havana, Faustino Pérez had returned to the Sierra early in March, and after conferences with Castro and other members, he joined him in signing a manifesto titled "Total War Against Tyranny," calling for the general strike. Pérez pushed for holding the strike as soon as possible, believing that Castro simply lacked "first-hand information on the existing conditions in Havana," and that it was his duty to provide it. After all, he saw himself as one of Castro's closest associates—he had been aboard the *Granma,* one of Fidel's two companions during the desperate days and nights after Alegría de Pío, and the first emissary from the Sierra to Santiago and Havana. He was certain that Castro would listen to his assessments.

The twenty-two-point Manifesto, from "the Camp of Column 1, General Headquarters of the Rebel Forces," started out by proclaiming that "the struggle against Batista has entered its final stage" and that "the strategy of the final blow is based on the general revolutionary strike, to be seconded by military action." The strike, it said, "will be ordered at the proper time," and it will continue along with the armed struggle "if a military junta should try to take over the government." Again, Castro was determined to prevent any deal behind his back by the military and the civilian opposition in the *llano*. In a further step to guarantee Castro's control, the new Manifesto reaffirmed the Movement's choice of Manuel Urrutia Lleó as the head of a provisional government, which would be formed after Batista's fall with the mission of preparing national elections. Castro had first proposed Urrutia in his December letter to the Cuban Liberation Junta, but now he and Faustino Pérez felt that the designation should be formalized as soon as possible.

Urrutia was the fifty-eight-year-old presiding judge of the Court of Appeals of Oriente province who had cast the dissenting vote when his two colleagues on the bench found guilty numerous *Granma* expeditionaries and Movement activists who had been arrested during an abortive uprising in Santiago on November 30, 1956. The verdict condemned the rebels to eight-year prison terms. Urrutia wrote later that to him the defendants were "models of dignity and patriotism" and that the basis for his dissent was "the 'sacred right of resistance against oppression' consecrated by Article 40 of the Cuban Constitution of 1940." He added that "I held that armed action by the accused men was legitimate because it was an attempt to end oppression in Cuba." Four years earlier, Urrutia was the investigating judge in Santiago who certified the deaths of Castro's rebels in the Moncada attack on July 26, and his sympathies with the Movement dated back to that bloody Sunday.

When Castro first conceived the notion of a provisional government, Raúl Chibás and Felipe Pazos, cosigners with him of the Sierra Maestra Manifesto in July 1957, were his prime candidates for president. But Chibás refused, and Castro concluded that Pazos was too much involved in traditional politics. He then thought of Urrutia as the perfect apolitical candidate, and in November sent emissaries to the judge in Santiago to offer him the presidency on behalf of the Movement. Urrutia agreed, resigned from the judiciary, and left for the United States at the end of December to await the call to serve when Batista fell. His mandate was "to lead our country to democracy, freedom, and a regime of law."

Urrutia, a liberal and an anti-Communist jurist, added prestige to the Castro cause—another of Fidel's immediate needs—and he was received in Washington by senior State Department officials dealing with Latin America. His concern at that point was to persuade the United States to halt arms deliveries to Batista, and he may have played a role in the Eisenhower administration's decision on March 14, 1958, to suspend such deliveries on the grounds that the weapons were being used for internal security and not for hemispheric defense as provided by law. In terms of winning the gratitude of the rebels in Cuba, however, the suspension was a belated gesture.

American-supplied bombs continued to be deployed against them, and on June 5, Fidel sent his famous note to Celia stating that it was his "true destiny" to fight against the North Americans to make them "pay dearly for what they are doing." Arms shipments were stopped, but Batista's bombers were allowed to refuel at the Guantánamo naval base between strikes at the rebels.

The "Total War" Manifesto declared that as of April 1, all highway and railroad traffic in Oriente province would be prohibited, and moving vehicles would be fired on. After that date it would be forbidden to pay any taxes in Cuba—to do so would be considered "unpatriotic and counter-revolutionary." At that stage, the Rebel Army had barely three hundred men under arms, but Fidel Castro's sense of drama and propaganda propelled him to announce that "from this moment, the country should consider itself in total war against the tyranny . . . the entire nation is determined to be free or perish!"

April 9 was secretly set as the general-strike date; Movement representatives were ordered to speed up their clandestine efforts, and the Rebel Army readied itself for coordinated attacks. The *Fidelistas* were strengthened by the arrival in late March of one of the first aircraft to land in the Sierra "Free Territory." A twin-engined C-47 transport flew in from Costa Rica with a load of arms and ammunitions—and two very important officers. One was Castro's principal military adviser, Pedro Miret, who had been under arrest in Mexico when the *Granma* sailed for Cuba, and was now finally able to join his companions, and the other was Huber Matos, the rice planter whose trucks took up volunteers from Manzanillo to the Sierra Maestra. Matos, who organized the flight and delivered a letter from Fidel to Costa Rica's president, was instantly given a troop command. Pedro Miret recalls that Castro was waiting for them in the meadow where the plane landed at twilight, smashing a propeller. Fidel named him to the general staff. Faustino Pérez, in the meantime, had gone back to Havana to coordinate the strike activities. Action groups in the capital were busy exploding scores of bombs every night throughout March to create a psychological climate for the general strike. This effect was enhanced by the spectacular kidnapping of world car-racing champion Juan Manuel Fangio of Argentina on the eve of a major Havana race. In an operation engineered by Faustino Pérez, Fangio was released the next day, announcing that he had been well treated. The regime was properly humiliated. In the 26th of July Movement, optimism was soaring.

Yet the strike was a terrible failure. Fidel Castro described it as the "hardest blow suffered by the Revolution during its entire trajectory" because "the people had never had as much hope as that day, and we had never had so many illusions as we had on that occasion." What happened was that for a long chain of reasons—poor planning, notably in Havana, a breakdown in coordination among various groups, bad timing, and less than adequate response and participation among the population—Castro's hopes of a precipitous revolution in the cities were skewered. The regime, which had again suspended constitutional guarantees and drafted an additional seven thousand men for the army, had been ready to smash the strike. At least one hundred Cubans were killed by the police throughout the island that day, and many hundreds were arrested. General Batista concluded that the tide had turned and that in the aftermath of the strike fiasco, the rebels in the Sierra Maestra would be demolished by a major military offensive.

One of the unresolved arguments about the general strike concerns the role of the Communist party and its relationship with the 26th of July Movement's leaders in Havana prior to the strike. Because of their basic opposition to the Castro rebellion, the Communists were charged with sabotaging the strike to provoke the downfall of the Movement, leading to a rise in their revolutionary influence and the adoption of their strategies. However, the Communist story is much more complex. No comprehensive account of the general-strike drama has ever been published—it remains an acutely sensitive topic in Cuba and therefore the reconstruction of this whole occurrence and its consequences is bound to be incomplete. Both ex-Interior Minister Ramiro Valdés and Faustino Pérez, representing different viewpoints, are still urging public silence in the controversy. There is no question, however, that the failed general strike marked a fundamental turning point in the history of the revolution. The political influence of the Movement's moderates in the *llano* vanished, and the assumption of total revolutionary power by Fidel Castro and his radical "militarists" in the Sierra became paramount.

As for the Communists, it was the Movement's National Workers' Front (FON) that avoided contact with the party's labor leaders, particularly influential among sugar workers, and for all practical purposes failed to include them in strike planning. FON did not give the Communists the date of the strike because it did not trust them, though the Communist party was certainly aware of what was going on. According to the Communists, the Movement's idea of a general strike was an appeal to the population to join the stoppage instantly and participate in armed attacks on preselected targets in Havana—without any preparation in plants, offices, and other workplaces. In early April the party concluded that the strike should be aborted if procedures were not improved at once, and it dispatched Osvaldo Sánchez, a Central Committee member, to the Sierra to inform Castro of the party's preoccupations. Sánchez reportedly told Fidel that the Movement leaders in Havana had overestimated their strength, failed to work at organizing the strike at work centers, refused to cooperate with the Communists, and trusted too much in a spontaneous response to strike appeals.

Six years later, on April 9, 1964, the Communist party newspaper *Hoy,* published for the first time a statement issued by Castro on March 26, 1958, to the effect that "in summoning the nation to the final struggle against tyranny, our Movement makes no exclusions of any kind . . . all Cuban workers, whatever are their political or revolutionary allegiances, have the right to belong to strike committees at their places of work. The National Workers' Front is not a sectarian organism. . . . The Front's leadership will coordinate with the workers' sections of political and revolutionary organizations that fight the regime, and with all the organized factions that struggle for economic and political revindications of their class, so that no worker can be severed from this patriotic effort." However, the Communists claim, the Movement never publicized Castro's instructions, and continued their unilateral activities. On April 2, the party's underground publication, *Carta Semanal,* called for a general strike, while noting that "the forces of disunity remain present."

In any event, Fidel Castro did not call off the strike. At ten o'clock in the morning on April 9, three Havana radio stations seized by the revolutionaries broadcast an appeal by the Movement for a "general revolutionary strike" to begin immediately. The appeal said: "Today is the day of liberation . . . all throughout Cuba at this very moment the final struggle which shall end in the overthrow of Batista has begun!" From the Sierra, the radio called out: "Strike! Strike! Strike! Everyone on strike! Everyone into the streets! . . ." Another broadcast urged patriots to "throw stones at all strikebreakers from your windows . . . throw ignited Molotov cocktails at the patrol cars . . ." But very little happened in the way of stoppages, mainly because workers were taken by surprise and the police were out in force. In Havana there were acts of sabotage, and heavy fighting developed in Sagua la Grande, a town in the central province of Las Villas. The following day, a message from Castro insisted that "all Cuba burns and erupts in an explosion of anger against the assassins, the bandits and gangsters, the informers and strikebreakers, the thugs and the military still loyal to Batista." But it was all over.

On April 13, Faustino Pérez and his Movement associates dispatched a communication to their committees in exile to acknowledge the failure and the errors committed by the urban leadership. They said the gravest errors had been to keep secret the date of the strike, then to broadcast an appeal to strike "at an hour when only housewives listen to the radio," instead of issuing the call forty-eight hours earlier. Fidel was much tougher: In a letter to Celia on April 16, he said that "the strike experience involved a great moral rout for the Movement . . . the Revolution is once again in danger and its salvation rests in our hands . . . we cannot continue to disappoint the nation's hopes. . . . No one will ever be able to make me trust the organization again. . . . I am the supposed leader of this Movement, and in the eyes of history I must take responsibility for the stupidity of others, and I am a shit who can decide on nothing at all. . . . I don't believe that a schism is developing in the Movement, but in the future we ourselves will resolve our own problems." Four days later, he told Celia: "We have not renounced the general strike as a decisive weapon against tyranny. . . . A battle was lost but not the war."

Now it was time for a final confrontation within the revolutionary ranks. Castro opened it with an April 25 letter to Raúl Chibás and Mario Llerena, leaders of the Movement's committee in exile in Miami, charging that "the Movement has failed utterly in the job of supplying us" and that "egotism, and at times trickery from other sectors, have combined with incompetence, negligence, and even the disloyalty of some comrades. . . . The organization has not managed to send us so much as one rifle, not one bullet from abroad. . . . But apart from all moral considerations, once again the task of saving the Revolution in one of its most profound crises falls on our men." Castro warned: "The danger of a military coup reaffirms the thesis that only the military can overthrow dictatorships, just as they first put them into power, a thesis that mires the populace in fatalistic apathy and dependence on the military . . . all this is now in the forefront as a result of the failure of the strike. *And the failure of the strike was a matter not only of organization but also of the fact that our own armed action is not yet strong enough* . . ."

All this set the stage for Castro's evisceration of the Movement as a political force outside his personal control. The Movement in the *llano* was dramatically accused of denying him resources from Cuba and abroad, although it is impossible to determine how accurate the charges were; for one thing, the first aircraft from Florida were beginning to land in the Sierra with war matériel. The twin specters of a military coup and a junta, Castro's constant fear, were raised anew. Finally, Castro washed his hands of any responsibility for the collapse of the strike he had approved in advance. The actual confrontation between the two wings of the 26th of July Movement came at a meeting of the National Directorate on May 3, at a farmhouse on the Mompié heights in the Sierra Maestra heartland. Lasting from early morning until two o'clock in the morning of the next day, the session was presided over by Fidel Castro and attended principally by the Movement's *llano* leaders—Faustino Pérez and Marcelo Fernández from Havana, the Santiago coordinator, René Ramos (Daniel) Latour, and the labor leader David Salvador—plus Fidel's closest personal associates in the Directorate,

Celia Sánchez, Vílma Espín from Santiago, and Haydée Santamaría. Che Guevara was not a member of the National Directorate, but he was invited at the request of Faustino Pérez and Latour because they had been targets of his violent criticism in the wake of the failed strike. It marked Guevara's formal entry into the top Cuban revolutionary policy-making circles. Guevara was also the best chronicler of what he called the "Decisive Reunion."

More honest and outspoken than his companions, Guevara wrote that "the division between the *sierra* and *llano* was real . . . differences of strategic concepts separated us." Prior to the strike, Guevara noted, "the comrades from the *llano* constituted the majority" in the National Directorate, and they were inclined "toward certain 'civilist' actions, a certain opposition to the *caudillo*, who was feared [to exist] in Fidel, and to a 'militarist' faction represented by us, the people in the *sierra*. As a result of the Mompié meeting, where Faustino Pérez, Latour, and Salvador were virtually on trial, the "*guerrilla* concept" of direct military action became "triumphant with the consolidation of the prestige and authority of Fidel [who] was named commander-in-chief of the forces, including the militias, that until now were subordinated to the *llano* Directorate." After what Guevara describes as "an exhaustive and many times violent discussion," Faustino Pérez, Latour, and Salvador were removed from their posts in the leadership of the Movement. He wrote that "politically, the National Directorate was shifted to the Sierra Maestra where Fidel took the post of Secretary-General, and a Secretariat of five members was created."

In Havana Marcelo Fernández was made the Movement coordinator under Castro, and Faustino Pérez was replaced as a national *llano* leader by Delio Gómez Ochoa, a Fidel military man. Faustino, Latour, and Salvador were transferred to the Sierra, where Castro gave the first two men important responsibilities. He is not a personally vindictive man, except when he senses betrayal, and he thought that Faustino's and Latour's talents and loyalty should be fully used. Faustino, who says that he was regarded at the time as part of the "right wing" of the Movement, has remained with with Castro in the revolutionary establishment for thirty years; Latour was killed in battle four months later that year. Urrutia, already designated provisional president of Cuba, was put on salary while awaiting the summons from Caracas, where he moved from the United States (he was the only salaried Movement leader). Haydée Santamaría was sent to Miami to coordinate fund-raising in the United States. Carlos Franqui was recalled to Cuba to take over Radio Rebelde, which was beginning to broadcast to the rest of the island from the top of the Sierra Maestra, where Castro's permanent headquarters were being set up. Radio Rebelde immediately became one of Fidel's most important psychological-warfare weapons. For Che

Guevara the most important outcome of the Mompié confrontation was that now "the war would be conducted militarily and politically by Fidel in his double capacity as commander-in-chief of all the forces and secretary-general of the organization."

But Guevara, despite his oft-repeated declarations of allegiance to Marxism-Leninism, remained unhappy with the Cuban Communists. In 1958 he wrote that there were "mutual fears" between the Movement and the Communist party, "and, fundamentally, the party of the workers had not perceived with sufficient clarity the role of the *guerrilla* nor Fidel's personal role in our revolutionary struggle." Che recalled having told a Communist leader that "you are capable of organizing cadres that are allowed to go to pieces in the darkness of a dungeon without saying a word, but you are not capable of forming cadres that can capture a machine gun nest in an assault." No matter what was said and what happened later, the top guerrilla leaders, with the possible exception of Raúl Castro, resented the Communists during most of the Sierra war. Whether or not this was part of his tactical line at that stage, Castro made a point of saying in the 1958 *Look* interview that "the Cuban communists . . . have never opposed Batista, for whom they have seemed to feel a closer kinship." He said that Americans "should know more about Latin American movements that are democratic and nationalist. . . . Why be afraid of freeing the people, whether Hungarians or Cubans?"

Fidel Castro would go on criticizing the Communists, sometimes savagely, in public and in private; but in reconstructing the events of 1958, in terms of his own political evolution, it appears that the trauma of the aborted strikes led him to make at least a preliminary decision to go the Marxist-Leninist route. The strike revealed the political unreliability of the 26th of July Movement liberals and moderates from a revolutionary as well as an organizational viewpoint. From men and women who lived that period with Castro, the consensus emerges that he had resolved that Marxist-Leninist strategies, procedures, and techniques would in practical terms be best suited to the future of his "real" revolution. That the traditional Communist party still denied him support, these witnesses say, mattered little to Castro, who was already thinking of fashioning his own Communist party, a pretension that was not any more absurd than his original pretension of overthrowing Batista single-handed. His early Marxist sympathies evidently played a role in turning him toward this alternative, as did his resentments against the United States. More to the point, however, he thought that the Communist's organizational talents and experience with mass organizations—unlike the 26th of July Movement—could be harnessed to the revolution, with him in command. In this sense, then, there was never an ideological struggle for Fidel's soul, as some commentators have suggested. And unquestionably the entire process was irreversibly set in motion as a consequence of the April 1958 general strike.

However, now it was General Batista who decided to launch a total war. Late in 1957, an offensive designed to isolate the Sierra Maestra from the outside produced no meaningful results, and the Havana high command elaborated a different strategy for the May "summer offensive." This concept provided for encircling the Sierra, gradually closing the circle, and then launching the final battle and the definitive blow to Castro at his headquarters in La Plata high in the heart of the Sierra Maestra. At first, this strategy left Raúl Castro's "liberated zone" to the northeast under little more than air attacks, the assumption being that once Fidel was liquidated, the younger brother's forces would collapse sooner or later. In June, however, Raúl came under powerful air and ground assaults. The Batista plan called for advances on the central Sierra Maestra from the south, where troops were landed on the coast, from the northwest, and from the north. As many as ten thousand men, totaling fourteen battalions, were deployed in Operation FF (for *Fin de Fidel,* or End of Fidel) in three battle groups; they were supported by artillery, helicopters, other aircraft, and navy frigates firing on the coast from the sea.

Fidel Castro's Column 1 in the center of the Sierra had 280 armed men, including Che Guevara's force to the east, and was later reinforced by Juan Almeida's several dozen rebels recalled from the easternmost "Third Front." Batista had crushing superiority in numbers and firepower, but Castro had turned the mountains into a fortress he could defend with a much smaller force, men, who knew every path in the forest, every turn of the road, and every peasant's house in the immensely complicated terrain. Both sides knew that this would be the decisive battle of the war, and Castro realized it could be touch-and-go. He was prepared to cede territory up to a certain line protecting his La Plata headquarters, then hit back with ambushes and fight to the death. "Every entrance to the Sierra Maestra is like the pass at Thermopylae," Fidel told Venezuelan newsmen, "and every narrow passage becomes a death trap."

When the government forces began to attack on May 20, Castro had entered what Guevara described as a "sedentary phase," expanding the La Plata headquarters, and consolidating and organizing the "Free Territory" administratively and politically. Not far from La Plata, Che Guevara set up a Rebel Army recruits' school at Minas del Frío where new volunteers were trained by a former Batista army captain. The headquarters, called *comandancia,* was in a large forest clearing on the crest of the Sierra Maestra that could be reached solely up a tortuous, narrow path, studded with rocks and boulders and usually covered with mud. It was an extraordinarily tough climb, and mules could be used only a part of the way.

Wooden structures erected in La Plata were concealed by tree branches above for protection from air attacks, and each structure was so cleverly located inside the forest at the edge of the clearing that one could not spot it even in daylight until coming upon it. Castro's house was built solidly against the side of a ravine over a creek, and it had an escape route down a long ladder to the creek if it were attacked through the entrance. The house consisted of a bedroom with a double bed for Fidel and Celia and a single chair, a room serving as an office for Celia, a deck where Castro often received visitors, and a kitchen. Nearby was the building were Faustino Pérez ran his civil affairs office for the "liberated territories," then a hospital, a guest house, and a structure for the women's combat regiment. In this building the women had a score of sewing machines which they used to make uniforms, but often they dropped their sewing, grabbed their rifles, and went into combat. The first battle of this Mariana Grajales detachment (named after the patriotic mother of Generals Antonio and José Maceo of the independence wars) occurred in September 1958, northwest of La Plata. The first house in the clearing was the office of a bearded dentist named Luis Borges who became a guerrilla officer. He was responsible for the men's teeth and for ammunition stores. Castro, who had bad teeth, was a frequent customer, and more than once received visitors and battle reports while the dentist worked in his mouth with a pedal-activated drill. Several hundred feet above the clearing was the studio and antenna of the rebel radio, which served both as a broadcasting station and as Castro's link with Caracas, Mexico, Miami, and points in Cuba. Finally, a field-telephone network was established between La Plata and a number of outlaying rebel positions, lessening the dependence on couriers for every message.

The Batista offensive lasted seventy-six days before being decisively beaten back by the *Fidelistas*. The rebels several times came very close to defeat when the army captured most of their positions and villages around La Plata. June 19, Castro said, was the "most critical day," as he gambled everything on preventing the Batista soldiers from dislodging his forces from the crest of the Sierra Maestra. So long as he held the long crest, the rebels could fire down on the enemy and keep them in check. Army battilions repeatedly attempted to cross a river near Santo Domingo and scale the range, but each time they were beaten back by the rifles, machine guns, and mortars of no more than forty rebels deployed along the crest. Castro, who raced with Celia and his aide, Arturo Aguilera, by jeep or on foot from spot to spot, was at times so close to enemy troops that he could count individual soldiers as he stared down at them with his binoculars. Fidel also applied psychological warfare for the first time in the Sierra war during this battle by installing loudspeakers that blared the national anthem, patriotic songs, and revolutionary exhortations at the exhausted Batista soldiers. He thinks it helped to sap their morale. In any event, the offensive ended.

It was a violent and vicious war, but Castro and some of the Batista commanders developed an old-fashioned gentlemanly relationship that Fidel, in particular, enjoyed. At the outset of the enemy offensive, he received a communication from General Eulogio Cantillo, the chief of the Batista forces in the region, inviting him, in effect, to surrender. Castro answered instantly: "I think highly of you. My opinion is not incompatible with my having the honor of recognizing you as an adversary. . . . I appreciate your noble feelings toward us, who are, after all, your compatriots, not your enemies, because we are not at war against the armed forces, but against the dictatorship. . . . Perhaps when the offensive is over, if we are still alive, I will write you again to clarify my thinking and to tell you what I think of you, the army, and we can do for the benefit of Cuba . . ." Che Guevara took a dim view of this exchange.

Earlier in July, before the rebels repulsed the Batista offensive, Castro defeated a battalion led by Major José Quevedo that had landed on the coast with orders to storm the Sierra up La Plata River. The force surrendered to the rebels at the battle of El Jigüe, but first Castro wrote the major, whom he knew as a fellow law student at Havana University. It was difficult to imagine, he said, "that someday we would be fighting against each other, despite the fact that perhaps we do not even harbor different feelings about the fatherland. . . . I have used harsh words in judging the actions of many, and of the army in general, but never have my hands or those of my companions been stained with the blood nor have we debased ourselves by the mistreatment of any soldier taken prisoner. . . . I write these lines on the spur of the moment, without telling you or asking you for anything, only to greet you and to wish you, very sincerely, good luck." Five days later, Castro sent Quevedo another message that "your troops are surrounded, they have not the slightest hope of being saved. . . . In this situation, I offer you an honorable, dignified surrender. . . . All your men will be treated with the greatest respect and consideration. The officers will be permitted to keep their weapons." But Quevedo kept fighting, and on July 19, Castro sent one of his men to ask him to surrender to save lives on both sides; he broadcast a radio appeal to the Quevedo battalion to give up, promising them special treatment; and then he went to meet the major. When Quevedo finally surrendered on July 21, the rebels took 220 prisoners, Castro entered the camp, and ordered Ramiro Valdés: "Have all the officers keep their side arms. Make sure that nobody tries to take them away from them." Then he sent a message to Guevara to "try to have them prepare lunch" for the prisoners at the Mompié hill before their release. The prisoners were turned over to the International Red Cross three days later, and Che Guevara arrived on his little mule to watch. Castro raced across the mountains to fight at Santo Domingo.

On August 12 both Fidel and Che were at Las Mercedes to observe the delivery of some one hundred prisoners from another Batista unit to the Red Cross. They met the army colonel representing his side at the ceremony, and had coffee together and chatted amiably. The colonel said he thought the rebels would win in the end, "but you will find a destroyed Cuba." When Castro showed interest in the colonel's helicopter parked outside, the officer invited the rebels to fly around with him. With Che, Celia, and a rebel captain, Fidel flew over the Sierra for fifteen minutes, having a marvelous time and spotting places he knew. It was his first time aboard a helicopter, however, and his horrified aides were concerned for his safety; "it was a Fidel-type thing," Colonel Aguilera said. From Las Mercedes, Castro returned to the *comandancia* to plan the rebel counteroffensive—which would become the final offensive of the war. As Aguilera remembers, "Celia never left him, Celia always was with him." The only times they separated was when Castro wanted her to solve some specific Rebel Army problem at La Plata when he had to make a dash to a unit caught up in sudden combat. Others remember that Celia's slacks and blouse pockets were full of Castro's and Rebel Army papers and documents; he would dictate anytime, anywhere, receive briefings whenever and wherever he could, and she operated what was, in effect, a portable office.

Batista lost the "summer offensive," and close to one thousand dead and wounded plus the four hundred prisoners, which the rebels kept returning as they captured them. Castro took over five hundred modern weapons, including two tanks, an extraordinary contrast with the moment in June at La Plata when he had only his telescopic-sight rifle and Aguilera, a shotgun. It was calculated later that 321 men beat back the huge Batista offensive, and politicians and senior officers in Havana also calculated that the regime could not last much longer. Castro broadcast on Radio Rebelde to all of Cuba every detail of the victories; his ringing battle reports were read by Violeta Casal, the first woman announcer in the mountain.

Late in June Raúl Castro kidnapped forty-nine American citizens in eastern Oriente in a desperate attempt to force the Batista air force to stop bombing his units in the "Second Front" and the peasant families spread throughout the Sierra Cristal war zone. Without consulting Fidel, who was fighting off enemy troops in the central Sierra Maestra, Raúl ordered the kidnappings when he obtained proof that Batista's planes were not only being refueled, but also loaded with bombs at Guantánamo; he had photographs of Cuban aircraft receiving ordnance at the U.S. naval base, and a rebel agent at the Cuban embassy in Washington had forwarded to the Sierra documents showing that three hundred rocket warheads had been delivered to the Batista command through Guantánamo. The Eisenhower administration's excuse three months after arms deliveries to Cuba were officially suspended was that the warheads were "replacements" for defective ones sent earlier; it was also explained that these replacements were done more easily in Guantánamo. In Washington nobody was thinking politically.

Raúl's forces were being badly bruised by the intensive bombings, rocket attacks, and strafing at a time when they were also nearly out of ammunition. In a lengthy operational report on June 2, Raúl informed Fidel of "our lack of all kinds of ammunition . . . this offensive worries me." Vílma Espín, who had joined the "Second Front" in April, wrote later that "Raúl took the offensive with the Americans because we were lost; we didn't have anything to push them back with. Our bullets arrived in the middle of the offensive. When the Americans left, [the army] attacked again, but now Fidel's offensive was going into action in the Sierra, which kept the army from attacking on other fronts."

It was on June 26 that Raúl's commandos raided the American-owned Moa and Nicaro nickel mines and the United Fruit Company's sugar mill at Guaro in the north of Oriente province, capturing twenty-five American managers and employees. Simultaneously, another commando on the south coast of the province took over a bus transporting twenty-four U.S. Marines back to Guantánamo. The rebels also commandeered tractors and trucks from the nickel mines "as strict war necessities." American consul general, Park Wollam, went from Santiago up to the Sierra to negotiate the release of the hostages with Raúl, and he was shown fragments of U.S.-manufactured bombs and the Guantánamo photographs. Earlier, the hostages were taken to see damage caused by the bombing and burn victims of napalm firebombing. Though Wollam and the other Americans were well treated and even given a Fourth of July party by the rebels, Raúl released the hostages only after Fidel ordered it in a broadcast from La Plata on July 3; even so, the last hostage was freed as late as July 18. During the entire time of the Americans' captivity, the "Second Front" was spared air attacks (presumably Washington suggested to Batista this would be a good idea) and Raúl was able to resupply and reorganize his guerrillas.

As for Fidel, he used the kidnapping incident in a way that combined an assertion of his war leadership, pleased the United States, and affirmed support of his brother's actions. Though his broadcast came a full week after the kidnappings, Fidel said that his headquarters had received no reports on it because of the distance and because Raúl's forces had no radio transmitters, but such an occurrence was "possible . . . as a reaction to the recent delivery of three hundred rockets from the North American naval base at Caimanera [the Cuban name for Guantánamo] to Batista's planes, with which civilian populations are being bombed in the territory occupied by the rebels." Then he grandly announced that "despite all this, today I am publicly ordering [the hostages'] release . . . the order should be received and carried out, if it is true that these North Americans are being held by some revolutionary troops, because I believe that these North American citizens cannot be blamed for the shipment of bombs to Batista by the government of their country. I am certain that no rebel forces would make hostages of United States citizens so that they could observe the results of the inhuman bombings of Cuban civilians with weapons sent by the United States. . . . The 26th of July Movement is fighting for the respect of human rights. We believe that individual freedom is one of the inviolable rights of every human being and therefore no one sould be arrested without a just cause. We hope that the United States government will, in like manner, respect the lives and liberty of Cubans. . . . This is the necessary condition for the continuation of the present friendly relations between the two countries."

Castro, behaving more and more like a statesman, handled the next incident with the United States, late in July, with equal diplomacy. As Batista had withdrawn army guards protecting the aqueduct from Cuban territory to the Guantánamo naval base (which depended completely on water from Cuba), U.S. Marines took over this responsibility, and the rebels protested it immediately as an act of American intervention. Rebel emissaries, including the journalist Carlos Franqui, entered into quiet negotiations with American diplomats in Santiago—reality increasingly forced American officials to deal with the *Fidelistas*—and rejected a proposal that the aqueduct area be regarded as a "neutral zone." Thereupon, Castro issued a formal declaration that "the presence of North American forces . . . is illegal and constitutes aggression against Cuban national territory," but "we are ready to give guarantees that the water supply will not be interfered with because our objective is not to attack that facility . . ." This message was both firm and conciliatory, and Castro's language carefully made the point that while he wished no disputes with the United States, he would make no concessions whenever, in his view, Cuban sovereignty was infringed by the *Yanquís*. This would remain his consistent policy toward the United States.

Apart from the fact that Raúl did not consult Fidel over the kidnapping of the Americans, which he probably could have, given his practice of sending long operational reports to rebel headquarters, there is no question that he was emphasizing Marxist ideological indoctrination in his zone much more than his brother. It is also a fact that the Communist party's contacts with the rebels in the "Second Front" were much stronger than with Fidel's units. The rosters of the Revolutionary Armed Forces, which Raúl organized into a modern army after 1959, demonstrate that the Marxist-Leninist hard core among the senior officers were the fighters of the "Second Front." Interestingly, while Fidel chose to deal with the Communists almost exclusively on a high policy level before and during the war years, Raúl concentrated on forging discreet but effective political and military links. This may have been a deliberate division of political labor between the brothers, aimed at protecting Fidel's democratic image before international public opinion, but the distinct ways in which their commands were run do suggest that Raúl took, or was given, the freedom to do what he pleased in this realm. A similar situation would develop later in Che Guevara's command in the final stages of the war. Even guerrilla dress differed: Raúl and Che wore black berets with a star for rank insignia; Fidel preferred his olive-green cap with no stars.

Raúl's first Communist link in the Sierra Cristal was José "Pepé" Ramírez Cruz, a party leader in the sugar workers' union in the Havana area, who was ordered early in March to travel to Holguín in the north of Oriente, then to join the "Second Front" when it was established during April. Most likely, Fidel knew about Ramírez's assignment, but the point is that he was assigned to Raúl, not to him. In the mountains Raúl instructed Ramírez to organize the peasants politically in the "Second Front" region, in preparation for a "Peasant Congress," and Communist party officials were sent up to help in the task. An Agrarian Bureau was added to Raúl's army. After kidnapping the Americans late in June, Raúl dispatched Ramírez to Havana to inform the Communist party leadership. The congress was held in September, and Raúl addressed the four hundred delegates in terms Fidel was not using at that juncture: "Reactionaries, backed by foreign capital, maintain a tyrannical and bloody Batista regime because they can enrich themselves at the expense of the people, even when this foreign backing means the strangling of our national economy." Since 1960, Pepe Ramírez has presided over the National Association of Small Farmers, a revolutionary organization in charge of cooperatives, and in 1986 he became an alternate member of the Politburo of the Castro Communist party.

Jorge Risquet Valdés-Saldaña, in 1986 a full Politburo member, was another young Communist whom Raúl put in charge of the Department of Revolutionary Instruction and the José Martí School for Troop Instructors at Sierra Cristal village of Tumbasiete. This school became the model for postrevolutionary military indoctrination centers as it taught selected fighters "ideological formation" along with other aspects of education. At Tumbasiete, Communist instructors introduced for the first time a Cuban history text based on Marxist interpretations. By the end of 1958, all the Rebel Army instructors in Raúl's territory were Tumbasiete graduates, and the Marxist influence spread to the peasants in the well-populated regions of northern Oriente. Raúl likewise organized a Corps of Rebel Intelligence Officers (a forerunner of the army's G-2 intelligence service) and a Committee of Revolutionary Peasants as an "information" branch.

When Carlos Rafael Rodríguez, one of the top Communist party leaders, finally traveled to the mountains late in July as part of the Communists' changing attitude toward the rebels, he stopped first to see Raúl. Rodríguez was subsequently quoted as saying that "in the Sierra Cristal, where Raúl Castro commanded, all was harmony with the Communists; but when I arrived at Fidel Castro's in the Sierra Maestra, harmony was converted into suspicion." This was Rodríguez's first trip to Fidel's headquarters; there would be others.

In the meantime, Fidel Castro was emphasizing unity and moderation. On July 20, while he was directing operations against the Quevedo battalion above the south coast, Radio Rebelde broadcast from the mountain the "Unity Manifesto of the Sierra Maestra," also known as the Caracas Pact (it was signed that week in the Venezuelan capital). This Manifesto was issued jointly by the 26th of July Movement and eight other Cuban opposition political parties and revolutionary and action organizations, including the rival Students' Revolutionary Directorate but not the Communists. Castro said later that this was due to objections to the Communists by all the other groups (which he accepted without public protest). Having long resisted pacts with other Cuban revolutionaries and politicians, Castro was evidently willing to sign the Caracas document because he was now dealing from a position of military strength. At the same time, the pact further improved his statesman's image: He was the indisputable leader of the revolution, and the other signers of the "Unity Manifesto," all of them political moderates, implicitly recognized this fact.

The Manifesto was not a revolutionary program but an agreement "to create a large revolutionary, civic coalition, made up of all of Cuba's sectors." It provided for "a common strategy to defeat the dictatorship by means of armed insurrection" and "the popular mobilization of all labor, civic, professional and economic forces, culminating in a great general strike on the civilian front." On the military front, it said, "action will be coordinated throughout the country." Concerning the future, only two points were made: "A brief provisional government will be formed to establish full constitutional and democratic rights"; and "a minimum governmental program will be formed to guarantee the punishment of those who are guilty of crimes, workers' rights, fulfillment of international agreements, public order, peace, freedom, as well as the economic, social, and political progress of the Cuban people." The Manifesto then asked the United States "to cease all military and other types of aid to the dictator." José Miró Cardona, the exiled president of the Cuban Bar Association, was the coordinator of the pact; it was understood that Manuel Urrutia, the president-designate who was in New York when the Manifesto was issued, would soon be flown to the Sierra Maestra.

It was to discuss post-Batista Cuba in the context of the Manifesto that Carlos Rafael Rodríguez had gone to see Fidel late in July, remaining until August 10, when he returned to Havana to report to the Communist party leadership. It is unclear whether Rodríguez had Fidel in mind when he spoke of the "suspicion" he encountered at La Plata, and there is no evidence to suggest that an actual agreement for subsequent collaboration emerged from these preliminary discussions. According to Rodríguez, Castro said that it would be a grave tactical error to alert the enemy prematurely by defining the revolutionary objectives with excessive clarity, but this did not necessarily commit him to an alliance with the Communists. Rodríguez was replaced at La Plata by Luis Más Martín, another ranking Communist leader and an old friend of the Castro brothers, who arrived on September 6. Más Martín told an interviewer later that Fidel, while reminiscing with him about obtaining books from the Communist party bookshop in Havana, remarked that "when the Revolution triumphs, we'll have Marxist books coming out of our ears." Again this is hardly a statement of political intent. Much more relevant was the Communist party's decision to maintain from then on a permanent presence with Castro, and his evident acceptance of it. Thus, Rodríguez came back to La Plata in mid-September, remaining with the rebels until the end of the war. No other Cuban political group sought or had such representation.

Although it is impossible to say with any precision when Castro made the final decision to strike a deal with the Communists, the likelihood is that it happened during Rodríguez's residence on the mountain, culminating a very pragmatic process that began with the collapse of the April general strike. This would explain the ease with which Castro slid into a relationship with the Communists immediately after the victory. It is important, however, to understand that Castro's idea was to use the "old" Communists for his instant needs in the absence of any other urban organization he could trust, including his own 26th of July Movement. The Rebel Army was the revolutionary vanguard, but it was not prepared to run the country for him when it came down from the Sierra. Chances are that except for Carlos Rafael Rodríguez, the "old" Communists did not comprehend Fidel's strategy until he finally swallowed them up; that is why they attempted to undermine him afterward. Raúl Chibás, who flew to the Sierra for a second visit in August 1958, says that his nearly daily discussions with Fidel over long weeks covered every subject except the question of communism; in exile from Cuba for over a quarter-century, he still thinks Castro is a *"Fidelista,* not a Communist."

Great insight into Castro's politics and ideology was expressed by Régis Debray, the French leftist intellectual who knew him better than any foreigner: "A Leninist is an opportunist with principles. Fidel is a Leninist. His principles remain firm, but the opportunities change. The unique thing about him is the combination of great realism in the evolution of the means available and the final goal." In his 1967 *Revolution in the Revolution?,* Debray remarked that the stronger the revolutionary nucleus in Cuba, "the more it could permit itself to seek alliances." He also underlines the Castro novelty the "old" Communists missed: "Eventually, the future People's Army will beget the [political] party of which it is to be, theoretically, the instrument: essentially the party is the army . . . [the party] already existed in embryo—in the form of the Rebel Army. Fidel, its commander in chief, was already an unofficial party leader by early 1959."

On October 10, 1958, Fidel Castro signed in the Sierra Maestra an agrarian reform law, known as Revolutionary Law 1, which preserved the moderate image. Drafted by the Rebel Army's advocate general, Humberto Sorí-Marín, but approved by Castro, the law turned over to tenants, renters, and squatters the land they worked; it said nothing about breaking up the great estates. This agrarian reform did not affect 58 percent of Cuba's land area held privately, though it benefited 64 percent of all the farmers. Coming after the bland Sierra Manifesto and the communication to the Cuban Liberation Junta in 1957, the revolutionary documents of 1958 were equally devoted to moderation despite (or because of?) the Rebel Army's victories.

It has been argued for over a quarter-century whether Castro and his movement were secretly more radical than they admitted in public, but historically it is an increasingly barren theme. Official Cuban histories cite Castro's "History Will Absolve Me" address in 1953 as an example of Marxist thought—and, taken at face value, even this evaluation is debatable—but no serious explanations are provided for the moderate voice from the Sierra, apart from Che Guevara's comment that the rebels had to live with "a minimum program." In the end, the public record must stand, and the Cuban exiled scholar Nelson P. Valdés is probably correct in saying that "the Cuban revolutionary leadership ached for truly radical credentials, but could not produce any." The truth may lie in the pragmatic opportunism that developed in the Sierra Maestra as the Rebel Army's final offensive began, and that was transformed into official radicalism after the victory a half-year later.

Karl E. Meyer, then an editorial writer for *The Washington Post*, visited Castro at La Plata for three days just as the rebel offensive was getting under way. He found Fidel relaxed, "sprawled across a bed," and expressing views that "as he described them to me are surprisingly moderate." Meyer wrote that "his social views are vague, but incline to a kind of welfare state liberalism." When Meyer arrived, Castro was reading *Kaput* by Curzio Malaparte, one of Mussolini's favorite journalists, and he quoted from Il Duce that "you can make a coup with the army or without the army, but never against the army." Flourishing his cigar, Castro observed: "We are proving Mussolini wrong. We are winning here in Cuba against the army." Castro complained that Israel was sending arms to Batista after the United States finally suspended deliveries. "Why should they do such a thing—we have nothing against the Jewish people." Meyer says that on his return to Washington, he passed this information on to an Israeli newspaper correspondent who used it in a story, "which in turn provoked a challenge in the Knesset and I believe brought an end to the shipments." Finally, Castro informed Meyer that "in three months, three fourths of the island will be in rebel hands." During Meyer's visit, a portrait of Fidelito in a gilt frame was delivered to Castro, and he exclaimed beamingly, "This is my son." Fidelito must have just returned to Havana with his mother after attending school for one year in Queens, New York, where they went shortly after the boy was kidnapped from his aunts in Mexico City.

In making his military prediction to Meyer, Castro was not too far off the mark. Camilo Cienfuegos, leading the "Antonio Maceo Column" with 82 men, was ordered to march from the Sierra Maestra to the westernmost province of Pinar del Río, and Che Guevara, at the head of the "Ciro Redondo Column" with 148 men, was assigned to take the central province of Las Villas, including the Escambray Mountains. They left in the third week of August. It must have seemed a demented plan, considering that this was the first time the rebels were leaving the safety of the Sierra for the lowlands where Batista still had tens of thousands of soldiers and policemen, but Castro was convinced that Cienfuegos and Guevara would triumph. They were to act as military governors as well as set up revolutionary authority while they advanced. Castro himself planned to attack Santiago later in the autumn.

Incredibly, Cienfuegos and Guevara made it to Las Villas, marching, fighting, and starving under the most adverse conditions—and creating new myths for the Rebel Army. It took the first column nearly six weeks to cross western Oriente and Camagüey provinces to reach Las Villas, walking through swamps and swimming through rivers, without food most of the time. The second column under Che made it in seven weeks. They had extremely few casualties, and it had never occurred to the commanders or the men to doubt the wisdom of Castro's orders to seize most of Cuba with a total of 230 men. But astonishingly, the army was pulling back, volunteers were joining the invaders from the Sierra, and the rebels kept advancing. The plan was modified for Cienfuegos and Guevara to work together to reach the north coast of Las Villas and thereby sever the island into two parts; the march on Pinar del Río was set aside. In the Escambray Mountains, the *Fidelista* chiefs encountered four other separate revolutionary forces: the Students' Revolutionary Directorate (DR), a spinoff from the DR known as the "Second National Front of Escambray," a Communist party unit, and a unit organized by former President Prío. Che Guevara wrote that it took "laborious negotiations" to create an approximately common strategy in regard to the enemy, the problems being political and ideological differences. The DR and "Second Front" fighters were convinced that Guevara and Cienfuegos were Communists, and at first refused all cooperation. Enrique Oltuski, the provincial coordinator for the 26th of July Movement, had a bitter clash with Che Guevara over ideological matters. The Communist party unit, the only fully Communist guerrilla group in the whole war, worked well locally with the Sierra rebels. Still, the fighting against the Batista troops went on during November and December for the control of the provincial capital of Santa Clara.

At the head of "Column 1, José Martí," Fidel Castro moved out of his mountain haven in mid-September, initiating an offensive designed to capture most of Oriente province, then surround Santiago and force its capitulation. Raúl's forces in the east fanned out to attack Batista forces from the rear. Castro now had military momentum going for him as well as the financial resources he had lacked in the past: Landowners, industrialists, and businessmen were contributing to the 26th of July Movement on such a scale that in October Fidel could instruct Major Juan Almeida to pay a dollar for a single semiautomatic rifle bullet, if necessary. The Rebel Army could not afford to run out of ammunition as it raced to crush Batista. On December 6 Castro's column won a tough battle for the town of Guisa, opening the way into the heart of Oriente. And on the same day, Manuel Urrutia landed secretly in a field on the western outskirts of the Sierra Maestra, ready to become Cuba's provisional president. On his flight from Caracas, Urrutia brought a shipment of arms and ammunition for the rebels from Venezuela's provisional president, Admiral Wolfgang Larrazábal. This was the first delivery of arms from a foreign government, but Larrazábal had ousted his country's dictatorship less than a year earlier, and he felt close to Castro; he was not the kind of traditional military officer the *Fidelistas* so feared at home as an alternative to Batista.

From Guisa, Castro kept up the pace, seizing the towns of Baire, Jiguaní, Maffo, and Contramestre, firing his rifle at the enemy, and issuing a daily broadcast of communiqués, proclamations, and ultimatums. His victories were particularly spectacular because at the outset his column consisted mainly of rough recruits from the Minas del Frío basic-training school. On December 19, Fidel established his command post outside of Jiguaní, receiving for the first time President-Designate Urrutia at a meeting also attended by Celia, Raúl, Vílma Espín, and Juan Almeida. Urrutia wrote later that Castro greeted him coldly, which seems unlikely in those euphoric days. The next morning, Colonel Aguilera, Castro's aide, had to hold Fidel back physically to prevent him from leading his column through enemy fire in the main street of the town they were attacking.

Palma Soriano, directly northwest of Santiago, fell on December 20, and Castro found himself retracing his steps of more than five years earlier as a conspirator en route to Moncada. At that point, Batista began to prepare to flee Cuba. People in Havana and the other cities as well as in the countryside were openly against him everywhere and, scenting blood and defeat, were assuming pro-Castro attitudes publicly. The ship was sinking. In Havana a Castro agent was contacted secretly by a faction of senior army officers proposing a peace settlement with the rebels by replacing Batista with a civilian-military junta. This junta would be composed of General Eulogio Cantillo, with whom Fidel had exchanged letters during the summer offensive, an anti-Batista army officer now in prison, Manuel Urrutia, and two other civilians chosen by the revolutionaries. The agent was further informed that the United States would immediately recognize such a junta. This was the military-coup trap Castro had suspected and feared all along, and he shot back a brief message: "Conditions rejected. Arrange personal meeting between Cantillo and me."

Despite subsequent denials, the United States was actively engaged in secret maneuvers to prevent Castro from taking power. The report of the board of inquiry on the failure of the Bay of Pigs invasion says that the CIA first sought to collaborate with Justo Carrillo's *Montecristi* group, which had army links, to set up a new regime that would block Castro. On December 8, 1958, the report states, former U.S. ambassador to Cuba, William D. Pawley, and the CIA station chief in Havana, James Noel, "approached Batista and proposed the establishment of a Junta to whom Batista would turn over the reins of government." The State Department supported these efforts, while the CIA seemed to have switched policies since financing the M-26 earlier in the year.

Castro and Cantillo met near Palma Soriano on December 28, and the rebel commander repeated his rejection of a junta; power, he said, had to be vested in the revolutionary army. With Santiago surrounded, Cantillo agreed to lead a rebellion on December 31, and turn over his troops unconditionally to Castro. Celia Sánchez and Vílma Espín were present along with Raúl Chibás and Major José Quevedo, who had just arrived at Fidel's new headquarters at the Oriente sugar mill outside of Palma Soriano. Quevedo was the superb Batista troop commander whom Castro had defeated in July and brought over to the Rebel Army (in the 1980s, Quevedo was the Cuban military attaché in Moscow). However, Cantillo broke his word, informed Batista of the junta plan, and gave him until January 6 to leave the country. Then, he asked Castro for a week's postponement of their deal, instantly arousing Fidel's suspicions. However, events moved very fast. Che Guevara finally captured Santa Clara on December 30, putting an armored train out of action, and the regime had nothing left to shore itself up. Just after midnight on New Year's Eve, Batista, his family, and closest associates drove to Camp Columbia's airfield and took off for the Dominican Republic. His final act was to appoint General Cantillo the head of the armed forces.

Fidel Castro spent New Year's Eve quietly at the sugar mill with Celia and his commanders, learning from radio broadcasts at dawn what had just occurred in Havana. Carlos Rafael Rodríguez was in Palma Soriano, too, but apparently he did not see Castro that night; Chibás, summoned by Celia to join Fidel in his room, says that Fidel seemed annoyed over Rodríguez's presence. It was not good for his image, Chibás thinks. (Incongruously, Errol Flynn was also in Palma Soriano at that exact moment, shooting a movie.) That morning in the capital, Cantillo formed a junta, chaired by Carlos M. Piedra, a Supreme Court judge, and Castro instantly swung into action. Speaking over Radio Rebelde from its new location in Palma Soriano, Castro first issued an ultimatum to the Santiago garrison to surrender by 6:00 P.M. or be attacked, saying that "the history of 1898 will not be repeated," an allusion to the fact that American forces had not allowed Cuban independence fighters to enter the city at the end of the war with Spain. Then, he issued a proclamation to the nation denouncing the junta as "accomplices of the tyranny," and calling a general strike the next day. He ordered Camilo Cienfuegos and Che Guevara to march on Havana to seize, respectively, Camp Columbia and the La Cabaña fortress. "The Rebel Army will continue its sweeping campaign," Fidel shouted. "Revolution, *yes;* military coup, *no!*"

In Havana the junta collapsed before nightfall. At noon a diplomatic delegation, which included the American ambassador, Earl Smith, met with General Cantillo at the presidential palace, but by the afternoon, Camp Columbia was turned over by its officers to Colonel Ramón Barquín, who had just been released from prison on the Isle of Pines where he was serving a sentence for conspiring against Batista. Barquín placed Cantillo under house arrest, and by midnight a plane arrived from the Isle of Pines, bringing other officers who had been imprisoned there, including Captain José Ramón Fernández and such civilian leaders as Armando Hart. Cienfuegos and Guevara would not reach Havana until the afternoon of January 2, so for three days the capital was without government or authorities; but there was no violence and no vengeance in the city, only joy, chanting, and singing into the night. Cubans had taken to heart Castro's broadcast appeals not to take justice into their own hands.

Fidel Castro and his entourage entered Santiago on January 2, surrounded by an explosion of popular happiness. Symbolically, he took possession of the Moncada barracks where he had launched the revolution on July 26, 1953. He named Manuel Urrutia as the provisional president of Cuba, declared Santiago the provisional capital of Cuba, and that evening delivered his first speech as the victorious chief of the revolution to a delirious, joyful crowd. Immediately, he set the tone for the future, the future that had already arrived: "The Revolution begins now. The Revolution will not be an easy task. The Revolution will be a very difficult undertaking, full of danger. This time, luckily for Cuba, the Revolution will truly come into power. It will not be like 1898, when the North Americans came and made themselves masters of our country. . . . At this moment we must consolidate power before anything else. . . . The Revolution will not be made in two days, but now I am sure that we are making the Revolution; that for the first time the republic will really be entirely free and the people will have what they deserve. . . . This war was won by the people! . . ."

IV

THE REVOLUTION

(1959–1963)

CHAPTER

1

Fidel Castro took over Cuba and launched his great revolution in January 1959, to the thunderous applause of an overwhelming majority of his compatriots. Within eighteen months he had guided the country to the threshold of being a Marxist-Leninist society and to an alliance with the Soviet Union. In the words of his closest Communist associate, Carlos Rafael Rodríguez, "the democratic-bourgeois period in Cuba really ended in August 1960," and Castro himself said at the time that "we are entering a new stage" of the revolution. He also explained that the revolutionary movement had to be the work of "new communists . . . because they were not known as such." The same month, the Central Intelligence Agency formally approved top-secret plans to assassinate him.

Castro set out from his first day in power to destroy every vestige of the old social order in Cuba. He accomplished it through the extraordinary procedure of operating for well over a year a "parallel" government in Havana, concealed from his own cabinet ministers, to say nothing of his fellow Cubans, until his revolutionary controls were fully consolidated. In an equally secret fashion, he instantly negotiated a pact with the "old" Communists, and, in a separate move, entered into the first conversations with a Soviet emissary over caviar and vodka in his INRA office in the autumn of 1959.

The inside story of the origins of the present Communist state in Cuba has never before been published, and its substance remains unknown to Cubans in general. It has been reconstructed here from interviews, held in Havana during 1985, with those who were personally involved in running Castro's "hidden government" and in such related activities as the creation of special schools where "old" Communists taught Marxism-Leninism to the "new Communists" among top *Fidelistas,* at first in total secrecy. Publicly, Castro savagely rejected domestic and foreign accusations that communism was creeping into his "humanist" revolution, and imprisoned on charges of treason those of his wartime companions who resigned their posts over this issue. Castro's Ariadne thread remained invisible while the "Maximum Leader (as Fidel was now known)," basking in national adulation, put together his "real" revolutionary structure.

In peace as in war, Castro was the master of strategy and tactics as well as the master of timing. Looking back at events of the first year of the revolution with the benefit of hindsight and inside knowledge, it is clear that he knew exactly what he was doing all along, that his apparent improvisations had been carefully thought out, and that nothing was left to chance. Castro understood above all else that his own personality was, as a purely practical proposition, the key to the success of his entire enterprise, and he exploited this factor to the utmost. Having always insisted that propaganda was vital in mobilizing the masses for a revolution, Castro immediately seized on television, which was already quite developed in Cuba in 1959, as the ideal vehicle to advance himself and the revolution. He was a natural television personality, and he literally sold the revolution on TV. Antonio Nuñez Jiménez, the erudite geographer and writer who was the coordinator of Castro's "inside team" and has remained at his side for nearly thirty years, says that in terms of a leader's relationship with his nation, "the case of Fidel is unique in modern history. . . . Lenin did not have radio and television, and, moreover, Lenin never had the ascendancy over the Soviet people that Fidel has had over Cubans."

Politically, Castro conducted himself with remarkable acumen from the moment power came within his reach. Having forced the collapse of the junta, which had attempted to replace Batista in Havana, by calling a general strike and threatening a military attack on the capital and Santiago, he kept up the political pressure for a full week. He maintained the work stoppage for two more days as a guarantee against new coup attempts (it also constituted a victory holiday with Castro as the center of national attention), and on January 3 he initiated a slow, triumphant march to Havana in the Roman manner. That morning, however, he first conferred with Camilo Cienfuegos, who flew from Havana to brief him on the security situation in the capital, and instructed Raúl to stay behind in Santiago to protect his rearguard. The advance to Havana lasted five days and nights, with Fidel, surrounded by his *barbudos,* riding atop a tank or in a jeep, receiving wild acclaim from the population, every step of the way relayed to the rest of the island by live television. His telescopic-sight semiautomatic rifle (now an American M-2) slung over his shoulder and his horn-rimmed spectacles perched over his Roman nose, Castro presented the image of the warrior-philosopher king. The famous beard, the cigar clenched in his teeth, and the olive-green combat fatigues (with a small medallion of the Virgin of Cobre on a chain around his neck conveniently visible under his open-collar shirt) were the symbols of the Fidel Castro personality, precisely the way he intended to be seen and remembered forever. Inching ahead through thick crowds, he halted almost every other minute to greet or embrace somebody he knew or recognized, to shout a slogan, deliver a few words, even to make a speech. When his final offensive began in the autumn, Castro commanded three hundred men and when he entered Santiago, he had a total of three thousand armed rebels. As he marched to Havana, he was like the Pied Piper, attracting more and more military followers. He says that when he reached Bayamo in western Oriente, the two-thousand-man regular army garrison with tanks and cannon joined his column of one thousand fighters (the balance of the Rebel Army was distributed in Santiago, Havana, Las Villas, and elsewhere). He welcomed these soldiers for the time being. The weeding out and the creation of the new army would come soon after.

It was unquestionably a risk for Castro, physically and politically, to be on the road for five days, but he had calculated that his trusted commanders throughout the island would assure that nothing untoward would happen; he was in permanent radio and telephone contact with them. Concerning the danger of assassination, Castro was always fatalistic, not prepared to sacrifice his relationship with the masses to security requirements. And the idea of the march to Havana down the length of the central highway represented such fantastic political exposure, and was such proof of his absolute sway over the nation and the adoration of him that was growing geometrically—thanks to television—that it simply could not be forfeited. Castro was even able to control the politics of the revolution from the road despite the maddening chaos and confusion that surrounded his progress. With Celia Sánchez firmly in charge of communications and of the whereabouts of people who mattered to him, Fidel managed to conduct private conferences with the men whom he wanted in key posts (for example, Raúl Chibás, the treasurer of the 26th of July Movement, flew from Santiago to Camagüey where he caught the Maximum Leader to say he was *not* interested in the Finance Ministry post being offered to him; others grabbed Castro elsewhere to discuss their futures). At the same time, Celia busily sent out messages around Cuba and abroad to summon old friends whom Fidel wanted to see as soon as possible.

In Cotorro, at the approaches to Havana, Castro finally encountered his son, Fidelito, whom he had not seen since leaving Mexico over two years earlier. Fidelito, now nine years old, was brought by relatives to meet his father; Mirta, his mother, evidently did not oppose the reunion and was already losing control over the boy. Soon, Castro took him out of a private Havana day school and put him in a public boarding school, making sure he could see him as often as possible. Soon, Mirta and her husband and children left for Spain for good, and Castro's old friend Naty Revuelta took it upon herself to find a school for Fidelito.

Whatever Castro was planning for the future of Cuba, his instinct had convinced him that a smooth and widely acceptable transition was politically advisable. For this reason, he let his hand-picked provisional president, Manuel Urrutia, select the prime minister and the cabinet in Santiago that first week, though naturally he kept a watchful eye on the process. For himself, Fidel reserved the title of military Commander in Chief that he already held in the Sierra Maestra, knowing that his real power lay in the blindly loyal Rebel Army. Urrutia, who had arrived in the Sierra the preceding December (but saw Castro only twice), named an exceptionally talented cabinet. Drawn mainly from the 26th of July Movement's moderate wing, the group included only three guerrilla companions as ministers, one of them being Faustino Pérez, a veteran mountain fighter but an ideological moderate. The other *barbudos* were Augusto Martínez Sánchez as defense minister and Humberto Sorí-Marín, who had drafted the Sierra agrarian reform law, as agriculture minister. From the founders of the 26th of July Movement after Moncada, only Armando Hart was invited to the cabinet and named education minister. The only ideological leftist was Osvaldo Dorticós Torrado, the minister of revolutionary laws, who had belonged in the late 1930s to the university committee of the illegal Communist party, joining the Castro movement in the late 1950s and serving for a time as chairman of the Cuban Bar Association. That same chairmanship had also been held by José Miró Cardona, whom Urrutia appointed as prime minister. Urrutia wrote later that he had proposed to Castro "the desirability of appointing a centralized cabinet representing all the revolutionary sectors, but Castro opposed it, asserting that the government should be as homogeneous as possible." At that juncture, Castro wanted this homogeneous group to be formed along the lines of the 26th of July Movement, which enjoyed worldwide respect.

More to the point, Castro needed instant competence. The Rebel Army, peasant based and overwhelmingly illiterate, could not provide administrators on any level, certainly not ministerial (Faustino Pérez, Martínez Sánchez, and Sorí-Marín were pre-guerrilla army intellectuals with university degrees, and all three were identified with the 26th of July Movement). While Castro explained later that the revolution had to turn to the "old" Communists because the Sierra rebels lacked government-management expertise, he could not actually do it for at least two years without triggering violent opposition from large segments of the Cuban population and from the United States. By the end of 1960, his police and political controls were strong enough to cope with domestic opposition, and his relations with the United States had deteriorated to such a point that he no longer had to take the American factor into account when formulating his defiant policies. At that point, Castro could afford to ally himself openly with the Communists of the Popular Socialist Party (PSP).

In keeping the Communists out of the cabinet in 1959 (Dorticós whose past Communist ties were generally unknown was easily accepted by Urrutia), the "Maximum Leader"—had to keep out the Students' Revolutionary Directorate as well maintain the "homogeneous" character of the provisional government; besides, the DR had challenged him by occupying the presidential palace in Havana and the university with armed guerrillas before the Rebel Army reached the capital. To avoid premature opposition, Fidel deliberately gave his brother Raúl and Che Guevara very low public profiles. Raúl had his military command in Santiago, and Che was the chief of the La Cabaña fortress in Havana, but their power and influence in real life greatly exceeded their job descriptions. They participated in all the secret revolutionary policy decisions, and they were instrumental in placing Communist-oriented Rebel Army personnel from their wartime commands in strategic middle-level positions throughout the country. With the impressive façade of the 26th of July Movement cabinet studded with such internationally recognized economic specialists as Rufo López-Fresquet as finance minister and Felipe Pazos as president of the National Bank, providing the new regime with respectability, Castro and his collaborators used the interval to discreetly construct the Marxist-Leninist edifice.

As usual, Che Guevara insisted on being frank and outspoken when the rest of the core leadership labored behind the façade. Enrique Oltuski, who was the regime's first minister of communications (and, at the age of twenty-three, the youngest minister) before being fired and imprisoned, later recalled this frankness in his encounter with Guevara during the Las Villas campaign in autumn 1959, when he was in charge of the 26th of July Movement in that province. As they discussed the future, and Oltuski urged caution in order not to provoke the United States, Che told him: "So, you are one of those who think that we can make a revolution behind the back of the Americans. . . . What a shiteater you are! We must make the revolution in a struggle to the death against imperialism from the first moment. A true revolution cannot be disguised . . ." In a Havana lecture on January 27 on the "Social Projections of the Rebel Army," Guevara went far beyond anything Castro was ready to say publicly when he declared that the agrarian reform law signed in the Sierra in 1958 was "not complete" without the seizure of large land holdings and that the "peasant mass" and the Rebel Army would impose a new law. At a time when Castro was pledging respect for foreign investments, Guevara informed his audience that "we are an armed democracy," that public utilities (which were owned by American companies) must be nationalized, and that "the entire Cuban nation must become a *guerrilla* army" to defend itself from aggression "by a power that is almost a continent." However, not much attention was being paid to the Argentine physician and his lectures before obscure groups—yet.

It was on Fidel Castro's public activities that the attentions of Cuba and much of the fascinated outside world were focused during these first months of the "Year of Liberation" (Castro liked the French Revolution's concept of designating years through reference to its own calendar, as if to erase the past), and he was mesmerizing. His entrance in Havana on January 8 was an apotheosis, marvelously staged. As Fidel drove into the ancient colonial city at the head of his Column 1, church bells tolled, factory whistles blew, and ship sirens sounded. The first stop was the harbor, and he stepped aboard the *Granma,* which was moored there—it had been recently brought to Havana—to the boom of cannon salutes from the navy frigates. The Castro motorcade then fought its way through the dense mob in the plaza in front of the presidential palace, where the Commander in Chief wished to call on Manuel Urrutia and the cabinet. Urrutia had been able to move into the palace earlier in the week when the Rebel Army persuaded the rival DR guerrillas to leave the building, but the armed students were still occupying the university, and now Castro was facing his first major crisis. He chose to solve it through rhetoric rather than through force, and this won him new acclaim and support.

Night had fallen on January 8 when Fidel reached the army's Camp Columbia headquarters in northwestern Havana to deliver his great victory speech before the tens of thousands of Cubans who had been waiting for him for long hours. His main theme was the Rebel Army's responsibility for the success of the revolution then being launched, and this led him to emphasize the need for revolutionary unity, and finally to bring out into the open the seizure of weapons by the DR. Turning to Camilo Cienfuegos, the Rebel Army's chief of staff and the second most popular revolutionary figure, Castro asked, "Am I doing all right, Camilo?" and Cienfuegos re-

plied, to the roar of the crowd, "You are doing all right, Fidel!"—and a
new revolutionary slogan was born.

Letting his voice drop, Castro announced he had a question for "the
people," thereby inaugurating a new approach to the art of governance: a
dialogue with the masses, through which they would affirm his policies by
chanting responses to his "questions." Soon, he would call it "direct democracy . . . of the marketplace," cleaner and more honest than the old-
fashioned corrupt electoral procedures of the past. But on that first night,
the questions were: "Why hide arms in different places in the capital? Why
smuggle arms at this moment? For what? . . . Arms, for what? To fight
against whom? Against the revolutionary government that has the support
of the whole people? [Shouts of: NO!] . . . Is it the same with Judge
Urrutia governing the Republic as it was with Batista governing the Republic? [Shouts of: NO!] . . . Arms, for what? Is there a dictatorship here?
[Shouts of: NO!] . . . Will they fight against a free government that respects the peoples' rights? [Shouts of: NO!] . . . Arms, for what, when
elections will be called in the shortest time possible? . . . Hide arms, for
what? To blackmail the President of the Republic? . . . Arms, for
what? . . . So I must tell you that two days ago, members of a certain
organization went to a military base and took five hundred weapons and six
machine guns and eighty thousand bullets!" (Shouts of: LET'S GET
THEM!)

"Arms, for what?" became the latest revolutionary expression, and later
that night the DR guerrillas, watching Castro on television, surrendered
the weapons to the Rebel Army, ending the crisis without bloodshed.
Shortly, key DR leaders would join Fidel's circle, becoming the mainstay of
his Security Services as he ushered in socialism in Cuba. At Camp Columbia, Castro consulted the crowd whether he should accept the "petition"
addressed to him by the provisional government to serve as Commander in
Chief of the land, sea, and air forces, and to reorganize the armed forces,
and received a unanimous "Yes!" Then he informed the nation that while
"decent soldiers who have not stolen and who have not assassinated" would
have the right to continue in the army, "I'm also telling you that those
who have assassinated will not be saved by anybody from the firing squad."
This was his way of serving notice that trials and executions of Batista "war
criminals" would soon start. Castro ended his first great public performance
as the leader of Cuba with these words: "For us, principles are above my
other considerations and we do not struggle because of ambitions. I believe
we have demonstrated sufficiently that we have fought without ambitions. I
believe no Cuban has the slightest doubt about it!"

As he finished speaking, the spotlights bathing him illuminated a pair of
white doves that suddenly came to rest on his shoulders. This astounding

symbolism touched off an explosion of "FIDEL! . . . FIDEL! . . . FIDEL!"
as the night was caressed by the first colors of the dawn. Cubans are a
people with powerful religious and spiritistic superstitions, going back to
the Afro-Cuban traditions of slavery, and that night in January confirmed
their faith: The dove in Cuban myths represents life, and now Fidel had
their protection. And, as it happened, doves again and again would alight
on Fidel's shoulder as he faced his people. The deification of Fidel Castro
became a phenomenon in Cuba in the aftermath of his victory, so greatly
had he touched the hearts and souls of the people. Soon, *Bohemia* magazine
published an immensely controversial portrait of the thirty-one-year-old
Maximum Leader with a Christlike halo subtly drawn around his bearded
countenance. Some Cubans thought this was overdoing political allegiance.
But Raúl Chibás, who traveled with Castro a part of the way from Oriente
to Havana, recalls that between Santiago and Bayamo, "elderly ladies embraced him as he went along. . . . Every five minutes, at every intersection
of the highway, women stopped him, the old women kissed him, telling
him he was greater than Jesus Christ."

Castro himself must have felt a profound kinship with Christ; in a pre-
Easter speech in March he intoned: "Because there are those who say they
are Christians and are racist. And they are capable of crucifying one like
Christ because one tells the truth to an insensitive and indolent society.
Because Jesus Christ—and I don't want to compare myself even remotely, I
don't want to compare myself in the least—because what I say is, why did
they crucify Jesus Christ? It is good that we should speak of this during
this Holy Week. They crucified Christ for something. And it was simply
because He defended the truth. Because He was a reformer within that
society, because within that society He was a whip against all that Phar-
isaism and all that hypocrisy. Because for Christ there was no difference of
race, and He treated the poor the same as He treated the rich, and the
black the same as the white. That society, to which He told the truth, did
not want to forgive His preachings, and they ended up simply crucifying
Him because He told them the truth."

Antonio Nuñez Jiménez says that later in 1959, in a "secret speech"
before officials of the new Agrarian Reform Institute, Castro said, "The
Revolution . . . ceased to be that romantic thing to become that in which
there is only room for those who are suffering the metamorphosis of conversion into revolutionaries, and are in accordance with that precept of
Christ when He said: 'Leave all that you have and follow me.' This is the
reality." In a televised speech in December delivered in defense of the revolution, Castro said he made a point of attending a Roman Catholic congress
in Havana because "our Revolution is in no way against religious sentiment
. . . our Revolution aspires to strengthen the noble desires and ideas of

men. . . . When Christ's preachings are practiced, it will be possible to say that a revolution is occurring in the world. . . . Because I studied in a religious school, I remember many teachings of Christ, and I remember He was implacable with Pharisees. . . . Nobody forgets that Christ was persecuted; and let nobody forget that He was crucified. And that His preachings and ideas were very much fought. And that these preachings did not prosper in high society, but germinated in the heart of the humble of Palestine . . ." Twenty-five years later, Fidel Castro continued to invoke Christ as his role model, and Christianity as the philosophical basis of the Cuban socialist revolution.

This socialist revolution was secretly set in motion through Fidel Castro's "hidden government" and his clandestine dealings with the "old" Communists within days of his arrival in Havana. Having ostensibly no governmental responsibilities during the first six weeks of the new regime, being "simply" the military Commander in Chief, Castro could engage in these enterprises without attracting undue attention. In any event, his whirlwind activities in and out of Havana were a perfect cover. During the month following his arrival in the capital on January 8, Fidel delivered at least twelve speeches, some of them major policy statements before huge crowds, and declarations to various groups; he held five major news conferences, mainly for foreign journalists, and made two lengthy television appearances (the speeches and press conferences were televised, too). At the end of January, he flew to Caracas on his first foreign trip as the victorious revolutionary leader to thank Admiral Larrazábal and the ruling Venezuelan government junta for dispatching arms to the Sierra in 1958; he also called on President-Elect Rómulo Betancourt despite the contempt he had for him as a reformist (and not a revolutionary) of the Latin American "Democratic Left." Venezuelans, liberated from dictatorship a year earlier, gave Castro a deliriously happy reception. He made a quick visit to Artemisa in Havana province and to Pinar del Río, the two areas from which most of his Moncada companions hailed, and spent four days in Oriente in the foothill towns of the Sierra Maestra, telling people about land reform. And on February 9 he announced the revolutionary regime's decision to declare Argentine-born Che Guevara a *native* Cuban citizen as an act of gratitude and as the legal step required to allow him to hold office in Cuba. All in all, it was hard to keep up with Fidel as he burst in and out of his penthouse suite on the twenty-third floor of the Havana Hilton Hotel, which served at the outset as occasional home and main office.

Meanwhile, his secret political operations worked on two levels simultaneously, that in the end became fused into one level, when socialist "revolutionary unity" was implanted in Cuba eighteen months later. On the first level were contacts and negotiations with the old-line Communist leadership of the Popular Socialist Party, growing out of Castro's lengthy discussions with Carlos Rafael Rodríguez in the Sierra Maestra in the latter part of 1958. On the other level was the "hidden government." That Castro made the decision to seek Communist collaboration before Batista's fall is corroborated by Fábio Grobart. Now in his eighties, the cofounder of the Cuban Communist party and presently its historian and the oldest member of the Central Committee, Grobart recalls that these consultations began "in the first days" of the new regime. However, the secret conversations between Castro, his associates, and the Communist leadership were not a short cut to a power-sharing deal, but highly complex debates on how a unified revolutionary party could be fashioned into a Marxist-Leninist force, and, in the meantime, how Communist talents could best be used in running the country and preparing the transition. Castro insisted from the outset that the "old" Communist party be incorporated into a new one under his leadership, thus requiring the actual delivery of the party to him, an unprecedented act in Communist history.

It goes without saying that Castro engaged in this process as part of his broader judgment that the moderate regime under Urrutia was transitory, unacceptable in the long run as an instrument of the revolution, and this is why he had to create a "hidden government" to move the nation rapidly along the revolutionary road while the unity concept with the Communists was being ironed out. Moreover, consultations had to be conducted in absolute secrecy because of immense ideological sensitivities shared by both the Communist party and the 26th of July Movement, and their deep resentments and mistrust of each other. Neither Castro nor the top Communist leaders meeting with him could admit that, in effect, they were involved in liquidating their respective political organizations in their present form. All these considerations existed quite apart from what might be expected from the United States if it became known that the Maximum Leader was in business with communism. This caution was expressed in a private wartime remark by Castro that "I could proclaim socialism from the Turquino peak, the highest mountain in Cuba, but there is no guarantee whatsoever that I could come down the mountain afterward."

As Fábio Grobart put it in a long discussion of the Castro-Communist relationship, "A process, taking months and years, was necessary to prepare public opinion for the necessity of having a unified Communist party, and that communism is not so grave, so dangerous, so bad . . ." But in 1959 the orthodox Communist party was not ready for Castro, either. On January 11 the party's Executive Bureau issued a declaration urging the defense of the revolution and the maintenance of revolutionary unity, but only in August of the following year did the party formally recognize its longstanding "errors" in minimizing and misunderstanding Castro's movement from the time of the Moncada attack. Pending this act of contrition, the top leadership had to tread very carefully, and Fábio Grobart says it was impossible to state publicly in the first months of 1959 that all these meetings were occurring. Even after the decision to form a unified Communist party under Castro was announced, some important old-line Communist leaders sought to sabotage it to the point where several of them were imprisoned for "conspiracy."

Castro held most of these clandestine encounters at a hilltop house in the fishing village of Cojímar, some ten miles east of Havana. This house was lent to him for an indefinite period in March by Agustín Cruz, a former *Ortodoxo* party senator, though the first meetings took place in private Havana homes. The large Cojímar villa, overlooking the sea from a distance, was under heavy Rebel Army guard, affording maximum privacy. Fidel used it as a residence, alternating with Celia's Vedado apartment and the Hilton penthouse, during the first years of the revolution. In his meetings with the Communists, Castro was always accompanied by Che Guevara, Camilo Cienfuegos, Ramiro Valdés, and often by Raúl, who commuted between Santiago and Havana. Cienfuegos, the Army Chief of Staff, appears to have been a "closet Communist" during the war, identifying with Marxism during the Las Villas campaign the previous autumn. His brother Osmany, an architect who sat out the war in Mexico, belonged to the party. Ramiro Valdés, veteran of Moncada, the prison, the *Granma,* and the Sierra, had been Che Guevara's deputy at the end of the war, and was now chief of the Rebel Army's investigations department, the G-2 (i.e., secret police). Valdés was an unabashed admirer of communism and the Soviet Union. Raúl Castro had been a party member since 1953, and Che Guevara was far to the left of all the Communist parties. Fidel, then, was the only one without an open Communist commitment. The party's negotiators were led by Blás Roca and its secretary general since 1934, and included Carlos Rafael Rodríguez and Aníbal Escalante of the Executive Bureau. All of them were considerably older than the *Fidelistas,* and were held in some awe by the young rebels—except by Fidel.

Blás Roca, who was seventy-seven years old and recovering well from a stroke when he agreed in 1985 to reminisce about the past, was the first to meet with Castro after the revolution, keeping up personal exchanges in addition to the group discussions. He says that "we began to hold meetings as soon as Fidel, Che, and Camilo arrived here," and remembers Castro, laughingly exclaiming, "Shit, now we are the government and still we have to go on meeting illegally." On another occasion, there was great merriment when Che remarked, "Yes, things have really changed now that we have an agenda before us." Blás Roca says that in those days the party's rank and file were not told that the top leadership had come to regard Castro as the principal revolutionary leader of Cuba: "We were not informing the militants, only a small group in the leadership." Likewise, according to Blás Roca, the party leadership refrained from informing the rank and file that Castro was regarded as a socialist and a Marxist because "the success [of the negotiations] was linked to the need of preventing the Americans from having a banner for an intervention, as they had done in Guatemala, and we had to go on maintaining the secret that had prevailed until then and had contributed to the success." However, top Communist leaders, he says, began educating party organizations to accept Castro's decisions on government appointments, stressing that party membership did not confer special rights during revolutionary periods, contrary to the belief of many activists. Blás Roca says that at labor-union conferences, he made a point of telling the workers that "a new leader of the Cuban working class was born, and this new leader is Fidel."

Fábio Grobart recalls that in time the meetings between the *Fidelistas* and the Communists became institutionalized. "There was," he says, "a coordination of activities and a collaboration. This was the beginning." Late in 1959 or early in 1960, Castro and the Communists concluded that the time had come to move ahead with the organization of a unified Communist party, but Fábio Grobart points out that the first step was setting up the Integrated Revolutionary Organizations (ORI) by bringing together the 26th of July Movement, the Popular Socialist Party, and the Students' Revolutionary Directorate. Each party maintained its identity and autonomy, he says, though Castro's overall leadership was recognized. In 1961 the three organizations were formally fused as a prelude to the establishment of the new Communist party in 1965. The real birth took place, however, in 1959, in the villa at Cojímar, the fishing village where Ernest Hemingway had found his old man and the sea.

Among the first decisions Castro and the Communists took together was to create special schools to teach Marxism-Leninism to the *Fidelistas,* particularly those with an obvious political future, in preparation for the ultimate transition to Communist rule on the island. They were called "Revolutionary Instruction Schools" (EIR), which at first disguised Marxist teachings behind a façade of showing officials how to run revolutionary institutions. In fact, they were the counterpart of the military political-education centers that were set up at the Havana commands by Camilo Cienfuegos and Che Guevara, and subsequently expanded to all the Rebel Army units. These centers grew out of the Troop Instructors School conducted by Communist officers in Raúl Castro's "Second Front" in 1958, and were run by members of the Popular Socialist Party and officers with membership in the party. This was consistent with Castro's principle that the Rebel Army must play a leading ideological role in the revolution, and the basic military text was the *Civic Preparation Manual* issued late in 1959. It was used in the Rebel Army's literacy program, and its language was generally Marxist, stressing "anti-imperialist struggles."

On the civilian side, the first Revolutionary Instruction School was established late in 1959, in a house on Primera Avenida in the Playa section of western Havana, and a full network of these schools was officially inaugurated in December 1960. The first director of the school was Lionel Soto, the head of Socialist Youth at Havana University in the 1950s and one of Castro's close friends, and the first teachers were Communist party intellectuals, like Raúl Valdés Vivó, and its top leaders, such as Carlos Rafael Rodríguez, Blás Roca, and Lázaro Peña. Fábio Grobart says that "the principal heads of the Revolutionary Instruction Schools came in the beginning from the Popular Socialist Party because they had the greatest experience in teaching and organizing such schools."

With the unveiling of the Integrated Revolutionary Organizations in 1960, the Marxist-Leninist schools became crucial in equipping the "new" Communists for their tasks in the unified party soon to be set up. In Grobart's words, "The person graduating from this school is a *cadre* prepared in multiple ways to be a political leader of the revolution." Over the years, these schools became centralized under the Ñico López Central School of the Cuban Communist party, which is in effect a Marxist-Leninist university offering everything from three-month basic courses to a five-year doctorate in the social sciences. All top Cuban officials must be Ñico López graduates (José Ramón Fernández, Cuba's vice-president and education minister, for example, went back to school in his fifties to earn this degree), and the curriculum includes scientific communism and atheism, party construction, ideological struggle, universal history, Cuban history, philosophy, the political economics of socialism, and the political economics of capitalism. By late 1961 over thirty thousand persons went through the indoctrination schools, but the elite was a class of fifty-three of the most promising young leaders who, starting in January 1962, were given exhaustive nine-hour-a-day courses in Marxism, economics, and philosophy. In March Fidel Castro came to the school to pick from this class a secret task force of twenty young officials to supervise a shift from the ORI to the United Party of Cuban Socialist Revolution (PURSC), a shift that just had been announced publicly and constituted the transitional stage to the new Communist party. By the end of 1960, there were still no Communists in the Cuban Council of Ministers, but Marxism-Leninism had made immense inroads in Cuba. This was certainly true of Fidel Castro's "hidden government" in 1959.

*   *   *

This group carried the innocent-sounding name of the Office of Revolutionary Plans and Coordination, in case questions should be asked, but its existence was virtually unknown outside the most intimate circle around Fidel Castro. Operating as a secret task force carrying out fundamental policy assignments for Castro, its chairman was Antonio Nuñez Jiménez and Che Guevara, Alfredo Guevara, Vílma Espín, Oscar Pino Santos, and Segundo Ceballos were members. Nuñez Jiménez, who knew Castro slightly from their university days, joined Che Guevara's column in Las Villas during the fall 1958 offensive, then, as a Rebel Army captain, became his deputy at the La Cabaña fortress. Ardently committed to Marxist-Leninist thought, Nuñez Jiménez was brought into Castro's personal entourage by Che Guevara, immediately turning into a full-time companion and trusted planner. As a geographer, geologist, and historian, he was (and is) very familiar with Cuba's problems—just the man Castro needed intellectually and ideologically for the transition.

Alfredo Guevara was Fidel's Communist friend from the university, his companion at the 1948 Bogotá uprising, and a victim of torture by Batista police in Havana in the last year of the war. He then went to Mexico, and Fidel had his sister Lidia summon him from Matanzas where he had just returned from exile, while still en route from Oriente to Havana during the first week of January. Alfredo Guevara, who had become a moviemaker, had hoped to launch a revolutionary motion-picture industry, but Fidel told him he was urgently needed for other tasks. Vílma Espín was an MIT-educated young woman from Santiago who joined Raúl Castro's "Second Front" in 1958, then married him in January 1959 at an Oriente wedding Fidel was too busy to attend. Raúl's presence with the army in Santiago was still required, but he was nevertheless deeply involved with the "hidden government" as he commuted between the two cities. Oscar Pino Santos, a Communist economist, and Segundo Ceballos, an elderly journalist who specialized in agrarian problems, were advisers but never participated in policy decision. Pedro Miret was Fidel's *aide-de-camp*, increasingly involved in the secret planning as the initial group evolved during 1959 into a full-fledged "hidden government." Celia Sánchez was Fidel's principal assistant.

The task force met at a house at the beach resort at Tarará, where Che Guevara was convalescing from illness and fatigue; two years of asthma attacks and malaria bouts during the Sierra war had ravaged him physically. Tarará was a half-hour drive from Havana, a few miles east of Cojímar where Fidel moved during March. The main assignment of the Tarará team was to draft, also in secret, a new agrarian reform law, much more drastic than the one Castro had signed in the Sierra the previous year, as well as additional revolutionary laws, and to become familiar with crucial areas in government operations in preparation for the ultimate takeover. Nuñez Jiménez says that "for two months, we held meetings during the night in Tarará where Che was recovering his health." Castro, he says, kept track of the drafting of the agrarian reform law, the centerpiece of the revolutionary legislation, "suggesting ideas and modifications." According to Nuñez Jiménez, the drafting was kept secret until Castro presented it to Revolutionary Laws Minister Dorticós for a review, bypassing the rest of the cabinet; Dorticós was a Castro ally.

Alfredo Guevara offers the best description of the latitude and mandate enjoyed by the task force. He says that "we met every night until dawn at Che's house, then Fidel would come and change everything" in the land-reform bill, "but we also prepared a merchant marine law, and we had to become specialists in the craziest things; for example, we began to work in the National Bank." Felipe Pazos was the regime's new National Bank president (he, Raúl Chibás, and Castro had drafted the first Manifesto from the Sierra in 1957), but, as Alfredo Guevara recalls, "Castro wanted us to start going to the bank, and we went there once a week. . . . Fidel kept saying, 'We don't know what a bank is, and we must know what a bank is.'" Subsequently, Che Guevara would replace Pazos as National Bank president.

"Nobody knew what we were doing," Alfredo Guevara says. "For example, the minister of agriculture did not know we were preparing the agrarian reform law, and nobody else knew it." The minister at the time was Humberto Sorí-Marín, who had drafted the Sierra law, and was later shot for conspiring against the Castro regime. But Guevara recalls, "We discovered in our discussions that nobody had any knowledge, that everybody favored agrarian reform, but nobody really had mastered what needed to be done in that field. . . . Merchant marine was another subject we had no knowledge about."

The activities of the "hidden government" changed and grew when Castro persuaded President Urrutia to obtain the resignation of José Miró Cardona as prime minister and to appoint him, Fidel, to this post on February 13, an easy undertaking. The circumstances did not allow Miró Cardona to be effective in the premiership, mainly because most ministers cleared their projects first with Castro privately at the Havana Hilton penthouse suite. Urrutia wrote before his death in exile that Castro had visited him several times early in February to say he would accept being prime minister, "but since he would be responsible for the policy of the government he would need sufficiently broad powers to enable him to act efficiently." Carlos Rafael Rodríguez, writing many years later about these events, put it more brusquely: "The government that emerged on January 1, could not be considered a true revolutionary government in the light of its composition or its procedures. . . . Revolutionary power at that moment resided outside of the government—in the rebel army headed by Fidel Castro. His designation as prime minister served to fuse together revolutionary power and the government." But Castro wanted to keep all the ministers in the cabinet for a time, and rejected Urrutia's offer of resignation. He was still proceeding step by step, although he obtained from Urrutia and the cabinet a change in the new Cuban constitution, giving the prime minister the power to direct government policy; that constitution had been approved by the cabinet only six days earlier on February 7.

Urrutia's power was reduced to signing laws, even though, as he wrote, Castro "conceded me veto power but asked that I use it as seldom as possible." From thereon, Castro began to preside over cabinet meetings at the presidential palace, with Urrutia in mute attendance, and with the Tarará task force acting as the invisible coordinator of policy. Alfredo Guevara says that Castro put him in charge of summoning the cabinet and helping him to run it. Shortly thereafter, Castro moved to Cojímar, where the task force set up its headquarters with still greater secret power. Che Guevara had recovered completely, and the Tarará site could be dropped. The first stage of the Cuban revolution had been completed with Castro's open assumption of total power. Now he and his teams prepared for the next phase.

CHAPTER

2

Yet, the immediate and overwhelming reality facing Fidel Castro's Cuba was the relationship with the United States, only ninety miles away. The antagonism between the new Cuban revolution and the Americans next door was instant, implacable, powerful—and inevitable. What Castro had touched off on both sides of the Straits of Florida was an explosion of nationalisms, historical resentments and misunderstandings, sharply differing perceptions of national interests, and an earth-shaking cultural shock for which neither side was even remotely prepared.

This antagonism, soon turning into open mutual hostility, predated the great Cuban-American clashes and confrontations stemming from Castro's choice of the Communist road to the revolution, the questions of ideology, and the ultimate Cuban-Soviet military alliance. Seen from a perspective of more than a quarter-century, it is evident that the antagonistic relationship was, in effect, foreordained by the forces of history, and there was virtually nothing either side could have done to avoid the collision course within the parameters of what was then politically possible. In a nutshell, Fidel Castro obsessively feared that his revolution would be stolen from the Cubans by the United States, as independence was stolen at the end of the Spanish-American War in 1898, while Americans (and not only the Eisenhower administration) saw in the *barbudos's* cry of defiance ominous threats to their national and economic interest, and were not always capable of distinguishing among real and imagined threats.

These fundamental attitudes, then, defined from the very beginning the behavior of the big and the little neighbor toward each other. It would be an enormous historical error, however, to conclude that had Castro and Eisenhower (and their political constituencies) acted in a less paranoid fashion the final outcome might have been different. The fact was that the United States could not tolerate a revolution it was unable to influence or control immediately south of Key West, Florida. Also, Castro, rejecting the traditional "geographic fatalism" that always made Cubans discard national decisions potentially objectionable to North Americans, was determined to assert his total independence. In his mind this freedom of choice already included implanting Marxism-Leninism in Cuba, although he was still far from ready to proclaim it. Therefore, no lasting compromise was ever possible, and it is demonstrably incorrect to believe that American actions pushed Castro toward communism or that, obversely, the United States resolved to try to oust him only after he had molded his revolution into an anti-American and pro-Communist instrument.

There is abundant evidence that Castro's aim in the immediate aftermath of victory was to forge revolutionary unity around his own Communist party when practicable, and there is no reason to doubt his comments on this theme twenty-seven years later: "We were carrying out our program little by little. All these [United States] aggressions accelerated the revolutionary process. Were they the cause? No, this would be an error. I do not pretend that the aggressions are the cause of socialism in Cuba. This is false. In Cuba we were going to construct socialism in the most orderly possible manner, within a reasonable period of time, with the least amount of trauma and problems, but the aggressions of imperialism accelerated the revolutionary process."

In Washington the mind-set was equally firm. Even before Eisenhower policy-makers began to understand what was happening in Cuba (and they never succeeded in really understanding it, as the Bay of Pigs invasion two years later would show), a top-level decision was made to get rid of Castro. Specifically, the secret agenda of the National Security Council meeting on March 10, 1959—two and a half months after Batista's defeat and with President Urrutia and a moderate cabinet still ostensibly governing Cuba— included as a principal topic the modalities of bringing "another government to power in Cuba." The Cubans had not yet seized or nationalized any American property on the island, and the United States had no reason thus far to complain about any Cuban actions. In fact, the official policy was to *appear* to be friendly to Castro; on January 7 the United States was the second country in the world (after Venezuela) to recognize the revolutionary regime. The Soviet Union had no diplomatic relations with Havana, and seemed to ignore altogether Castro's struggle against Batista. Philip W. Bonsal, a U.S. career diplomat with a liberal reputation, an excellent knowledge of Latin America, and fluency in Spanish, was immediately named ambassador to Cuba to replace Earl T. Smith, the friend of Batista. Bonsal (who had served in Cuba as a young diplomat and whose father was a war correspondent in 1898), met Castro at Cojímar on March 5, the day after presenting his credentials to Urrutia, and their first conversation was pleasant. The ambassador wrote later that "I was encouraged to believe that we could establish a working relationship that would be advantageous to both our countries," and "Castro had gone out of his way to express a warm desire for frequent meetings with me." The following day, Castro spoke during a television speech of his "cordial and friendly conversations with the ambassador of the United States," and disclosed his plans to visit the United States the next month as a guest of American newspaper publishers (but in December 1961, Castro offered a different version in another TV speech, accusing Bonsal of "a style of someone who came to deliver instructions").

It remains a mystery why the National Security Council discussed Castro's liquidation within five days of his first encounter with the American ambassador without giving diplomacy a chance. The council did not know that Castro was not amenable in the long run to a cordial relationship with Washington since he undoubtedly realized that it was impossible in the light of his revolutionary plans, so the whole discussion made little sense—unless the Eisenhower administration had a secret second plan to derail the revolution as a matter of principle. A similar approach had worked in Guatemala in 1954, under the same White House and CIA leadership, and it may have seemed easy to rerun the operation in Cuba five years later. That such a policy undermined Ambassador Bonsal, who was not informed of it, posed no problems to the policy-makers: it was par for the course in this type of situation.

However, the Intelligence Community was split on Cuba. Late in March, for example, a special panel on Cuba in the CIA's Board of Estimates concluded in a secret review that Castro was not "a Moscow-oriented Communist" at that time, a correct estimate in which Ambassador Bonsal had concurred. The full board rejected that conclusion under pressure from CIA Director Allen W. Dulles, the architect of Guatemala, but as late as November 5, 1959, General C. P. Cabell, the Agency's deputy director, testified before the Senate Internal Security Subcommittee that even the Communists in Cuba did not consider Fidel Castro "a communist party member, or even pro-communist. . . . We know that the communists consider Castro as a representative of the bourgeoisie, and were unable to gain public recognition or commitments from him during the course of the revolution." In retrospect, it is obvious that these opinions were culled by the Havana CIA station either from low-ranking party members, unaware of the secret talks between the *Fidelista* command and their leadership, or from those in the know who wished to continue the impression that Castro was unsullied by Marx and Lenin. It was not until March 1960 that the United States government formally decided to mount a paramilitary operation against Castro, the CIA had begun to arm the first anti-*Fidelista* guerrilla bands appearing in the Escambray Mountains in central Cuba in the latter part of 1959. These guerrillas were mainly ex-Batista soldiers, wealthy peasants, and a sprinkling of rightist ideologues and adventurers; just as these bands started posing a problem for Castro, the CIA abandoned them as incongruously as it had initially encouraged them.

In the spring of 1959, as Fidel Castro prepared to go to the United States in triumph, the real tensions developing between the two countries had to do with an endless series of daily irritants, feeding on each other, rather than with broad government policies. Again, emotional and psychological factors colored the relationship—and the gulf of cultural misunderstanding grew. The problem was that Americans insisted on judging Cuban attitudes by American standards, and Cubans responded by judging Americans by Cuban standards—in a quickening and damaging vicious circle. Nowhere was this phenomenon more obvious than in the case of the trials of the Batista "war criminals" by revolutionary courts.

Reduced to its simplest terms, the issue in the eyes of Fidel Castro and masses of Cubans was that while the United States government never protested the killing and torture of thousands of the old regime's opponents by the Batista police and soldiers (and U.S. public opinion hardly took notice of it), Americans were now indignant that the victorious revolutionaries were punishing culprits with executions and lengthy prison sentences. Neither Castro nor a great many other Cubans who were not normally bloodthirsty could comprehend why, in effect, a double standard was being applied to them, and why the United States was so roundly and vociferously condemning the revolution for enforcing its notion of justice. The climax in this tragedy of misunderstandings came when Wayne Morse, the great liberal from Oregon, rose on the Senate floor to denounce the Cuban "bloodbath" and urge that executions be halted "until emotions cool." Castro, who saw a threat to his revolution in any and every criticism of Cuba, shot back that justice would proceed "until all criminals of the Batista regime are tried," and that "if the Americans don't like what's happening in Cuba, they can land the Marines and then there will be

200,000 *gringos* dead . . ." That was an off-the-cuff remark to a crowd of newsmen, and Castro had the sense to apologize for it, but the threat made worldwide headlines, and the atmosphere became even more poisoned. It was only January, less than a month after the revolutionary triumph.

There is no question that Castro used the trials issue, an immensely emotional one to thousands of families whose sons, brothers, and husbands had been mutilated and killed, to orient Cuban public opinion against the United States. It obviously could not have been planned that way, but this whole question served to accelerate as well as justify the extreme radical course he was giving his revolution. In this sense, Americans were playing into his hands, but it is just as important—historically—to place the episode of the trials into perspective. The first point is that in the normally accepted meaning of the expression, no "bloodbaths" occurred in Cuba after Batista fell. In other words, no vengeful crowds took revenge into their own hands, and former Ambassador Bonsal, a most objective observer, wrote afterward that "thirty years earlier, the hirelings of the Machado regime deemed guilty of similar crimes were simply ferreted out by the mob and killed. . . . The Castro procedure of setting up special tribunals to try the cases of people who, on the basis of the Nuremberg principles, were accused of serious crimes could have been an improvement over the earlier method. . . . These special courts were subject to all sorts of pressures including those generated by the circuslike atmosphere in which many of them were conducted."

Castro acknowledges that around 550 of these Batista "criminals" were executed after summary courts-martial and then by special revolutionary tribunals in 1959 and 1960 (the special courts were abolished after the first six months or so, then reactivated late in 1959 to deal with "counterrevolutionaries" who were beginning to emerge, sometimes with CIA assistance). Whatever can be said of the procedures before these tribunals, defendants were not picked at random, but because they were believed to have committed crimes and brutalities on a large scale, and so were punishable under the provisions of revolutionary laws proclaimed from the Sierra in 1958. Cuban revolutionary trials, then, bore no resemblance to the real bloodbaths that followed the Mexican, Russian, and Chinese social revolutions in the twentieth century—or to the vengeance-in-the-streets that erupted in Cuba after Machado, in France, and other Nazi-occupied nations after liberation in World War II, or in Venezuela following the deposition of the dictator Pérez Jiménez in 1958 and in the Dominican Republic after Dictator Trujillo was murdered in 1961. By the same token, the Cuban revolution refrained from institutionalized mass killings such as those perpetrated against hundreds of thousands ethnic Chinese in Indonesia in the aftermath of the 1965 army anti-Communist coup, or those thousands attributed to Chilean military authorities when they overthrew the Marxist president, Salvador Allende Gossens, in 1973. Considering that in the first few days of revolution, public order in most of Cuba was assured by 26th of July Movement local militias, Boy Scouts, and advance units of the Rebel Army, it is quite remarkable that violence-prone Cubans remained so unviolent.

Castro's worst mistake, however, was to hold a "show trial" in Havana's sports stadium in January for three exceedingly brutal Batista ex-commanders. He may have thought that such public televised proceedings would both defuse the temptation for a private enforcement of justice and show the people that Batista's crimes were being swiftly judged. From the viewpoint of international public opinion, however, it was an unmitigated disaster and created the legend of "circuslike atmosphere" to which Bonsal alluded (in fact, only one trial was held in the sports stadium). The principal defendant was ex-Major Jesús Sosa Blanco, charged with scores of murders in Oriente and famous for brutality, but this was not a kangaroo court. The three judges were Agriculture Minister Humberto Sorí-Marín, a Catholic lawyer who was the Rebel Army's judge-advocate (later shot for counterrevolutionary activities), Raúl Chibás, the moderate treasurer of the 26th of July Movement (who fled Cuba two years later), and Castro's wartime companion, Universo Sánchez. Sosa Blanco and his codefendants had defense attorneys. However, Sosa Blanco contributed to the circuslike atmosphere by shouting that he was subjected to a "Roman circus" procedure. He was sentenced to death, then resentenced later at a more tranquil trial in a military courtroom.

Raúl Chibás, who at Castro's personal request had served as judge in other trials, says a quarter-century later at his exile home in the United States that these procedures were totally justifiable. "Sincerely, I was in agreement with these trials," he says. "I was in agreement beforehand, and I discussed with Fidel {in the Sierra} the need for carrying out justice after Batista's fall . . . on the grounds that if there is no justice, people would enforce it as in the time of Machado when mobs dragged corpses through the streets. . . .When one tries to do it legally, adverse reaction sets in. Unfortunately, this is people's hypocrisy, and we talked about it with Fidel in the Sierra. . . . They {Batista officers} had been forewarned because we issued proclamations from the Sierra that justice would be applied to all those who robbed and murdered the peasants, and I thought we had to act accordingly." Chibás thinks that the idea of holding the trial in the stadium was a "boomerang," but "Fidel wanted to demonstrate that these were assassins, that they had committed over 100 murders, that they had raped women . . ." Castro's next error was to demand a new trial in March when a revolutionary court in Santiago acquitted forty-four aviators from the Batista army for bombing peasants in the Sierra. He announced on television that "revolutionary justice is not based on legal precepts, but on moral conviction." Major Manuel Piñeiro Losada, a founder of the Castro secret police, was named chairman of the tribunal, at which all the fliers were sentenced to varying prison terms.

There remains a philosophical aspect to the question of the trials, which were decisive in poisoning the Cuban-American discourse. The central point is that Castro believes that the revolution had every right to hold them on the basis of wartime revolutionary laws, while American opinion in 1959, a sanctimonious view that ignored Cuban emotions and demanded Anglo-Saxon legal procedures, denied the Cubans this right. Apart from the fact that Cuban law is based on the Napoleonic Code, which (unlike the English common law) places on the defendant the burden of establishing his (or her) innocence, Castro takes the view that "in the Sierra Maestra, when we were an embryo state, we wrote a penal code to punish war crimes . . . when the Revolution triumphed, the courts of the land accepted these laws as applicable laws, validated by the victorious Revolution, and the tribunals tried many war criminals who could not escape . . ." Castro says that "this started the first campaigns abroad against Cuba, especially in the United States, which realized quickly that we had here a different government, not a very docile government, and which began furious campaigns against the Revolution." Rufo López-Fresquet, the moderate and pro-American treasury minister during the first fourteen months of the Castro regime, has written that "the foreigner, especially the North American, put his emphasis on the legal aspects of the revolutionary trials," but "the Cuban was interested in moral justice . . . When a man who has boasted of killing dozens of men while protected by his Batista uniform was executed, the Cubans believed justice was served. The rest of the world concentrated on criticism of the revolutionary judicial process. Perhaps both were right, but they were miles apart. Not many calm voices dedicated themselves to explaining these differences . . ."

(The trials should not have surprised Cubans or Americans: Early in February 1958, a year before the war ended, *Look* magazine in New York and *Bohemia* in Havana published a photo-reportage piece on "Justice in the *Sierra*," showing Castro sitting informally on the ground while interrogating prisoners charged with murder and rape before a "revolutionary tribunal," trials lasting twelve days, and Raúl Castro commanding a firing squad.)

Consequently, anti-Cuban campaign did begin to shape up in the American government, in Congress, and segments of the media, and Castro let nothing escape his attention. Convinced that a "Plattist mentality" still existed among many Cubans and Americans—the belief in the applicability of the long-abrogated Platt Amendment granting the United States the right to intervene in Cuba—he saw conspiracies and his self-fulfilling prophecies coming true. When Americans condemned the trials, Castro reacted with accusations that the United States had granted asylum to the worst Batista "war criminals," which was true, and that they would plot against the revolution, which also was true.

From his first week in Havana, Castro used every speech to tell the United States as plainly as possible that it no longer had a say in Cuba. Before the Lions Club on February 13, he reminded his hosts that "the Platt Amendment is finished," that the revolution was already being attacked in the United States, and that Cubans had the right to trace their own destiny and "do things better than those who spoke of democracy and sent Sherman tanks to Batista." Two days later, he told the Rotarians that "nobody can intervene here because sovereignty is not a favor that is granted us, but a right we deserve as a nation." The following afternoon, he touched off approving roars of hundreds of thousands attending a rally at the presidential palace when he warned that if the United States wished to have good relations with Cuba, "the first thing they have to do is to respect her sovereignty." At a mass rally at the palace five days later, Castro reopened the question of the validity of the trials by demanding the huge crowd to raise their hands "if you agree that the murderers must be executed"—a forest of hands rose over the plaza, and Fidel said: "The jury of one million Cubans of all ideas and all the social classes has voted! . . ."

During March Castro engaged in political activities that were contradictory, to say the least, if he truly wanted a favorable shift in United States public opinion about the Cuban revolution. Accepting an invitation from the American Society of Newspaper Editors to speak to their annual meeting in Washington (thus allowing him to visit the United States without an invitation from the Eisenhower administration) and launching "Operation Truth" for the benefit of American newsmen willing to see Cubans as official guests, Castro appeared to be courting Americans. But at the same time, he had "intervened" in the American-owned Cuban Telephone Company, which meant that the regime took over management in order to investigate its operations. This was consistent with Castro's statements dating back to Moncada about Cuba's need to operate its own utilities, but the political timing was awful in terms of his approaching United States trip. Next, he publicly embarrassed José Figueres, the former president of Costa Rica and one of his early supporters, by accusing him of intolerable "imperialist" tendencies for having suggested at a rally in his presence that in the Soviet-American struggle, there could be only one place—alongside the United States. With that, Fidel Castro departed for the United States.

Most new Latin American leaders made a pilgrimage to Washington as soon as possible to curry official favor and seek emergency economic aid.

However, Fidel Castro was an exception in his refusal to ask for money, or even talk about it. That threw American officialdom off balance. Rufo López-Fresquet, the finance minister, who was among the hundred-plus Fidel entourage on the trip, recounted in his memoirs in exile this conversation with Castro: "I don't want this trip to be like that of other new Latin American leaders who always come to the U.S. to ask for money. I want this to be a goodwill trip. Besides, the Americans will be surprised. And when we go back to Cuba, they will offer us aid without our asking for it. Consequently, we will be in a better bargaining position." López-Fresquet replied that "the reasoning was not completely illogical," although at that point the government's cash reserves were below $1 million. What Castro did not confide in his minister, however, was his determination to demonstrate to Cubans and other Latin Americans that he could appear in the United States on his own terms (and without an official government invitation), act as an equal, and not tarnish himself with money discussions in which he, as the poor party, would be the *supplicant*. The American visit, in Fidel's mind, had to underscore the absolute independence of the Cuban revolution from the United States, mainly because he already saw himself as the great hemispheric leader. Nevertheless, he brought along all his top cabinet economic and financial experts.

Castro's visit to the United States, which was primarily an exercise in public relations and an attempt to educate North Americans about Latin American nationalism, began on the evening of April 15, amid the chaos and confusion characterizing all political activities. He was two hours late in boarding his special Cubana Airlines plane for Washington, keeping everybody waiting on both sides (he told a nervous adviser: "We are going to be in the United States fifteen days; what difference does an hour or two make?"). Clearly, Fidel was savoring the moment. His last visit to the United States had been nearly four years earlier, when, as an impecunious and rather obscure revolutionary, he asked Cuban communities for cash contributions to finance the promised war against Batista, and had to humbly request an extension of his visitor's visa. And for all his anti-Americanism, Castro had come to seek, consciously or not, North American approval for the person and the deeds of Fidel Castro; this is presumably why, even thirty years later, he remains exceedingly sensitive to what is being written about him in the United States, down to the smallest detail.

In the United States, Castro was lionized and continually lectured on the dangers of communism and the beauties of democracy by American government officials, congressmen, and editorial writers in that patronizing and irritating fashion Americans apply to foreigners. Having a marvelous time, he went along with this game, and being the modest charmer, he said what he knew Americans wanted to hear. He basked in the applause and the huge national attention he was commanding, drawing additional sympathy from the presence of nine-year-old Fidelito, whom he took almost everywhere, and saving his private contempt for his hosts in comments to friends in his personal bodyguard. President Eisenhower arranged to be playing golf out of town during the five days Castro spent in Washington, which was not a slight inasmuch as the Cuban was not a head of state and not even an official guest, although it might have been a useful gesture to have invited him to the White House. Nonetheless the olive-green-fatigue-clad Castro enjoyed red-carpet treatment in Washington: a lunch given by Acting Secretary of State Christian Herter; a two-hour-and-twenty-five-minute private meeting with Vice-President Nixon at his office in the otherwise deserted Capitol (they met there Sunday afternoon after Fidel refused to go to the vice-presidential home) that was notable for an absolute lack of mutual understanding and for Nixon's conclusion that the Cuban was controlled by Communists; a lunch at the National Press Club, where he treated newsmen to twenty-minute answers to questions in his surprisingly fluent if heavily accented English; a *Meet the Press* television interview; visits with key senators and congressmen; and hosting a reception at the Cuban embassy on Sixteenth Street, where he was staying. On that occasion, Castro appeared for the first time in a formal uniform with tie and jacket, and had a brief conversation with the Soviet ambassador, Mikhail A. Menshikov. It was his first known encounter with a Soviet official.

In his free time, Castro went to see the Lincoln and Jefferson memorials, and walked for an hour around the grounds of Washington's home at Mount Vernon. At the Jefferson Memorial he was asked whether he thought governments should ever be overthrown, and he was ready with appropriate words: "I am not an advocate of frequent changes in laws and constitutions, but laws and constitutions must go hand in hand . . . this *is* a revolutionary principle for . . . progressive changes of institutions as the minds of men change." Saturday night, after the embassy reception, Castro playfully reverted to his Havana practices, disappearing for four hours from the surveillance by the enormous American security apparatus to tour Washington in a private automobile with five Cuban companions. He dined in a downtown Chinese restaurant, debated with a group of university students at nearby tables, and finally returned home at three o'clock in the morning.

Politically, Castro was most adroit in Washington, although in retrospect it is clear that he had engaged in deception, which in his mind had continuing "historical justification." On the issue of communism in Cuba, endlessly raised with him in Washington, he repeated time after time that "we are not Communists," that if there happened to be any Communists in his government, "their influence is nothing," and that he did not agree with communism. It has been argued subsequently that at the time of his 1959 United States visit, Castro had not yet fully resolved his ideological allegiances, and therefore was being truthful in his responses about communism. However, this argument did not take into account the fact, not known then, that Fidel had initiated secret coalition talks with the old Communists at least three months before his American trip. Presumably to reassure Americans during the postvictory transition period—pending ultimate consolidation—Castro announced that Cuba would not confiscate foreign-owned private property (which meant mainly American-owned concerns), and would seek additional investments to provide new jobs. Addressing the luncheon meeting of the newspaper editors who had invited him to the United States, Castro said: "The first thing dictators do is to finish the free press and establish censorship. There is no doubt that the free press is the first enemy of dictatorship."

It was in Washington that Castro for the first time publicly ruled out elections in Cuba in the foreseeable future, telling television interviewers that four years would have to elapse before the revolutionary regime could "establish conditions for free elections." Heretofore, Castro had been saying that the delay before holding elections would be no more than two years, and he had not attached conditions to the timing. But the first indications of a fundamentally changed political line were voiced in a Havana speech on April 9, a week before flying to Washington: "We want that when elections come . . . that everybody be working here, that the agrarian reform be a reality . . . that all the children have a school . . . that all families have access to hospitals . . . that every Cuban know his rights and his duties, that every Cuban know how to read and write. . . . Then, we can have truly democratic elections! . . ." What Castro was proposing was a generational delay, but in Washington he chose to say that the conditions would be ripe within four years and, in the meantime, "real democracy is not possible for hungry people."

Almost imperceptibly, Castro had changed the rules of the Cuban political game as he smiled disarmingly before television cameras in a Washington studio. The "conditions" for free elections—and Castro certainly was not saying there would be no elections—sounded eminently sensible, and after all he had the trust of 90 percent of the Cuban population. Advancing along several fronts, including some not visible, the Prime Minister of the Revolution had the overall strategy and tactics well in hand. Very soon, the slogan of "Revolution First, Elections Afterward!" began to be heard and read in Cuba, and Fidel could say he was simply responding to *vox populi,* and by midyear, elections (like anticommunism) became a counterrevolutionary theme.

Secure that revolutionary politics at home were under control with Raúl

Castro fully in charge (though Defense Minister Martínez Sánchez was the acting prime minister), Fidel could enjoy his North American tour and quietly prepare to extend his travels to South America. He also found himself in the midst of an extraordinarily bizarre episode, which was also a great game of bluff with neither player knowing enough about the other. In fact, to this day Castro may not know the whole truth; only old-time CIA insiders do. Either to test his non-Communist protestations or to profit from his experiences with the Communists, the CIA arranged to have Castro receive its supposedly leading expert on communism in Latin America, a cigar-smoking wartime German refugee named Gerry Drecher (afterward often mistakenly identified as "Droller") who used the pseudonym of "Frank Bender." He had no Latin American experience whatsoever. Castro had refused to see him in Washington but relented and agreed to a meeting at his hotel suite in New York, where he arrived by train (after giving a speech at Princeton University) on April 21.

The contact was arranged through Finance Minister López-Fresquet (urged on by his American friends) who wrote later that Castro and Bender talked privately for more than three hours. Then, Bender "returned to my suite in a state of euphoria . . . he asked for a drink, and with great relief exclaimed, 'Castro is not only not a Communist, but he is a strong anticommunist fighter.'" Bender next informed López-Fresquet that "he had arranged with Castro an exchange of intelligence information on the activities of the Communists and that I was to be the Cuban contact." López-Fresquet wrote that "the following month, during a reception at the French Embassy in Havana, a high U.S. official approached me and gave me an oral message for Castro from Mr. Bender. At the next Cabinet session, I gave Castro the intelligence. He didn't answer me, and he never gave me any information to pass on to Mr. Bender, the 'expert' on communists." What López-Fresquet did not mention was that Bender-Drecher was already preparing for his role as one of the main architects of the Bay of Pigs invasion that would take place two years later; he would be the chief of political action of the operation. It was a strange encounter: Did Bender really expect Castro to be his informant on communism, and was he a dupe (he had no way of knowing about Castro's Cojímar sessions with the Communists)? Did Castro suspect that he was receiving an arch-plotter-to-be (probably not)? And to López-Fresquet, who still had his illusions, Castro said in Washington, "Look, Rufo, I am letting the Communists stick their heads out so I will know who they are. And when I know them all, I'll do away with them, with one sweep of my hat."

In New York Castro spent four days as a conquering hero, touring the United Nations, addressing a nighttime crowd of thirty thousand in Central Park, visiting the Coffee and Sugar Exchange and City Hall, and speaking at luncheons and dinners to publishers, businessmen, and financiers. He made a superb impression, and little attention was paid to his remark at the United Nations that Cubans were "unanimously" opposed to immediate elections because they would risk the return of "oligarchy and tyranny." Likewise scant attention was paid to a page-one dispatch in *The New York Times* from its Havana correspondent, Ruby Hart Phillips, on April 23 that "the communist Popular Socialist Party is organizing every town and village . . . the communist influence in unions is rising . . . leaders of the 26th of July Movement are combatting these communistic efforts, but the youthful rebels are amateurs at organization compared with the communists." From New York Castro traveled by train to Boston, gave a speech at Harvard University, then went on to Montreal, where he said that he would go immediately to Buenos Aires to attend an inter-American economic conference.

Fidel Castro's decision to fly to South America at the end of his North American tour resulted from his determination to reach out instantly for Latin American leadership and statesmanship—it was his first Bolívarian gesture—and to emphasize his independence from the United States. At the same time, Castro was evidently keen to clarify his stand on Latin American revolutions and any possible Cuban assistance given them in the light of his own pronouncements and a series of odd events occurring in Cuba during his stay in the United States.

Castro's principal statement in this context was that Cuba would provide "hospitality," the opportunity to work, and help to exiles from Latin American countries who hoped to overthrow dictatorships at home. This was, in effect, his first declaration of what has become known as Castro's "internationalism," but, as he would keep repeating for the next thirty years, "the Cuban revolution was not for export" because revolutions must stem from internal conditions. Nonetheless, he must have been aware of nervousness in the region following reports coinciding with his presence in Washington and New York that Nicaraguan, Panamanian, and Haitian rebels were on the verge of invading their countries from Cuban bases. In fact, he was clearly embarrassed by the reports and there were strong suspicions at the time that these expeditions had been authorized by Raúl Castro in Fidel's absence. Thus on April 18, over one hundred Nicaraguans were arrested at a camp in the province of Pinar del Río, and their arms seized, by the provincial Rebel Army commander, who declared that Fidel had forbidden invasions from Cuba. That same day, a Panamanian opposition leader named Ruben Miró said in Havana that his armed groups would land in Panama within a month. On April 21, Dame Margot Fonteyn, the ballerina, was arrested in Panama and expelled, and her politician husband,

Roberto Arias, went into hiding after their yacht had circled off the coast in a suspicious fashion. López-Fresquet, who was in Boston with Castro, remembered overhearing an angry telephone conversation between Fidel and Raúl on the day the Panamanians announced the capture ashore of three rebels, two of them Cubans; Fidel was apparently chastising his brother.

When Fidel stopped in Houston, Texas, en route from Montreal to South America on April 27, Raúl flew up from Havana for a half-hour face-to-face conference with him at the airport. It was never explained publicly why Fidel had summoned Raúl, but the next day, in his aircraft flying over Cuba, he broadcast a denunciation of "irresponsible" Cubans landing in Panama, damaging the prestige of the revolution. At the Organization of American States in Washington, the new Cuban ambassador, Raúl Roa García, also denounced the Panamanian venture—and Fidel was widely applauded for being so statesmanlike. True to his "internationalism," however, he never turned his back on any Latin American revolutionaries. It was Castro's talent for having the best of all worlds.

Castro devoted ten days to his South American journey, another triumph for him and his revolution, attracting vast crowds, wild applause, and the riveted attention of local leaders. In Port of Spain he was greeted by Prime Minister Eric Williams; in São Paulo he announced that "our aspirations are the same as those of all Latin America," and then flew to the site of the future capital of Brasília to confer with President Juscelino Kubitschek; in Buenos Aires he met President Arturo Frondizi; in Montevideo he was welcomed by the Uruguayan government and spoke at a huge street rally; back in Brazil he conferred again with Kubitschek, addressed a mass rally, and appeared on the television program *This Is Your Life,* loving the exposure. But the high point of the journey was the economic conference in Buenos Aires where Castro sat in his olive-green fatigues among the hemisphere's ministers of economy (the United States sent only an assistant secretary of state, a sad counterpoint) to proclaim that "the hour has come for the people of Latin America to make a daily effort to find a true solution to the root of our problems that are economic in character." In this May 2 speech before the "Committee of 21," Castro urged the United States to grant Latin America $30 billion in economic aid over a decade, an idea immediately derided in Washington as ridiculous and demagogic. But less than two years later, President John F. Kennedy would offer $25 billion for Latin American development under the Alliance for Progress program, over which Castro later chuckled as an attempt to steal his thunder. To be sure, Castro had an uncanny sense of Latin American needs and moods that no United States administration in the decades to come could or would comprehend. And the contrast between his conquering voyage and the rock-

and-spit assaults on Vice-President Richard Nixon across South America exactly one year earlier underlined dramatically the moods of the region.

Mixing with crowds everywhere he went—in the United States, in Canada, and in South America—Castro was always an easy target for assassination. Yet, no public attempt was ever made against him anywhere, and Fidel displayed the *guerrillero* fatalism. In New York when shown a headline about an "assassination plot," he smiled and said, "I'm not worried. I will not live one day more than the day I am going to die."

Castro returned to Havana on May 7, and the following day launched the next major phase in the Cuban revolution. Always conscious of the need for contact with the masses, he convened a rally of tens of thousands on the vast civic square (now called Revolution Plaza) to pledge that "the Revolution will never renounce its human principles . . . nor the existence of social justice in Cuba." Then he drove off in a motorcade of Oldsmobiles to his seaside house at Cojímar to present his ministers with the text of the agrarian reform law that the "inside team" had secretly drafted previously at Che Guevara's residence. On May 17 the law was signed by the entire cabinet just below his wartime La Plata headquarters in the Sierra Maestra, and Castro announced on television from there that "Cuba is beginning a new era."

The main provision of the new law (replacing the 1958 Sierra agrarian law) limited land ownership to 966 acres per individual; sugar, rice, and cattle holdings were allowed to be as large as 3,300 acres. Private land ownership was not established as a matter of principle, and indemnifications were promised for nationalized lands. For practical purposes, however, the *latifundium* was abolished in Cuba as an economic and political phenomenon. Chiefly, the law was intended to lead to the consolidation by the state of nationalized lands so that great plantations and pastures could go on operating efficiently. In this sense, agrarian reform marked the first real revolutionary milestone; as Castro said afterward, "It truly established a rupture between the Revolution and the richest and the most privileged sectors in the country, and a rupture with the United States, [and] with transnational companies." As Castro noted, some American companies each owned as many as 480,000 acres of Cuba's "best land." Not surprisingly, the law was immensely popular among Cubans, especially the peasants, and the already finely honed internal propaganda apparatus instantly produced the slogan "The Agrarian Reform Works!" which was repeated endlessly on radio and television, and by every telephone operator in the country answering a call.

Politically, the law handed Castro his greatest instrument of unchecked power by creating the National Institute of Agrarian Reform (INRA), which, in effect, provided the structure for his "hidden government." Castro became the president of INRA in addition to his premiership and his post as Commander in Chief, and Antonio Nuñez Jiménez, the coordinator of the Tarará secret task force, was named executive director. A nexus was immediately established between INRA and the Rebel Army, the latter executing the economic and political decisions of the former: It was a logical step in the unfolding of Castro's governance concept that a strong and modernized rebel army be the revolution's power center and its vanguard. INRA and the Rebel Army became, in fact, indistinguishable. Castro rarely attended the ministerial cabinet at the presidential palace, since it was no more than an adornment.

Looking back at that first year, Nuñez Jiménez says that under Castro, INRA "was the bastion where the Revolution occurred in those initial months . . . the organism that dealt the real blow to bourgeoisie and imperialism." He remarks that "it was not tactical to change suddenly the Council of Ministers . . . our people were not yet prepared ideologically for the outbreak of an open battle between the revolution and the counter-revolution within the government itself." Therefore, Nuñez Jiménez recounts, "Fidel duplicated in INRA the most important functions of the Revolutionary Government." One of these functions was the creation within INRA of a department of industrialization, headed by Che Guevara, which was a de facto ministry of industries. When Guevara subsequently became president of the National Bank, Nuñez Jiménez was named its executive vice-president to assure INRA's voice there; the old Commerce Ministry was transformed into INRA's Department of Commercialization. Gradually, ministries and INRA departments overlapped "until the Council of Ministers," Nuñez Jiménez says, "was formed totally by revolutionaries." Next, INRA created its own armed 100,000-man militia units with the aid of Raúl Castro, and eleven million pesos in INRA funds were used to organize an army artillery school and the first antiaircraft and antitank artillery units. Likewise, INRA financed the construction of most of the highways in Cuba in the first revolutionary period, built peasant housing and tourist resorts. Rebel Army officers were given the power to seize private land and to run farms as cooperatives under INRA, and Che Guevara decided that the new industrialization projects should be directed by Rebel Army companions. There was, indeed, a "hidden government," but strangely most Cubans, even those in senior positions, were unaware of it.

Using the tall INRA building (erected by Batista as Havana's city hall) as his principal office, Castro fueled the revolution in periodic meetings with the Institute's top officials and field administrators. In what Nuñez Jiménez has called Fidel's "secret speeches," the prime minister announced late in 1959, for example, that all sugarcane fields would be seized from individual small owners after the winter 1960 harvest and turned into co-operatives; this action violated the May agrarian law, but what counted were Fidel's orders. And at that stage he proposed to go on running Cuba from INRA, declaring in another "secret speech" that the Institute "is a political instrument and the apparatus for activating the country's masses to carry out a task and to defend this task . . . INRA will be gigantic apparatus with an extraordinary power of mobilizing [the people], espe-cially if we organize peasants in social groups and military groups. . . . An armed people is the definitive guarantee of the Revolution, precisely be-cause it is armed."

And Castro said, "In this whole situation INRA is the decisive apparatus and it has a tremendous revolutionary responsibility. INRA and the army are the two decisive apparatuses for the Revolution, and we must rally around them all the forces that we can marshal for the battles that await us in the future. . . . We shall consolidate what has been won. Let us not look for more enemies that at a given moment we must find. Remember that in the war we first attacked small towns, then big towns . . ."

Late spring of 1959 was a moment of danger for Castro's revolution, internally as well as externally. Still, as he kept reminding his companions, the revolution had to move carefully and gradually while he consolidated power from his INRA "bastion." At home, Castro and the revolution re-tained their immense popularity, but Fidel's sensitive political antennae were attuned to disturbing signals in the air. He was uncertain of the revolutionary loyalty of his own 26th of July Movement, particularly of its "bourgeois right wing," as he called it, and this was why he set up the secret INRA government to bypass the cabinet and President Urrutia—until he was ready to discard them. As Pedro Miret, one of Fidel's closest associates, recalls that period, "It was necessary to dilute the 26th of July." Though Castro had finally succeeded in bringing the Students' Revolution-ary Directorate to his side after the past rivalries and to attract DR's chief, Faure Chomón, to the government, he still felt uncomfortable with this middle-class organization. Then, there was the problem of the strutting *barbudos* (some of whom did and some did not actually fight in the moun-tains) demanding rank and privilege, an attitude that drove Fidel to parox-ysms of fury. In a rage, he denounced "the stupidity, the demagogy, the opportunism, and the politicking of all those who today give the appear-ance of being more revolutionary than anyone else."

And, finally, there were the Communists. Though he continued to hold long-range unity talks with the party's leaders, Castro was aware that the Communist leadership was split, and many old-timers resented and envied him. The party as a whole had still failed to pay him fealty (possibly be-cause Moscow remained unconvinced) and recognize his revolutionary genius in triumphing his way instead of the orthodox Marxist-Leninist way. Pragmatic as they may have tried to be, key party leaders found it hard to swallow Castro's denials in the United States of Communist influence in the Cuban revolution, and his occasional private outbursts of impatience with the assertiveness and arrogance of some party members. A polemic over Communist support for Batista in the 1940s developed between the Movement's official organ, *Revolución,* and the Communist daily, *Hoy.* For a time, only Carlos Rafael Rodríguez kept the channels open at the top level; Conchita Fernández, who was Castro's secretary at INRA, recalls that Rodríguez was the only Communist of importance to join the inner circle there in those days.

Externally, Castro increasingly feared an American intervention, especially after the signing of the land-reform law. Bonsal, the American ambassador in Havana, wrote later that "in the spring of 1959 Castro believed it probable that the Cuban revolution as he envisaged it would sooner rather than later come into irreconcilable conflict with American interests on the island and that the United States government would respond with a full-scale invasion of Cuba." But Bonsal also commented that "Castro's scenario at this time did not contemplate the massive help in the form of economic aid and weapons that he later received from the Soviet Union . . . [he] became oriented toward dependence on the Soviet Union only when the United States, by its actions in the spring and summer of 1960, gave the Russians no choice other than to come to Castro's rescue." This is a complex proposition, and set forth by a diplomat of Bonsal's acuity, it is worth examining.

If Bonsal's assessment is correct—and, in the end, only Fidel Castro knows the truth, free of ideological and mythical embellishments—then my own impression is that the American determination to prevent Cuba from acquiring arms anywhere in the world during 1959 and 1960 was a decisive factor, even before Eisenhower launched economic and covert-action warfare against the revolutionaries. It is certainly arguable that Castro might have chosen domestic Marxist solutions without becoming wholly dependent on the Russians economically and militarily if the United States had not threatened his survival; Yugoslavia and China are not such farfetched analogies, and are not a world away from the Soviet Union. Moreover, at least a year elapsed before Soviet assistance began arriving on the island, and Washington did not have to close off all alternatives. Finally, a careful study of the record will show that Castro had begun to move headlong toward a socialist revolution long before Moscow promised or gave him help. And all of this history is consistent with the Castro claim that American "aggressions" simply accelerated the socialist process of the revolution.

Immediately after victory, Castro made the modernizing of the Rebel Army—and subsequently of the new people's militia—a top priority. The militia was particularly important because Castro, much as he believed in the Rebel Army as the chief instrument of revolutionary power, did not want the army to become an elite organization that even theoretically could some day turn into a political threat, especially in the absence of a ruling political party. But Castro also thought that it would be too costly to maintain an oversized standing army even if he daily feared an American invasion, and that a well-trained militia, which could be mobilized for action within hours, would be the perfect complement to a smaller though heavily armed professional army. Finally, he considered that the militias as "an armed nation" represented an additional dimension of popular revolutionary commitment. In mid-1959, in his "secret speeches" Castro was telling INRA associates that he wanted one hundred thousand peasant militias (in addition to workers' and students' militias) to be trained in the use of weapons, including .50-caliber tripod machine guns, at a rate of one thousand every forty-five days. What Batista had left behind, however, did not suit the modern mobile forces Castro had in mind, so by mid-May, trusted emissaries fanned out overseas to try to buy arms.

Rufo López-Fresquet, the revolution's first finance minister, recalls being summoned to Castro's hotel penthouse late in January and asked if there were funds available for "an immediate purchase of arms." Castro told him he was concerned about an invasion from the Dominican Republic, where Batista had found haven with Trujillo, and that Cuba was disarmed because "the generals stole the money, and, in spite of the heavy expenditures on armaments, these are nowhere to be found." López-Fresquet was able to track down $5.3 million of Cuban military funds in European banks, and the regime purchased 25,000 light automatic FAL rifles (including 2,000 weapons with grenade-throwing attachments), 50 million rounds of ammunition, and 100,000 grenades from Belgium. The United States had not begun pressing foreign governments to refuse Cuba arms, and the first FAL shipment reached Havana in September, just in time for the fresh militia units (though the militias' creation was only announced by Castro on October 26).

José Ramón Fernández, the career army officer imprisoned by Batista for plotting against him, was one of Castro's principal arms buyers as well as the chief organizer of the militias. Now a vice-president of Cuba and an alternate member of the Communist party Politburo, Fernández recalls shopping for arms in Italy, Switzerland, West Germany, and Israel in mid-1959, but was only able to obtain them in Italy. Cuba took delivery of nearly one hundred .81mm-caliber mortars, two batteries of 105mm howitzers (made in the United States), and one hundred heavy machine guns, flamethrowers, and light weapons, but a subsequent dispute between Rome and Havana blocked further purchases. In Israel, Fernández met with Golda Meir, then foreign minister, but after touring Israeli defense plants, he could not find suitable weapons for the Cuban Army; Israel was willing to sell him Uzi submachine guns, but these weapons did not fit into the Cubans' plans. Soon, however, the Europeans, under pressure from Washington, declined to sell arms to Cuba. Britain, honoring an American request, refused to supply Hawker Hunter jet fighters to the Cuban Air Force. Yugoslavia, approached secretly by the Cubans in 1959, decided on its own not to sell arms to Castro, presumably to avoid antagonizing the United States, with which it maintained a delicate relationship.

Announcing in October the creation of the militias because, as he put it, of growing counterrevolutionary dangers and attacks from the United States, Castro told a million Cubans at a presidential palace rally that "if we cannot buy planes, we shall fight on land, when the time comes to fight on land . . . we shall immediately start training peasants and workers . . . if they don't sell us planes in England, we shall buy them wherever they will sell to us; and if there is no money for warplanes, the people will buy the warplanes . . ." Turning to Army Chief of staff Juan Almeida, Castro exclaimed: "And right here, I hand you a check from the President of the Republic and the Prime Minister as a contribution for the purchase of the planes."

José Ramón Fernández says that the first arms from Czechoslovakia and the Soviet Union began arriving late in 1960, months after the final breach in Cuban-American economic ties; the first jet aircraft were delivered in mid-1961, at the earliest. As Castro had said, Cuba would buy arms wherever possible (and accept them free), but the record suggests that neither Havana nor Moscow were interested in arms deals during 1959. Castro had tried Western sources first, and Ambassador Bonsal is right in saying that the revolutionaries were forced into dependence on the Soviet bloc. According to Fernández, the first Czech shipments were automatic M-52 rifles, BEZA-792 machine guns that could be used for antiaircraft fire, and 82mm mortars. Czech instructors accompanied the weapons. Soviet arms came next.

If Castro feared an invasion from the United States in 1959, he also feared an attack from the Dominican Republic by his archenemy, the generalissimo. Trujillo, who remembered Fidel's involvement in the abortive Cayo Confites expedition against him twelve years earlier, was just as concerned with the threat of a Cuban invasion, directly or through Haiti. Both rulers therefore devoted themselves to preparations for preemptive blows, though neither seemed to understand the political situation in the other's country. Trujillo started out by organizing a "foreign legion," including Caribbean mercenaries, anti-Castro Cubans, Fascist Blue Division veterans from Spain, Germans, and right-wing Croatians. It was never clear if Trujillo planned to use this "legion" to defend Haiti from Cuba or to attack Cuba, but Castro, who had been training anti-Trujillo Dominicans in his camps, struck first. At dusk on June 14, a C-46 twin-engine transport plane provided by Venezuela landed in Constanza in the central mountains of the Dominican Republic with fifty-six rebels aboard. Ten of them were Cubans, and the commander was Major Delio Gómez Ochoa, the former 26th of July Movement coordinator in Havana. The Dominican Army rapidly destroyed the invading force, capturing Gómez, and six days later Trujillo's air force and warships sank two yachts filled with additional Dominican rebels trying to land at Puerto Plata on the north coast. This marked the end of Castro's only attempt to tangle with Trujillo, and it is probably irrelevant if he was again practicing "internationalism" or simply hoping to discourage the old dictator from assaulting him.

In any event, Trujillo would not give up, and immediately after Constanza, his top intelligence agents resumed plotting with two Rebel Army *comandantes* whom they considered ready to betray Castro. One was an American mercenary named William Morgan and the other was a Spaniard named Eloy Gutiérrez Menoyo who had participated with his brother Carlos in the Students' Revolutionary Directorate attack on Batista's palace in 1957. Carlos was killed, and in the attack the following year Eloy joined the DR guerrillas in Escambray where he met Morgan who seemed to be fighting there strictly for the money. Trujillo supposedly offered Morgan and Gutiérrez Menoyo $1 million for starting a rebellion in the Escambray Mountains that would then be supported by coastal landings by Cuban exiles and the Dominican "legion." Inasmuch as there was bad blood between the DR factions and Castro, Morgan and his Spanish friend appeared to the Dominicans to be perfectly plausible traitors, and the American mercenary actually collected $500,000 as a down payment.

What Trujillo did not know was that Morgan and Gutiérrez Menoyo had informed Castro of the conspiracy, which allowed the Cubans to tape radio traffic between the two *comandantes* and the Dominican capital. It is entirely possible that the two men were playing a double or triple game, depending on the most favorable outcome, but on August 12 they radioed Ciudad Trujillo on Castro's instructions that "rebel guerrillas" had taken the port of Trinidad and were now awaiting reinforcements. Castro and his bodyguards sat inside a mango grove off the airfield listening to the radio exchanges all night, then they watched a Dominican plane land munitions and depart, and finally came out from behind a mango tree to capture ten Cubans disembarking from a second Dominican plane at dawn on August

13. It was his thirty-third birthday treat. The area was surrounded by several Rebel Army battalions, but Fidel proudly relived the guerrilla days by personally capturing the pilot, Lieutenant Colonel Antonio Soto, who had flown Batista to exile on January 1. It is strange that Trujillo had fallen into the Castro trap, considering that only four days earlier the Cuban security forces had rounded up around a thousand ex-Batista soldiers and others with Dominican connections, but then logic was never a prevailing wind in the Caribbean. As for Morgan, he was shot two years later for being involved in a "counterrevolutionary" conspiracy, and Gutiérrez Menoyo was captured in 1961, after landing clandestinely from a CIA boat from Florida. He was still in prison in 1986, despite entreaties by Spain's socialist government.

For Fidel the Trinidad incident was a welcome and exhilarating distraction from the rigors of managing a revolution, living behind a microphone, and juggling all his political balls. His temperament required constant movement and change, and he succeeded reasonably often in pleasing himself. In Havana he had the choice of working at either of his offices at INRA (he disliked the presidential palace and never set foot at the nearby official quarters of the prime minister), at the hotel penthouse suite, at the seaside house in Cojímar, at Celia Sánchez's apartment on Eleventh Street, or at another house set aside for him next to the Chaplin theater in Miramar. He also had his secret military command post in the Vedado residential district. Though Castro commuted continually among all these places, he was ever restless. Occasionally, he turned up at the Cerro stadium in the evening to pitch a few balls at the Sugar Kings' batters; or he would appear at Carmelo's terrace café in Vedado, across the street from Alicia Alonso's ballet school, so that Fidel could eat ice cream and chat with the patrons; or he would drop in at a foreign embassy reception. One night, late in March, after a midnight steak dinner in the kitchen of the Habana Libre Hotel (the Hilton's new name), he took Herbert Matthews and me to a suburban beach in the small hours to have Cokes and show us the resorts the revolution had built for the people. Afterward, we chatted in the hotel cafeteria until midmorning, Fidel being the only one who was not sleepy. Sometimes, he spent several days at a military camp, usually in Oriente, chatting with the troops, reading, and writing. Caught by three days and nights of rain in one such camp, he avidly played dominoes and chess for hours on end, strangely showing no impatience. His torso bare, he did chin-ups, shot baskets, or played with his German Shepherd, Guardián. Celia Sánchez was usually along on these trips, but Castro seemed to enjoy the military camaraderie with his friends and bodyguards: his physician, René Vallejo, his chief bodyguard, Jesús Yañez Pelletier (the Santiago jailer who had saved his life), until he was fired for improper behavior, his friend Nuñez Jiménez, and Raúl Castro and Che Guevara if they happened to be around. Fidelito visited his father at the Cojímar house, but he was a boarder at school and their contacts were not very frequent. During the spring of 1959, Fidelito was in an automobile accident near Matanzas, perforating his spleen, and it took hours for Castro to be located and brought to the hospital.

Always looking for the new and the unusual, Fidel Castro discovered and invaded the Bay of Pigs two years before the Americans thought of it. Moreover, when the CIA decided to surprise him in the great southern swamp at that time, they had no idea that Fidel knew the Ciénaga de Zapata like the back of his hand, making it even easier for him to win the battle. It was one of those coincidences of history. The Ciénaga de Zapata is an immense, virtually uninhabited marshland stretching far inland along the southern coasts of Havana and Matanzas provinces—the domain of quicksand, charcoal men, crocodiles, and mosquitoes. Probably because it was the only geographic challenge left for him in Cuba after conquering the Sierra Maestra—and assuredly because it was even poorer than the Sierra— Castro became fascinated with the Ciénaga, and in March 1959, he began visiting the area regularly. At least twice, he nearly lost his life in the immensity of the treacherous swamp.

The Ciénaga was a classic example of Castro's enthusiastic, sweeping, generous, and more often than not impractical approaches to social-economic development projects. His idea was that great drainage works would turn the 480,000 acres of western Ciénaga into an immense rice granary, that the destitute charcoal men's families would prosper through rice cultivation and the expansion of tourism, and that canals, roads, and resorts should be built at once to make it all possible. More than a quarter-century later, however, no huge rice plantations followed Castro's original experimental station, tourism has remained marginal, though crocodiles and sea cows have been saved. The usual plague of insufficient resources, know-how, and follow-up have diluted the revolution's most ambitious plans.

On one of Fidel's first visits to the swamp, the vessel taking him, Celia Sánchez, and Nuñez Jiménez and his wife down a canal suddenly went down, and only the agile Commander in Chief managed to jump to a bank. When his wet companions joined him, they found Castro relaxing, reading a copy of Giovanni Papini's *The Remote Past* he happened to be carrying in his pocket. The next few days brought greater dangers. First, the pilot of Castro's helicopter, Air Force Commander in Chief Major Pedro Luis Díaz Lanz, told Fidel he lacked sufficient fuel for the planned tour of the Ciénaga. He left Castro, Nuñez Jiménez, and Pedro Miret on Playa Girón, one of the landing beaches for the future Bay of Pigs invasion, to pick up gasoline at a sugar mill in the north, and he was to return within a few hours. The three men spent the afternoon in gun practice on the beach, but Díaz Lanz did not return, and they had to sleep in a fisherman's shack. In the morning, Castro and his companions walked ten or more miles to a military post in order to telephone the sugar mill—and learned that the helicopter had never made it. After summoning search aircraft from Havana, Castro climbed into a single-engine light plane with a pilot, and took off in driving rain to look for Díaz Lanz; returning an hour later, he reported having spotted the crashed chopper with no sign of survivors. Meanwhile, Raúl Castro, Che Guevara, and other top rebel commanders arrived aboard Fidel's personal plane, *Sierra Maestra,* from Havana, followed by two helicopters. Fidel took off in one of the helicopters, with the storm still raging, and four light planes, Raúl in one, went after him in a search pattern. While Fidel's helicopter landed near the crashed craft, the light plane with Raúl vanished in the rain clouds. Díaz Lanz was found late that day at the far end of the swamp where he had reached help, but Raúl was still missing. The following morning, Raúl's plane was located pancaked in the mud near the coast. Raúl and two pilots were lost in the marshland, but were rescued by a search party and placed aboard a navy flying boat on the beach. On the way to Havana, the Catalina's landing gear collapsed, and the pilot made a successful crash landing at the airfield.

The Ciénaga's pitfalls and adventures did not discourage Fidel. The marshland had become his pet project, and he kept referring to it rhapsodically in his speeches. "We have rediscovered the Bay of Pigs, broad and deep," he said in December in a speech the CIA must have missed, "and we have one more bay, the one of Pigs. Cuba is a country that rediscovers a bay!" On Christmas Eve Castro and Nuñez Jiménez drove from Havana to the Laguna del Tesoro resort in the center of the Ciénaga, then flew by helicopter to the charcoal men's village of Soplillar, joining local families at a dinner of roast pig and revolutionary songs. A contemporary photograph shows a smiling Fidel, his rifle in his left hand, looking down at a table surrounded by children and holding a machine gun on a tripod, an ammunition belt, bottles of wines and rum, glasses, and candy. After Christmas he had a small aluminum house-and-office structure erected at the Laguna del Tesoro: for a time, it was Castro's favorite hideaway and a place to receive special guests.

In Havana, in a rhythm of point and counterpoint, the Castro revolution and its enemies, foreign and domestic, were turning increasingly confrontational. In June 1959, Castro felt powerful enough to force the resignation of Foreign Minister Roberto Agramonte, a pro-American politician of the old school, replacing him with Raúl Roa, the vociferously revolutionary ambassador to the Organization of American States who only two years earlier had bitterly denounced the Soviet intervention in Hungary in a series of articles. But Roa had joined the *Fidelista* camp. Four other "moderate" ministers were also dismissed, and Castro now prepared to remove President Urrutia. The basic issue on which the internal battle had focused was communism: The charges from the moderates that communism was taking over the revolution, and the response from the Castro camp that anticommunism was the principal instrument of the "counterrevolution," mainly forged in Washington. This was exactly two months after Castro had proclaimed in New York, "I have said in a clear and definitive fashion that we are not Communists. . . . The doors are open to private investments that contribute to the industrial development of Cuba. . . . It is absolutely impossible for us to make progress if we do not understand each other with the United States."

By June, however, not even Castro and Ambassador Bonsal seemed to understand each other. Following the signing of the new agrarian law, Bonsal had sent a note to the Cuban government, pointing to the problems it posed for American investors in Cuba, and the next day, June 12, the ambassador was invited to visit Castro in Cojímar. It was their first formal meeting since Bonsal's get-acquainted call on the prime minister in March despite Bonsal's repeated requests. Bonsal has written that Castro was "cordial" and replied with "an emphatic affirmative" to the ambassador's question if American interests still had a role in the development of revolutionary Cuba. According to Bonsal, Castro also denied in press comments that "my approach to him [was] . . . a proconsular attitude." In a speech in December 1961, however, Castro had a different recollection of his overall relations with Bonsal: "From the first moment, shocks began over criteria and viewpoints, and these meetings became so intolerable that . . . he kept asking for an interview for three months until, in the end, there was no way of not granting it, for the most elementary protocol norms. Why? Because this gentleman's statements were simply intolerable . . ." Twelve years later Bonsal's comment in his book was that "[it] is far from a unique example of the manner in which the Maximum Leader's flexible memory permits him statements contrary to the truth of the event he is recalling." All this was regrettable, principally because Castro did not seem to realize that Bonsal was virtually putting his own career on the line in arguing in Washington for maintaining a dialogue with Cuba.

The removal of President Urrutia in July was dictated by the logic of Castro's revolutionary politics: The naïve, patriotic judge from Santiago he had named in the Sierra to be provisional president was now turning out to be an obstacle to the revolution. Having been restricted since February to signing laws prepared by Castro or the cabinet, by mid-spring Urrutia made himself even more vulnerable by staying away from cabinet sessions chaired by Castro and by delaying law-signing. Even López-Fresquet, the finance minister, wrote that Urrutia's actions were "a mistake" and "his conduct disrupted the functioning of the government . . . when Castro had an interest in the legislation in question, his animosity toward Urrutia increased . . . by his actions, Urrutia succeeded only in alienating the cabinet, which later offered no opposition to his overthrow." Then, Urrutia began to make anti-Communist speeches and television statements as if to force a confrontation with Castro he could not possibly win. Playing blindly into Castro's hands, he offered to go on a "leave of absence"—from which he did not propose to return—on the day in June when the first batch of moderates were fired from the cabinet. Urrutia did not realize that Castro believed a resignation was an attack on him and the revolution, and so let himself be talked out of leaving late in June. Fidel intended to control events himself.

On June 29, the air force chief, Major Díaz Lanz, who had piloted Castro's helicopter in the Ciénaga the previous month, defected to the United States, appearing before a Senate subcommittee in Washington to denounce communism in Cuba, and thereby quickening the crisis in Havana. That same week, Urrutia went on television to say, "I do not believe in communism and I am ready to debate these questions with anyone." Castro, now ready to go for the jugular, responded publicly, "I consider it not entirely honorable that if we are to avoid being called Communists we must embark on campaigns against them; no honorable government would do this . . ." On July 13, Urrutia again went on television to feed Castro more rope: "The Communists are inflicting terrible harm on Cuba." In retrospect, it is clear that Urrutia would have accomplished much more in his confrontation with Castro if he had resigned at the moment of his choice over the communist issue, which seemed to concern him so, but politically he was no match for the chief of the revolution.

When Fidel Castro was ready to strike, he marshaled the full panoply of revolutionary drama. On the evening of July 16, Cuban radio and television announced that Castro had resigned as prime minister (but not as Commander in Chief), because Urrutia had blocked approval of revolutionary laws and other government measures. He dropped out of sight for the next twenty-four hours. As it happened, Castro had ordered that all the Cuban peasants who wished to do so be brought to Havana for the celebrations of the sixth anniversary of the Moncada assault, and they began arriving en masse in the capital as the Urrutia guerrilla theater began to unfold. On July 17 the cabinet was summoned to the presidential palace, surrounded by troops and a growing crowd, but Urrutia was not to be seen. Defense Minister Martínez Sánchez appeared to be in charge as the ministers awaited further developments. In the evening, Castro went on television for a two-hour speech to explain that he had to resign because of Urrutia's attitude as president, and then he went on to link him with the Díaz Lanz defection, exclaiming, "This has come close to treason, companions; we have been on the brink of treason!" He charged Urrutia with fabricating a Communist "legend" in Cuba in order to provoke foreign aggression, and mentioned that the president had bought a house "for thirty or forty thousand dollars."

With crowds outside chanting demands for Urrutia's resignation, the president signed it while Castro was still on television. The ministers accepted the resignation and named Osvaldo Dorticós, the minister for drafting revolutionary laws, to be the new president. The proposal was made by Education Minister Armando Hart, one of Castro's most devoted associates, whereupon Fidel arrived at the cabinet meeting room to be "informed" of the change. Dorticós would be Castro's loyal collaborator. The drama, however, was still running its course, since Fidel required an overwhelming national show of support. So he attended a textile workers' rally at the sports stadium to hear pleas for his return to the premiership, and on July 23, revolutionary labor unions called an hour-long general strike to insist on his return. Castro had carried out a coup d'état by television, and now *Habaneros* and the visiting peasants, wearing straw hats and carrying sharp machetes inside brown leather sheaths, paraded around the city to hail their chief, songs about Fidel and the agrarian reform rising in the warm evening air. On July 26, Castro addressed a million of his supporters on the Civic Plaza, agreeing to be prime minister again and warning that "to attack Cuba is to attack all of Latin America."

In the meantime, however, Castro had to continue mending revolutionary fences at home, to defuse rising criticism of the Communist inroads in his regime and the Rebel Army. As in Urrutia's case, he considered all politically motivated resignations by important personages as acts of treason, and he acted accordingly when Major Huber Matos Benítez, military governor of Camagüey province and a top Sierra fighter, prepared to resign over the Communist issue. Tipped off on October 20 that Matos had sent him a resignation letter and that twenty officers would likewise resign, Castro dispatched Rebel Army Commander Camilo Cienfuegos to arrest the major, and himself rushed off to Camagüey. On the morning of October 21, he led a revolutionary crowd in a march on the provincial military headquarters to preempt whatever rebellion may have been brewing. The Matos episode held immense dangers for Castro in assuring the unity and loyalty of the Rebel Army, and potentially it was his worst political crisis since he assumed power. Because Díaz Lanz, the former air force commander, had flown over Havana that same day in a light plane from Florida dropping anti-Castro leaflets and allegedly machine-gunning the city, Fidel immediately linked him to the Matos affair as part of a larger American-based conspiracy also involving the hapless Urrutia. On October 26, Castro gathered once more an enormous crowd—around one million—in Havana to protest these counterrevolutionary plots, to announce officially the creation of the armed militia, and to consult his supporters if the revolution should restore the death penalty and revolutionary tribunals. The crowd shouted its approval, chanting, *"Paredón! Paredón!* (To the wall! To the wall!)."* In December, after Castro's impassioned, prosecutorial speech, Matos was sentenced to twenty years in prison. Castro would not trust anyone else with the defense of the revolution at such a crucial time.

At the same time, he continued to consolidate revolutionary power and controls. The day before the Matos incident, Raúl Castro was named minister of revolutionary armed forces, making official his sway over the army. On November 26, Che Guevara took over the presidency of the National Bank while remaining in charge of national industry. Guevara's wartime deputy, Ramiro Valdés, became the head of DIER, the military intelligence and secret police, assisted by Manuel Piñeiro Losada, the red-bearded ideologue. But late in October, the revolution lost its second most popular leader after Fidel when Camilo Cienfuegos disappeared on a solo flight aboard his light plane from Camagüey to Havana. Despite a land and sea search personally coordinated by Castro, no trace of his aircraft was ever found, and Camilo joined the ever-expanding pantheon of revolutionary heroes upon whom today's myths still repose.

The Cuban revolution's growing orientation toward socialism or communism was not matched at the outset by Soviet support or ties. This fact does away with demonological views that the Castro rebellion was planned all along with the Russians (or the Chinese) to create a Communist presence at the doorstep of the United States. On the contrary, there are convincing indications that Castro had to sell the Soviets on himself and his revolution when it became obvious that he needed them to survive economically and militarily in the light of United States antagonism. It can even be argued that Castro, taking advantage of the convergence of various international situations, pushed the Russians into the relationship faster and to a greater degree than they desired. If nothing else, the Kremlin long shared the Cuban Communists' view that Fidel Castro was an unreliable customer, ideologically and practically. By the end of 1959, however, Castro had the Russians pretty much where he wanted them—as an antidote to the Americans.

The Cuban revolutionary government was recognized by the Soviet Union on January 10, 1959, three days after the United States did so, but this was basically meaningless because Batista had broken diplomatic relations with Moscow in 1952, and neither side proposed to restore them at that stage. Fidel had never been an ardent and uncritical admirer of the Soviet Union like Raúl, Nuñez Jiménez, and Che Guevara (until he soured on the Russians), and relations with Moscow were not a priority. Nevertheless, in the course of a television interview on February 19, he said Cuba was prepared to sell sugar to the Soviet Union, not a startling idea considering that the Russians were traditional buyers anyway. During 1959 Moscow contracted to buy 500,000 tons of sugar, roughly the volume purchased in 1955; China bought 50,000 tons. Together, this was less than 10 percent of the Cuban harvest.

A Soviet labor-union delegation had been invited to attend May Day celebrations in Havana, but visas were not issued in time. Three Soviet trade unionists did appear, however, in November, though the CIA station in Havana was much more interested in the visit in May and again in October of one Vadim Vadimovitch Listov, who was reported to be a high-ranking Soviet intelligence operative. At the same time, the Americans learned that four top Cuban pilots had gone secretly to Czechoslovakia to prepare a MiG jet training program; their leader was Captain Victor Pina Cardoso, a wartime Royal Air Force flier.

Officially, Moscow was not encouraging a close relationship with the Castro regime during 1959, possibly not to interfere with the "Spirit of Camp David" resulting from Premier Nikita Khrushchev's visit to President Eisenhower in September. Deputy Premier Anastas I. Mikoyan, a powerful Politburo member, had gone to New York and Mexico to open Soviet trade fairs, but nothing was suggested about hopping over to Havana to meet Castro. A more subtle method to make contact was devised instead. On October 16, ten days before he disappeared with his plane, Camilo Cienfuegos told Nuñez Jiménez that he had had a long conversation at a Havana hotel with Aleksandr Alexeiev, a correspondent for TASS, the Soviet news agency, who had come to Cuba after waiting eight months for a visa and would like to meet Castro. It is not known how Alexeiev was able to meet Cienfuegos, the army commander.

Nuñez Jiménez passed the word on to Castro, who informed Alexeiev that he would receive him for a "friendly" conversation at the prime minister's office on the top floor of the INRA building. Wearing a black suit and a gray necktie, the Soviet correspondent was brought from his hotel by two bearded Rebel Army soldiers. Finding Castro and Nuñez Jiménez in olive-green fatigues, Alexeiev greeted Castro and handed him a package wrapped in a Moscow newspaper: it contained a bottle of vodka, several cans of black caviar, and an album of photographs of Moscow. Opening the conversation, the Russian told Castro of the "great admiration" the Soviet people had for him and the Cuban revolution. The Soviet government and the Communist party, he said, held in great esteem his work for Cuba's social progress. Castro responded pleasantly that his revolutionary regime would be disposed to enter into trade relations with the Soviet Union at the proper time.

Nuñez Jiménez mentioned having met Mikoyan at the Soviet trade fair in New York in July, and a conversation developed about such an exhibit being brought to Havana, too. Castro then said that it would produce a "great impact" if Mikoyan came to Cuba to open the fair. At one point, Alexeiev noticed a silver medallion around Castro's neck, and the prime minister told him, "Don't worry, it's the image of a Christian saint that a little girl in Santiago sent me when I was in the Sierra Maestra." Then Fidel said, "So long as you've brought the caviar and the vodka, let's taste it." Conchita Fernández brought crackers, and the three men leaned back to enjoy the feast. Turning to Nuñez Jiménez, Fidel remarked, "What good vodka, what good caviar! Nuñez, I think it's worth establishing trade relations with the Soviet Union? What do you think?"

"Very well, Fidel," Alexeiev said, "we can now count on reestablishing economic relations, but what about the most important one, the diplomatic relations?" Castro replied, "Ah! . . . I can see why you came dressed so formally. . . . But it's better that we go on talking. We'll have to do it this way for the moment because we need time to create the [proper] conditions. Do you remember an article by Lenin in which he was saying that to apply a new policy or to introduce new ideas it is necessary to persuade the masses, make them participate in these decisions? We shall do that. . . . The idea of bringing the trade fair is excellent. . . . It's an opportunity to show the Cuban people the progress of the Soviet Union. Until the present, everything said about the Soviet Union is negative, and we shall see to it that this type of information does not continue. The fair and a visit by Mikoyan could be a successful beginning, don't you think? We've already started with caviar and vodka." The three men clinked glasses, and Castro said: "The fundamental thing now is not diplomatic relations. The most important thing is that Cubans and Russians are already friends."

Within three months, Mikoyan was in Havana to launch the immensely far-reaching relationship with Cuba. Aleksandr Alexeiev would become the second Soviet ambassador to Cuba. But five successive Soviet rulers—from Khrushchev to Gorbachev—would also learn in the ensuing quarter-century the massive frustrations of dealing with Fidel Castro as an ally.

CHAPTER

3

Fidel Castro and Nikita Khrushchev, two of the greatest political actors of our time, warmly embraced in a Harlem hotel in New York on September 20, 1960, sealing an alliance that two years later would push the world to the brink of a nuclear war. The sixty-six-year-old Soviet Communist party chairman's visit to the thirty-four-year-old Cuban prime minister, whom he dubbed a "heroic man," was his first public appearance after arriving to attend the United Nations General Assembly, thus creating the impression that Cuba was Russia's number one foreign-policy priority. So delighted was Castro with this attention that later that week he was the first delegate to leap to his feet to applaud Khrushchev's address to the General Assembly fervently. The Castro-Khrushchev New York act was magnificent political theater, leaving no doubt that Cuba had exchanged the American influence sphere for that of the Soviets, with all its attendant political implications in the Western Hemisphere and in East-West relations.

The immediate backdrop for the cordiality between Khrushchev and Castro was the catastrophic deterioration in Cuba's relations with the United States; the two nations, separated by ninety miles of blue Caribbean water, were now on an accelerating and unavoidable collision course. During the summer, Castro had nationalized $850 million worth of United States property on the island, from sugar mills and cattle ranches to oil refineries and utility companies, while the Eisenhower administration deprived the Cubans of their vital quota in the high-premium American sugar market. Moreover, the United States had taken two secret parallel decisions: to train and equip a Cuban exile force to invade Cuba, and to have the American Mafia assassinate Fidel Castro on the CIA's behalf.

At the same time, the general international situation had changed radically during 1960, making Cuba and its *Fidelista* revolution much more attractive and interesting to the Soviet Union. Because Khrushchev and Castro were calculated-risk gamblers, they joined forces to exploit the new state of affairs, though it must be assumed that, as a matter of principle, they had a healthy distrust of each other. From the Kremlin's viewpoint, the previous September's "Spirit of Camp David" was replaced by ugly tensions, beginning in May when the Soviets shot down a CIA U-2 spy plane over their territory, and the Khrushchev-Eisenhower summit meeting was canceled. Simultaneously, the long-simmering ideological feud between Moscow and Peking had finally boiled over in public view, and the two great Communist nations became rivals for the affections and allegiance of the emerging Third World. In this suddenly changed context, Cuba acquired a desirability for the Russians it had theretofore lacked (up till that time, the Soviets had diplomatic relations in Latin America only with Mexico, Argentina, and Uruguay; the region was one of comparatively minor interest to them. Khrushchev was not about to let China become the champion of the Cuban revolution. The Soviets must have been aware of the attraction China had initially had for many young Cuban revolutionaries, seeing parallels between their "peasant" revolutions, and even for ranking PSP members, such as the party's secretary general, Blás Roca, who was cordially received by Mao Zedong as late as April 1960. It was from Peking that Roca went to Moscow to meet Khrushchev for the first time (up to that moment, no senior Cuban leader had appeared in Moscow).

The Chinese were courting the Cubans while deliberately ignoring Khrushchev. For example, the official Chinese news agency, Hsin Hua, covered Khrushchev's two-hour General Assembly speech in only one paragraph in its daily world report. At that juncture, however, Castro obviously had already made the logical observation that the Soviets had the means to assist him greatly in his revolutionary endeavors, which impoverished China did not—although for the next two years he strove to navigate between the Communist giants, even coming up with the astonishing notion that he could mediate in that vast Marxist-Leninist dispute. Castro and his principal advisers acknowledge privately that the seriousness of the Sino-Soviet split in 1960 had to be among the overriding elements in Khrushchev's decision to go all-out for Cuba, going far beyond the economic accord Mikoyan had signed in Havana earlier that year, and following the caviar-and-vodka session between Castro and Alexeiev. Observing the rising hostility between Cuba and the United States, Khrushchev also reached the conclusion that Cuba was a priceless strategic asset for the Soviets in those days of relatively limited nuclear arms range and technology. After the U-2 incident with *its* strategic overtones, Castro's island offered too much potential military opportunity for the Soviet chairman—with his penchant toward adventurism—to resist the temptation of a full commitment. For Khrushchev, in effect, this was the moment of "the buying of Cuba," as a senior Cuban official commented privately and with certain bitterness not long ago, and the chairman presumably knew what he was doing. It is less clear, on the other hand, if Castro fully understood the geopolitical process upon which he was embarking in his bear hugs with Khrushchev in New York. Yet, it was he who pushed for the marriage, oblivious of the potential consequences.

Castro and Khrushchev were among the heads of state and government who had decided to attend the 1960 session of the United Nations General Assembly, marking the twenty-fifth anniversary of the birth of the world organization. There was an unprecedented gathering of notables—President Tito of Yugoslavia, President Nasser of Egypt, Premier Jawaharlal Nehru of India, Prime Minister Macmillan of Britain, and President Eisenhower were there—but the tall, bearded Cuban and the rotund, bald Russian instantly emerged as the stars, even aside from their personal eccentricities. Khrushchev's visit to Castro at the Theresa Hotel in Harlem, where in pique the Cuban and his entourage had moved from a midtown Manhattan hotel earlier in the week, was designed to crown their new alliance publicly.

This alliance took visible shape when First Deputy Premier Mikoyan, one of the few surviving "old Bolsheviks" of the Russian Revolution, alighted in Havana on February 4, 1960, for a nine-day visit. Mikoyan came with plenipotentiary powers to strike a complex and far-ranging deal with Fidel Castro, one that has never been made public in its entirety. Castro and Che Guevara, who greeted Mikoyan at the airport, were the principal Cuban negotiators, the Argentine in his capacity as National Bank president. After Mikoyan inaugurated the Soviet Science, Technology, and Culture exhibit at Havana's Bellas Artes Museum, the ostensible reason for his trip to Cuba, he delivered a speech to the government-controlled labor-union confederation (incongruously handing Castro a little check for ninety thousand dollars to purchase planes and arms.)

Then he was taken to see land cooperatives in Pinar del Río, Camagüey, and Oriente, and he visited Fidel's beloved Ciénaga de Zapata and the Isle of Pines, where he was shown the prison cell once occupied by the chief of the revolution. On February 13 a joint Cuban-Soviet communiqué was issued, emphasizing that the consolidation of world peace depended largely "on the inalienable right of every nation to decide freely its own political, economic and social road," which was the first Soviet commitment, indirect as it may have sounded, to the protection of the Cuban revolution. Another political aspect of the communiqué was the agreement to discuss, when convenient, the resumption of Cuban-Soviet diplomatic relations; it was obviously convenient very quickly because this resumption "on the level of embassies" was announced on May 8. Cuban officials say that Castro sought to obtain some form of explicit security guarantees from Mikoyan, but the communiqué does not touch upon this theme. It is known, however, that the deputy premier informed Castro that Soviet arms would be delivered to Cuba as soon as possible; indeed, they began arriving late that year, immediately after the initial Czech shipments. In this broad sense, a strategic understanding was reached.

Economically, the results—as outlined in the communiqué—were not particularly impressive, but this relationship had to develop gradually in terms of Cuba's ability to absorb Soviet aid. The arrangement worked out by Mikoyan with Castro and Guevara was called a "trade agreement," a term that would always be applied in the future to all economic dealings between the two countries; "assistance" was not a word the Cubans desired to see in print. Specifically, the Russians agreed to buy 425,000 tons of Cuban sugar during the balance of 1960, in addition to the 345,000 tons they had already bought during the spring (the total exceeded somewhat the 1959 purchases, but it was less than one fifth of the current Cuban crop). Because the communiqué did not mention the price, it was assumed that Moscow was paying the very low prevailing world price, roughly one half the subsidized price per pound the United States was paying under its quota system, in which Cuba still participated. Additionally, the Soviets committed themselves to buy one million tons of sugar annually (the 1960 total was 770,000 tons) for the next four years, certainly not enough to meet Cuban foreign-exchange requirements inasmuch as it was basically a barter transaction with payment in kind. Thus, for the 1961–1964 period, the Soviets offered Cuba $100 million in credits "for the acquisition of equipment, machinery and materials" and for technical assistance.

Contrary to Cuban assertions, this was not a phenomenal deal for Havana—actually, it was a better one for Moscow—but it did involve the Soviets even more deeply in the survival of the Cuban revolution. If this was Castro's and Guevara's long-term objective, they did outfox Mikoyan (who was of Armenian-descent) who could not possibly predict how Cuban-

American relationships would subsequently develop. On the other hand, the Cuban explanation of the agreement showed that their leaders either did not understand world economics or were willfully misleading the citizenry. In speeches a month after signing the Mikoyan accord, Castro and Guevara argued that it did not matter that only 20 percent of the Cuban sugar harvest going to the Soviets under the trade agreement would generate actual dollars because, as he put it, "the dollar is nothing more than an instrument for purchasing, the dollar has no value other than its purchasing power, and when we receive manufactured goods or raw materials [from the Soviets], we are simply using sugar as if it were dollars." A quarter-century later, when 80 percent of all Cuban exports went as barter to the Soviet bloc, Castro was desperate for dollars because the high technology he now needed could be obtained only in the West and only for dollars he did not possess. But on March 18, before the Mikoyan deal was four weeks old, Castro received a helicopter as a present from the Soviet Union for his personal use. It was handed over to him ceremoniously by the chief of the Soviet trade mission, which had already installed itself in Havana.

(Curiously, one person Mikoyan specifically asked to meet in Cuba was Ernest Hemingway, who had lived on and off since 1939 at Finca Vigía in San Francisco de Paula, a southeastern Havana suburb. Hemingway had been in Cuba during the first half of 1958, the last year of the war, and his ample published correspondence shows a certain interest in the revolution. For example, in a letter to his son Patrick written from Idaho in November, he said that "Cuba is really bad now . . . living in a country where no one is right—both sides atrocious—knowing what sort of stuff and murder will go on when the new ones come in—seeing the abuses of those in now—I am fed on it . . ." Still from Idaho, he wrote a New York editor late in January 1959 that "the officer commanding Havana Garrison is an old San Francisco de Paula boy who used to play ball on local team I used to pitch for . . . I knew Phil Bonsal the new ambassador when he used to work for I.T.T. . . . He is a very sound able guy but will naturally be working our interests. . . . Castro is up against a hell of a lot of money. . . . If he could run a straight government it would be wonderful . . ." Hemingway was away from Cuba all of 1959, but shortly before returning to the island early in 1960, he wrote his friend General Charles T. Lanham that "I believe completely in the historical necessity of the Cuban revolution. I do not mix in Cuban politics but I take a long view of this revolution and the day by day and the personalities do not interest me. . . . In the present situation there is nothing I can say that would not be misinterpreted or twisted. I have a terrible amount of work to do and want to be left alone to do it." Nevertheless, Hemingway found time to entertain Mikoyan at lunch at Finca Vigía; a photograph shows him smilingly pouring a drink for the deputy minister and Vladimir Bazikin, then the Soviet ambassador in Mexico. Castro did not accompany Mikoyan, and the only time Hemingway and Fidel did meet was in May at Havana's Barlovento Yacht Club where both attended the Hemingway Fishing Tournament. Castro had won the individual championship that day, catching the biggest blue marlin. It is intriguing why Castro never sought out Hemingway while he still lived in Cuba in 1960, given his admiration for the writer and Hemingway's support for the revolution; this may be the only time when Fidel was uncomfortable imposing on someone, in this case a man he regarded as a genius. Gabriel García Márquez has written that Castro knows Hemingway's work "in depth, that he likes to talk about him, and knows how to defend him convincingly." In a 1984 interview, Castro said lamely that he never got to know Hemingway "because those early days of the Revolution were very busy ones, and no one thought that [he] would die so quickly." He added: "What I like most about Hemingway are his monologues.")

After the Mikoyan visit and the establishment of Soviet-Cuban diplomatic relations (Sergei Kudryatsev, a specialist on Latin America, was the first ambassador to Havana, and Faure Chomón, the former Students' Revolutionary Directorate leader, went to Moscow), the new alliance's growth was directly proportionate to the increasing hostility between Cuba and the United States. In retrospect, it does appear that at this stage both Castro and Khrushchev were keen on exploiting this hostility—Castro probably more than Khrushchev—and that the United States did not quite know what to make of the emerging Havana-Moscow axis, other than being scared of it. Ambassador Bonsal wrote that the Mikoyan trade agreement in his opinion "did not in itself jeopardize the American economic position in Cuba," but that "the pleasantries exchanged by Mikoyan and Castro in February had been given the most alarming significance in some Washington quarters. The economic arrangements between Cuba and the Soviet Union seemed intolerable to people long accustomed to a dominant American position in Cuba."

Then a ship blew up in the Havana harbor, destroying the last chance for an accomodation between Fidel Castro and the Americans. The vessel was *La Coubre,* a French freighter that docked in Havana's inner harbor on March 4, with seventy tons of ammunition and explosives from Antwerp, which was the balance of the matériel purchased from Belgium the previous year. The first explosion, killing and maiming mainly crewmen and stevedores, came around 5:00 P.M., and a second one about an hour later killed and injured Cuban soldiers, militiamen, and firemen. The death toll was eighty-one. Castro arrived and directed rescue operations while accusing the United States of "sabotage." No proof of sabotage was actually produced by the Cubans, and the cause of the explosion was never officially established. The United States angrily rejected Castro's charges, pointing out that docking an explosives-laden ship inside a busy harbor violated international safety rules; however, the French captain never explained publicly why he had tied up his ship there. In any event, the *La Coubre* incident rallied the Cuban masses around Castro at a time when he was beginning to face growing internal political problems and there were signs of erosion in his popularity. At funeral services the next day at Colón cemetery, Castro delivered an immensely emotional oration, declaring that "today I saw our nation stronger than ever, today I saw our revolution more solid and invincible than ever" and vowing that "Cuba will never become cowardly, Cuba will not step back, the Revolution will not be detained. . . . The Revolution will march ahead victoriously! . . ." He ended by using for the first time his great revolutionary slogan *Patria o Muerte, Venceremos!* (Motherland or Death, We Shall Win!). Since that day, every revolutionary speech in Cuba ends with this phrase, which the audience picks up and repeats with rising fervor.

La Coubre was a milestone in many ways. For a majority of Cubans, the explosion confirmed Castro's predictions that the United States was determined to stamp out the revolution; they believed that this tragedy was the work of "enemies of the revolution . . . who do not wish us to receive arms for our defense." For Castro it was another useful confirmation of his premise that external danger strengthens the revolution because of the powerful impact of nationalism on popular reactions; the United States, of course, obliged him from the first year of the new regime to the 1983 invasion of Grenada when armed Cubans fought an overwhelming force of armed Americans.

For the United States, Castro's accusations over *La Coubre* became, in the words of Ambassador Bonsal, the official argument that "perhaps . . . tipped the scales in favor of Washington's abandonment of the policy of non-intervention in Cuba." Likewise, 1960 was an election year, and, as Bonsal put it, "the American posture of moderation in the face of Castro's insulting and aggressive behavior was becoming a political liability." Bonsal was never informed of it officially, but he believed that "the new American policy . . . was one of overthrowing Castro by all the means available to the United States short of the open employment of American armed forces in Cuba."

Indeed, on March 17, President Eisenhower approved a basic policy paper on the subject: "A Program of Covert Action Against the Castro Regime." Developed by the CIA and the White House "Special Group" (the deputy under-secretary of state, the deputy secretary of defense, the director of the Central Intelligence Agency, and the special assistant to the president for national security affairs), the plan's principal feature was "the development of a paramilitary force outside of Cuba for future guerrilla action." Before long, the concept of the paramilitary force for guerrilla warfare grew into a full-fledged invasion brigade. It is now known that Vice-President Nixon was the chief advocate of the March 17 plan. After meeting Castro in Washington the previous year, he concluded he was dealing with a Communist. And Nixon was also a presidential candidate in a race in which the question of Cuba loomed large.

However, it is not really credible that the Cuban reaction to *La Coubre* was crucial in the U.S. decision to attack Castro. Such a program could not have been elaborated in only twelve days. In fact, the report of the board of inquiry on the Bay of Pigs invasion emphasized that the Special Group had drafted the program for covert action against Castro at a meeting on *January 13* for Eisenhower's signature on March 17. There is ample evidence that the initial idea to remove him already constituted top-secret policy in March of *1959,* and almost any excuse would have served for the presidential approval of the anti-Castro "Program.". The Cuban regime, and most notably INRA's Rebel Army officers controlling land reform, were seizing American property in absolute disregard of Castro's own law, anti-American propaganda was unbridled, and the revolutionary government felt much more secure now that Mikoyan had given it Moscow's blessings. Washington's responses, produced in the heat of approaching elections and in incomprehension of Cuban realities, played into Castro's eager hands and rewarded him with the fulfillment of his prophecies.

The astonishingly patient Ambassador Bonsal had concluded in mid-spring that Cuban-American relations had no future, though he neither favored violent action against Castro nor was he being kept posted by the Eisenhower administration on its latest secret plans. In a private conversation one afternoon, he remarked, "You know, it's a no-win situation. . . . It's damned if we do, and damned if we don't. . . . You just can't please Castro." Bonsal and Castro had not met since the previous September (their third and last encounter), and official Cuban propaganda criticized the ambassador frequently. For example, in January he was attacked in the press for going to the airport to say good-bye to the Spanish ambassador, Juan Pablo de Lojendio, whom Cuba had ordered expelled on a twenty-four-hour notice. Lojendio was ejected because he had burst into a TV studio as Castro was speaking to protest Castro's attacks on the Franco regime made earlier in the broadcast. The Spaniard had watched the televised speech at his residence and, full of indignation, rushed to the station to confront Fidel. The ambassador was a short, balding man, and there was a touch of the grotesque to his brief shouting duel with the towering Castro before he was bodily removed.

Before *La Coubre,* there was a final attempt by the State Department at basic Cuban-American negotiations, but they were frustrated by Castro's insistence that as a first step the United States should take no steps against Cuba while the talks lasted. He had wanted to make sure that the U.S. Congress would not cut or reduce the Cuban sugar quota in the interim— the administration had requested Congress in January to grant it executive discretion on the quotas, but the State Department decided to turn him down on the technically correct grounds that it could not speak for the legislative branch. Had there been goodwill, however, a way to negotiate might have been found, but basically Eisenhower and Nixon were sick and tired of Castro: They had other solutions in mind. Since Castro was equally correctly convinced that the term "historical fatalism" would be brutally applied to him, he was in no mood to compromise, either. This mind-set on both sides would—with brief periods of relaxation and mutual probing—define the Cuban-American relationship in the next quarter-century.

Economic warfare broke out in earnest in May when Che Guevara informed two American and one British company, each owning a refinery in Cuba, that henceforth they would have to process crude petroleum imported from the Soviet Union. The companies had traditionally shipped to Cuba the oil they were producing in Venezuela as part of the worldwide production-shipping-marketing-refining system then practiced by the multinationals. Guevara argued that Cuba had the sovereign right to import the cheaper Soviet crude (the Soviets chose to make it cheaper and, besides, Cuba would be paying for it with sugar instead of dollars, under the Mikoyan accord), and warned the companies that the government would not pay them $50 million owed for earlier imports. Acting on the direct advice of the Treasury Department (the State Department not having been consulted), the companies decided to reject the Cuban demand without any serious negotiations. On June 29, Castro seized the three refineries while the Soviets assembled enough tankers to transport all the crude Cuba needed from Black Sea ports, and Soviet technicians adjusted the plants for processing the new oil. The arrival of the *Andrey Vishinsky,* the first tanker, is now celebrated as a revolutionary anniversary in the country. Castro portrayed Cuba as an aggrieved party in the dispute, and Bonsal wrote that "the Cuban Revolution had won a great victory and had had a powerful ally thrust into its arms." That day, a U.S. aircraft carrier sailed past Havana, and sent two of its jets roaring over the refineries. Both Eisenhower and Khrushchev were, in effect, doing Castro's bidding in their apposite fashions.

On July 6, Eisenhower made the next move in the economic war with Castro when he announced that the United States would not allow the importation of the balance of the Cuban sugar under the 1960 quota, roughly one quarter of the annual total of some three million tons. Eisenhower also made it clear that the United States would not buy *any* Cuban sugar until further notice. The White House acted immediately after Congress authorized the administration to allocate and reallocate the sugar quotas of foreign producers, a law that Fidel Castro described as the "Dagger Law," the dagger in the back of the revolution. Bonsal, who had opposed ending the quota, rejects the suggestion that Eisenhower acted in reprisal for the oil refineries' seizures: "The suspension of the sugar quota was a major element in the program for the overthrow of Castro." As for Castro, he had warned in a television speech two weeks earlier, when the "Dagger Law" was about to be passed, that "if we lose our entire sugar quota, they could lose all their investments in Cuba"—plus the huge annual trade surplus. The loss of the quota, Castro said in a five-hour speech, which was a treatise on the history of sugar and Cuban-American trade relations over a century, "would cost Americans in Cuba down to the nails in their shoes."

However, Castro waited a full month before striking back. Always concerned with strategy and tactics, he first sought to extract maximum political advantage from the sugar affair by portraying Cuba as the victim of "economic aggression" by the United States in violation of the charter of the Organization of American States (for which he otherwise had no use), and then to obtain overwhelming Soviet public support for the Cuban cause.

Fidel had always understood the immense importance of Third World solidarity with his revolution, and as early as mid-1959, he had dispatched Che Guevara to Africa and Asia to look for friends. Early in May 1960, President Achmed Sukarno of Indonesia, a virtual ally of the Soviet Union, became the first foreign chief of state to visit Cuba; the Marxist prime minister of British Guyana, Cheddi Jagan, came twice that year. At the end of August, the Cuban delegation walked out of an inter-American foreign ministers' conference in San José, Costa Rica, where a resolution criticizing Cuban meddling in Latin America was approved; on September 2, Castro convened one million Cubans in Havana to hear his denunciation of the San José conclave and to "approve" the First Declaration of Havana, condemning man's exploitation by man in the impoverished world. He was already reaching out for Third World leadership, knowing that his natural allies were there, but in those days Washington was not taking the Third World very seriously; postwar decolonization was still very much under way.

Nikita Khrushchev came through with flying colors. As soon as Eisenhower chopped off the sugar quota, Moscow announced it would buy (though at world prices) the 700,000 tons that would have gone to the

United States, in addition to the 770,000 tons the Soviets had already purchased in 1960. This meant that over one fourth of that year's sugar harvest was going to Russia. Though Castro and Guevara still could not make up their minds whether Cuba should remain a major sugar producer or concentrate (as Che urged) on industrialization, the Soviets were becoming, in effect, the guarantors of the island's economic viability. Khrushchev clearly had already taken the strategic decision to form an alliance with Cuba, but it is difficult to understand the intellectual processes in the Eisenhower administration that helped to smooth the way for this new friendship. Khrushchev's most extraordinary commitment to Cuba came on July 9 when he declared in a speech that "the Soviet Union is raising its voice and extending a helpful hand to the people of Cuba. . . . In a figurative sense, if it became necessary, the Soviet military can support the Cuban people with rocket weapons . . ."

This comment may have been meant symbolically—Khrushchev did use the word "figurative"—but Castro chose to manipulate it for his own purposes, in the same way in which the origins of the 1962 Cuban missile crisis are still being obfuscated by him. In a display of high drama, Castro spoke on television from his sick bed the next day (his illness was never explained) to thank the Soviets for their expressions of support, but he repeatedly stated that the Khrushchev offer was "absolutely spontaneous"— although he also threw out heavy hints that Moscow had offered "real" rockets. Typically, Fidel appeared to be trying to avoid the impression in Washington that *he* had requested Soviet "rockets," while at the same time magnifying the scope of the Soviet commitment to suit himself. Khrushchev, who was beginning to learn that life with Fidel was not always easy, must have objected to Castro's public interpretation of his words because the entire reference to "rockets" has disappeared from modern Cuban official texts. But in July 1960 it was crucial for Castro to be able to print prominently in his newspapers the Soviet promises of "rockets," even though a joint communiqué issued in Moscow when Raúl Castro met with Khrushchev early in August mentioned no rockets of any kind.

In any case, Fidel was in fine form on July 26 when he returned to the Sierra Maestra to celebrate the seventh anniversary of Moncada and to proclaim that "we shall continue to make our fatherland an example that will make the Andes mountain range into the Sierra Maestra of all America." He was ready now to punish the United States for taking away his sugar quota. The Cuban cabinet passed a Law of Nationalization right after Eisenhower had acted on the sugar question and on August 5 Castro was ready to implement it. He called it the "Machete Law" in retaliation for the U.S. "Dagger Law." Much land, chiefly from the United Fruit Company, had already been seized, but now Fidel moved on the bulk of American investment on the island. Conchita Fernández, his secretary at INRA, recalls that late in the evening he called her into his fourth-floor office and told her: "Call Che and call that Mexican named Fofo, because I'm going to nationalize now all these foreign companies: Shell, Standard Oil, Esso . . . right now, at midnight." Castro instructed her to summon Carlos Franqui, then editor of *Revolución*, so that the announcement of the nationalizations could be published the following morning. Conchita Fernández says that Castro had remained at his INRA office three days and nights preparing the nationalizations with Che Guevara, and then he suddenly announced that he was ready to sign the documents. He told Conchita: "Right now, we'll give the back of the hand to these imperialist companies." Then, he went to a Latin American youth meeting at the Havana stadium to announce what he had done. In all, Cuba nationalized thirty-six American-owned sugar mills, two oil refineries, and two utility companies (two nickel mines, one belonging to the U.S. government, were taken in October). It was the end of an era—and the start of a new one for Cuba, at home and internationally.

In the context of a nearly total break with the United States and an emerging strong Soviet alliance, Fidel Castro flew to New York in September to address the United Nations, and, as much as anything else, to meet Nikita Khrushchev on American turf. It was the kind of irony in which Fidel delights. He felt very strong and self-assured, having successfully weathered his great confrontation with the United States and consolidated his domination at home. Finance Minister Rufo López-Fresquet, the last moderate in the cabinet, had resigned in March, and Castro no longer needed to rule with a "hidden government." He and his companions were fully and openly in charge of all that mattered in Cuba, and what remained of relatively independent life—such as culture—would soon be regimented as well. The Cuban Workers' Confederation (CTC) had tried to keep Communist leadership at bay, but Castro put an end to that in November 1959 when he forced the CTC convention to drop an elected non-Communist slate; the force of his personality always won the day when all else failed. The CTC's secretary general, David Salvador, a top 26th of July Movement leader, was dismissed and joined the anti-Castro underground before being arrested.

Propaganda and the control of public opinion are fundamental Castro concerns, and special attention was given them as soon as sufficient revolutionary consolidation was attained. Superficial observation of Castro's seemingly wildly uncoordinated activities and his chaotic life-style had suggested that he was running Cuba from whim to whim. In retrospect, however, it is evident that Fidel knew exactly what he was doing—and how and when. Every step logically followed the previous one; the timing was impeccable because Castro sensed what the nation was prepared to accept, his sermonlike speeches fit perfectly into the concept of revolutionary indoctrination, and his improvisations were in reality carefully thought-out chessboard moves. When outsiders commented on Castro's "rantings," they failed to perceive that he was engaged in a campaign to educate the masses in his beliefs and political and economic analyses.

It was logical, then, to subordinate the media to the revolution. During the first year, he was basically served by *Revolución,* the organ of the 26th of July Movement, and the Communist party's newspaper, *Hoy,* resumed publication shortly after the victory; Carlos Rafael Rodríguez, Fidel's principal Communist ally, was the editor. From the outset, Castro acted as *Revolución*'s supereditor, visiting its offices, conferring almost daily with Carlos Franqui, who was the editor and chief propagandist, and seeing to it that the revolutionary line was phrased precisely the way he wanted it to be every time. The two principal TV channels, CMQ and *Mundo,* were put instantly under official control, though for a time their private ownership was left untouched. Radio stations were joined in a network called FIEL, always available to broadcast Fidel's words.

Early in 1960 it was time to curtail the independent press. Castro invented the notion of the *coletilla,* a postscript appended at the end of every article, news dispatch, editorial or photograph that happened to disagree with the official line on anything. Castro-controlled journalists' and printers' unions were charged with drafting and publishing these *coletillas,* or, told simply to refuse to run what they disliked. It was a lethally subtle form of censorship-cum-intimidation, all in the name of the revolution, and by mid-1960 the so-called "bourgeois press" went out of business because editors could not control the content of their publications, and the rapidly shrinking private sector of the economy could no longer provide the required advertising. Castro's view was that only the revolution brought real freedom of the press to Cuba to replace the right-wing biases of the bourgeoisie. His propagandists went the absurd extra mile to ban Santa Claus as a Christmas symbol. A revolutionary figure named "Don Feliciano" took its place, and a popular American tune reemerged equipped with new lyrics: "Jingle Bells, Jingle Bells, Always with Fidel!" It did not catch on (nor did that other revolutionary jingle, "Ping-Pang-Poong! Viva Mao Tse Tung!").

Astonishingly, it was the "old" Communist party that went on giving Castro trouble, notwithstanding Nikita Khrushchev's fervent July support of the Cuban revolution. Castro and his group and the old-line Communist leaders had begun meeting secretly eighteen months earlier to plan an ultimate merger, but the party still resisted the "exceptionality" of the

Fidelista triumph outside Marxist dogma. Only Castro himself was trusted by the PSP old guard, to the extent they trusted anyone, and Pedro Miret recalls that on many occasions when Raúl Castro made a proposal, Blás Roca, the PSP secretary general, would ask, "but have you cleared it with Fidel?" At the PSP's Eighth Congress in August 1960, Blás Roca seemed out of synch with reality when he described the revolutionary regime as a power that "represents and executes the policy of the coalition of the pro-letariat, the peasantry, the petty bourgeoisie, and the advanced sectors of the national bourgeoisie." Aníbal Escalante, another top PSP leader, op-posed the confiscation of all private property because of the "national bour-geoisie's strong fear of the revolutionary changes," adding that "we maintain the strategy of the alliance of classes with which the revolution originated." This was historical nonsense, but as subsequent events would show, the "old" Communists appeared to be holding out for key posts in the regime, a step Castro was not about to take.

Only Carlos Rafael Rodríguez was Castro's unconditional ally (he seemed to be better attuned to Moscow's thinking than his colleagues), and he was largely responsible for the October 1960 merger between the party's So-cialist Youth and the 26th of July Movement's youth division. Recalling those days, Pedro Miret says that the first step for the *Fidelistas* was to "dilute" the Movement without most Cubans realizing it, because "we had to create our own little group." Castro was setting the stage for the Inte-grated Revolutionary Organizations, the ORI mechanism he regarded as a preparation for his own Communist party, but through most of 1960, the "old" Communists were dragging their feet, often insisting that they repre-sented the views of the Kremlin. Finally, Aleksandr Alexeiev, back in Cuba ostensibly as the permanent TASS correspondent (while Kudryatsev acted as ambassador), delivered a private message from Khrushchev to Castro to the effect that the Soviet government considered "there is no intermediary party," between them, and that Fidel was the "authentic leader" of the revolution. Anticipating meeting Castro in New York, Khrushchev was careful not to make him feel like the head of an Eastern European satellite country.

Unlike Cuba's old-line Communists, Khrushchev also evidently had the sense to accept Castro for what he was at that stage in history, although he learned later that Fidel could not easily be maneuvered into going along with the Kremlin dogma. In attempts to reconstruct Castro's ideological evolutions—and to dispel the persistent impression in successive American governments that he always was a Soviet tool—it is useful to consider his comments on Marxist interpretations in conversations with the French writer Régis Debray in the mid-1960s. Castro said: "I am accused of heresy. It is said that I am a heretic within the camp of Marxism-Leninism.

Hmmm! It is amusing that so-called Marxist organizations, which fight like cats and dogs in their dispute over possession of revolutionary truth, accuse us of wanting to apply the Cuban formula mechanically. They reproach us with a lack of understanding of the Party's role; they reproach us as heretics within the camp of Marxism-Leninism." Debray explains that the great difference between Castro and the dogmatics accusing him of heresy was his belief that Marxist-Leninist parties are not necessarily the only or the best "vanguard" leadership in launching revolutions. The *Sandinista* victory in Nicaragua in 1979 would prove Castro right, but in the meantime he had to keep protecting his heresies.

Fidel Castro arrived in New York on September 18, the day after signing decrees nationalizing three American bank branch offices in Cuba. Four days before he reached the city, a CIA-organized plot to assassinate him was formally set in motion at a meeting at a New York hotel with a key Mafia figure whom the Agency wanted to handle the assignment. The decision to murder Castro was an outgrowth of the "Program" against him approved by Eisenhower in March. The assassination was to coincide with intensive guerrilla warfare to be triggered on the island; the first guerrilla operations had, in fact, already begun in the Escambray Mountains.

An internal CIA memorandum states, "In August, 1960, [Deputy Director] Mr. Richard M. Bissell approached Colonel Sheffield Edwards to determine if the Office of Security had assets that may assist in a sensitive mission requiring gangster-type action. The mission target was the liquidation of Fidel Castro." Bissell, who was also in charge of preparing the Bay of Pigs invasion, briefed CIA Director Allen Dulles, who "gave his approval." What remains unknown is whether Dulles acted on his own authority in ordering the assassination of a foreign head of government or whether he obtained President Eisenhower's explicit assent. In the 1960s, however, the White House practiced the concept of "plausible denial" in risky and potentially embarrassing operations to protect the prestige of the president, and usually the CIA director and the national security adviser took it upon themselves to keep Eisenhower in the dark about specific enterprises so that he could claim ignorance without actually lying if the Agency were caught red-handed. Authority for such operations was derived from overall policy directives by the president—such as the March decision to oust Castro. "Plausible denial" was invoked for Eisenhower in May when the U-2 spy plane (another Bissell undertaking) was shot down over the Soviet Union, and chances are that the same principle was applied to the Castro assassination plot.

Castro says that he had assumed all along that he was on an American hit list, but he had also concluded that he would be safer in the United States than at home; he did not think the CIA would risk the awesome political fallout of murdering him in Manhattan. According to the CIA memo, a former agent of the Federal Bureau of Investigation, Robert A. Maheu, was asked if he could develop "an entree into the gangster element" as the first step toward organizing the murder. Maheu therefore met at the Plaza Hotel in New York on September 14 with Johnny Roselli, described as "a high-ranking member of the 'syndicate.'" An offer of $150,000 to kill Castro was conveyed on behalf of "businessmen" who had suffered financially in Cuba because of the revolution, Roselli put Maheu in touch with Mafia chiefs Salvatore "Momo" Giancana and Santos Trafficante in Miami. On the gangsters' suggestion, the CIA's Technical Services Division developed and produced pills with "elements of rapid solubility, high lethal content, and little or no traceability." The CIA says that "several attempts without success" were made to have Castro take the pill in some fashion, and "the project was cancelled shortly after the Bay of Pigs episode." Subsequent testimony at Senate hearings into anti-Cuban intelligence operations disclosed that the murders of Raúl Castro and Che Guevara were also contemplated because, in the opinion of the chief of the CIA's Western Hemisphere Division, unless the three top leaders "could be eliminated in one package—which is highly unlikely—this operation can be a long, drawn-out affair and the present government will only be overthrown by the use of force."

This was the attitude prevailing toward him in the United States government when Castro materialized on short notice in New York to lead the Cuban delegation to the U.N. General Assembly. The great unwitting irony of his reception was that the U.S. government confined him and his entourage to the island of Manhattan "to insure his personal safety," and that during his ten-day stay in the city, the special 258-man police detail guarding him was nearly as large as Castro's Rebel Army of 300 when it launched its final offensive in 1958. The Police Department obviously had no way of knowing that another official United States agency was planning to kill elsewhere the man they were guarding in New York, and the State Department (also unaware of the CIA plot) was desperately searching for a city hotel willing to accept the Cubans. To compound matters, the Cubana Airlines plane that brought Castro to New York had to race home to Havana to avoid being seized by American creditors; another Cubana airliner had been placed under lien at the airport two days earlier. Castro's presence in New York triggered a mass of misunderstandings, plenty of American harassment, dangers to Fidel from Cuban exiles, and—secretly—the shadow of official assassination. Castro may not have known about Maheu, Roselli, and poison pills, but he was thriving on the rest of it. In Havana, where Raúl Castro was acting prime minister, the American ambassador's movements were restricted to the Vedado residential section in reprisal.

Nikita Khrushchev had planned to meet Fidel Castro even before he boarded the liner *Baltika* in Kaliningrad for New York. Arkady N. Shevchenko, then a junior diplomat traveling as an adviser, recalls that the chairman began talking to him about Cuba one day on the *Baltika's* deck as she sailed north of the island. (Shevchenko, who later rose to the rank of Under-Secretary General of the United Nations, defected in 1978 and wrote his memoirs in *Breaking with Moscow*.) Shevchenko recorded Khrushchev's remarks: "I hope that Cuba will become a beacon of socialism in Latin America" and "Castro offers that hope, and the Americans are helping us." Khrushchev said that the United States was trying to drive Castro to the wall instead of establishing normal relations with him, adding, "That's stupid, and it's a result of the howls of zealous anti-Communists in the United States who see red everywhere, though possibly something is only rose-colored, or even white. . . . Castro will have to gravitate to us like an iron filing to a magnet."

In New York, Khrushchev took immediate steps to accelerate this gravitation by visiting Castro in Harlem. Fidel's choice of black Harlem as his headquarters made the visit even more interesting politically to the chairman. The day Khrushchev landed in New York, Castro had abandoned the Shelburne Hotel on Lexington Avenue and Thirty-seventh Street, a comfortable establishment near the United Nations, in protest against what he described as the management's "unacceptable cash demands" for deposits; he may not have realized that the Shelburne had rented the Cubans the twenty suites (at twenty dollars each) only because the State Department had pleaded with the hotel to house the Cuban delegation. The next day Castro and fifty Cubans rushed eight blocks in the descending dusk to the United Nations Secretariat Building on First Avenue to confront Secretary General Dag Hammarskjöld. Dressed in olive-green combat fatigues, Castro piled into a black Oldsmobile with seven companions, followed by other Cubans in cars or on foot, by police, and by hundreds of newsmen. Once arrived, Fidel informed the mild-mannered Hammarskjöld that his delegation would stay there until the housing problem was solved. He said the Cubans were prepared to march to Central Park, remarking that "we are mountain people, we are used to sleeping in the open air."

Turning down an offer of free accommodations at the fairly luxurious Commodore Hotel, just three blocks from the United Nations, Castro then directed his wild-looking motorcade to the eleven-story Theresa Hotel at Seventh Avenue and 125th Street in Harlem, where the management evidently awaited the Cubans. They took forty rooms, finally moving in at

12:30 A.M. Having totally overshadowed all other United Nations-related activities with his nocturnal thrusts around the city, Castro announced that he had thought all along about staying in Harlem because blacks would be more sympathetic to the Cuban revolution. Then, he stayed up most of the night, receiving black journalists and the Black Muslim leader Malcolm X.

Khrushchev's account of these events appears in his memoirs, *Khrushchev Remembers,* in which he reports his indignation upon learning that the Cubans "were thrown out" of their hotel and invited to Harlem. The next morning, he drove to Harlem to "shake Castro's hand as a gesture of sympathy and respect" after phoning the Cuban that he was on his way uptown from the Soviet U.N. residence. Khrushchev wrote that Castro had offered to call on him instead because "he thought that the Soviet Union was being a great country and his being a young revolutionary government representing a small country, it would be proper for him to pay a visit to me first." But, Khrushchev went on, "I felt it would be better for me to make the first visit, thereby emphasizing our solidarity with Cuba, especially in the light of the indignation and discrimination they were being subjected to. . . . By going to a Negro hotel in a Negro district, we would be making a double demonstration: against the discriminatory policies of the United States of America toward Negroes as well as toward Cuba."

At high noon on Tuesday, September 20, Fidel Castro greeted Khrushchev at the entrance of the Theresa Hotel in one of the most improbable diplomatic encounters of the postwar years. Khrushchev himself described it best: "He made a deep impression on me. He was a very tall man with a beard, and his face was both pleasant and tough at the same time. His eyes sparkled with kindness toward his friends. We greeted each other by embracing. When I say 'embrace,' I'm using the word in a rather specialized way. You have to take into consideration my height as opposed to Castro's. He bent down and enveloped me with his whole body. While I'm fairly broad abeam, he wasn't so thin either, especially for his age." The two men went up to Castro's ninth-floor suite for a twenty-two-minute conversation through interpreters. Khrushchev wrote that the Cuban "expressed his pleasure at my visit, and I repeated my sentiments of solidarity and approval of his policy. The meeting was very brief; we exchanged only a few sentences. . . . You can imagine the uproar this episode caused in the American press and elsewhere as well." *The New York Times* reported that "it was the biggest event on 125th Street since the funeral in 1958 of W. C. Handy, who wrote 'St. Louis Blues.'"

Back at the Soviet residence on Park Avenue and Sixty-eighth street, Khrushchev told reporters he was "very much pleased with the conversation I've had with Dr. Castro," whom he described as "an heroic man who has raised his people from the tyranny of Batista and who has provided a better life for his people. . . . I salute Fidel Castro and wish him well." But this was not the end of Khrushchev's courtship of the revolutionary chief. At the afternoon session of the General Assembly, he walked from his seat, almost at the rear of the hall, to the front, across the rostrum, and over to the front row of the other side where Castro sat with his diplomats. Castro then rose to his feet, and the two men embraced repeatedly and beamingly for the benefit of photographers. Three days later, Castro dined with Khrushchev at Soviet headquarters on Park Avenue for four and a half hours, which was evidently when they held their substantive talks—including discussion of the scope of Moscow's military backing for Cuba. After the meeting broke up at midnight, Khrushchev walked arm in arm with Castro to the Cuban's car. Their next meeting would be three years later at the Kremlin when they tried to make up after a bitter dispute over Soviet nuclear missiles in Cuba.

Shevchenko writes that on his return from Harlem, Khrushchev told his staff that "he had found that Castro wanted a close friendship with the U.S.S.R. and asked for military aid . . . moreover, he got the impression that Castro would be a good communist. While Khrushchev was enthusiastic, he also added that it would be necessary to be cautious. 'Castro is like a young horse that hasn't been broken,' he said. 'He needs some training, but he's very spirited—so we'll have to be careful.'"

Apart from his meetings with Khrushchev, the Cuban prime minister saw only Czechoslovakia's president, Antonin Novotny, and Bulgaria's Premier Zhivkov, and, from the neutralist group, President Gamal Abdel Nasser of Egypt, Prime Minister Nehru of India, and President Kwame Nkrumah of Ghana. Yugoslavia's Marshal Tito declined to see Castro despite efforts by Foreign Minister Raúl Roa. Fidel treated the General Assembly to a four-and-a-half-hour speech on September 26, speaking from only a single sheet of notes, accusing the United States of "aggression" against Cuba. Khrushchev sat through the entire address, often interrupting with smiling applause.

While in New York, Castro also chaired by telephone a cabinet meeting in Havana to decide on diplomatic relations with the People's Republic of China and North Korea. He spent much time in his suite at the Theresa Hotel, presumably because there were not enough diplomatic contacts available to fill his ten-day stay in New York. He brought from Havana Major Juan Almeida, the Army Chief of Staff, who is black, to meet with black United States leaders, but little came out of it. He worked with Celia Sánchez and Captain Nuñez Jiménez to keep track of developments in Cuba, and many of his meals were chicken-and-rice dishes delivered to the hotel from a nearby restaurant. One evening, he invited the Theresa's black employees to a steak dinner at the hotel with him and Major Almeida; on another evening, Castro received the leaders of the Fair Play for Cuba Committee, including the poets Langston Hughes and Allen Ginsberg. After a few days, the novelty of his presence in Harlem wore off, the crowds thinned out, and only the police maintained a heavy presence. A police horse named Bangle collapsed in front of the Theresa from a kidney ailment resulting from exhaustion.

On September 29, Castro finally flew home aboard a Soviet Ilyushin-18 turboprop airliner, so as not to risk the seizure of a Cuban plane by American creditors. At the airport, before leaving, he declared that "the Soviets are our friends. . . . Here you took our planes—the authorities rob our planes—Soviets give us planes." Back in Havana, greeted by 150,000 cheering fellow citizens, he told them the United States was a "cold and hostile nation" and New York was "a city of persecution." But on balance, Castro evidently achieved what he came to seek in New York: a great deal of public attention and a solid understanding with Nikita Khrushchev concerning all forms of assistance for Cuba. It would be very urgently needed—and very soon.

On his return to the island, Fidel Castro instantly discovered a plethora of problems. Most of them were related to American pressures now that the Eisenhower administration was absolutely determined to liquidate his revolution. On October 18, Ambassador Bonsal was recalled "on extended consultations," knowing perfectly well that this marked the end of his Cuban assignment; the United States no longer wanted any dealings with Castro. The following day, the administration banned exports to Cuba of any American goods—except nonsubsidized foodstuffs, medicines, and medical supplies. This was the embargo, or "blockade" as Castro calls it, and it was still in force twenty-six years later, complicating the Cuban economy, having made Cuba totally dependent on the Soviet Union but failing to demolish the revolution.

The internal-security situation was a dangerous problem on another level, but it too was linked to American efforts to oust Castro. By September, he had to accept the reality that the rebel guerrilla groups in the Escambray Mountains in central Cuba (and to a much lesser extent in Oriente) could no longer be tolerated. Castro knew from his own experience how tough it is for a conventional army to deal with guerrillas—and now *he* had a conventional army—if they are allowed to expand. Moreover, he was aware that the bands in Escambray were air-supplied by the CIA, even if most of the drops never actually reached the rebels, being recovered instead by the Rebel Army and the militia.

Castro's principal advantage was that these rebels had no centralized command and no nationally identifiable leader—since *he* had been the undisputed leader in the Sierra Maestra—and therefore no coordination. On September 8, shortly before Castro flew to New York, the Rebel Army organized special battalions to launch the "Escambray Clean-up Operation," surrounding the mountains with as many as one hundred thousand militiamen in a massive search-and-destroy effort. Just before Fidel's return to Havana, the regime announced it had scored a victory over a rebel unit in Escambray—the first public admission that there was fighting in the central Sierras. At the same time, the first heavy Soviet arms—82mm mortars and 122mm howitzers—began arriving to bolster the island's defenses. The first Soviet tanks came early in 1961. Castro and his advisers believed late in 1980 that the Escambray uprisings were linked to a subsequent invasion, and that therefore it was urgent to destroy the guerrillas. Fidel was quite right because the CIA had considered a linkage between the Escambray forces and the Bay of Pigs landings (the mountains are adjacent to Zapata); when the exiles came ashore, however, the Escambray had already been neutralized.

Vice-President Fernández, then one of the few professional military commanders serving under Castro, says that the most dangerous moment in Escambray was between December 1960 and February 1961. That was why Castro had assumed personal command of the operations on his return from the United Nations, spending days and nights with the militia battalions and showing himself to the troops and the local population. Again, his *guerrillero* experience came in handy. Fernández recalls that the Escambray strategy devised by Castro was to place a militiaman permanently every forty or fifty yards along a road or a ridge, to live in his trench, and have food delivered three times a day. In this fashion, the mountains were completely sealed off, and other units moved into the hills to pursue the rebels. The rebel bands, Fernández says, never had more than twenty men each, moving rapidly from spot to spot, but never forming a strong group. Castro's and Fernández's best estimates are that the total guerrilla numbers may have reached five thousand at one point; in a single operation, for example, five hundred rebels were captured. They were a mix: small landowners who feared agrarian reform, ex-Batista soldiers, disgruntled Rebel Army fighters, and simple adventurers; much of the combat by then was in the name of anticommunism.

According to Fernández, the last band in Escambray was liquidated as late as 1965, which meant that the militiamen were tied down there, in greater or lesser numbers, for over five years. To deal with these uprisings without denuding other Cuban defenses, the militiamen were given twenty-one days of basic training before being sent off to do battle. Around six thousand men were trained simultaneously at any given time, and in time the militias became the backbone of Cuban defenses. They were crucial, for example, at the Bay of Pigs.

Another vital Castro weapon was his intelligence service under the command of Ramiro Valdés and Manuel "Redbeard" Piñeiro. The Rebel Army's intelligence arm had infiltrated to an astonishing degree inside rebel and antirevolutionary groups, making it possible for Castro to hold all his enemies at bay for nearly three decades, including the CIA. Castro says that at one stage there were as many as three hundred "counterrevolutionary organizations," each expecting American support, and that "we knew more what they did than they knew themselves" because of the infiltration. He claims that "a moment came, almost at the end, when our people were the chiefs of almost all these counterrevolutionary organizations." According to Castro, his security files were so complete that when a plotter was arrested, say in 1962, the regime knew everything he had done and whom he had seen in 1960 and before.

Full-time security organizations, however, were not considered sufficient, so on September 28, the day he came back from New York, Castro announced the creation of the Committees for the Defense of the Revolution (CDR) as a system of the people's "collective vigilance." The CDRs were Castro's invention because nothing on such a scale exists even in the Soviet Union, and their immediate function was to keep the police and Security Services informed of strangers appearing in their neighborhoods (there is a CDR for every urban block and in every plant and farm), of citizens voicing criticisms of the regime, and so on. Castro estimated in 1986 that 80 percent of the population belonged to the CDRs, an unparalleled security network. But nowadays the CDRs are also responsible for the vaccination of children and other community tasks.

For all practical purposes, the first major phase of the Cuban revolution was completed at the end of 1960, a two-year period. In the words of Carlos Rafael Rodríguez, this was the finish of "capitalism" in Cuba with the nationalization of foreign companies as well as Cuban industry, farms, and businesses. Ties with the United States had been broken, and an alliance with the Soviet Union created as a basic new relationship. Internal security was firmly in hand with the militias and the new CDRs, and the revolutionary government was free of "liberal" or "moderate" influences.

On October 13, Castro ordered the expropriation of 382 large industrial and commercial companies "belonging to the Cuban bourgeoisie," and all Cuban and foreign banks (except Canadian banks). On October 15 he went on national television to say that his revolutionary program outlined in "History Will Absolve Me" had been fulfilled—nationalization of foreign companies and agrarian reform—and that the revolution had entered a new stage. In the same speech, however, Castro pledged that the revolution "has no need" of liquidating small private businesses, such as shops and small factories. To set up "People's Stores" in the cities, he said, would create "an obstacle to the revolution," and the revolution has no interest in retail distribution mechanisms. In 1968, when Castro came up with still another revolutionary stage, he went back on his word to nationalize everything from the corner coffee shop to taxis in the cities. Not even ambulatory vendors were permitted to exist when Castro was experimenting with "pure communism," one of his greatest errors, as he would say much later.

When Castro presented the ideological program to the First Congress of his new Communist party in 1975, he declared that during the second half of 1960, "the Cuban Revolution entered its era of socialist construction." In October 1960, of course, he had omitted the word "socialist" in announcing the new revolutionary phase. Cuba was not quite ready yet for the "construction of socialism." But on December 31, 1960, Castro ordered general mobilization in Cuba to defend the nation from an imminent military attack "by the troops of Yankee imperialism," which he insisted President Eisenhower would carry out as his last act at the White House. Fidel Castro did not know he would be dealing with John F. Kennedy, now the president-elect and already partially briefed on the Bay of Pigs plans, and not with Dwight D. Eisenhower. In fact, the new phase in Cuban life would be more complex and explosive than Castro seemed to realize as he was ordering his New Year's Eve mobilization.

CHAPTER

4

Fidel Castro and John F. Kennedy belonged to the same generation. Both had superb minds and a vision of history. They had never met but were fascinated by each other as adversaries and as national leaders. I know it from discussing Castro with Kennedy in the short years when they were simultaneously in power in their respective countries, and discussing Kennedy with Castro over a quarter of a century later. Each was intensely interested by everything concerning the other. There was an intellectual respect between them. Historically, they had an immense impact on one another and their nations: Castro's existence pushed Kennedy into the tragedy of the Bay of Pigs and, in a strangely contradictory manner, into launching the Alliance for Progress programs for Latin America. Kennedy's presence, still feared by Castro after the Bay of Pigs, led to the Cuban request for Soviet military guarantees and to the subsequent installation of Soviet nuclear weapons on the island, causing the great missile crisis of 1962, the closest the world has ever come to a nuclear confrontation.

In this sense, Castro and Kennedy had a common destiny. To this day, Castro believes that had Kennedy lived, they would have, sooner or later, settled intelligently the basic Cuban-American dispute. To Castro, himself an assassination target, Kennedy's death came as a tremendous blow, and he frequently returns to the theme. Whether Kennedy would have wished a settlement that left Castro in power is a question without an answer; American historians hold differing opinions on this subject.

What is beyond question is that Castro played a central role in the history of the Kennedy presidency, and that Kennedy and his policies were crucial in Castro's ultimate attitudes toward the United States as well as the Soviet Union. Finally, Castro was the issue that forced the two superpowers to rethink and revise their fundamental nuclear strategies; in the case of the Soviet Union, the fiasco of Khrushchev's Cuban nuclear adventure resulted in the decision to accelerate efforts to attain at least nuclear parity with the United States, if not to gain superiority. In other words, the history of the Cuban revolution in the 1960s is to a very large degree the story of Fidel Castro's and John Kennedy's impact on each other.

Nineteen sixty-one, denominated by Castro the "Year of Education," was the third year of the revolution—and its ideological turning point in terms of openly embracing the Marxist-Leninist doctrine. For the United States, it was the year when Fidel Castro was to be liquidated as the Kennedy administration proceeded with the invasion plans secretly formulated by the Eisenhower administration.

Eisenhower's farewell gesture toward Cuba was to break diplomatic relations on January 3, the day after Castro demanded his speech on the anniversary of the revolution that the American Embassy in Havana be reduced to eighteen diplomats—the same number the Cuban Embassy had in Washington. The United States had over sixty diplomats in the Cuban capital, principally because American interests on the island were so sizable in every way, and Eisenhower chose to regard Castro's request as a provocation and as justification to sever diplomatic ties. Now the United States felt even freer to pursue every conceivable course of action to oust Castro. Even as John Kennedy was being inaugurated in Washington, Cuban exiles were being trained for the invasion at the CIA's secret camps in Guatemala. Castro, a believer in the strategy of crying wolf, used the anniversary celebrations to warn that "imperialism" remained a mortal danger despite the change in administrations in the United States, and he underlined it by presiding over the first revolutionary military parade in Havana. Rebel soldiers and militia units, no longer ragtag *guerrilleros,* happily displayed new Soviet, Czech, and Belgian weapons.

This display, of course, also symbolized the nature of Castro's essential dilemma of striving to take the revolution beyond rhetoric and of the simple dismantling of the pre-1959 economic and social order. Because he really did face a lethal danger from the United States, he was forced to divert much manpower, resources, energy, attention and leadership to defense—and away from "revolutionary construction." On the other hand, as he has often said himself, external danger was necessary to keep alive the revolutionary spirit, particularly when tough realities of daily life replaced the romantic euphoria of the victory days. To compound matters, Castro had no clear economic policy in the first years of the revolution, and then he went on to alter, rewrite, and replan as required by changing circumstances. Moreover, he lacked economists and managers on the policy level after dispensing with the "moderates" by the mid-1960s: For years, the Cuban economy was conducted from INRA by Castro personally with advice from Che Guevara, Captain Nuñez Jiménez who was a geographer and an ideologue, and, later, Carlos Rafael Rodríguez, a charming intellectual and the best politician Cuban Communists had produced.

The events in Cuba directly involving Castro, during just one week at the start of 1961, illustrate the pressures on him and the way in which he allowed himself (or forced himself) to be pulled in every imaginable direction. During 1960, Fidel had made up his mind that Cuba's principal priority had to be to make the population literate if it was to enjoy any progress, the fact of the revolution notwithstanding. When the revolution triumphed, roughly 40 percent of the six million Cubans were illiterate (as was most of the rebel army), and Castro calculated that there was a shortage of ten thousand elementary school teachers in the rural zones. In the course of its first two years in power, the Castro regime had added ten thousand classrooms to the educational system, but it lacked teachers for them. The notion of a crash literacy campaign was born from these realities, and beginning in mid-1960, the government formed "literacy brigades" of university students and high school seniors in the cities who fanned out all over the countryside. Training teachers would be the next step, and Castro's plan was to make 1961 the "Year of Education," himself directing the literacy campaign along with defense and the economy. This, then, was Fidel's week:

On New Year's Eve, he dined with ten thousand elementary teachers at the old army Camp Columbia in Havana, converted earlier into a school complex, to launch the literacy campaign, but to warn, at the same time, that "imperialist" agression was imminent. A few hours earlier, he had gone downtown to direct firemen in efforts to put out a fire that destroyed the vast *La Época,* and that in his opinion was an act of sabotage. He had also ordered that day general mobilization against an American invasion. On January 1, Castro officially inaugurated the National Literacy Campaign, declaring that Cuba needed "a revolutionary conscience that it is shameful not to know how to read and write." On January 2, he watched the military parade on *Plaza Cívica,* and demanded the cutback in the staff of the American Embassy. On January 4, Fidel donned his battle fatigues and a brown beret to join, rifle in hand and compass on his left wrist, the militiamen fighting the bands in the Escambray, having first signed in Havana documents creating a National Culture Center. On January 5, word reached Castro in Havana that Conrado Benítez García, a volunteer student teacher participating in the literacy campaign, had been assassinated by "counter-revolutionaries" in an Escambray village where he had just begun working with the local peasants.

Benítez now also belonged to the pantheon of the martyrs of the revolution, and his murder, too, was put at America's door by Castro because the Escambray hands were CIA-supported. Wherever he now went, Fidel heard the chant *"¡Cuba sí, Yanqui no!,"* the latest revolutionary slogan. Then came "Fidel, For Sure, Hit the Yankees Hard!," which schoolchildren still chanted in 1985 to greet Castro. He sheepishly explained to an American friend when they went together to visit a Pioneers children's camp: "They don't know that *you* are American. . . ." A teacher said, "Yes, we're supposed to teach it to the kids," but, strangely, there was no touch of hate in the children's voices or in the teacher's explanation. It was simply revolutionary folklore, turning into living history twenty-five years later; the teacher had not been born when the slogan was first sounded in Cuban streets.

But slogans and revolutionary fervor could not shore up the economy as it foundered in 1961, unguided and improvised from day to day, and brutally severed from the American economy of which it had always been an appendage. The central problem was that Castro desired to refashion the Cuban economy rapidly but had no real concept or blueprint for it. As René Dumont, a leading French agronomist whom Castro consulted in the early 1960s, wrote later, "Cuba in 1959–1960 confusedly searched for truly original socialism," but in the next decade it committed "an impressive series of economic errors . . . that did not happen by chance." (After Dumont published critical but well-meaning conclusions in a 1971 book, he was classified, in effect, as an enemy of the revolution because Castro does not take kindly to criticism. Dumont has since died.) This "truly original socialism" never really acquired an identifiable shape inasmuch as it degenerated into ill-conceived imitations of orthodox Soviet central planning, administrative and managerial chaos, new vested ideological and bureaucratic interests, and brusque course corrections. Castro lost his way in implementing his pledge that in every sense "the Cuban revolution will be as Cuban as the palm trees."

In stating late in 1960 that the Moncada Program, contained in the "History Will Absolve Me" discourse, had been fulfilled, Castro led his nation to mistake the decision to carry out the objectives of "History" for actual results. Nationalizations and the agrarian reform laws meant that the revolutionary regime had embarked on a new approach to the industrial, trade, banking, and farming segments of the Cuban economy, but it was not spelled out what this approach would represent in terms of production, the distribution of resources, and so on. Che Guevara wrote, for example, in the October 1960 issue of the army publication *Verde Olivo* that "the laws of Marxism are part of the development of the Cuban revolution" and that between the landing of the *Granma* and the victory of the revolution there had been "ideological transformations" of its leaders. But he offered no clue as to what these Marxist laws were supposed to be or what they portended for the third year of the revolution. For one thing, Guevara, running as usual ahead of the more cautious Fidel, was the first top leader to use publicly the word *Marxism* to describe the revolution. In proclaiming that same month that the Moncada program had been completed, Fidel did announce the start of a new revolutionary era, but he eschewed ideological adjectives.

In terms of fundamental economic policy, which should have been practical as well as ideological and political, Castro and Guevara, the principal planners, long could not make up their minds about the role of sugar in the Cuban economy. Though sugar had always been Cuba's mainstay, the revolutionary chiefs developed during 1959 and 1960 the daring notion that the island should end forthwith its dependence on it—presumably because it was a bitter reminder of the American-dominated "colonial" past. Consequently, production dropped sharply in the harvests between 1962 and 1964, plantings having been reduced starting in 1959. Then, economic realities—such as the fact that Cuba had only sugar to pay even in part for the mounting Soviet aid and that nothing had yet replaced canecutting as a gainful rural occupation at a time of endemic national unemployment— forced Castro and Guevara to abandon the idea that industrialization could overnight be the new fountainhead of wealth. In 1965, therefore, the Cubans (possibly under Soviet pressure) returned to sugar as their main production priority, then went to the other extreme of shooting for record harvests, which also turned out to be a painful error.

Fidel Castro, ever the great teacher, kept the nation informed of this shifting course through a formidable array of statistical data, lengthy projections, and interpretations. He always concluded on the note that whatever decision was being announced or explained by him on a given day represented the collective wisdom of the revolution, and the adoring crowds cheered and chanted, "We Shall Win! . . ." Fidel's word was never questioned by the masses, and there was not enough expertise around to sort it all out. Colossal errors, as Castro later acknowledged, were committed by INRA's military administrators of nationalized lands, especially on the cattle ranches where their absolute inexperience and politically directed demand for immediate increases in meat production combined to destroy the herds and the industry.

Castro had launched from the outset a series of grandiose but nevertheless rational and promising food-production and diversification schemes, ranging from much higher rice yields and a huge increase in vegetable farming (as China had done successfully at about the same time) to a poultry industry on a national scale to provide rapidly cheap and abundant protein supplies from chicken and eggs. In 1959—as in 1986—Castro was convinced that perfect, state-of-the-art technology in agriculture or anything else could alone solve great economic problems. But he lacked resources, personnel, patience, and the art of the follow-up to transform his dreams into reality. Thus nothing was accomplished along those lines while the rest of the economy deteriorated, and Fidel rushed restlessly from a rice experimental station to a new stand of eucalyptus trees, exploding with bursts of impatience and bursts of new ideas the length and the width of his island.

This was still the romantic period of the revolution, but above all it was the obsession of Fidel Castro to do away with human, social, and economic underdevelopment in Cuba. To understand Castro and his revolution, it is essential to comprehend this concept of underdevelopment as a crucial psychological attitude in Cuba, probably to a greater degree than elsewhere in the Third World. To Fidel and the revolutionary generation, underdevelopment means illiteracy and disease, economic inadequacy, dependence on the West under the shackles of "neocolonialism," and the thinking patterns of people in poor countries. To him and his disciples, underdevelopment is the shame of a society's cultural and economic poverty, it is a mental prison, it is third- or fourth-class citizenship in the world. In Cuba since the birth of the revolution, this word is used constantly since the birth of the revolution in speech and print by Castro and everybody else: It expresses degradation, a justification for failures and insufficiencies of the system and of individuals, and even a sense of defiance in response to criticisms. The expression appears in endless articles, novels, films, and conversations as Cubans keep alive the psychology of underdevelopment year after year and decade after decade in the life of their revolution. In fact, one of the most popular—and one of the best—Cuban post revolution movies was *Memories of Underdevelopment,* directed by Tomas Gutierrez Alea, dwelling on this theme.

To eradicate underdevelopment in all these meanings was, indeed, Fidel Castro's magnificent obsession from the beginning, and this was what the social-justice aspects of the Moncada program were all about, long before he chose to apply Marxist language to it. Once in power, social-improvement goals were paramount to Castro, and from the first day they commanded an enormous share of the revolution's time and attention. Clearly, decisions on sugar and industry were vital, but such was the mood in those early years that what really captivated him as a human being were the literacy campaign, the classrooms in the Sierra Maestra, the creation of a public-health network (including forty-five new hospitals built in one year) that showed instant results, especially among children, decent peasant housing, country roads, and that new Cuban sense of pride that really meant more to Fidel than anything else. He spoke of racial equality and of the "new Cuban man (or woman)," before he turned to "the new socialist man (or woman)," and naturally the masses adored it. Through the first agrarian reforms, Castro gave peasants land (before, in effect, taking it away again), and through urban reforms he halved the rents and banned ownership of more than one dwelling per person (before the state went massively into the city landlord business in a subsequent revolutionary paroxysm). It was all marvelous, but it cost money, and it required a functioning economy that Cuba simply lacked at that revolutionary stage.

Castro, therefore, turned to improvisations—and to an ever-increasing dependence on the Soviet Union for the economy and for defense. He was both economically and militarily vulnerable, and he knew that a moment of awesome confrontation with the United States, now under President Kennedy, was both inevitable and imminent. To put his house in order in anticipation of the great clash, Castro tried to tighten up the economy at home through the creation in February of three new ministries: Industry under Che Guevara, who gave up the presidency of the National Bank (after delightedly signing new banknotes with his nickname "Che"), Foreign Trade, and Internal Trade. Guevara had just returned from a lengthy visit to Eastern Europe and the Soviet Union—his first trip to Moscow—where he joined the Soviet leadership at the Kremlin in reviewing the October Revolution military parade and signed a new trade agreement providing for Russian purchases of one half of the Cuban sugar crop (2.7 million tons) at a price above the world market. Guevara sold another 1.3 million tons to other Eastern European countries, also Cuba's new allies. Six months earlier, Nuñez Jiménez had gone to Moscow, heading the first Cuban trade mission to the Soviet Union, and returned with a commitment that thirty Soviet industrial plants would be installed in Cuba. This was the Cuban leadership's concept for industrializing the island—but sugar had to be grown to pay for the plants and oil and everything else being sent by Moscow.

Soviet military deliveries to Cuba were formalized when Raúl Castro met Khrushchev in Moscow in July 1960, and their joint communiqué restated that the Soviets would use "all means not to consent to an armed United States intervention against the Cuban republic." This assured Fidel he could adequately equip the FAR (Revolutionary Armed Forces) and the militias to resist a conventional forces' attack from the outside; (the deal Guevara had signed in Moscow at the end of 1960 gave him a breather in trying to keep the economy afloat the deal included the assignment of 189 Soviet industrial advisers to Cuba). And Castro made no secret of his dependence on Soviet aid: In an interview on February 1, 1961, with Jiřy Hochman, the correspondent of the Czech Communist party organ, *Rude Pravo,* he remarked that "if it were not for the intervention of imperialism, the Cuban revolution would have developed without difficulties. . . . The solidarity and the aid given revolutionary Cuba in this situation by the socialist countries have played a decisive role for the definitive victory of our people. If the socialist camp and its attitude did not exist, we would pay very dearly for our revolutionary laws. . . . Thanks to arms that we have received from socialist countries, we were able to create a defensive force capable of arousing the respect of the mercenaries and the respect of the aggressive circles of imperialism. . . ."

The stage was now set for the Cuban-American confrontation. Castro felt secure that he could defend Cuba from anything short of a full-fledged American attack, though, as he said later, he doubted that Kennedy would engage United States forces in support of the exiles' brigade whose existence was amply known to Havana. In Washington, the new president was being urged by the CIA not to delay the invasion because, according to the CIA Havana Station, Cuban pilots in Czechoslovakia were about to complete their training and would be returning home any day to fly the MiG jets that the Soviets were expected to provide (none had yet arrived).

Still concerned about the Escambray bands that he suspected of having been instructed by the CIA to support in some fashion the exiles' invasion, Castro went back to the mountains on March 1 to satisfy himself that the situation there was really under control. Again, he appeared in battle gear, going up some hills, chatting with militiamen and their officers, joking with local peasants—all these activities being televised, photographed, and reported in detail. He was orchestrating everywhere his preparations for the battle, though in the case of Escambray, the CIA had lost interest in the bands there in terms of the planned landing on the shores of the Bay of Pigs. They were no longer considered an asset, but Fidel could not be sure of this and was taking no chances. In the meantime, the United States had banned travel to Cuba by Americans, which Castro took to be another sign of approaching hostilities. However, on February 14, he made a point of telling United Press International correspondent Henry Raymont that Cuba wished to have relations with the United States provided that Washington halted its active support for the "counterrevolution."

But now events were picking up momentum. On March 2, the Kennedy administration announced it planned to prohibit all imports from Cuba. On March 11, Kennedy blocked the sales of U.S. farm products to Cuba. And in the days and weeks that followed, Kennedy and Castro engaged in a chain of simultaneous activities affecting each other's countries—sometimes in surprising and contradictory ways. March 13 was such a day.

Speaking in inspired tones, President Kennedy invited Latin America to join the United States in an "Alliance for Progress" to ameliorate peoples' lives and achieve economic advancement, pledging $25 billion over a decade as the American contribution to this goal. The irony, which Castro suspected, was that even as Kennedy spoke in Washington, the invasion-brigade training was being perfected in the Guatemalan camps. Still that same day, rockets were fired from a high-speed boat at the oil refinery in Santiago, killing one sailor and injuring a militiaman; this was part of the CIA-organized pre-invasion softening up of Cuba. And in Havana, Castro went to the great stairs of the university to commemorate with a fighting speech the fourth anniversary of the assault on Batista's palace by the Students' Revolutionary Directorate.

We now know that Kennedy approached the Alliance for Progress with much greater enthusiasm than the supposedly secret plans for the Bay of Pigs. Both enterprises, however, had roots in the Castro revolution, one to prevent its repetition elsewhere in Latin America, and the other to liquidate that revolution. The rationale in the administration, to the extent that a coherent one existed at all, was that the Alliance would build a bright future for Latin America on the ruins of the Cuban revolution. It evidently had not occurred to the planners that in the aftermath of the Bay of Pigs, the great gesture of the Alliance might have been regarded in Latin America as something other than pure altruism. Having turned a deaf ear for long years to Latin American urgings for a large-scale economic-development program, the United States, under a younger man's administration, finally chose to do exactly what Castro had proposed in Buenos Aires two years earlier, but it still completely underestimated how the Cuban revolution had captured imaginations in the hemisphere.

Discussing the Alliance many years later, Fidel Castro easily admitted that there was a direct nexus between his revolution and the Kennedy initiative. He made the point that the Alliance sought the same goals as the first phase of the Cuban revolution under the Moncada program: concepts of agrarian reform, social justice, a better distribution of national wealth, tax reforms, and so on. Castro thought that the Alliance was an "intelligent strategy," but it was a hopeless attempt because Latin American establishments would not have allowed real reforms. In his view, Latin American conditions could be approached in only two ways: social reform or "political repression," and in this sense the Alliance was a "politically wise idea," though even reforms were not enough to meet the needs. Basically, Castro said, the Alliance sought to avoid revolutions. In a long, thoughtful discussion of Kennedy in January 1984, he remarked that the Alliance emerged with "the political objective of acting as a brake on the revolutionary movement, and I think it was a merit of Kennedy to have really understood [and] recognized that there was an economic and social situation that sooner or later had to become translated into revolution."

Castro was also impressed with Kennedy's idea to create the Peace Corps, albeit this was not sufficiently revolutionary for his taste. What he saw in it was an American version of what he called "internationalism," that is, direct involvement in the development process of other countries through assistance teams or individuals. In a way, Cuba's assignment of thousands of Cuban doctors, nurses, teachers, and technicians in Africa and Nicaragua is an imitation on a vast and political scale of Kennedy's Peace Corps: Kennedy and Castro both understood the need for establishing person-to-person contacts in the Third World as a factor of influence. So aware was Castro of what he calls "Third World internationalist solidarity" that in the spring of 1960, when Cuba herself lacked resources, he dispatched a shipload of medicine, foodstuffs, and clothing aboard the freigther *Habana* for the victims of a terrible tidal wave that had hit Chile. As it happened, Senator Salvador Allende Gossens, the Marxist physician who ten years later would be elected Chile's president, was a frequent guest in Havana, and it was to him that Castro made the first offer of aid. Also in 1960, Castro lent $5 million to Cheddi Jagan's leftist British Guyana.

To be sure, both Castro and Kennedy thought in guerrilla terms, too, concerning the Third World. From the outset, the Cubans proceeded to train young Latin Americans and Africans for future guerrilla operations in their countries (aid shipments to Chile included copies of Che Guevara's *The War of Guerrillas,* his manual on rural insurrection and warfare). There was Cuban support for Venezuelan urban and rural guerrillas in 1960, and the leaders of the movement that grabbed revolutionary power in the British colony of Zanzibar in 1963 had been trained in Cuba where they maintained an office since 1960. (Zanzibar later merged with Tanganayka to form the republic of Tanzania whose leadership has cordial ties with Cuba.) Much more such guerrilla training came later, and it continued into the late 1980s. Kennedy responded to the threat of the Cuban subversion by creating a special warfare school at Fort Bragg, North Carolina, the home of the U.S. Army's 82nd Airborne Division, now known as the John F. Kennedy Special Warfare Center, and Guevara's *War of the Guerrillas,* hurriedly translated into English, became one of the principal textbooks in the art of insurgency and counterinsurgency. U.S. Army Special Forces Detachments (or the Green Berets) as they now exist were born at the JFK School, and one of the first units assigned overseas was sent to Panama to aid Latin American armies in counterinsurgency. Ironically, it is quite probable that American advisers who helped the Bolivian army track down Che Guevara and kill him in 1967 had studied Guevara's manual.

The United States and revolutionary Cuba competed for Latin American influence on the highest political level as well. After President Eisenhower toured South America early in 1960, Castro sent Osvaldo Dorticós, Cuba's nominal president, to make a similar journey in May. Kennedy made two trips to Latin America during his brief presidency, and Castro returned to the region in the seventies. On May 1, 1961, Radio Havana inaugurated its short-wave international service in scores of languages (from English to the Andean Indian Quechua language) with one of the world's most powerful signals as part of Castro's "internationalist" offensive. The *Prensa Latina* international news agency was started about the same time. Somewhere along the line, Fidel seriously proclaimed that the Indian chieftain Hatuey, born on the island of Hispaniola in the Age of Discovery and later the head of a tribe in Cuba, was the "first internationalist." This was his colorful way of saying that in Latin America, leaders moved from one country to another from the start of history as "internalists" because to them frontiers had no meaning.

Between the end of March and mid-April 1961, tensions went on rising. Kennedy ordered the suspension of the Cuban sugar quota for the year (the law gave the president the authority to "suspend" or cut quotas, but not abolish them altogether), and on April 3, the administration issued its "White Book" on Cuba that decried the denial of democracy under the revolution. It was, in effect, the intellectual justification for the coming invasion, and this was certainly the way Castro read it when he received news-agency reports on it that afternoon. Now he addressed Cubans daily with warnings of an imminent attack: He spoke to thousands of construction workers who had organized Committees for the Defense of the Revolution at the Public Works ministry, to labor-union delegates preparing May Day celebrations, to another labor rally to protest the flight abroad of Cuban workers, intellectuals, and technicians, and to the nation over television to discuss revolution and education. He also found time to dispatch a message to Khrushchev to congratulate the Soviet Union on the first manned space flight.

That an attack was about to happen was no secret to Castro and his highly efficient intelligence apparatus, and in his April exhortations he blended sarcasm with dire warnings to his foes. Speaking of the Democratic

Revolutionary Front (FRD), the CIA-constructed political organization in whose name the invasion was being prepared—and its chairman, José Miró Cardona, who was the first revolutionary prime minister—Castro asked, "Are those the men who will come to overthrow the armed people? Don't make us laugh! . . . This mercenary government will not last twenty-four hours in Cuba. . . ." The next day, he taunted the exiles: "The people here have asked many times, when will they come? People, too, are impatient. . . . Those who have illusions and vain hopes in those plots, they have no alternative but to come here, sooner or later. . . . It is inconceivable that they do not understand that a struggle against the revolution could be started here, but it would never end, never in the fields, never in the cities. . . ." Fidel was determined to have the nation psychologically ready for the attack when it came, and he gave this preparation as much importance as he gave his military planning.

And his intelligence and security networks were functioning with perfection. Ramiro Valdés, who was Interior Minister until 1986, and who had created the security services as one of the first endeavors of the revolutionary regime in 1959, says that Cuban intelligence was able to track invasion preparations step by step, from Miami to the training camps in Guatemala. "It was an open secret," Valdés adds. Because of the traditional indiscretion of Cubans, he recollects, the "Little Havana" district of Miami was so full of invasion talk that Cuban security's biggest problem was to sort out truth from rumors in the mass of information streaming from the mainland. (Miami, of course, was full of infiltrated Valdés agents at the time.) He says that the flow of intelligence was so heavy that there were moments when he suspected that the CIA had mounted a campaign of "disinformation" or a psychological warfare campaign to keep Cuba off balance. "When reports of invasion reach you every day, on the other hand, you catch the wolf on the day he finally arrives."

Based on information from Miami, Central America, and from within Cuba, State Security Services began on April 1 to round up people suspected to be CIA-linked or involved in clandestine antiregime activities. As Valdés says, "We knew who everybody was, what weapons they carried, how much ammunition they had, where they were going to be, how many of them, at what time, and what they proposed to do. . . . We were very seriously infiltrated in the counterrevolutionary bands." The CIA's justification for not informing the anti-Castro underground beforehand about the date and place of the invasion was that the information could fall into the hands of Cuban intelligence, and according to Valdés, the Agency acted correctly. However, the invasion made absolutely no sense without instant and massive support from the underground, such as it was, and for this reason the invasion was doomed to total failure. Meanwhile, the Security

Services arrested members of six separate antiregime groups in the first week of April, mainly men and women involved in CIA-aided sabotage acts. In several instances, shootouts preceded the arrests. And neighborhood Revolution Defense Committees in the cities were crucial in ferreting out suspicious persons. Nevertheless, Havana's El Encanto department store was torched and burned. But in Oriente, 145 members of "counterrevolutionary bands" were captured early in April.

On April 12, President Kennedy offered assurances at a Washington news conference that United States forces would not intervene in Cuba. Castro reasoned—correctly—that Kennedy would not make a point of excluding an American participation unless an attack, presumably by exiles, was in the offing. He took Kennedy's comment as a confirmation that an invasion was imminent, and he based his defensive strategy on the assumption that he would be fighting only exiled Cubans (but American planners did not realize Castro had reached that conclusion). Fidel naturally had alternative contingency plans if Kennedy, after all, did send Americans into combat (which nearly happened in terms of air support). He was not certain exactly where and when the main landing would be made, but he thought it would come in the south, around Cienfuegos or Trinidad, with simultaneous attacks in the south and the north of Oriente and in Pinar del Río in the west. On the eve of the battle, Castro had the relative planning advantage of being ready for a number of alternatives, whereas the CIA and its Pentagon advisers, as it soon turned out, had no real idea what they were facing; a board of inquiry discovered subsequently that the planners believed Castro had "no doctrine" of any kind. Once more, he had the luck of being underestimated.

To defend the island with his clear doctrine of main-force deployments and mobility by tactical units, Castro had the regular rebel army of about 25,000 well-trained and equipped men plus around 200,000 militiamen organized in battalions and stationed throughout the country in strategic areas. The army was divided into three tactical regional commands, and Castro's battle plan provided for Raúl Castro to command forces in the east (Oriente and Camagüey), Che Guevara in the west (western Havana Province and Pinar del Río), and Chief of Staff Major Juan Almeida in the center with headquarters in Santa Clara. Fidel, as Commander in Chief, would coordinate all operations from a secret command post in the Nuevo Vedado section of Havana (he also commanded directly the troops in the capital), although he was prepared to move rapidly from place to place as required. As he explained later, "Every time there was talk of an invasion from the United States, we divided ourselves in the country." Under this defensive doctrine, no main-force units would move from their respective assigned areas unless battle developments made it necessary; early tactical operations were assigned to the militia battalions. Thus Castro would not be trapped in diversions and deceptions the CIA had planned.

About two weeks before the invasion, Castro had paid one of his periodic visits to the Ciénaga de Zapata, and at one point strolled along the Girón beach at the entrance to the bay to see how a tourist-village construction was coming along. Suddenly, he turned to a Cuban journalist accompanying him, and pointing to a one-story concrete house, he said, "You know, this is a great place for a landing. . . . We should place a fifty-caliber heavy machine gun there, just in case." The day before the invasion, Juan Almeida had toured the Bay of Pigs area and decided to send a militia company there, dividing it among three principal beaches because he thought communications were inadequate. At that moment, the invasion fleet was already sailing from Puerto Cabezas in Nicaragua (where the brigade had been moved from Guatemala) under a U.S. Navy escort. The first ship of the invasion force left on April 11, and the last one on April 13: The landings were set for post-midnight hours of April 17.

Fidel Castro spent the night of April 14–15 awake at his emergency command post known as "Punto Uno" (Point One), a two-story house at Forty-seventh Street in the residential district of Nuevo Vedado in Havana, near the Zoological Gardens. Major Sergio del Valle, his Chief of Staff, who had been a guerrilla physician in the Sierra Maestra, and Celia Sánchez were with him. Castro had had so many signals in recent days that "something" was about to happen that he made "Punto Uno" his temporary home; it was principally a communications center with the rest of the country. Specifically, Fidel was informed on Friday, April 14, that a suspicious-looking ship had been spotted off the coast of Oriente, not too far from Guantánamo, and he had to conclude that it could be the vanguard of a major invasion force. Oriente *was* the traditional gateway to "liberations" of Cuba. Castro's information was correct, but the ship, a freighter named *La Playa,* was carrying a diversionary force—not the main one. The CIA's idea was that diversionary activities at various points on the island forty-eight hours before the Bay of Pigs landing would confuse Castro and force him to desperately improvise his defense. *La Playa* therefore was to land 164 men commanded by Major Nino Díaz at the mouth of the Mocambo River, some thirty miles east of Guantánamo, in the small hours of April 15, after steaming three days from Key West, Florida. Nino Díaz had fought under Raúl Castro in that same area of Oriente, but he had turned against the revolution and now was willing to fight there again. However, a reconnaissance party from *La Playa* aboard a rubber boat reported back that there were militia units on the beaches, and Díaz decided to abort the landing, sailing for home.

Fidel Castro had no way of knowing about the fiasco of this planned diversion, but another deception almost literally exploded in his face a few minutes after six o'clock on the morning of Saturday, April 15. This was the sight of two B-26 light bombers with his air force's FAR insignia rocketing from low altitude the runway at *Ciudad Libertad*. Here the former Camp Columbia army camp had been transformed into a school, but a military landing strip was still operational. It was less than a half-mile from Castro's command post. Then the planes came back for bombing and strafing runs, hitting houses in the densely inhabited neighborhood around *Ciudad Libertad*. Within minutes, Castro was informed that air-force bases at San Antonio de los Baños, his principal base, near Havana, and at Santiago in Oriente had been attacked simultaneously by two B-26s each. Castro's first thought was that the aerial attacks signified the start of the invasion, but there was no follow-up, and a few hours elapsed before it became clear what had happened.

The air strike at the three Cuban bases was designed to destroy Castro's aviation to assure full control of the air on invasion two days later, but the CIA believed it could perpetuate the deception that the B-26s belonged to the FAR, and that their pilots were defecting to the United States after destroying other aircraft on the ground. The planes had actually flown from Puerto Cabezas in Nicaragua, and they were part of the CIA's exile brigade, but the agency thought that there were political and propaganda advantages in making the world believe that Castro's pilots were defecting; Castro obviously knew within seconds that there were no such defections. The strike was as ill-conceived as everything else in the invasion: Whereas sixteen B-26s were supposed to have flown the mission, Kennedy had the number cut in half the day before (he was uneasy about the whole proposition, and this was a compromise). Out of the eight bombers in the sortie, only six participated in the attacks (one being shot down by teenage militiamen firing Czech-made quadruple-barrel antiaircraft guns at San Antonio) and two flew directly to Florida, running out of fuel. The others returned to Nicaragua. Finally, the political deception collapsed when news photographs of a B-26 that had landed in Miami were presented that night to the Political Committee of the U.N. General Assembly meeting in an emergency session in New York at Cuba's request. They showed that this particular B-26 had a solid-metal nose whereas the B-26s Castro had inherited from Batista had Plexiglas noses, a detail the CIA had forgotten but which the Cubans immediately called to the Political Committee's attention.

Castro's tiny air force lost five planes on the ground in the raids, including two B-26s, an AT-6 propeller trainer, a DC-3 transport, and one T-33 jet trainer—leaving it with four British-made Sea Fury light attack bombers, one B-26, and three T-33s as the only operational aircraft. As Fidel remarked later, he had eight planes and only seven pilots, but the CIA could not be sure just what the Cubans could still do in the air. And it understood even less Castro's special talent for turning seeming defeats into triumphs. Apart from the fact that the Saturday attacks warned him of the real imminence of an invasion—all the military units went on high alert and the seven pilots took turns sitting inside cockpits or sleeping on cots under aircraft wings—Fidel was handed a great political victory. Seven persons were killed and fifty-two wounded as a result of the raid on Havana, and he turned the funeral into a stirring act of patriotic and nationalist mourning and revolutionary defiance. A quarter of a century later, Castro was still emotional and indignant about the attack, telling a foreign visitor that "one of those who were dying there, a wounded man, was bleeding away, and he wrote my name with his blood on a wall. . . . It reflected the attitude of the people: A young militiaman who is dying, and his protest was to write a name with his blood."

In his funeral oration at Colón cemetery on Sunday, April 16, Castro compared the air raid to Pearl Harbor, except, he said, that it was "twice as treacherous and a thousand times more cowardly." He proclaimed that "the attack of yesterday was the prelude to the aggression of the mercenaries" paid by the United States, and that the American government could be called "liars" for the deception it practiced by pretending that the attacking pilots were defectors. He reminded the audience of soldiers and militia men and women that the previous year the Eisenhower administration had also initially lied about the U-2 spy plane shot down over the Soviet Union, and went on to compare the "admirable" Soviet achievement of putting man into space with the American achievement of "bombing the installation of a country that has no air force." His voice rising to a roaring climax, Fidel Castro then issued his ideological indictment: "Because what the imperialists cannot forgive us . . . is that we have made a socialist revolution under the noses of the United States . . . and that we shall defend with these rifles this socialist revolution! . . ."

This was the first time Castro had publicly described the Cuban revolution as a "socialist revolution," marking another watershed in its history. But, again, he left it unclear as to when and how he had concluded that time was ripe for socialism in Cuba—again interpreting history to suit his needs at that particular moment. In a long discussion of ideology in January 1984, Castro told me that even though he already considered himself a Marxist-Leninist, "not at the time of Moncada nor at the triumph of the Revolution did we consider the intention of the development of a socialist revolution in Cuba as an immediate question. . . . I don't want to say that I didn't dream, that I wasn't convinced that in the long run the type of revolution that should be made in our country was socialism, but this was not a question that could be considered an immediate objective at that stage, thinking of the realities of our country, the level of political culture in our country, the level of preparation of our people, the enormous objective difficulties that we would have encountered if we tried to push ahead with this type of revolution."

If that was his judgment, the question arises why the CIA air raids in April 1961 suddenly made Cuba ready for socialism. In interviews in 1985 with Frei Betto, the Brazilian Dominican friar, Castro explained that "faced with an invasion organized by the Yankees, our nation is already struggling for socialism. . . . If since 1956 [the nation] is fighting for the constitution, for the overthrow of Batista, for an advanced social program, but not yet a socialist one, at this moment it is fighting for socialism. And this has great symbolism because tens of thousands of men were disposed to face what may come. . . . I proclaimed the socialist character of the Revolution before the battles of Girón. . . ."

Castro and other Cuban leaders have said on many occasions that he had to unveil the revolution's socialism on April 16 because men had the right to know for what they would be dying in the new inevitable confrontation. This would imply, however, that socialism was being created, in effect, behind the backs of most of the population—which was true—and it would be an admission that Castro had been misleading Cubans, especially when he fulminated against the "lies about communism." And 1961 was the year when Castro was busy quietly organizing the fusion of his revolutionary movement with the Communists. There is a school of thought propounding that it was American pressure, culminating in the Bay of Pigs, that pushed Castro into openly espousing socialism (or communism) before he was really committed to it. But such top leaders as Armando Hart, a member of the party's Political Bureau, and Blas Roca, former secretary-general of the "old" Communist party, told me in separate conversations in mid-1985 that Castro had all along planned to proclaim the socialist revolution in his May Day speech. In this case, the proclamation came *two weeks* ahead of schedule. Under the circumstances, Castro's course was absolutely logical: The patriotic passion aroused by the air raids created the perfect conditions to make the new official ideology fully acceptable to a nation on the threshhold of the battle for its survival. As Fidel always said, Martí and Marx are inseparable in the Cuban revolution, and he told the irate cemetery crowd that Cubans would defend this revolution "of the humble, by the humble, and for the humble to the last drop of blood."

And even as Castro spoke in Havana on Sunday, April 16, John F. Kennedy gave the final authorization for the invasion in a telephone call from the Virginia estate of Glen Ora to the CIA headquarters in Washington. But he simultaneously signed the invasion's death warrant when he forbade air strikes by Cuban exiles' B-26s against the remainder of Castro's aviation in support of the landings on D-Day. The President feared that such missions, which would have to be flown from Nicaragua or Florida, would publicly compromise the United States in the eyes of the world. Cubans, he told the CIA, could fly combat missions the moment they secured a field on Cuban territory. The invasion might have failed even if Kennedy had allowed the D-Day strikes: Without them it did not have a prayer of success because Fidel had a secret weapon and quite a few surprises for his foes.

The secret weapon and the other surprises were produced by Fidel Castro, the master at the game of letting his enemies trap themselves in defeat, within five hours or so of the first landing of the exiles' invading force on the beaches of the Bay of Pigs. The men of Brigade 2506 (so named after the serial number of its first volunteer to die during training in Guatemala) began coming ashore first on Playa Larga ("Red Beach") deep inside the bay and then on Playa Girón ("Blue Beach") at its eastern entrance about 1:15 A.M. of Monday, April 17, from landing craft launched from the ships that had brought them from Puerto Cabezas. The force of approximately fifteen hundred men was under the command of José Pérez "Pepe" San Román, a young career officer who was trained in the United States, fought in the Batista army against Castro, then turned against the dictator toward the end of the war. As the Havana regime kept emphasizing later, the Brigade included nearly two-hundred ex-Batista officers, soldiers, and officials, sons of rich or middle-class families, and over one hundred of what it contemptuously called "lumpen." But the Brigade was well trained and equipped, and Castro took the landings with the greatest seriousness when word reached him just before 2:30 A.M.

Militia patrols had spotted the invaders immediately and began the first fire exchanges, but it took messengers in jeeps almost an hour to reach the town of Jagüey Grande to the north where the nearest telephone was located. From there, reports on the landings were phoned to the "Punto Uno" command post in Havana and instantly relayed to Fidel who was spending the night at Celia Sánchez's Eleventh Street apartment, less than ten minutes away. Castro's instinctive reaction was that the Bay of Pigs was the principal invasion area, and he proceeded at once to unveil his strategic surprises: on the land and in the air—where his secret weapon awaited action.

The first telephone call Castro made was to his trusted friend Captain José Ramón, "Gallego" Fernández at his commander's quarters at the army cadets' school in Managua, just south of Havana, to pass on the fragmentary information he had received, and to order him to the Bay of Pigs. Castro told him to pick up on the way, in Matanzas, the elite Militia Officers' School Battalion headquartered there—870 men—and assume operational command in the whole battlefield region; the other unit immediately available to Fernández was Militia Battalion 339 from Cienfuegos, which had detachments throughout the great swamp. At that stage, Castro had no idea of the size or composition of the invading army, but his political judgment was that he had to prevent at all costs the consolidation of a beachhead large enough to allow the exiles to set up a provisional government and request international recognition. As it happened, this was exactly the centerpiece of the CIA's master plan. While Fernández was racing south in his green Toyota jeep, Castro was ordering artillery units from Managua and Havana to move urgently to the battle zone and having Soviet-built T-34 tanks placed on flatbed trucks to be transported there as well. He was careful, however, not to denude Havana of its rebel army and militia troops: he could not yet rule out landings on the north coast.

Now operating from "Punto Uno," Castro made his next telephone call to the air force base at San Antonio de los Baños where the exiles' B-26s had attacked on Saturday. Poised on the runway were two Sea Furys, two B-26s, and three T-33 jet trainers. And the T-33s were Fedil's secret weapon in the context of his very imaginative overall aerial strategy. He had decided that the rocket-equipped Sea Furys would concentrate on attacking the eight-vessel invasion fleet to sink as many as quickly as possible while the three T-33s (one was destroyed on Saturday) were to neutralize enemy aviation. CIA and Pentagon planners had assumed that Castro's small air force would be destroyed on the ground, and it had never occurred to them that whatever planes were left would be used against the ships. The vessels therefore had no antiaircraft weapons. As to the jet trainers, the Americans never suspected that Castro had them armed with two .50-caliber machine guns each. With this armament and a jet's maneuverability and speed, the T-33s were vastly superior to the lumbering B-26s that began at first daylight flying exhausting round-trip combat missions between Nicaragua and the Bay of Pigs. In this fashion, the little jets played a key role in Castro's victory, depriving the brigade of air support, and allowing the Sea Furies to go unchallenged after the ships. Testifying in May before a presidential inquiry board, the Air Chief of Staff, General Thomas White, remarked, "Well, I really believe that the Cuban air force had a whale of an effect on the bad outcome. . . . I was surprised to find that [the T-33s] were armed." He acknowledged that the Joint Chiefs of Staff did not consider the T-33s to be "combat aircraft." National Security Adviser McGeorge Bundy wrote in a formal statement to General Maxwell Taylor, chairman of the inquiry board, that "one startling omission . . . is the failure of any of the President's advisers to warn of the danger of the T-33s."

Castro made a point of personally urging his pilots to find and destroy the ships, impatient over every minute elapsing. He knew the pilots from earlier visits to the base, and he wanted to impress on them how vital it was to deprive the invaders of the ships still carrying weapons, ammunition, food, and stores; this was consistent with his goal of isolating the Brigade on the beachhead and then smashing it. His greatest concern was that the exiles might succeed in establishing positions on the high ground in the Zapata swamps beyond the beachhead and controlling the three paved highways linking the Bay of Pigs with firm land in the north and east. If they achieved it and continued to be supplied from the sea, it might have become impossible to dislodge them. Castro correctly guessed the enemy plan, and acted accordingly.

At 4:30 A.M., Fidel called the air base and demanded to talk to Captain Enrique Carreras, the senior pilot, who had been sitting strapped inside the cockpit of his single-seat Sea Fury. Carreras raced to the phone to hear Castro tell him, "*chico,* you must sink those ships for me! . . ." At the first light, minutes before 6:00 A.M., Carreras took off in his Sea Fury armed with rockets and four 20mm cannon, followed by another Sea Fury and a B-26. Reaching the Bay of Pigs, Carreras saw landing craft moving toward the beaches, and a large freighter approaching Playa Larga. He missed on the first pass, but on the second pass his rockets hit the freighter, which was the *Houston* carrying the Brigade's Fifth Battalion and its equipment, and seconds later the second Sea Fury scored, too. At 6:30 A.M., the *Houston* went aground five miles south of Playa Larga, and the battalion could never land to join the fray. *Barbara J,* a landing craft infantry (LCI) serving as a CIA command vessel, was damaged by Sea Fury machine gun fire, began to take water, and fled for the open sea. Carreras returned to San Antonio to refuel and rearm, and he was back over the bay at 9:30 A.M., this time hitting and sinking the freighter *Rio Escondido,* which carried ten days' supply of ammunition for the Brigade and essential communications gear. At that juncture, other invasion ships steamed out of the bay, leaving some 1,350 Brigade men stranded ashore. Not even Castro understood it yet, but the battle of the Bay of Pigs was won by the Sea Furys eight hours after the first landings. Captain Carreras was hit in an engine by fire from a Brigade B-26, but he limped home to San Antonio where ground crews rushed to repair the plane. Then the T-33s took to the air to cope with the exiles' B-26s, shooting down four of them during the day. Castro lost two Sea Furys (not Carreras's; twenty-five years later he is still flying, now as an airline pilot out of Havana) and two B-26s, but strategically it no longer mattered. His aerial secret weapons had accomplished their mission.

*   *   *

Proceeding south from Matanzas with his militia officers' battalion, Fernández halted in Jovellanos, where he was intercepted by a phone call from Castro, who wanted to check on his progress. Fernández and his battalion reached the Australia sugar mill on the outer perimeter of the Zapata swamps around 8:00 A.M., and now Castro ordered him to take Palpite, a village just three miles north of Playa Larga, one of the two Brigade beachheads. Palpite was some twelve miles from Australia by paved highway, and advancing in buses and trucks under enemy air bombardment, the militia battalion occupied the village by noon while another unit raced a few miles southeast to take the village of Soplillar where there was a landing strip. Fernández was just in time because Brigade units from Playa Larga were trying to reach Palpite and the south-north highway. He was helped by the fact that paratroops earmarked for this mission were dropped too far away to be of any assistance, and soon they were captured by the militias. Another paratroop unit took San Blas to the east of Girón, but it had no strategic importance at that point, even the men came immediately under fire from the Cienfuegos battalion.

When Fernández in Australia telephoned Castro in Havana to report the capture of Palpite, Fidel exclaimed, "We've already won the war! . . . Aviation has sunk three or four ships, and it continues in action. . . . You are to attack Playa Larga with the militia battalion. . . ." Castro had triumphantly shouted "We've won the war" when he and the two other survivors of Alegría de Pío encountered Raúl Castro and his few companions after the December 1956 debacle at the foothills of the Sierra Maestra, and this time his assessment was equally precise. With the invasion fleet eliminated and the invaders bottled up on the beaches, Castro faced what was essentially a cleanup operation, although much heavy combat still lay ahead.

Thus Fernández's attack on Playa Larga by five hundred militiamen armed with mortars, machine guns and rifles was repulsed at 2:00 P.M. with considerable casualties for the battalion. But Castro now felt that his presence might energize his troops in general (and anyway he wanted to be in the thick of the action), and he arrived by car at Australia at 3:15 P.M. —a three-hour drive at breakneck speed. He did not take a helicopter because it would be vulnerable to air action in the skies all around. At the sugar mill, he informed Fernández that artillery, antiaircraft guns, and tanks were on their way, and that the attack on Playa Larga must be resumed promptly. He wanted to go to the front at Palpite, but Fernández talked him out of it. His first photograph in the combat area—walking in battle dress and brown beret, rifle in the left hand, and cigar clenched in his teeth as he listened to Fernández a step behind—was taken there and immediately circulated around the world. The other famous Castro war picture, showing him jumping off a tank at Playa Larga, was shot the following day; it has since been made into millions of heroic posters.

That Castro materialized in the war zone the first afternoon also served to emphasize that he felt militarily and politically secure. The troops and militias in Havana were totally loyal, the seaside Malecón boulevard was ringed with artillery and antiaircraft guns, and in the capital alone about thirty-five thousand persons suspected of antiregime sentiments were detained during the morning by State Security, the police, and the Revolution Defense Committees. Havana's Roman Catholic auxiliary bishop was among the detainees. Sixteen thousand of them were placed in Havana prisons, ten thousand in the sports stadium, and four thousand in the huge Blanquita theater. If the CIA had really ever hoped for anti-Castro Cubans to rise in support of the invasion, Fidel made sure that they did not have any leadership anywhere. And before leaving Havana, he drafted an appeal to "The People of America and the World," signed by him and President Dorticós, for "solidarity" with Cuba in her struggle against "the imperialism of the United States" and its "mercenaries and adventurers who have landed at a place in the country."

There was tough combat for two more days as the Brigade fought the rearguard actions well, even in Castro's opinion, notwithstanding the hopelessness of the situation. Despite heavy artillery bombardment, the Brigade defended Playa Larga until the morning of the next day, April 18, when additional militia units arrived. That morning Fidel returned to Havana because of reports that enemy troops were landing in Pinar del Río in the west, but it was another CIA deception involving electronics gear aboard small boats well offshore, and he was back on the battlefield early on Wednesday the nineteenth—in time to see final victory. He was with Pedro Miret's artillery batteries east of Girón when San Blas was retaken and the noose tightened around the "Blue Beach." Brigade B-26s, which inflicted heavy casualties on Fernández's militias as they moved toward Girón on Tuesday, could no longer fly because their Cuban pilots were totally exhausted from the seven-hour round-trip flights from Nicaragua. Four CIA-recruited American pilots from the Alabama Air National Guard died when their two B-26s were shot down by Castro's T-33s. Kennedy had authorized one-hour morning sorties by American jets from the carrier *Essex* on Wednesday to help protect the evacuation from Girón beach, but unexplainedly they appeared too early to be of any help. The last efforts at resistance ended at 5:30 P.M.

Considering that he had no radio or telephone communications within the vast battle region of the Zapata—handwritten messages were rushed by jeep or motorcycle or even on foot—Castro maintained a remarkable control over the events as they unrolled for three days. Celia Sánchez was at the

Havana command post, keeping him informed by microwave radio and telephone of developments elsewhere. And, above all, his strategic instinct was unerring. When Fernández reported to him on the nineteenth that two U.S. destroyers were approaching the beach at Girón in what might be the prelude to a new landing, Castro replied, "What you're seeing are not landings but evacuations. . . ." He wanted the rubber boats with escaping exiles pinned down, but cautioned Fernández not to fire on the destroyers even though they were within Cuban territorial waters. He assumed correctly that the warships would fire back, and a sudden clash with the United States would erupt with unforeseeable consequences.

Castro spent Thursday the nineteenth on the beach in Girón satisfying his curiosity by inspecting the enemy positions and talking to the prisoners. There were so many prisoners that quite a few still had their weapons with them as they surrounded Fidel to answer his questions. Fernández, who knew most of the Brigade officers personally from the old days, says that the prisoners first feared immediate execution by their captors, then "were surprised that they were treated with total correctness . . . human dignity was scrupulously observed." It took several days to round up the remnants of the Brigade; in the end 1,189 prisoners, including the entire high command, were taken to Havana and interned at the Naval Hospital near La Cabaña fortress. Castro lost 161 dead in the battle, and the Brigade lost 107.

Castro's victory at the Bay of Pigs defined for the future the basic relationship with the United States as well as with the Soviet Union. In the first instance, Castro proved that he had an extremely high capacity for military defense, and in the second instance, the invasion demonstrated that the Kremlin was prepared to act in Cuba's defense. As Khrushchev said in a note to Kennedy on April 18, "We shall render the Cuban people and their government all necessary assistance in beating back the armed attack on Cuba." Still, eighteen more months would elapse before Khrushchev showed how far he was really prepared to go: Unquestionably the Bay of Pigs affair led directly to the Cuban missile crisis in 1962.

There is also little question that Soviet and Czech weapons, including artillery and tanks, made Castro's triumph quicker and easier, even though it was his use of air power that was the decisive factor from the outset. The weapons were so new that, Fernández says, some of the tank crews were learning how to fire the weapons en route from the depots to the battlefield. There is nothing to suggest, on the other hand, that the Soviet advisers who began arriving in Cuba with the equipment late in 1960 had anything to do with the victory. General Aleksei Dementiev, the first Soviet officer of this rank to be stationed on the island, arrived several months after the invasion to be chief of the military advisory group, and no one of lesser caliber would even dare to second-guess Castro. The revolutionaries won because Castro's strategy was vastly superior to the CIA's; because the revolutionary morale was high; and because Che Guevara as the head of the Instruction Department of the Revolutionary Armed Forces, in charge of the militia training program, and Fernández as commander of the militia officers' school had done so well in preparing 200,000 men and women for war. Moreover, the battle at the Bay of Pigs brought a new sense of unity to Cubans.

The report of the Taylor board of inquiry, parts of which remained classified twenty-five years later, freely recognized that Castro had been totally underestimated. It said that his forces' "operational doctrine was estimated to be virtually non-existent, and yet they seemed to have a very clear understanding of the value of control of the air." Richard Bissell, the CIA's architect of the invasion, testified that among the wrong judgments was "the underestimation of Castro's capability in certain specific respects, mainly his organization ability, speed of movement and will to fight. . . . Contrary to our opinion, the T-33s were armed and flown with skill, loyalty and determination." Finally, the board agreed that the notion that defeated Brigade troopers might join guerrillas in the Escambray (an idea pushed by the CIA) was pure fantasy.

To Ambassador Bonsal, the Bay of Pigs "was a serious setback for the United States. . . . It consolidated Castro's regime and was a determining factor in giving it the long life it has enjoyed. . . . It became clear to all concerned in Washington, in Havana, and in Moscow that for the time being, the Castro regime could be overthrown only through an overt application of American power." A quarter of a century later, a huge billboard overlooks the Girón tourist beach, proclaiming: GIRÓN—THE FIRST IMPERIALIST DEFEAT IN AMERICA! On a sunny day in summer 1985, a group of Soviet sailors from a visiting warship in Cienfuegos had their pictures taken in front of the billboard; most of them had probably not yet been born when that event occurred.

Having "defeated imperialism," Fidel Castro lost no time in extracting from the victory every drop of glory for the revolution and political and ideological advantage for his revolutionary regime—and, as always, for the Cuban sense of national pride. He did so by staging an extraordinary revolutionary passion play, running from spring to fall. On April 23, victory Sunday, he opened it with a four-hour appearance on *Popular University,* his favorite television program, to narrate the saga of the invasion and revolutionary triumph, using maps, a pointer, and captured documents. Fidel blended humor, sarcasm, scorn, defiance, and rousing explanations of revolutionary strategy to narrate the tale of Girón, and his voice soared in crescendo when he proclaimed that "our men know how to die and they have demonstrated it to excess in recent days!" Practically the entire population of Cuba watched Fidel on television that Sunday. Streets and plazas and parks were deserted, and his popularity seemed even greater than on the day of his first victory, the victory of 1959. Girón had unified the nation behind him, and, as Castro would say later, "Our Marxist-Leninist party was really born at Girón; from that date on, socialism became cemented forever with the blood of our workers, peasants, and students."

At the same time, Castro began producing captured prisoners on television to explain their backgrounds and involvement in the invasion Brigade. In what became a veritable television serial for days on end, the exiles repented their participation in the Bay of Pigs expedition (or were maneuvered into repentance), or stolidly confirmed having volunteered to invade their own country. Under questioning, fourteen of the prisoners confessed to murders and other crimes while serving in the Batista forces after the 1952 coup—and again the "imperialists" were shown up as the sponsors of assassins of the Cuban people. But some were defiant, asking why Castro held no elections if he was so popular. The following week, Castro had the prisoners assembled at Havana's Sports Palace where, also on television, he questioned the men, and argued and discussed with them—another act in the passion play.

Castro had decided while the battle was still progressing that the prisoners would not be harmed, and he did not wish to damage the image of revolutionary purity and generosity with brutality, summary executions, or a mass trial. Much more subtly—and practically—he held out for ransom from the United States for the overwhelming majority of the men, including all the commanders, and had revolutionary courts try in September only the fourteen prisoners accused of prerevolution crimes. Five of them were executed, and nine sentenced to thirty years in prison. Seven of the nine had their terms shortened, and in 1985, only two were still imprisoned.

The negotiations for the release of the Brigade prisoners took twenty months—Castro had first asked for five hundred tractors in exchange for them, and the Kennedy administration had balked at it—and Fidel came up with another sensation in his drama by allowing ten of the men to fly to the United States on May 20, to support his demands. The prisoners were personally told by Castro that ten of them, chosen by the whole group, could go to Miami and Washington for seventy-two hours, or longer if required for the negotiations, on their word of honor as officers and gentlemen that they would return to Cuba. He stunned them with this offer when he visited the captives just after being awarded the Lenin Peace Prize at the Soviet Embassy, a touch of Fidel's sense of ironic timing. And the other touch was that they traveled in their Brigade camouflage battle uniforms, clean and shaved, a gesture of military honor Castro always applied to his foes in war. They brought mail for the families of those left behind when they flew back to Havana a week later, and they were allowed to carry 660 pounds of gifts for their fellow prisoners. Though Eleanor Roosevelt agreed to chair a "tractors' committee" (and Castro immediately announced his willingness to negotiate with "the widow of the great president"), no accord could be reached, in part because the United States refused to let the Cubans make the exchange appear as a form of indemnification of Cuba for the invasion. Only on December 23, 1962, were the prisoners released against the delivery of 53 million dollars' worth of medicines and food.

On two widely separated occasions, I had the opportunity of discussing the Bay of Pigs—and President Kennedy—with Fidel Castro. The first time was in June 1961, less than two months after the invasion, when I joined Castro on a tour of the battlefield. He was the guide, the victorious commander, and the military historian as he pointed to sites where crucial events had occurred. Describing the second day's battle for Playa Larga, he said emphatically, "The attack was incessant, and we attacked them incessantly." On Girón beach, Castro rested his boot on the wreckage of a Brigade B-26 bomber, and said, expansively waving his long cigar, "They underestimated us and they used their own forces incorrectly." He went on to say that the invaders were lethally surprised by his mastery of the air, and that they should have engaged in multiple landings, instead of a single one.

"That was their first error," he expounded. "And because they had established a large beachhead, it became an urgent political problem for us to oust them as quickly as possible so that they could not establish a government here." The second major error, Castro said, continuing his critique, was the Brigade's failure to prevent tanks on flatbed trucks from reaching the Zapata area from Havana; the failure was due to the fact that paratroopers were dropped too late on D-Day morning to cut off the road communications already controlled by the militiamen. The paratroopers, he said, were used "too conservatively," but when he was asked how *he* would have used them, he wagged his finger, laughingly remarking, "I am not going to tell *you* that." Furthermore, Castro commented, attacking troops were not landed fast enough after the first wave came ashore at Playa Larga, allowing the *Houston* to be sunk with a whole battalion still aboard.

"Their problem," Fidel said, clearly enjoying the lecture, "was that they did not have a guerrilla mentality, like we do, and they acted like a conventional army. We used guerrilla tactics to infiltrate their lines, while attacking steadily from the air and on the ground. You must never let the enemy sleep." He thought, however, that the Brigade had first-rate equipment and excellent firepower. His error, he added, was to let a militia battalion advance on the second day on the open road that rises above the quicksand of the marshes where they were easy targets for enemy aviation. Yet the CIA planners' greatest miscalculation, Castro concluded, was to believe that the air strike on the three Cuban bases on April 15 had destroyed most of his planes. This was the key to the "imperialist" defeat and his victory.

The next time Castro spoke to me of the invasion was at his office at the Palace of the Revolution in Havana late in January 1984—almost twenty-three years later. We had not seen each other in the intervening period, and the conversation picked up where, in effect, we had left it off on Girón beach. I mentioned to him that in November 1961, seven months after the Bay of Pigs, I had been summoned by President Kennedy to the Oval Office for a private discussion of Cuba, and that I had been stunned when he asked me, "What would you think if I ordered Castro to be assassinated?" I told Castro that my reply was that the United States should not be involved in political assassinations, and the President said, "I agree with you completely." Kennedy had added, I informed Castro, that he was under pressure from some of his advisers to have the Cuban leader killed, but was "glad" that I opposed the idea because, indeed, he felt that for "moral reasons" the United States must not be party to assassinations. Richard N. Goodwin, then Assistant to the President, who was present at this conversation, testified in 1975 before a Senate committee that when he asked Kennedy several days later about this discussion, the President replied, "We can't get into that kind of thing, or we would all be targets." I made a point of mentioning this to Castro, too, and my story started him on the subject of John Kennedy. Castro, of course, knew in 1984 that numerous attempts had been made by the CIA in the 1960s to murder him, but, as he said, he could never bring himself to believe that President Kennedy would have authorized them.

"Well, what you tell me is really very interesting, and I had never heard it before," Castro said. "This is very illustrative for me because, in the first place, there is a great coincidence between what you have mentioned . . . and Kennedy's idea of having a dialogue [with Cuba]." Castro told me that on November 23, 1963, the day Kennedy was assassinated in Dallas, he was meeting in Cuba with Jean Daniel, a French magazine editor, who had brought him a secret message from the President. This was a year after the Cuban missile crisis. Kennedy, according to Castro, had asked Daniel to come to Havana and ask him how he felt about "discussing and having a dialogue with the United States . . . to find some channel of contact, of dialogue, to overcome the great tensions that had existed." Castro said that the radio flash about Kennedy's death came just as he and Daniel were discussing over lunch the presidential message to him. "For this reason," Castro said thoughtfully, "I have always kept the impression that Kennedy was meditating about the question of relations with Cuba."

Castro then alluded to a speech Kennedy had made in the spring of 1963, at American University in Washington, proposing nuclear-arms-control negotiations with the Soviet Union. "It was really a peace speech," Castro said, "and, in my opinion, it marked a change in the position of the United States in relation to international problems." Coming back to the subject of Cuba, he remarked, "I always had the impression that Kennedy was capable of rectifying that policy [toward Cuba]. . . . I have considered him to have sufficient valor to rectify that policy. For this reason, I consider that for us, for Cuba, and for relations between Cuba and the United States, the death of Kennedy was a great blow, an adverse factor."

Turning again to the question of his proposed assassination, Castro said that my account of the Kennedy conversation "fits logically with the idea I have of Kennedy's character . . . with the concept I have of Kennedy." But, he went on, "I judge Kennedy at the root of everything that has happened with relation to Cuba, beginning with Girón. I do not hold Kennedy responsible for Girón because the idea of Girón had appeared much earlier. . . . Since we proclaimed the Agrarian Reform Law in May 1959, the United States had taken the decision of liquidating the Cuban revolution, one way or another. It is possible that, initially, it was thought that the economic blockade, with the suspension of sugar purchases, with the halt of sales of equipment and spare parts from the United States, would be sufficient to force the collapse of the Cuban revolution, to produce a huge commotion in the people—and even to cause serious internal dissension that would force back the revolutionary process. I understand that after my conversation with Nixon at the Capitol in 1959—what apparently had been a frank and amiable conversation . . . in which I spoke very frankly to Nixon about the problems of Cuba—Nixon was convinced that I was a Communist and that it was necessary to liquidate the Cuban revolution. In reality, from what I said to Nixon, it could not be deduced that Castro was a Communist."

Two years later, Castro continued, there was the convergence of the Bay of Pigs and the Alliance for Progress. But, he said very emphatically, "Kennedy had inherited the whole plan of Girón from the Eisenhower government. At that time, Kennedy was, from my viewpoint, unquestionably a man full of idealism, of purpose, of youth, of enthusiasm. I do not think he was an unscrupulous man, I don't have that concept of Kennedy. He was, simply, very new, you might say—besides, very inexperienced in politics although very intelligent, very wise, very well prepared, with magnificent personal qualities. I can speak of experience and inexperience in politics because when we compare ourselves now with what we knew then about politics—the experience we had in 1959, 1960, and 1961—we are really ashamed of our ignorance at that time. Twenty-five years have elapsed [for us], and Kennedy only had a few months in the government."

Addressing the question of the invasion, Castro said: "I am convinced that he had doubts, for all the reasons of judgment I can assemble, about all that—but he did not decide to cancel it because many forces were committed: prestigious institutions in the United States, the Pentagon, the CIA, the tradition of [keeping] government decisions. So, with many doubts, he decided to move ahead with the invasion of Girón. And even though he launched the invasion, I think that Kennedy had great merit. If it had been Nixon, he would not have resigned himself to the defeat of the invasion, and I am convinced that there would have been escalation, and that in this country we would have been trapped in a very serious war between North American troops and the Cuban people, because the people, without any doubt, would have fought. The Revolution, without any doubt, would have resisted. At that moment we already had tens of thousands of weapons, we had them distributed in the mountains, we had them everywhere. We had hundreds of thousands of men, even if not yet very well trained . . . because when we saw that a military threat was hanging over us, we did everything possible to acquire arms, especially infantry weapons, and there would have erupted in our country a war that would have cost us tens of thousands, hundreds of thousands of lives. And I think that the man who had the personal qualities, who had the personal courage to recognize that a great error had been committed, to calm himself and to brake himself—this man was Kennedy. So, if we had an invasion that was prepared by Nixon and Eisenhower, we also had the luck that a Nixon was not elected to the presidency, and that at that moment it was a Kennedy, who was ethical, who was President. . . . Thus we recognize that Kennedy had the moral valor to assume the responsibility for what had been done. . . .

"For this reason, I place this idea of my physical elimination in the same context as Girón—something he had inherited. I have not had an attitude of resentment about it, even if he had considered the idea, because it could have been done indirectly, it could have been the interpretation of an expression at a given moment 'to get rid of Castro,' which could mean through a counterrevolution, through an invasion like Girón, through economic blockade, a direct invasion, or the physical elimination of Castro. . . . But it is hard for me to believe that Kennedy would have ever given a direct order of that nature because of the opinion I have of him—

not because he is dead, but analyzing it with calm and cold blood. I can add that I really felt a profound pain the day I received the news of his death—it shocked me, it hurt me, it saddened me to see Kennedy brought down."

And Fidel Castro was remembering John Kennedy thus not only in the light of the Bay of Pigs but also the missile crisis that had occurred the following year.

CHAPTER

5

The Cuban Missile crisis of October 1962 was a historically inevitable consequence of the Bay of Pigs events, even though, as Fidel Castro has observed, President Kennedy chose to refrain from launching an actual American invasion of the island. The dynamics on both sides of the straits of Florida had to force a new confrontation—it was like a law of physics—and this time the conflict was escalated beyond just Cuba to the superpower level. Never before or since had the world come so near a nuclear war. But Cuba and Castro would have been simply the pretext and the trigger, a very crucial point that this globally minded revolutionary leader could not grasp at the height of the crisis. This failure on Castro's part to understand global power realities served to fit him into the Soviet power system, though he would resist it for over six years. It would be a subtle and silent struggle between him and the Russians, much of it still shrouded in secrecy.

No sooner had the blood dried at the Bay of Pigs than a new process of polarization developed in both Washington and in Havana, John Kennedy and his advisers and Fidel Castro sharing equally the blame and responsibility for the results. In mid-June, Kennedy received the conclusion of the Taylor board of inquiry on the Bay of Pigs disaster that "there can be no long-term living with Castro as a neighbor" and that the Cuban situation be reappraised "and new guidance be provided for political, military, economic and propaganda action against Castro." Studies on how best to execute such action were provided by the National Security Council staff and the CIA, and on November 30, Kennedy sent a memorandum to Secretary of State Dean Rusk to inform him of his decision to "use our available assets . . . to help Cuba overthrow the Communist regime." In the meantime, Castro proclaimed that after the April invasion "the struggle in this country is for socialism," and he proceeded to accelerate this process in every possible way, presumably quite aware that he was inviting fresh violent responses from the United States. But for Castro, it was a matter of revolutionary principle—now that it was no longer "counterrevolutionary" to regard the revolution as the socialist stage in the march toward communism. Before Girón, men went to prison for saying it was so.

The postinvasion acceleration of the revolutionary process spotlighted the immense transformations in Cuban society and its way of life as the country experienced the third and approached the fourth year of Castro's rule. The new realities were the very considerable amelioration in the living, health, and education conditions of the population, especially in the impoverished countryside; a painful deterioration in the national economy for reasons ranging from chaotic planning to defense requirements and the United States' economic warfare; and a nearly total regimentation of Cuba in terms of its political, intellectual, and cultural life. Put more crudely, a revolutionary dictatorship had been implanted by the end of 1961, because, as Fidel Castro emphasized in his 26th of July speech that year, "this is a life-or-death struggle that can only end with the death and destruction of the Revolution or of the counterrevolution."

Having declared that this was a socialist revolution and that the next step would be the creation of the United Party of the Cuban Socialist Revolution (PURSC) as the ruling political organism under his guidance, Castro moved to eradicate all forms of opposition (loyal or not) and every vestige of independent thought. Always insisting that all these measures aimed at the absolute national unity required for the revolution's survival, he urged Cubans to join him in joyfully greeting the advent of the new age of regimentation. In the course of his 26th of July speech at Havana's Revolution Plaza, Fidel asked those who belonged to the militia, to the Revolution Defense Committees, the Revolutionary Workers' Confederation, and the Cuban Women's Federation to raise their hands—and a hurricane of applause and shouted approval swept the huge crowd. It was another "consultation" with the masses, and Castro exclaimed, "The Revolution has organized the people! . . . Even children are being organized in the associations of Rebel Pioneers!" And Cubans now addressed each other as *compañeros* in this brotherhood of the revolution.

Notwithstanding the Bay of Pigs invasion and the threat of new American thrusts, the Escambray antirevolution guerrillas, who managed to regroup in mid-1961, and economic adversity, Fidel never deviated from his blueprint for step-by-step political consolidation and control. In fact, the greater the pressures and dangers, the more determined he was to carry out methodically and uncompromisingly his fundamental program. And nothing could disturb his sense of priorities or interfere with his painstaking and often convincing rationalizations for enforcing his policies.

Castro's handling of Cuban intellectuals, writers, and artists, forcing the country's cultural community into an ideological straitjacket and depriving it of the last ounce of freedom in the sense accepted in the nontotalitarian world, was one of his masterful exercises in power, intimidation, and manipulation. Because the Revolution's wisdom could not be questioned and because of Fidel's skills, Cuba's best and richest minds gave him an ovation when he had concluded the lethal surgery on them. He, in effect, created a cultural wasteland in Cuba, where a quarter-century later, creativity is not even beginning to revive—at least not visibly. In terms of Castro's logic, this operation was consistent with his goals of political philosophy, but the resulting intellectual and literary climate in Cuba is a blend of Cervantes, Kafka, and Orwell. Cervantes lives in Castro's magnificent and constant use of the Spanish language; Kafka is represented by the element of nightmare in Cuban intellectual life; Orwell symbolizes the terrifying efficiency of the all-powerful state and its great leader in dictating what culture may be allowed to exist.

The events that occurred in the conference room of the José Martí National Library in Havana on three successive Saturdays in June 1961 brought about this state of affairs, and they provide a special insight into the workings of the amazing mind of Fidel Castro. Familiar with the writings of Antonio Gramsci, the Marxist thinker who propounded the view that the control of popular culture is needed to win (or retain) the support of the masses, Castro approached the Cuban problem from this premise. Much earlier, he had acquired full control over all the mass media, establishing at the same time a high-quality cinema institute, directed by his old friend Alfredo Guevara, to produce feature films, documentaries, and newsreels with heavy revolutionary content. Now came the turn of the writers, poets, journalists, artists, composers, movie directors (as individuals), playwrights, and ballet masters. Actually, the June crisis was precipitated by a more or less ideological dispute involving a mildly controversial twelve-minute documentary film and the decision by the Castro-created National Culture Council not to exhibit it, but it was instantly clear that what was at stake was the definition of cultural freedom under the revolution.

Incongruous as it may appear, the real issue was the unhappiness of Castro and Edith Garcí Buchaca, the council's head and ranking member of the Communist party, with the weekly literary supplement of the newspaper *Revolución,* which was still the official organ of Fidel's own 26th of July Movement. Published on Mondays and therefore called *Lunes de Revolución,* it had been launched in March 1959, and it was probably the best and most interesting literary publication in Latin America. Its problem, however, was defined in its first editorial's statement that while the revolution had broken "all the barriers of the past," *Lunes* "has no defined political philosophy although we do not reject certain systems [such as] dialectic materialism, psychoanalysis and existentialism." *Lunes* therefore published what its editors (who were leading writers, most of whom had lived abroad during the Batista era) considered interesting: It ran the full gamut from Raúl Castro's and Che Guevara's war diaries to articles about Marx and Lenin (annoying moderate *Fidelistas*), Trotsky and Djilas (annoying the Communists), Proust, Chekhov, Hemingway, and American "beatnik" writers. At the outset, nobody seemed to mind, and Fidel occasionally dropped in at *Lunes* offices late at night for *café con leche,* one time bringing along Jean-Paul Sartre and Simone de Beauvoir. This was still the romantic phase, and Fidel enjoyed being the bohemian intellectual. In 1961, however, the revolution was no longer romantic, Castro had publicly chosen a dogmatic ideology, and hard-line Marxists ran the Culture Council. *Lunes* no longer belonged.

At the National Library, Castro patiently listened for two Saturdays to lengthy debates among the writers and the artists on the meaning of cultural liberty under the revolution; the sessions were made even livelier by a few favor-currying writers accusing some of their best friends in Fidel's presence of being "counterrevolutionaries." On the third Saturday, June 30, Fidel Castro laid down the intellectual and cultural law of the revolution in one of his most important speeches, known as "Words to Intellectuals" (two hours' worth of them). Apart from Gransci's dictum about popular culture, Castro defined in the clearest fashion to date the philosophy of the revolution and the limits of its tolerance, or, better said the biological rigidity of its intolerance.

Too intelligent and sophisticated to spout Marxist slogans to such an audience (he refrained from mentioning socialism even once in his "Words"), Castro put it very plainly: "We believe that the Revolution still must fight many battles, and we believe that our first thought and our first preoccupation must be what to do to make the Revolution victorious . . . The fear expressed here is whether the Revolution will drown this liberty [or expression], suffocate the creative spirit of writers and artists. . . . The most polemical point is whether there must exist, or not, liberty of contents in artistic expression. . . . [But] the revolutionary places something even above his own creative spirit; he places the Revolution above all else, and the most revolutionary artist would be disposed to sacrifice even his own artistic vocation for the Revolution. . . . Writers and artists who are not revolutionary must have the opportunity and liberty to express themselves within the Revolution. This means that within the Revolution [it is] everything; against the Revolution—nothing. Against the Revolution nothing because the Revolution also has its rights, and the first right of the Revolution is the right to exist. . . . What are the rights of writers and artists, revolutionaries or not revolutionaries? Within the Revolution: everything; against the Revolution, no rights."

In this manner, Castro established the policy principle that the revolution (or the revolutionary bureaucracy or himself) would interpret—arbitrarily—what *is* and what *is not* "within the Revolution," with no right of argument or appeal. It implied more or less subtle censorship of ideas, and it encouraged self-censorship—presumably on the lowest-denominator level of averting ideological risks. Should writers or artists have doubts, however, about what belongs "within the revolution," Castro offered them the National Culture Council as "a highly qualified organ to stimulate, promote, develop, and orient—yes, orient—this creative spirit. . . ." Reassuringly, Fidel added that "the existence of authority in the cultural realm does not signify that there is reason to be preoccupied over abuse of that authority."

Unveiling the revolutionary concept of cultural freedom, Castro proceeded to implement it by calling a congress to organize an association of writers and artists—one more revolutionary unity organization—and by proposing that this association publish a "cultural magazine" open to all, instead of multiple literary and artistic publications that tend to provoke disunity. The message was understood clearly: on November 6, 1961, *Lunes de Revolución* published its last issue. As one of its editors put it later, *Lunes* vanished unlamented in the climate of revolutionary conformity and submission. The supplement, as well as the more obscure literary supplement of the Communist newspaper *Hoy*, was presently replaced by the new association's "cultural magazine." It is as uninspiring as the rest of the Cuban press today.

Culture is a matter of perception as well as definition. The traditional Western concept of cultural liberty was erased by the Castro revolution, which, instead, offered Cuba mass culture. Fidel, the only truly cultured mind in the guerrilla leadership, would surely and irately reject the notion that his island was transformed into a cultural wasteland. (Che Guevara, who was another cultured mind, was cynical enough to have acknowledged it had he lived to see the ravages.)

Castro, however, has convinced himself that true culture for the people does exist in Cuba because of the near-miracle of making the whole nation literate, because by 1986, over *fifty million* copies of books were being printed annually, and because ballet, music, and quality theater and cinema were available to the masses. It does not seem to matter in the context of the revolution that no fresh ideas are germinating in a country ruled by a man endowed with an extraordinary intellect; to Castro this is not a contradiction even in terms of his own place in history. It does not seem to trouble him that between 1968 and 1976, the best Cuban writers of his own generation were blacklisted without explanation by Cuban publishers, and that self-censorship is now a revolutionary institution as well. It may be escapism, but twenty-five years after Castro delivered his "Words to Intellectuals," the most widely read authors are Hemingway (so admired by Fidel) along with Mark Twain, Dashiell Hammett, and Raymond Chandler.

Finally, in its cultural travail, the revolution is responsible for the persecution of homosexuals—among them some of the most talented writers and artists with the freest minds—that reached its apex in the 1960s and 1970s, when hundreds or more were forcibly enrolled (along with common criminals) in the so-called Military Units for the Support of Production (UMAP). These forced-labor units have since been abolished, but there is no explanation of why a man of Fidel's humanistic and intellectual orientation tolerated them in the first place as if to assert the *machismo* of the fiery fighters of the Sierra Maestra.

To proclaim his revolutionary defiance at home and abroad regardless of the possible consequences of his attitudes was part of Fidel Castro's nature as a permanent *guerrillero*. Indeed, he loved to taunt his enemies, the "imperialists," as if anxious for another confrontation and another test of his militarized society. In his speech on July 26, 1961, he preached the necessity of always keeping weapons in hand "because the imperialists do not want to forgive us our successes, and the more we become organized, the more they become filled with ire over the conquests of our people." He spoke after awarding the newly created Order of Girón to his guest of honor, the Soviet cosmonaut Yuri Gagarin, whom he told, "You can fly twice around the world while I'm delivering my speech."

Castro's defiance included his willingness to accept the awesome brain drain affecting the Cuban economic structure with the exodus of tens of thousands of middle-class physicians, engineers, managers, and professors. As many as 250,000 Cubans out of a population of 6 million fled in the first three years of the revolution, but Fidel saw them as "parasites" and potential counterrevolutionaries, and he preferred them in the United States. It was a calculated long-range risk to do without a management class he did not trust anyway, and to educate his own revolutioniary elites, and from his viewpoint Castro was undoubtedly right. He has been vindicated now, twenty-five years later, when Cuba has trained enough physicians to assign thousands of them to work throughout the Third World, after achieving at home one of the world's highest ratios of doctors to the population.

Fidel Castro's moment of supreme challenge came on the evening of December 1, 1961, when he informed Cuba and the world that the new united revolutionary political party would have "a Marxist-Leninist program adjusted to the precise objective conditions of our country," that this would no longer be "a secret," and that "today we shall see to it that to be a Communist is a merit." This culminated three years of the revolution, and it crystallized once and for all Castro's political and ideological identity. At the age of thirty-five, Fidel Castro had attained the ultimate definition of his personality in every observable sense: As a man and as a political figure he would not appreciably change in the years and decades to come. Obviously, great maturity and experience came with age, but it was not very different to converse with Fidel in 1985 from what it was after the Bay of Pigs in 1961.

In the conversations in 1961, Fidel offered his special mix of earnestness and good-humored informality when I asked why and how he had become a Marxist. We were sitting in the Riviera Hotel restaurant in Havana, and after I raised the ideological question, he twirled his cigar, and said, "Men evolve politically. You mustn't forget that." Then his eyes twinkled, and Castro added, "You know, Dante once wrote what was essentially a novel, but he insisted on calling it *The Divine Comedy.* We made a revolution that we first called 'humanist,' but we now call it 'Socialist.' . . . I came to believe in socialism when I discovered that capitalism means the exploitation of man by man, when I saw the cyclical crises of capitalism, when I realized that imperialism was doomed. . . . Now, you should not be surprised that I came to this conclusion. We all read the same books, don't we?"

In December, Castro typically chose to announce the advent of Marxism-Leninism on the *Popular University* television program, and the format was not a formal speech but a chat about the history of his revolutionary movement, back to his own youth, and badinage with others on the panel. He speculated aloud about how little Lenin and Marx could have done if they lived in the eighteenth century, to make the point that nothing exists in a vacuum, and that the Cuban revolution was made when the necessary conditions developed. Of Marx and Lenin, he said that "one cannot be the intellectual of a class that does not exist or the creator of the doctrine of a revolution that cannot occur." Describing the early phases of his movement, Castro acknowledged that at the time of Moncada "certain proposals were made with the hope of not damaging the scope of the revolutionary movement. . . . If we hadn't drafted this document carefully, and had made it more radical, the revolutionary movement of struggle against Batista would not have acquired the amplitude that made victory possible."

Rhetorically, he asked himself, "Do I believe absolutely in Marxism? I believe absolutely in Marxism. . . . Did I understand it [in 1953] as I understand it today after ten years of struggle? No, I didn't understand it then as I understand it today. . . . Did I have prejudices concerning communists? Yes. Was I influenced by the propaganda of imperialism and reaction against communism? Yes. . . . Did I think communists were thieves? No, never. . . . I always thought communists were honorable, honest people. . . ." And twenty-five years later, addressing Latin American participants at a Havana conference—and in private conversations—Castro handled the topic in almost identical phrases and formulations. He had become an experienced juggler of politics and ideology, with dozens of balls in the air at the same time, each intended to entice and fascinate a different viewer in the audience. He always felt in control.

In declaring Cuba to be embarking on the Marxist-Leninist road, Castro knew that he was increasing by a vast order of magnitude the risk and likelihood of new United States intervention in one form or another. His announcement hit the Kennedy administration like a bombshell. And even as he explained before the television cameras why Cuba must move toward communism, he was publicly warning that "the entire military science of the Pentagon will smash itself against the reality in which live the people of America." Secretly, he was pondering how best and how fast he could protect Marxist-Leninist Cuba from the Pentagon—with his own resources and those of his new faraway allies. The great crisis was less than a year away, but neither Kennedy in Washington nor Khrushchev in Moscow had yet comprehended where Fidel Castro was pushing them.

And the Soviets may also have had difficulty reconciling Castro's overwhelming need and desire for greater economic and military assistance, especially as a new crisis with America was obviously approaching, with his ruthless treatment of the "old" Communists with close Moscow ties, who had dared to challenge his leadership at home. Fidel had survived and succeeded in the last decade because he never compromised in the attainment of his goals—not at Alegrí de Pío and not at the Bay of Pigs—and he was not prepared to abandon his principles in dealing with local Communists or with the Russians. This power battle was over the issue of "sectarianism" that erupted early in 1962, and Castro dealt with these Communists as if they were the "bandits" of Escambray.

After Castro had sealed his alliance with the "old" Communists in the secret Cojímar meetings early in 1959, and decided to merge the 26th of July Movement, the Students' Revolutionary Directorate (DR) and the Communists' Popular Socialist Party into a single political organization, it became necessary to hammer it all together. As the first step, the three groups were joined in the ORI (Integrated Revolutionary Organizations) in preparation for the emergence of the United Party of the Cuban Socialist Revolution (PURSC) with Castro as the secretary general. The "new" Communist party would be the final step. In accordance with Castro's respect for the "old" Communists' supposed political skills, Aníbal Escalante, an old-line leader, was given the task of organizing the ORI. It seemed like a fine idea until Castro indignantly realized that Escalante and his associates were packing the whole organization with their own party people on every level, and, in effect, carrying out a classical Communist takeover from the inside. It was a quiet attempt at a coup d'état, which would have placed the revolution in the hands of orthodox Communists, and presumably turned Fidel into a magnificent figurehead.

Castro himself has not suggested publicly that this astonishing notion had been elaborated in Moscow, but there are valid reasons not to exclude this possibility. The Russians still regarded Fidel as a loose cannon, and may have wished to exercise some form of control over a regime whose upkeep was becoming increasingly onerous. It is also possible that at least a faction among the "old" Communists may have concluded that they should run Castro, and not the other way around, and that they might find allies among ideologically sympathetic rebel officers who would agree to elevate him to the rank of an exalted statesman of the revolution, but with reduced powers of governance. The only certainty is that Aníbal Escalante could not have invented it all by himself; too many other Communist old-timers were involved. How these Communists came to believe that they could capture Fidel Castro is still a mystery. Their conspiracy was elegantly called "sectarianism"—in revolutionary Cuba it was a superiority complex that many 26th of July Movement officers as well as "old" Communists tended to display in relation to each other and everybody else—but to Castro is was pure and simple counterrevolution.

With his superb sense of timing, Castro waited for the right moment to strike. Thus he said nothing publicly when the composition of the twenty-five-man National Directorate of ORI was announced—evidently with his approval—on March 9, 1962, with him, Raúl Castro, and Che Guevara leading the list as chiefs of the 26th of July Movement. But of the twenty-five members, ten were "old" Communists (who played no serious role in the war), and among the thirteen men from Fidel's Movement, at least three had strong Communist leanings. The Students' Directorate, which fought against Batista in Escambray, received only two seats. Theoretically, then, as Fidel knew, the orthodox Communist faction could assemble a leadership in the ORI leadership. Four days later, Castro exploded in unprecedented public fury when at anniversary commemorations at the university of the Directorate's 1957 attack on the Batista palace, a Communist party orator omitted the invocation to God in the text of the "testament" of the student leader José Antonio Echevarría. He shouted that the memory of a dead companion was being censored and falsified, and it was entirely in character for Castro to do so as a matter of principle.

Nevertheless, those who knew him well thought that Fidel was sending the Communists a message. If he was, they missed it altogether. Castro waited two more weeks, and on March 26 (his favorite day of the month in his private superstitions), he staged one of his great dramatic television productions. In his television appearance earlier in March, Castro had announced a depressingly sharp program of food rationing (agriculture under the revolution was not keeping up with consumer demand and needs), but now he focused the national attention on the "sectarian" plot. Aware that political dramas must be personalized in order not to be abstract for the masses, Castro singled out Aníbal Escalante for some of the most withering accusations in the formidable arsenal of sarcastic invective. He informed Cubans that Escalante had created "a counterrevolutionary monstrosity" in ORI, that he had built up his own "machine" to take over the party and the government, and that *Fidelista* veterans from the Sierra Maestra were losing troop commands to Communist officers (who may have been part of the Escalante affair). Escalante was suffocating Cuba, Castro said, to the point that when "a cat had four kittens, one had to go to the ORI office to resolve the matter" of the kittens' fate.

However, Fidel had the political sagacity not to declare war on all the "old" Communists, having made his point with the public political execution of Escalante. Several other party leaders were purged along with Escalante, including the cultural chief, Edith García Buchaca, and her husband Joaquín Ordoqui, an executive bureau member of the PSP. Others in the party leadership realized that they could not confront Castro, and they were delighted to let Aníbal Escalante be the principal sacrificial lamb. (His brother César Escalante, joined in the rites and remained in the leadership.) Once more, Carlos Rafael Rodríguez rose to the occasion to mediate between his friend Fidel and his old comrades, negotiating from a position of power toward the party inasmuch as Castro had named him president of INRA only the month before. With the farm crisis raging, Castro preferred to concentrate on other matters and resigned from the INRA presidency, leaving agricultural problems to Rodríguez and Che Guevara. Others fell instantly into line. The PSP's secretary general, Blas Roca, wrote in the party newspaper *Hoy* that Castro was "the best and the most efficient Marxist-Leninist in our country" and Marxism-Leninism's "insuperable guide and chief," remarkable praise for a man who had formally embraced this doctrine only five months earlier. A quarter-century later, Escalante's "sectarianism" remains a sore topic with the "old" Communists; both Blas Roca and Fabio Grobart uncomfortably called it "simple measles" in separate interviews in 1985.

The Soviets chose to stay publicly out of this internal affray, preferring to let nature take its course and not assuming positions prematurely. That Castro was angry at them for unspecified reasons was made obvious when he refused to receive Soviet Ambassador Kudryatsev for a farewell audience (apparently he had requested his removal, and had said to friends that Kudryatsev "tires no more than Bonsal did") in a typical display of displeasure. Still, Moscow chose to maintain the friendship with Havana, and it is probable that the mounting schism with China played a role in the decision to swallow Castro's manifestations of independence. Thus Aleksandr Alexsiev, the young "journalist" who drank the vodka toast with Fidel in 1959, was named as the new ambassador to Havana to the Cubans' great satisfaction. Alexsiev accompanied Mikoyan in 1960 for the signing of the first trade treaty, and now he came from his ambassadorship in Argentina. *Pravda* wrote an editorial on April 11 praising Castro and denouncing the unfortunate Escalante, and Deputy Foreign Trade Minister I. I. Kuzmin turned up in Havana to conclude a new trade treaty for 1962. It increased the two-way trade (in reality, Soviet deliveries to Cuba) from $540 million to $750 million.

A trade accord with China, though on a much smaller scale, had been signed a month earlier, and the Peking official journals hailed Castro for opposing "sectarianism." Khrushchev obviously did not think the Cubans would go over to the Chinese side, but even the thought of evenhandedness toward China was troublesome to him in the midst of the rising Soviet battle for the control of the world Communist movement as well as the sympathies of the increasingly important Third World. Castro had his way, and now he was in the best possible position to extract Soviet guarantees of military protection for Cuba against the United States. In a June speech, he was happy to describe Khrushchev as "that great and dearly beloved friend of Cuba." Again, Fidel was demonstrating that strategy and tactics must never be confused.

And foremost on his mind was the threat from the United States. There were no plans afoot by the Kennedy administration, as far as is known, to launch an American invasion of Cuba, although Castro (and possibly Khrushchev) believed that a direct attack was in the offing. This was a misperception that produced dire results, but it is just as true that Castro could not afford to rule out an invasion, and that he had to be ready for one. His subsequent behavior and accords with the Soviets are consistent with his overall precautionary stance, even though it remains unclear a generation later how exactly Castro and Khrushchev arrived at the notion of deploying Soviet nuclear weapons on the island. Nor is it comprehensible how the two of them could have expected that the United States would fail to discover the presence of these weapons—and stand still for it.

Castro, however, was absolutely right in assuming that the Kennedy administration was attempting by all means short of invasion to remove him from power, and his intelligence services were providing him with growing evidence of such subversive activities. Under the circumstances, Castro had to conclude that an invasion could well be the next logical move against him. What he was witnessing, starting early in 1962, was the launching of "Operation Mongoose," authorized by the President the previous November "to help Cuba overthrow the Communist regime."

General Edward Lansdale, a counterinsurgency specialist, was named to head Mongoose, and the operational plan he presented at the White House in mid-January 1962 called for a six-phase effort by the entire United States government to undermine Castro from inside. It was designed to culminate—ironically, as it turned out—in October, "with an open revolt and overthrow of the communist regime." An invasion might indeed have come around that time inasmuch as the Lansdale plan included planning for "use of U.S. military force to support the Cuban popular movement." Such a movement was most unlikely to materialize, but Castro certainly had reason to be very much worried about the Americans.

As it happened, Operation Mongoose never came close to attaining any of its anti-Cuban goals (although four hundred CIA officers in Washington and Miami were attached full time to this enterprise). The best it could do was to run minor intelligence-infiltration missions, carry out minimal sabotage, and, as far as the CIA was concerned, revive plans to assasinate Castro. The agency had attempted an assassination plot through Mafia figures in 1960 and, acting on its own, dusted off those plans when Mongoose was created. Richard Helms, at the time the CIA's deputy director for plans (covert operations), testified before a Senate committee in 1975 that he had assumed that the "intense" pressure exercised by the administration to oust Castro had given the agency authority to kill him, albeit assassination was never formally ordered. In this sense, Helms used almost identical reasoning, from his standpoint, as Castro did in his conversation with me in speculating that plots for his demise may have stemmed from overzealous interpretation of broad policy instructions. Thus Helms said: "I believe it was the policy at the time to get rid of Castro and if killing him was one of the things that was to be done in this connection, that was within what was expected."

Historians have been debating for many years whether or not the Kennedy administration's efforts to oust Castro after the Bay of Pigs fiasco were *directly* responsible for the Soviet deployment of nuclear weapons in Cuba. All the available evidence, including Castro's own judgments, suggests that an affirmative answer is possible only if it is accepted that the Cubans and the Russians did really believe that an actual invasion, as distinct from Mongoose, threatened the island in 1962. Since it is now known that Kennedy entertained no invasion thoughts, the debate over American responsibilities for the *nuclear* crisis is couched in false terms. The nuclear issue in 1962 was strictly a matter of Soviet strategic decision-making once Khrushchev was convinced (or convinced himself *and* Castro) that there would, indeed, be an invasion. Naturally, Castro had every right to seek maximum Soviet protection because he was facing great uncertainties, but it may never be known whether Fidel used Khrushchev to escalate the nuclear crisis with the United States—or vice versa. The existing record is fragmentary and contradictory.

There is no question, however, that Cuban-Soviet negotiations over significant Kremlin support for the revolutionary regime began in earnest in the spring of 1962, just as Castro completed liquidating the Escalante "sectarian" Communist challenge to his leadership. Not only did Castro see imminent military dangers from the United States, but in January the Organization of American States expelled Cuba from membership, under Washington's pressure, at the hard-fought foreign ministers' conference at Punta del Este in Uruguay. This was seen in Havana as political preparation for an invasion, the idea being to present Cuba as the Communist enemy of the entire Western Hemisphere, not only of the United States.

Castro fought back politically by issuing the "Second Declaration of Havana" in an exceptionally emotional speech on February 4, charging that at Punta del Este "Yankee imperialism gathered the ministers together to wrest from them—through political pressure and unprecedented economic blackmail in collusion with a group of the most discredited rulers of this continent—the renunciation of the national sovereignty of our people and the consecration of the odious Yankee right to intervene in the internal affairs of Latin America." Always on the offensive, Fidel responded to the expulsion from the OAS with the kind of challenge calculated to make Washington even more hysterical about him. Addressing the vast crowd at Revolution Plaza, he said: "The duty of every revolutionary is to make the revolution. . . . The revolution will triumph in America and throughout the world, but it is not for revolutionaries to sit in the doorways of their houses waiting for the corpse of imperialism to pass by. The role of Job doesn't suit a revolutionary."

At home, Castro also had cause for intense concern over the reappearance of guerrilla bands in the Escambray Mountains and, to a lesser degree, elsewhere in Cuba, after the successful "cleanup" operations late in 1960 and early in 1961. That "clean-up" had neutralized the Escambray as an area of operations in support of the Bay of Pigs landing, but already in 1962, the guerrillas were again a serious problem. Strangely, the CIA provided no help to these bands in the context of Mongoose, but the situation was grave enough for Raúl Castro to describe it as "the second civil war." In any event, Escambray was a costly drain on Cuba's defense resources, adding to the regime's overall vulnerability. The Castro brothers and men like Che Guevara realized that the existence of the guerrillas was intolerable to the regime—they knew it from their own experience as mountain rebels—and they watched with great concern the bands add up to approximately three thousand men by mid-1962. This was ten times as many men as Fidel had had in the Sierra at the end of the war, but, fortunately for him, there was no unified leadership and no real leader, and the guerrillas were split up into scores of groups without communication among them.

Politically, it was embarrassing for Castro to admit that so many men were up in arms against him, even if they did not constitute a coherent force, and practically nothing on the subject was published in Cuba well into the 1960s, when the last remnants of the guerrillas were finally destroyed. Most of the *guerrilleros* were small landowners—the Cuban rural middle-class—and former estate managers, foremen, and workers, but there were also rural merchants and quite a few ex-officers of Castro's rebel army and the Students' Revolutionary Directorate. This was a group to whom revolutionary socialism had nothing to offer, and, in fact, these men had lost much privilege in the wake of the revolution. Unlike Castro in the 1950s, they had no support in the cities, and they did not present ideological or political positions. According to a Cuban specialist on this subject, the Escambray bands "were not a true danger for the revolution if we acted in time and in silence, and we did not convert it into a national preoccupation."

At a strategy meeting in Havana, Fidel suddenly interrupted one of his commanders, who was referring to the new rebels as *guerrilleros,* to say, "Don't ever again say that they are *guerrilleros*—they are bandits." The counterinsurgency units that were then being trained were consequently given the names of "Battalions of Struggle Against Bandits," and the description of "bandits" caught on. Castro had understood the psychology of the situation facing him, and he never allowed the problem to get out of hand. Many years later, Castro explained at great length in a confidential speech before Angolan officers, that he had defeated the "bandits" through a combination of throwing great military resources against them and secret negotiations with guerrilla chieftains for surrender on generous terms and for encouraging others to surrender. These included bands totaling over five hundred men in the Oriente mountains as well. Still, these were costly operations, and Castro's casualties reached more than three hundred killed in Escambray alone; economic losses were calculated at around $1 billion in ruined crops, burned houses, destroyed rolling stock, roads, and bridges— and the military cost of the "antibandit" operations. Castro, despite his other concerns, was always in overall command of this fight; on one occasion, he personally captured a number of "Bandits" when with several soldiers he climbed a wooded hill above the Cienfuegos highway, where, he was told during a trip, a guerrilla band was hiding. But the last of the Escambray "bandits" were not caught until 1966.

Things could not have been worse for Castro than in the spring of 1962, when he turned to Moscow for military protection. There were guerrillas in the mountains, "old" Communists were trying to undermine him, the farm economy was collapsing, and the Americans were after him with Operation Mongoose on one level and diplomatic isolation in Latin America on the other. There is no single event that seems to have triggered Castro's requests to Moscow, but indications are that he raised the issue for the first time when S. R. Rashidov, an alternate member of the Soviet Politburo, came to Havana late in May. He was the highest-ranking Soviet government official to visit Cuba since Mikoyan's trailblazing presence two years earlier. There must have been continuing secret talks during June because Castro disclosed in a subsequent speech that the negotiations for "the strengthening of our armed forces and the dispatch of strategic missiles to our country" occurred that month. They may have been conducted through Alexsiev, who had just arrived as the new ambassador, or through Carlos Olivares, the new Cuban ambassador in Moscow. But a secret channel is more likely.

At this stage, Fidel turned once more to public drama. On June 15, he departed with great fanfare for the Sierra Maestra, wearing his battle fatigues and carrying his rifle, and he spent eight days there, reliving the glories of the guerrilla, and being ostentatiously absent from Havana. Castro's mountain meanderings were covered in detail in the press and on television, including his statement that "once more, I have raised the banner of rebellion." The symbolism must have been meant for Kennedy and Khrushchev as well as for Cubans who had to be prepared for a new crisis. On July 1, the Communist newspaper *Hoy* started publishing daily reports on U.S. "violations" of Cuban air space and territorial waters (though *Revolución,* still the regime's official organ, failed to do so).

This sequence of events continued wtih a two-week visit to Moscow early in July by Raúl Castro and a delegation of his officers. Official announcements reported that Mikoyan and Marshal Rodion Malinovsky, the Soviet defense minister, received Raúl on July 3, but the Soviet historian Roy Medvedev writes that the military talks lasted a week. Medvedev, who always had access to official Soviet sources, also adds in his biography of Khrushchev that "he attended the talks on 3 and 8 July." And according to Medvedev, these discussions centered "on the provision of military aid to Cuba and the secondment of a number of Soviet military specialists," and "it was presumably during that week that the decision was taken to send to Cuba medium-range missiles with nuclear warheads and bombers capable of carrying atomic bombs." This is most likely accurate although there may have been followup conversations through secret channels between July and early September when the Soviets began secretly to ship the missiles to Cuba.

There may also have been secret contacts before Raúl Castro's *official* visit to Moscow in July. Thus Fidel said to a visitor a year after the crisis that in June 1962, my brother Raúl and Che Guevara went to Moscow to discuss ways and means of installing the missiles." If Fidel's dates are correct—and despite his prodigious memory he occasionally has problems with precise dates when he talks about past events involving his activities or interests— Raúl and Che would have gone to Moscow secretly even as he was waving the rebellion banner in the Sierra Maestra.

There is no known record of an actual Soviet-Cuban agreement on the missiles, and it is possible that nothing was ever put on paper in order to protect Khrushchev as well as Castro in the future. Raúl Castro mentioned in a little-noticed speech late in August that "Soviet troops" had begun to arrive in Cuba, but he offered no details, and it was unclear whether he referred to advisers, combat units, or rocket forces specialists who came to prepare missile sites. Simultaneously, MIG jet fighter-bombers began arriving for the Cuban air force. Che Guevara and Major Emilio Aragonés, who was very close to the Castro brothers, visited Khrushchev at his vacation *dacha* in the Crimea, but this was not announced. Guevara was in the Soviet Union for the ostensible purpose of signing an agreement on modernizing Cuban steel plants, and he may also have finalized the missile arrangements with the premier.

A Soviet-Cuban military agreement on September 2 announced that as a result of "imperialist threats," Cuba had asked the Soviet government to "help by delivering armaments and sending technical specialists for training Cuban servicemen," and Moscow had responded affirmatively. However, this text had to have been designed to explain publicly the arrivals of Soviet military personnel and conventional equipment, and even possibly to allay any American fears over missiles. That same week, Anatoly Dobrynin, the new Soviet ambassador, conveyed to Attorney General Robert F. Kennedy a message from Khrushchev to the President to the effect that no "offensive weapons" were among the equipment being sent to Cuba. On September 12, TASS issued a statement to stress that the Soviets had "no need" to deploy retaliatory "defensive weapons" in any other country, adding "Cuba, for instance." This was in reply to a Kennedy warning that the United States would not tolerate the installation of ground-to-ground missiles in Cuba. Khrushchev had now set in motion his incredible attempt to deceive the United States about nuclear weapons in the Caribbean.

Twenty-five years after the Cuban missile crisis, it remains a matter of debate whether it was Khrushchev or Castro who first proposed the Soviet nuclear deployment on the island. There are also subtleties in various accounts concerning the fashion in which the two leaders decided on this risky course of action. However, Castro is still most anxious for his interpretation of the events of October and November 1962 to form the historical record. In the course of an all-night conversation in his office at the Palace of the Revolution in Havana, I touched on the missile crisis, and Castro said, "Are you interested in my opinion about that moment?" He then proceeded to narrate for hours the origins, the development, and the aftermath of the crisis in the context of his relationship with Khrushchev. For reasons of clear chronology, I have rearranged here the order in which Castro told me the story—it was a conversation with many questions and answers and backing and filling—but this account is based on a transcript of our taped Spanish-language talk.

I asked Castro where and how the idea of deploying the Soviet missiles in Cuba had emerged, and he replied:

"Look, I shall tell you with much precision how the idea arose. After Girón, the United States government indubitably was very irritated, very dissatisfied with what had happened, and the idea of solving through force, of liquidating through force, the Cuban Revolution was not abandoned. But it was not considered possible to go back and repeat the Girón experience, and the idea of a direct invasion of Cuba was being seriously considered and analyzed. And we, through various sources, had news of the plans being elaborated, and we had the certainty of this danger."

Castro said that at the meeting between Kennedy and Khrushchev in Vienna in June—two months after the Bay of Pigs—the Cuban question was discussed, and "Kennedy spoke with much irritation there." Castro went on to say that "from the terms in which Kennedy expressed himself, it could be deduced that he considered he had the right to use the armed forces of the United States to destroy the Cuban Revolution. He referred to different historical events, [and] on that occasion made a reference to Hungary. Having received information about that conversation, we reached the conclusion, as did the Soviets, that the United States persisted in the idea of an invasion."

Evidently the Soviets were Castro's source for accounts of the Khrushchev-Kennedy sessions, and the suggestion has been made by many historians that they had set out to convince the Cubans that Kennedy was planning an invasion, although they knew it was not true, in order to provoke Castro into demanding far-reaching military protection. In another conversation, Castro has said that Kennedy had pointedly reminded Khrushchev that the United States remained neutral when the Soviets invaded Hungary—and that this should be considered a hint that, in terms of reciprocity, the Russians should not interfere if there were an American attack on Cuba. In his own memoirs, Khrushchev does not even mention Cuba in the chapter on the Vienna meeting with Kennedy, but this does not necessarily disprove the Castro version. What matters is that Fidel chose to accept Soviet reports on Vienna as confirmation of his own suspicions about a U.S. invasion, and he behaved accordingly. Finally, the decision on the missiles may have been triggered by Khrushchev's impression in Vienna that Kennedy was indecisive—he allowed himself, for example, to be defeated at the Bay of Pigs—and therefore would live with the accomplished fact of missiles in Cuba once they were emplaced.

"We were then in discussions with the Soviets," continued Castro, narrating the 1962 events. "At that moment, the Soviets were already quite committed to us. They were giving us maximum help, they had responded with the purchase of Cuban sugar when [our] market in the United States was totally closed, they supplied petroleum when all the sources of petroleum supply were suspended, which would have annihilated the country. They, I say, had acquired a great degree of commitment toward us, and now we were discussing what measures should be taken. They asked our opinion, and we told them exactly—we did not speak of missiles— that it was necessary to make it clear to the United States that an invasion of Cuba would imply a war with the Soviet Union. We told them: It is necessary to take steps that would imply, in a clear manner, that an aggression against Cuba is an aggression against the Soviet Union. These were the statements we made . . . as a general concept."

"It was then," Fidel Castro told me, "that they proposed the missiles. As a result of all these discussions, our position was that steps were needed to demonstrate that an aggression against Cuba was equivalent to an aggression against the Soviet Union—it could be a military pact, it could just be that. And, then, among the measures that were analyzed, the installation of medium-range missiles was analyzed. We were thinking fundamentally about the political inconveniences this had. At that time we were not thinking so much of actual dangers because, you know, we had come down from the mountains, we had come from a war, we were very irritated with all the things that had happened, all the aggressions of which we were the victims, and so we analyzed the political inconveniences we faced."

Speaking of the missiles' deployment, Castro said, "We analyzed [the fact] that this, besides being convenient for us, could also be convenient to the Soviets from a military viewpoint. This is to say that we analyzed what advantages there were for us and what advantages for them . . . from the strategic viewpoint—we understood this. We reached the conclusion that this [the missiles] was mutually beneficial." Castro then explained at length that it would have been "morally incorrect" for Cuba to "expect for a country to support us, even to the point of going to war, but—for reasons of prestige or to avoid military-type commitments, or for strictly political reasons—for us to fail to do what could also be convenient for the other side." Therefore, he said, "it seemed to us to be really equitable, it seemed to us just, it seemed to us to be basically reciprocal to accept these measures that implied safety even though they also implied a cost from the political viewpoint—the political fact that these missiles would be installed here."

"Then, after analyzing it with truly serious, just, and honorable criteria, we took the decision of communicating to the Soviets that we were in accordance with the installation of the missiles here," Castro said. "That is, it was not pressure by them—it does not emerge as the consequence of their coming over to us one day and saying to us, 'We want to install the missiles because it is convenient for this and for that.' Really, the initiative of soliciting measures that would give Cuba an absolute guarantee against a conventional war and against an invasion by the United States was ours. The idea of the missiles, in a concrete way was Soviet."

Castro's words convey the impression that he was easily swayed by the Soviets to accept the missiles as a quid pro quo for the Soviet military protection once Khrushchev made the case that, as he put it in his memoirs, there was "no other way of helping [the Cubans] meet the American threat except to install the missiles." But Khrushchev himself indicates that he had to overcome considerable Castro resistance to the missiles. He wrote, "When Castro and I talked about the problem, we argued and argued. Our argument was very heated. But, in the end, Fidel agreed with me. Later on, he began to supply me with certain data that had come to his attention. 'Apparently what you told me was right,' he said. That in itself justified what we then did." Khrushchev's reference to their "heated . . .

argument" also raises the question of *where* they argued, unless he meant it figuratively as exchanges through Raúl Castro, for example, or some other channel. The record does not show any Khrushchev-Castro meetings after New York in 1960. It is possible, of course, that Castro secretly visited Khrushchev in the Soviet Union during June (when he was supposedly raising rebellion flags in the Sierra Maestra and was out of sight for several days), but this has never been mentioned even as speculation.

Most historians tend to agree that, indeed, Khrushchev was the author of the missiles idea although its evolution into a military agreement between the two governments is sketchy. Castro has been generally consistent from the beginning in claiming, as he did in his immensely detailed interview with me, that the proposal had come from the Russians; the only known exception is his statement to Herbert L. Matthews in October 1963 that the idea came from him, "not the Russians." In November 1964, however, he denied it to C. L. Sulzberger of *The New York Times,* stressing that "Cuba took the responsibility for the presence of the missiles here . . . both Russia and Cuba participated." Inasmuch as an agreement did exist, the point of joint responsibility is self-evident.

More interesting is the fact that the official Soviet line at the time of the crisis, both for Kennedy's benefit and for Soviet public opinion, portrayed Castro as the party asking for the missiles. In his letter to Kennedy on October 26, at the acme of the crisis, Khrushchev insisted, concerning Soviet missiles in Cuba, that "all the means located there, and I assure you of this, have a defensive character, are on Cuba solely for the purposes of defense, and we have sent them to Cuba at the request of the Cuban government." On November 6, Deputy Premier Aleksei N. Kosygin told a gathering of Soviet Communist party leaders that the Cubans had requested the missiles to protect their national security. This speech was printed in *Pravda,* but strikingly not in the Cuban press—most likely because it also justified the withdrawal of the missiles under American pressure and without consulting Castro. Khrushchev repeated it before the Supreme Soviet on December 12. But, then, the record shows that Khrushchev had lied to both Kennedy and Castro on numerous occasions during the crisis period.

Absent from official Soviet and Cuban pronouncements on the origins of the October crisis is any hint of how Khrushchev and Castro imagined Kennedy would react to the missile deployment. The assumption is that Khrushchev had hoped to get away with completing the installation before being discovered, and then make Kennedy blink with a Soviet ultimatum. Nowhere are there any indications of any consideration of the fact that no matter when the missiles' presence surfaced, Cuba would be much more vulnerable to a United States attack than before the introduction of these weapons. Therefore I asked Fidel Castro how he had thought Kennedy would react to the deployment.

"I was convinced," Castro replied, "that a very tense situation would be created, and that there would be a crisis." Then he offered me a rationale for the missile gamble that was pure Castro in terms of his calculated-risk instinct blended with his sense of principle—and which in the light of his own history is entirely believable. Contrary to common belief, Castro's gamble succeeded better than Khrushchev's: He won a noninvasion guarantee from the United States at no cost to himself, whereas the Soviets were simply humiliated by Kennedy. But it took Castro a few months to realize it in his fury at Khrushchev for making a missile-withdrawal deal with Kennedy behind his back. Being Fidel, he had wanted from the crisis more than just survival.

In the meantime, he said to me, "put yourself in our place—between a situation of impotence facing a very powerful country that could at any moment decide to invade Cuba, costing the lives of millions of Cubans who would have resisted, and a situation of running a risk from a more secure position, a risk of a world nature, but not a risk of conventional war." Castro's calculation was that, in effect, the threat of nuclear conflict would save him from a nonnuclear attack by the United States.

"We preferred the risks, whatever they were, of great tension, a great crisis," he said, "to the risks of the impotence of having to wait, impotently, for a United States invasion of Cuba. . . . At least they gave us a nuclear umbrella, and we felt much more satisfied with the response we were giving to the policy of hostility and aggression toward our country. From the moral point of view, I never had and I shall never have doubts that our attitude was correct. From a strictly moral as well as strictly legal viewpoint, as a sovereign country we had the right to make use of the type of arms we considered gave us a guarantee. And in the same way that the United States had missiles in Italy and Turkey, in the same way as the United States has bases in all parts of the world around the Soviet Union, we, as a sovereign nation, considered we had the absolutely legal right to make use of such measures in our own country."

Thus the only relevant issue was "political," Castro said, and the United States "acted according to political positions and formulas and according to force" to prevent the Soviet deployment. However, he volunteered the comment that at the time "the correlation of forces in the nuclear realm favored the United States," adding that it was a fact "I ignored" in 1962. Castro was thus saying that he had engaged in the missile confrontation without being adequately informed by the Soviets of the superpowers' relative nuclear strength. Too elegant diplomatically to spell it out, Castro was nevertheless giving me a very clear impression that Khrushchev had led him astray with his missile proposals—leaving one to wonder what Fidel would have done had he known the truth. This was unprecedented insight into the secrets of Soviet-Cuban relations.

"At that time," Castro said, "I ignored how many nuclear weapons the Soviets had and how many nuclear weapons the North Americans had. I ignored it, and it did not occur to me to ask the Soviets about it; it did not seem to me I had the right to ask, 'Listen, how many missiles do you have, how many do the North Americans have, what is the correlation of forces?' We really trusted that they, for their part, were acting with the knowledge of the entire situation. We did not have all the information to be able to make a complete evaluation of the situation, we only had part of the information available." As it turned out, Castro never really forgave the Soviets for keeping him in the dark, or misleading him, about the world nuclear balance of forces when he put Cuba's life on the line on the basis of Khrushchev's assurances. So confident were the Cubans of the Soviets' good judgment that President Dorticós told the United Nations General Assembly in New York on October 8, when the missiles were already being clandestinely deployed, "We warn that if an error is committed, aggression against Cuba can become transformed, to our great regret and against our desires, into the start of a new world war."

Among other continuing controversies about the 1962 nuclear crisis is the question of how close the world really *did* come to war, atomic or, at least, conventional. The accepted wisdom is that war might have begun if Soviet ships bound for Cuba had not halted on October 24 without crossing the quarantine line drawn by Kennedy. Otherwise they would have been forcibly stopped by U.S. warships and planes—an act of war presumably followed by a Soviet response, and then unpredictable rounds of escalation. The danger still persisted until the morning of October 28, a Sunday, when the crisis was settled through the final Kennedy-Khrushchev exchange of letters. A nuclear duel was certainly possible during that week inasmuch as Soviet medium-range missiles in Cuba had become operational on October 23, and it had to be assumed that they were armed with nuclear warheads.

However, as Fidel Castro recounted the history of the crisis, Saturday, October 27, was a most critical day, not only because Soviet surface-to-air (SAM) batteries had shot down a high-flying Air Force U-2 reconaissance plane that morning, but because the Cubans were trying to shoot down low-flying American aircraft. Castro insisted that contrary to published allegations, the Russians and not the Cubans had brought down the U-2 because they had the exclusive control of the SAMs, but he was just as vehement in saying that he was determined to destroy any American plane

584 / F I D E L

his antiaircraft artillery could reach—regardless of consequences. For him, it was again a matter of principle and sovereignty, and he had ordered that American aircraft that appeared over Cuba a few days after the crisis was settled be fired upon—even if it reopened the confrontation. Castro was aware when he spoke to me that Kennedy had resolved to bomb Cuba if a second American aircraft was shot down after the loss of the U-2.

"The [SAM] rockets were in the hands of the Soviets, and the antiaircraft batteries—all the conventional ones—were in our hands," Castro recounted. "We had hundreds of batteries. In those days, this type of rocket could not fire below one thousand meters. It was effective above one thousand meters, but in those days of the crisis the North Americans began flights at a very low level, in addition to the U-2s. They began flights at two hundred or three hundred meters. I realized that the SAMs as well as intermediate-range [ground-to-ground] missiles were threatened with destruction simply by low-level attacks when the SAMs were absolutely impotent."

Castro continued: "Then I ordered the deployment of all the antiaircraft batteries we had, some three hundred batteries. I submitted to the Soviets that we could not permit the low-level flights and we were going to use the batteries. We installed all these batteries around all the SAM bases and around all the missiles, and that day we issued the order to fire. It was we who gave the orders to fire against the low-level flights. This is rigorous historical reality."

On the morning of October 27, Castro said, "a couple of planes, or several couples of planes, appeared in low-level flight over different places, and our batteries began to fire." Official U.S. records confirm that on that morning two low-flying reconnaissance aircraft were fired upon, but not hit, around 10 A.M., when the U-2 was shot down by a SA-II rocket. Castro said that "the inexperience of our artillerymen, who had recently learned to operate these pieces, probably made them miss as they fired on the low-flying aircraft." When the U-2 was crossing Cuba and flew over Oriente Province, Castro said, "a Soviet surface-to-air rocket battery fired on the plane and hit it."

Fidel told me, "It is still a mystery how it happened; we had no jurisdiction, no control over Soviet antiaircraft batteries." He said: "We had simply presented our viewpoint to [the Soviets], our opposition to low-level flights, and we ordered our batteries to fire on them. We could not fire against the U-2. But a Russian there—and for me it is still mystery, I don't know whether the Soviet battery chief caught the spirit of our artillerymen and fired, too, or whether he received an order—did fire the rockets. This is a question that we do not know ourselves, and we didn't want to ask much about this problem."

I commented that the downing of the U-2 could have triggered a world war. Castro replied, "I don't know what would have happened if the U-2s had flown over again, but I am absolutely certain that if the low-level flights had been resumed, we would have shot down one, two, or three of these planes . . . with so many batteries firing, we would have shot down some planes. I don't know whether this would have started a nuclear war." He said the planes did not return the following day because the Soviet-American agreement was reached, but they did come back several days later in low-level flights.

Castro said that Soviet Deputy Premier Mikoyan was already in Havana "to explain all this to us, and we warned him that we did not accept low-level flights under any circumstances. We told the Soviets that although the accord was reached, we would fire against low-flying aircraft, and we gave orders to our batteries. But that day, a contact between the Soviets and the Americans may have occurred, and they suggested to them not to fly. That day I was at the San Antonio air force base, where we had some batteries, and that was where every day, at ten A.M., these planes were flying over. I went there, and I waited for the planes at ten A.M. I knew that there would be a counterstrike, and that possibly we would have many casualties, but I thought it was my duty to be there, in a place that surely would be attacked, but the planes did not come that day." Fidel sounded almost wistful. But he lived up to his concept of military honor: Four days after the U-2 was shot down by the Soviet battery, Castro ordered that the body of the pilot, Major Rudolph Anderson, be returned to the United States for dignified burial at home. Ironically, Major Anderson was one of the two U-2 pilots who, on October 14, had brought back the first photographs of the Soviet missile deployment in Cuba; the crisis was set in motion by these photographs.

Fidel Castro never concealed his "irritation," as he put it, with the Soviet Union for having struck a deal with the United States to repatriate the missiles without consulting him. The rancor was still there when he was telling me the crisis story twenty-two years later, remarking that "it had really never crossed my mind that the option of withdrawing the missiles was conceivable." Although Castro said that in the end he understood why Moscow had removed the missiles—because of the Soviet nuclear inferiority of which he had not been sufficiently aware—and that the Russians had been right, "we were irritated for a long time." He volunteered the remark, never before publicly uttered by him, that "this incident, in a certain way, damaged the existing relations between Cubans and Soviets for a number of years—many years elapsed." Despite displays of surface friendship and mutual high-level visits, these relations remained tense, difficult, and fragile until 1969—six long years—and they have never really returned to the situation of trust Castro had in the Soviet Union prior to the October crisis. Soviet failure to react meaningfully to the United States invasion of Grenada in October 1983 (exactly twenty-two years after the missile crisis), where American and Cuban troops found themselves in combat for the first time in history, produced a new explosion of Castro's rage against the Russians.

Recalling his shock over the crisis settlement, Castro said, "I never considered the withdrawal solution. Perhaps in the revolutionary fervor, passion, fever of those days, we did not consider as conceivable the removal of the missiles once they were established here." Yes, he added, "there certainly were communications between us and the Soviets over different situations, but in the last two days, after the plane was shot down, events moved so rapidly that it was not possible for a prior exchange to be held between us and the Soviet government about the proposition of withdrawing the missiles. And we were really very irritated over the fact that an agreement was reached without our participation, or without a consultation with us. . . . We were informed when the accord had already virtually been concluded." Castro told me that had he been consulted, "I think we would have understood the necessity of finding a solution, but we would have demanded at least three things: the cessation of aggressions against Cuba, the cessation of the [economic] blockade by the United States, and the return of the Guantánamo Naval Base [to Cuba]. These would have been conditions that were perfectly understandable and acceptable."

What Castro did not know at the time—as he told me—was that a secret part of the Kennedy-Khrushchev deal in 1962 provided for the removal of U.S. Jupiter nuclear missiles from Turkey in exchange for the Soviet repatriation of the missiles from Cuba. All Castro knew was that Khrushchev was taking away his nuclear weapons to avert a global war and because he had received guarantees from President Kennedy that the United States would not invade Cuba. Inasmuch as the entire missile adventure was undertaken by Khrushchev and Castro supposedly to protect Cuba from an American invasion, they had achieved their objective. Fidel nevertheless took the view that as long as his fate was being negotiated, he was entitled to political bonuses and as he told me, "an honorable accord could have resolved once and for all the questions that continued during a long time to poison relations between the United States and Cuba." Castro's point was well taken, but the Soviets at that stage became more interested in getting the Jupiters out of Turkey and were not prepared to press Kennedy for more concessions.

Castro said that all the terms of the October agreement, including the secret clauses on the Turkish missiles and a formal noninvasion guarantee for Cuba, are contained in writing in documents exchanged between Kennedy and Khrushchev. But, he remarked, "nothing was said about the Turkish missiles in the weeks following the crisis. It was kept behind a mantle of silence, and we did not know that the withdrawal of missiles from Turkey was part of the accord."

I asked how long he was kept in the dark, and Castro replied that it was until he had gone to visit the Soviet Union in late April 1963. He said: "One day, Nikita is reading to me all the documents that had been exchanged between the United States and the Soviet Union. Then he says, 'In the American document they say they have made such and such commitments, and besides, we have made the commitment of withdrawing the missiles from Turkey.' And this was when I learned that the missiles in Turkey were in the agreement, and that they had actually been removed. But this was never discussed because the Americans had asked the Soviets not to make public this part of the compromise." Castro was right. The Jupiter missiles, which were obsolete, were removed from Turkey during late April 1963 without any public announcement. Over the years, the pullout became generally known, but the U.S. government did not declassify the pertinent documents until 1985—nearly two years after Castro told me the October story. The full exchange of documents with the Kremlin on the settlement of the crisis is expected to remain classified indefinitely.

In the meantime, there is nothing on the public record to confirm the existence of an *explicit* commitment by Kennedy not to invade Cuba, although no subsequent U.S. administration has negated it in so many words. Castro insists that the commitment is "explicit, not implicit," and that the proof of its validity is that there never has been an invasion attempt against Cuba. Under the Kennedy-Khrushchev agreement, the withdrawal of the missiles from Cuba was to be subject to U.S. inspection, but Castro furiously rejected this on the grounds it violated Cuban sovereignty. U Thant, the U.N. secretary general, was unable to budge Castro during a special trip to Havana, and, in the end, the inspection was carried out by U.S. aircraft from Guantánamo flying low over Soviet ships displaying the missiles on their decks. Castro's refusal to accept U.S. inspection led the Kennedy administration to become publicly vague on the noninvasion pledge; as Arthur M. Schlesinger, Jr., the historian of the Kennedy era, has written, "The guarantee never went into formal effect." After Khrushchev brought back home IL-28 bombers from Cuba in November (another source of anger for Castro), Kennedy stated that "if all offensive weapons are removed from Cuba and kept out of the hemisphere in the future, under adequate verification and safeguards, and if Cuba is not used for the export of aggressive Communist purposes, there will be peace in the Caribbean."

While the missiles were being shipped out of Cuba, Castro contrived to balance his private ire at Khrushchev with public assurances that all was well between the two countries. His economy was in appalling shape (there had even been antiregime demonstrations over food shortages in Matanzas earlier in 1962), he needed to raise very substantially the sugar production, and he had to have fresh resources for the first four-year economic plan that had just been launched with a commitment to invest $1 billion in industrial and farm development. Khrushchev's deal with Kennedy brought Castro unsolicited support from China, Cuba being portrayed as a victim of Soviet betrayal, but it was on Moscow that he depended for basic assistance. Therefore, Castro went before television cameras on November 1 to proclaim, "We are Marxist-Leninists. . . . There will be no breach between the Soviet Union and Cuba." But Anastas Mikoyan, who arrived in Havana on November 2 to pacify Castro, received the brunt of bitter recrimination from the Maximum Leader. Fidel greeted him at the airport but refused to see him for weeks, until the elderly deputy premier had to plead for a meeting so that he could return to Moscow for his wife's funeral. He stayed in Havana for over three weeks. And so unhappy was the whole mood that Ambassador Alexeiev, who considered himself Castro's friend, was said to have wept during conversations with him.

Although the relationship with the Kremlin had to be the centerpiece of Castro's foreign policy, he never lost sight of the possibility of some form of accommodation with the United States—at some point and if no exaggerated conditions were placed at the start of an eventual negotiation. Castro has said that the settlement of the missile crisis could have included adjustments in Cuban-American relations, and indeed it may have been a lost opportunity. In our 1984 conversations, he remarked that "the removal of the missiles should have permitted Kennedy to make these small concessions to Cuba that would have eliminated the great obstacles that have remained in the way of relations between the United States and Cuba."

Speaking of the message from Kennedy brought to him by Jean Daniel on the day of the President's assassination, Castro told me, "Kennedy would not have received a rebuff from us. . . . I was meditating and thinking very much and I was considering giving him a constructive and positive answer. . . . At that moment, we could have perhaps begun a dialogue, an exchange of impressions. . . . After the 1962 crisis, Kennedy had the authority, he had the will. . . ." And Fidel's judgment was quite perceptive. The crisis led to the dismantling of Operation Mongoose (though the CIA went on for years attempting to assassinate Castro), and Kennedy maintained until his death a latent interest in a possible opening to Cuba. Even before Jean Daniel went to Havana, the President was considering sending a secret emissary of his own to see Castro, and the National Security Council staff was exploring possible "channels of communication with Castro." As Arthur Schlesinger has observed, in 1963, "the White House [was] drifting toward accommodation." But John Kennedy's death in November froze prospects for Cuban-American diplomacy for nearly fifteen years. Through an incredible irony, the CIA officer in charge of planning the latest attempt to kill Castro had met with the prospective murderer in Paris on the day Kennedy was assassinated in Dallas. The President had never been apprised of the plot against Fidel.

If the Soviet Union was vital for Cuba's economic survival, the Russians needed the Cubans politically—and even militarily, despite the October nuclear fiasco. In the light of the deepening ideological conflict with China and its impact on Soviet standing in the Third World, the Kremlin could not afford a break with revolutionary Cuba. The removal of the missiles had already caused the Russians considerable embarrassment internationally, and a reaffirmation of friendship with Cuba was now a top priority for Khrushchev. The marriage of convenience between Castro and Khrushchev had reached such extraordinary proportions that in the name of "internationalist and socialist solidarity" they staged an unprecedented spectacular around the Cuban leader's spring 1963 visit to the Soviet Union, a visit that lasted forty days, unquestionably a record in the annals of international travel by chiefs of governments.

Left behind in charge of Cuba were Raúl Castro and the Revolutionary Armed Forces, now better trained and equipped than ever. A military consolation prize for the Cubans after the departure of the nuclear missiles was SAM batteries (enabling them to shoot down marauding U-2 spy planes, if they wished) to which the United States did not object because of their defensive character. What the United States did not know at first (and forgot afterward) was that a Soviet Army combat brigade remained on the island following the repatriation of some twenty thousand troops who had accompanied the missiles. The brigade was meant as a symbol of continuing Soviet commitment to an active defense of Cuba, and Castro described it to a visiting American journalist in 1964 as "a solid Russian combat force."

Actually, this force oscillated between four thousand and five thousand men (in addition to military advisers, also in the thousands), and nobody seemed to pay much attention to it until the brigade was "discovered" by the Carter administration in 1979. This caused such disarray that the President grew nervous about submitting the new Strategic Arms Limitation Treaty (SALT II) with the Soviets to the Senate for ratification. (It has never been submitted.) Cuba seemed to affect American policies in every imaginable way, and when I asked Castro about the brigade during one of our conversations in Havana in January 1984, he told me with a chuckle, "This is the same brigade the Russians left behind in 1962." It must have been the second or third generation of Soviet soldiers stationed in Cuba.

To set the tone for Castro's first trip to the Soviet Union, Khrushchev delivered a speech in Moscow at the end of February declaring that an "imperialist" attack on Cuba or any other socialist country would mean the start of the Third World War. On April 26, Castro and a vast entourage boarded the giant Soviet TU-114 turboprop airliner for the nonstop flight from Havana to Murmansk, the northern Russian port. This was a regular Aeroflot route inaugurated in January, but this time Fidel almost did not make it. When the TU-114 reached the Murmansk area the following day, the fog was so dense that the airport had to be closed. There was not enough fuel left to look for alternate fields, and the pilot was forced to land in zero visibility with inadequate instrument-landing-system facilities at the Murmansk airport. But he did it perfectly, and, once more, Fidel Castro embarked on a new adventure in a dramatic fashion. Mikoyan, for whom Cuba and Castro had become practically a full-time occupation, was on hand to welcome the visitor—and Fidel, now wearing a fur hat, went directly from the plane to a mass rally downtown as the most natural thing in the world. The speech in Spanish was short—he said, "We've encountered temperatures to which we are not used: lots of cold outside, but plenty of warmth inside the heart!"—and the people of Murmansk naturally loved it.

During his forty days in the Soviet Union, Castro visited fourteen cities, from Central Asia to Siberia, from the Ukraine to Georgia, and from Moscow to Leningrad; inspected the Northern Fleet and a strategic forces rocket base; delivered countless speeches at sports stadiums, factories, battlefields, and town squares; reviewed the May Day Parade from the top of the Kremlin Wall; received the title of Hero of the Soviet Union, the Order of Lenin, and the Gold Star; attended the Bolshoi Ballet in Moscow and an open-air concert; and spent scores of hours publicly and privately with Nikita Khrushchev. The two of them spoke at a special mass rally in Red Square, an unusual honor for a foreigner, but Castro also had the opportunity of quietly meeting other top Soviet leaders. At a *dacha* near Moscow, he joked and chatted with Leonid Brezhnev, the man who would oust and replace Khrushchev the following year—in part because of the Cuban missile humiliation. A photograph at the *dacha* shows Khrushchev and Brezhnev in stodgy suits and ties, with rows of medals on their lapels, and Castro in his usual olive-green battle fatigues.

The tour was a fantastic success: No foreigner had been received in such grandiose fashion since the Great Patriotic War, the crowds and television audiences were fascinated by the romantic guerrilla fighter from the faraway

Cuban mountains, and they all adored the spectacle. Khrushchev accompanied Castro to Murmansk for the flight home at the end of May. Fidel had never in his life enjoyed such attention and on such a scale, and as far as the world was concerned, it was a love feast between the Cubans and the Soviets. But Castro never lowered his political guard either, quietly conveying to his hosts his economic-aid requirements, insisting on his independence, and neatly keeping the revolutionary factions at home off balance. On the one hand, he had made sure that not a single "old" Communist accompanied him on the Soviet tour (not even the ever-mediating Carlos Rafael Rodríguez), as if to enhance the *Fidelismo* of the revolution. On the other hand, he launched a savage public attack on *Revolución,* the organ of the *Fidelista* movement, for the way it had reported his Soviet trip. He criticized it for an article comparing him to Lenin in terms reminiscent of the adulation he found in the Soviet Union, and for reporting "amusing" details that he thought detracted from the significance of the visit. Clearly, it was a warning to the remaining "moderates" of the revolution grouped around the newspaper, and his message was clear when Castro stated in all seriousness that *Pravda* was the best newspaper in the world. The following year, *Revolución* was merged with the Communist newspaper *Hoy* to become the regime's chief mouthpiece under the name of *Granma* and under hardline Communist editors. *Gramna* now rivals *Pravda* in quality, especially in the generous use of red ink on its pages.

After five years of the revolution, Fidel Castro had established himself as the undisputed, powerful, ruthless, imaginative, and unpredictable leader of Cuba as the country now edged toward a Communist system of life and governance. He continued to command great national following—often national adoration—as he emerged supremely victorious from the Bay of Pigs confrontation with the United States and essentially undamaged from the October missile crisis. As at the time of the exiles' invasion, Cubans rallied around their country's flag—and their Maximum Leader—at the moment of danger in 1962, when American warplanes roared over the countryside and antiaircraft batteries fired their volleys.

Above all, Castro knew to keep the nation's faith alive and engender hope in the midst of frightful hardship and sacrifice. When Hurricane Flora, one of the worst hurricanes of the century, flattened and flooded the island during 1963, Fidel Castro was everywhere, on the front lines as usual, directing rescue operations, taking chances, leading and inspiring. Despite massive Soviet aid, the economy was performing appallingly below standard, but what Castro made his people remember was that they were being educated, fed, housed, protected, and cared for by the revolution as Cubans had never been before in history. Therefore, they were in the main loyal to the revolution, grateful to the Commander in Chief, and ready to ferret out counterrevolutionary "worms," patrol the cities and the farms at night, and be indoctrinated to the point of numbness in the mysterious phrases of Marxism-Leninism.

When, at the end of 1963, Castro told Cubans that not only would the economy be geared again to sugar production but that a 10-million-ton harvest was planned for 1970 (the 1963 production was 3.8 million and the all-time record 6.7 million), the people trusted his judgment. It would mean "voluntary" work on weekends for students and city dwellers to help their peasant brethren, but it would be for the good of the revolution, and therefore it was the proper thing. Besides, it was unwise to complain or shirk revolutionary duty: One's fellow revolutionaries had sharp eyes and keen ears.

As the first quinquennium of the revolution drew to a close, Castro and Cuba were essentially set on a firm course. Fidel, now thirty-seven years old, knew precisely where he was leading the nation even though it was a hard, uphill march—as in the Sierra Maestra. Having survived the Bay of Pigs and the missile crisis, the revolutionary regime was as secure from external threat as any government in the world—or more so. Survival in the long run was, in effect, guaranteed by the Soviet Union because history and Fidel Castro left the Russians no political alternative in the rivalry-ridden world.

His own life was still that of a *guerrillero,* restless, impatient, eager to conquer, unorthodox, sharp-shooting. Celia Sánchez was succeeding in making Fidel's activities somewhat more orderly, but not too much so because he despised schedules and hated administration. He was still hugely enjoying himself. Fidelito, at sixteen, was preparing for the university and then for studies in the Soviet Union. In August 1963, Fidel's mother, Lina Ruz de Castro, died at their family home in Birán, and the clan gathered for the funeral. Afterward Castro addressed a rally on Birán's main street.

In the next two decades, Fidel Castro would maintain the rhythm of revolutionary life he had imposed on Cuba, but with few great surprises ahead in terms of his own behavior and attitudes—or in organizing the new society. He would strive to consolidate Cuba ideologically under the banner of a ruling Communist party that he would head. He would continue with reforms where anything was left to be reformed (the third agrarian reform in October 1963 limited individual holdings to 168 acres, with the state owning 70 percent of the land), he would practice "internationalism" throughout the Third World by helping revolutionaries everywhere and dispatching Cuban combat troops across the seas, and he would seek for himself the role of a world statesman. And he would lose Che Guevara.

V

THE MATURITY

1964–1986

CHAPTER

1

The death of Ernesto Che Guevara in the Bolivian jungle on October 8, 1967, climaxing the destruction of his guerrilla movement there, was the central drama in the history of the Cuban revolution in the decade of the 1960s. Apart from the myriad mysteries, never resolved, surrounding Che's presence and death in Bolivia, his ultimate disappearance had a profound impact on the evolution of Fidel Castro's domestic and international policies. It even helped Castro to settle his disputes with the Soviet Union, festering since the 1962 missile crisis, inasmuch as in the name of romantic revolutionary purity, Che had become a severe critic of the Russians for their "internationalist" timidity and their "imperialist" way of managing economic aid to the Third World. Finally, Che Guevara was the last totally independent mind and spirit in the increasingly rigid power structure built by and around Castro. He had vanished from Cuba in the early part of 1965, for reasons never adequately explained, and therefore did not participate in the final stages of formalizing the establishment of Communist rule in Cuba through the creation of the new party under Castro.

There must be powerful reasons for Castro's refusal, even nearly twenty years after the fact, to throw any light on Che's presumably spontaneous decision to sever his ties with Cuba and therefore with Fidel himself. The only explanation—such as it is—is contained in Che Guevara's letter to him where he says, "other hills of the world demand the aid of my modest efforts" and in which he renounced all his positions in Cuba as well as his Cuban citizenship. This Castro read before a stunned assembly of "new" Communist leaders in Havana on October 3, 1965. The occasion was the presentation of the membership of the Central Committee of the freshly organized Cuban Communist party—the creation of the party had been announced only the day before—and there was an awkward silence after Fidel said that the only deserving name missing from the roster was Che's and proceeded to read the letter. In Cuban newsreels Castro seemed immensely ill at ease and clearly unhappy as he read; this was the last thing Fidel said publicly about Guevara's whereabouts until October 15, 1967— two years later—when he went on television to announce that the news of Che's death was "unfortunately true."

That Guevara had left Cuba was confirmed by Castro as early as April 20, 1965, when he tersely told foreign newsmen that "*Comandante* Guevara is where he is most useful to the Revolution." Guevara had returned on March 15, from his latest tour of Africa and Asia, and was met at the airport by Castro and President Dorticós, but he was not seen again, and journalists began inquiring about him. It is not known exactly when Che left Cuba or when he wrote his farewell letter to Fidel. However, there is no reason to doubt the authenticity of this missive with its assertion that "I feel that I have fulfilled the part of my duty that tied me to the Cuban Revolution in its territory, and I bid farewell to you and the *compañeros,* your people who are already mine"—and that "my only shortcoming of some gravity was not to have trusted in you more from the first moments in the Sierra Maestra and not to have understood with sufficient celerity your qualities as a leader and a revolutionary." Not only is the style pure Guevara, but it was also in character for him to mention, in the revolutionary spirit of self-criticism, his Sierra "shortcoming." This was his doubt, which he had expressed in a wartime letter, about Castro's honesty in dealings with exiled anti-Batista politicians late in 1957.

Simultaneously, Che had written to his parents, who then lived in Buenos Aires, that "once again I feel beneath my heels the ribs of Rosinante . . . I return to the road with my lance under my arm." He reminded them that over ten years earlier he had written "another letter of farewell." On leaving Argentina he wrote: "Here goes a soldier of the Americas"—and now "my Marxism has taken deep root and become purified. . . . I believe in armed struggle as the only solution for those people who fight to liberate themselves. . . . Once in a while, remember this small *condottiere* of the twentieth century." In 1985, Ernesto Guevara Lynch, Che's father, told me in Havana (where he moved after his wife died and he remarried) that he had taken the letter at face value, understanding his son's restlessness. But he said that was all he knew, not having seen Che since 1961. Most probably, Fidel Castro is the only person who knows the full truth.

Castro knows, for example, why Che chose Bolivia for his new guerrilla enterprise, improbable as the choice was, given the fact that he would fight in terrible mountain and jungle territory where he did not speak the local Indian languages (Andean peasants seldom know Spanish), instead of Salta province in his native Argentina where he had originally hoped to start a revolution. The Bolivian mission was much more suicidal than the *Granma* landing a decade earlier, but Fidel evidently went along with this idea, at least to the extent of assigning Rebel Army fighters to Che's guerrilla detachment, equipping and financing the expedition, and maintaining regular radio contact with Guevara until the end. Perhaps Castro should have guessed that Che would indeed be betrayed by the Moscow-oriented Bolivian Communist party, which virtually delivered Guevara to the Bolivian Army rangers and their CIA advisers.

In retrospect, it is understandable that there was really not much left for Che to do in Cuba after the consolidation of the revolution. He was minister of industry, but on fundamental issues of Marxist economic development, he was very much at odds with Castro. In an oversimplified fashion, their differences centered on Che's idealistic belief in moral incentives for the population to produce and Castro's more practical conclusion that material incentives, such as higher wages and bonuses, were more likely to stimulate work. For a time in the early 1960s, Che's view prevailed, but then Fidel in effect overruled him (though in 1986, Castro reinvented moral incentives). None of this, of course, has ever been publicly debated or reported in Cuba, yet Castro provided the best clue to the ideological convulsions of those days in the reply he gave me in 1984 when I asked him what errors the revolution had committed. He was surprisingly candid, during that Havana dawn in his office, about the trajectory of the revolution, and the awesome problems of creating a new society. He was pensive, stroking his beard slowly as he spoke.

"At the outset of the Revolution," he said, "when we had to assume all the functions of the state and all the functions of the economy . . . we began this task without experts, just ignorant people who did not know what had to be done. . . . Our economic development was sustained development, it had highs and lows, reaching an average of four point seven percent in twenty-five years. It was slow in the first years, when our objective was fundamentally to survive more than to develop, but it was more accelerated in the subsequent years. We passed through different stages. We suffered the consequences of different errors. Let us say that one error we committed was to want to jump stages, wanting to arrive at Communist forms of [wealth] distribution, jumping over the socialist form of distribution—and it is impossible to jump stages. Our own history demonstrates that we wanted to go too far and establish Communist forms of distribution, when the correct course really is to follow socialist distribution forms, in which distribution is made according to the work of the people. . . . The Communist formula is: Each must give according to his capacity and receive according to his needs. The socialist one is: Each must give according to his capacity and receive according to his work. . . . We were marching too rapidly toward Communist formulas when it was really premature. It was a jump, and it created problems. But we rectified it in time."

Castro then went on to explain how, in the Eighties, a blend of communism and socialism existed in Cuba, and "many things are distributed in a Communist form." This was his definition of Cuban Marxism-Leninism: "I think that salaries are paid according to the work and the capacity. But education, for example, universal and generalized, is free, and it is a service of the Communist type. The same possibilities of education and the same education are received by the son of an engineer and the son of a worker, the son of a father who works hard, and the son of a father who does not work much. . . . Medical services are received in an egalitarian form by the whole population, that is, a Communist-type service. . . . We start from the concept that according to the ideas of Marx, Engels, and Lenin, communism is the final objective, but there is the stage of socialism. This formula is not applied in a chemically pure fashion. . . . In salaries, what is paid a worker is *not* according to the needs he has. There are workers who have one or two children, and they earn twice what is earned by a worker who has three, four or five children because their contribution through work and their capacity may in a given case be greater. So, this is not Communist distribution. The Communist system would be to pay more to the worker who has seven children than the worker who has one child. . . . In the socialist formula, the worker who has more capacity, even though he has less necessity, is paid more. The one who has less capacity, although he may have more need, is paid less. In a harbor a worker who can load, say, twenty tons in a ship receives more than the worker who can load ten tons. But there was a moment when we paid the same to the worker who carried twenty tons and the one who carried ten or five tons. This was an error. It doesn't really stimulate work."

Castro never suggested that Che Guevara might have inspired such errors, but they were "rectified," as he put it, in the mid-1960s. This coincided with Che's gradual withdrawal from economic policy-making, and his growing concentration on contacts with the Third World, evidently in concurrence with Castro. Guevara seemed to enjoy this mission, seeing it through the prism of revolution making, and the contributions the Cubans could offer in this realm. This was the impression I gained spending hours with him—for the last time—in December 1964 in New York when he came to address the United Nations General Assembly. After we appeared together on a television interview program, we chatted for several hours about Cuba and the Third World; the next day Che left for Africa and Communist China in what would be his farewell world tour on behalf of the Cuban revolution.

This was a time of continuing—as well as additional—tensions between Fidel Castro and the Soviet Union, and the "old" Communists at home. Despite Castro's regal reception in the Soviet Union in May 1963, he went on resenting the Russian "betrayal" of the October missile crisis, and pressing them for more and more economic assistance. In fact, he had developed the rationale that inasmuch as Cuba was dealt out of the crisis settlement by Khrushchev and Kennedy, thereby winning no concessions, the Soviets owed his country even more aid than before. It was Fidel's ever-successful way of turning defeat to advantage.

Nevertheless, a public display of fraternal love was maintained by both sides. In December 1963, Nikolai Podgorny, a powerful member of the Soviet Politburo, came to Cuba for a two-week visit, the highest-ranking Kremlin figure to appear in Havana up till then. On January 12, 1964, Castro flew with Podgorny to Moscow, where he conferred at length with Khrushchev and Leonid Brezhnev: Ten days later, the Russians signed an agreement with him to purchase the bulk of the Cuban sugar harvest for the next five years with a guaranteed minimum price—above the world market. Then Che Guevara signed an accord on technical assistance, and the two Cuban leaders went home, apparently satisfied with their achievements.

Two months later, however, Fidel Castro orchestrated the trial of Marcos Rodríguez, the pro-Communist student who in 1957 had betrayed four of his companions in the conspiracy against Batista. They had been murdered by the Batista police in the famous "Humboldt Street Crime," and it developed later that Rodríguez was in effect being protected by key members of the "old" Communist party. It was Faure Chomón Mediavilla, formerly the leader of the Students' Revolutionary Directorate to which the four assassinated youths belonged, who produced the evidence against Rodríguez, a fact rendered especially interesting because he was currently Cuba's ambassador to Moscow. In terms of internal Cuban politics, the Rodríguez trial inevitably revived all the *Fidelista* resentments against the Communists and the Russians, and Castro chose to let this happen. He may have decided to use the Rodríguez affair as another warning to the "old" Communists not to try to repeat their previous "sectarianism" as the new party was about to be launched; in any event, the trial, held during March 1964, was a Byzantine event, highlighted by an immensely confusing courtroom speech by Castro. He succeeded in clearing the party of actual guilt, but left enough of its officials besmirched; Rodríguez was sentenced to death twice (the second time on appeal, suggested by Fidel) and executed.

During the summer of 1964, Castro threw out hints—mainly through newspaper interviews—that he was interested in improved relations with the United States, inaugurating a pattern that was to continue for the next twenty years of always leaving the door open to some accommodation with Washington while keeping a reliance on the Soviets for military and economic survival. He always believed in multiple facets in foreign policy, emphasizing or minimizing one or another, depending on changing circumstances. In mid-1964, for example, Fidel was showing new interest in dealing with the United States even though he was simultaneously investing in revolutionary conspiracies around Latin America (though not very successfully), and being the target of harsh pressure by the Johnson administration. American U-2 spy planes were again overflying the island, and the CIA was back in the assassination business with a vengeance.

(Ramiro Valdés told me in an interview in mid-1985, when he still served as interior minister, that most of the CIA assassination attempts on Castro, perhaps as many as thirty, had occurred during 1964 and 1965. He said the only attempt that nearly succeeded was in 1964 when Fidel stopped, as he often did, for a milkshake at the cafeteria at the Habana Libre Hotel. The CIA had discovered this habit, and had suborned a cafeteria employee to try to slip a cyanide capsule into the milkshake. The next time Castro came to the cafeteria and ordered the milkshake, the employee took the capsule out of the refrigerator, where he had kept it, to put it into the drink. But, Valdés said, "The capsule was frozen and it broke, and the man couldn't slip it into the milkshake. It seems he was very nervous. And, you know, cyanide is lethal poison; it would have instantly killed Fidel. . . . This was the closest it ever came.")

Castro's overtures to the United States led nowhere, but in October 1964 he was suddenly faced with a new set of partners in Moscow. Just as President Dorticós was visiting Moscow, Nikita Khrushchev was fired by the Central Committee and replaced by Leonid Brezhnev as general secretary of the Communist party, Aleksei Kosygin as premier, and Nikolai Podgorny as chief of state. In November Che Guevara flew to Moscow for what turned out to be his final visit as well as the confirmation of his growing suspicions that the Soviet Union no longer stood for real revolution. But for

Castro, who had the immediate responsibility for keeping Cuba afloat, the question was whether the Brezhnev leadership would continue Khrushchev's attitude of supporting the Cubans at all costs, and he reasoned correctly that there was no reason for any change.

While failure to consult his Politburo colleagues adequately during the Cuban missile crisis was one of fifteen formal charges against Khrushchev, and the whole Cuban adventure was one of the real reasons for his removal, Castro remained an extremely valuable ally and client in terms of Soviet strategic interest, and it would have made no sense to penalize him. Fidel remarked once in a private conversation that the supreme irony was that it was Khrushchev's humiliation by Kennedy in Cuba that led to the Soviets' decision to catch up with the United States by embarking on a crash nuclear armament program so that they would never again be defeated thus. Khrushchev wrote in his memoirs that "the experience of the Caribbean crisis also convinced us that we were right to concentrate on the manufacture of nuclear missiles. . . . When we created missiles which America and the whole world knew could deliver a crushing blow anywhere on the globe—that represented a triumph in the battle . . . in defending the security of our homeland." Thus unwittingly Fidel Castro contributed to a fundamental shift in the superpower nuclear balance.

Meanwhile, Castro decided to demonstrate his independence by not rushing to Moscow to congratulate Brezhnev on his elevation to power; such hurry befitted Communist satellite leaders in Eastern Europe and Asia but not the proud Cuban revolutionary chief, who had many disagreements with the Russians. In fact, Fidel waited eight years before his next voyage to Moscow. He told me that many years elapsed before Cuban-Soviet relations were restored to the cordiality that preceded the missile crisis. Instead, Raúl Castro was given the mission of conducting high-level contacts at the Kremlin; he saw Brezhnev for the first time on April 2, 1965, and went on visiting Moscow on the average of twice a year. President Dorticós went there four times between 1964 and 1971, while Premier Kosygin visited Havana in 1967 and 1971. But after 1965 permanent contact beteween Cuba and the Soviet Union was established at the level of deputy premiers; Carlos Rafael Rodríguez had had this responsibility for over twenty years, and Deputy Premier Vladimir Novikov replaced Anastas Mikoyan in what became virtually a full-time job of dealing with the Cubans. More recently, Deputy Premier Ivan Arjipov inherited this strategic headache.

Throughout the 1960s, the underlying issue of contention between the Cubans and the Soviet Union was that of Third World revolutions—apart from the aftermath of the missile crisis, permanent economic-aid problems, and the role of Moscow-influenced "old" Communists in Castro's new Cuban communism. Che Guevara played a crucial intellectual and inspirational role in this area, convinced that the responsibility of a triumphant revolution was to spawn revolutions elsewhere. (Che was too subtle, however, to imitate Trotsky's "permanent revolution" rhetoric.) Guevara and Castro shared this view from the time they first met in Mexico, and both sought to put it into effect as soon as they had seized power.

To Guevara the concept was ideological, romantic, and mystical—the ultimate ideal embodied in his Bolivian self-sacrifice. To Castro the idea was considerably more practical in that he knew the Cuban revolution would be more secure if it were successfully repeated elsewhere in Latin America: He had coined the Guevara-like phrase that the Andes would be the Sierra Maestra of the Americas. And when he solemnly declared on December 2, 1961, that "I am a Marxist-Leninist and I shall remain it until the last day of my life," he was committing himself to the revolutionary propagation of Marxism-Leninism throughout the Third World. In the early 1960s, this was the meaning of Castro's support for the emerging guerrillas in Venezuela (after the failure of his earlier revolutionary expeditions around the Caribbean), and of Che Guevara's advising the leftists in Congo (Brazzaville) and the Frelimo guerrillas of Mozambique.

None of it, however, pleased the Soviet leadership (neither during Khrushchev's time nor Brezhnev's), which, thinking in cautious superpower terms, preferred the traditional "united front" coalitions of local Communists with the "progressive bourgeoisie" in the Third World rather than unpredictable revolutions. Curiously, the Soviets were applying moderate "Euro-communism" political tactics to the Third World, being careful always not to challenge the United States excessively. Moreover, they were concerned that sudden successful revolutions might be captured by China, the Kremlin's newest archenemy. In the case of Cuba, therefore, the Soviets discouraged revolutionary impulses in the direction of Latin America or Africa; most ironically, Moscow disliked Cuba's "export of revolutions" as much as Washington feared it. They regarded the Cuban promotion of revolutions to be just as "putschist" and "adventurist" on an international scale as Castro's Moncada and Sierra war undertakings in Cuba earlier.

To show their displeasure, the Russians harassed Castro with long delays in signing annual economic-aid agreements (they were called "trade agreements") and other accords as well as in delivering such vitally needed commodities as petroleum. The Soviets believed this was the way to bring Castro back to his senses if not to his knees—and Fidel responded with gestures of independence, such as the creation of new international mechanisms for encouraging revolutions and his refusal to visit Brezhnev in Moscow. But it was Che Guevara who took it upon himself to deliver a public attack on Soviet behavior. In his last public speech, on February 24, 1965, to the Economic Seminar of Afro-Asian People's Solidarity Organization in Algiers, Guevara, in effect, accused Moscow of being as bad as the "imperialists" in the treatment of the struggling new countries.

Che's first point was that "the development of the countries which now begin the road of liberation must be underwritten by the socialist countries," and he took the Russians to task for imposing excessively harsh terms on the recipients of their largesse. "There should be no more talk of developing mutually beneficial trade based on prices that the law of value and unequal international trade relations imposed on backward countries," Guevara said. "If we establish that type of relationship between the two groups of nations, we must agree that the socialist countries are, to a certain extent, accomplices in imperialist exploitation. . . . It is a great truth, and it does not do away with the immoral character of the exchange. The socialist countries have the moral duty of liquidating their tacit complicity with the exploiting countries of the West. . . . There must be a great change of concepts on the level of international relations. Foreign trade must not determine politics, but on the contrary, it must be subordinated to a fraternal policy toward the people."

This was not language that the Soviets would normally accept from a client state, but they could not afford an open break, and Guevara received no public response to his charges. As it happened, Raúl Castro and Osmany Cienfuegos, then construction minister, arrived in Moscow on the same day that Che Guevara spoke in Algiers, to attend a preparatory meeting for a conference of world Communist parties being planned for March. Cuban support was important in the Soviets' rivalry with China, and clearly they were not going to make an issue of Che's maverick attitudes. Whatever Fidel thought of Che's speech, he made a point of greeting him at the airport in Havana on March 15, which was a gesture for the Soviets' benefit too.

At this point, of course, the thread was broken and Che was never seen again—except dead, two and a half years later in Bolivia. It is an absolute mystery what happened, whether there was a breach in Castro and Guevara's friendship, as has been alleged, and whether Che really left of his own volition. Given Fidel's character, it is unlikely that he would turn for political or ideological reasons on so loyal and intimate companion as Guevara. Che certainly was not a rival. It is true, however, that in 1965 there really was no more place for him in the Cuban revolution. Fidel rebuffed his plans for the economy and at the same time resumed personal control over Cuban agriculture by returning to the presidency of INRA. Internationally, Che was losing his usefulness to Castro because the Communist community now regarded him with suspicion. In the absence of conclusive evidence to the contrary, it must be assumed that they jointly agreed on Guevara's permanent departure from Cuba. Apparently, it happened on April 1. And indeed it would be difficult to imagine him today as an aging revolutionary, approaching the age of sixty, in a subordinate position to Fidel and Raúl. Such a thought must have crossed Che's very proud and sensitive mind.

It would be wrong, however, to conclude that with Guevara's disappearance Fidel renounced their revolutionary objectives or become more tractable to Moscow and its local friends. In November 1964, for example, he had fired Joaquín Ordoquí, an "old" Communist leader, from the post of deputy defense minister, and arrested him and his wife, Edith García Buchaca. Both had been involved in the "sectarian" purge in 1962, and in the Rodríguez trial in March 1964, but Castro struck at them immediately after the change of leadership in the Kremlin as if to remind the new men they were not to interfere with Cuban Communist politics. The trade agreement for 1965 was signed on February 17, well behind schedule, a week before Che Guevara's denunciations in Algiers. Guevara was still abroad when Castro engaged in his own criticism of the Russians for not helping North Vietnam more effectively, and for not reacting more forcefully against the American bombings that had begun in February. "We are in favor of giving Vietnam all the aid that may be necessary, we are in favor of this aid being arms and men, we are in favor of the socialist camp running the risks that may be necessary for Vietnam," he said in a speech on March 13. And Castro abandoned his pro-Soviet stand in the Moscow-Peking dispute to protest this split, saying that "even the attacks on North Vietnam have not had the effect of overcoming the divisions within the socialist family."

Fidel naturally was helpless when the United States landed troops in the Dominican Republic in April 1965, intervening in the civil war there on the side of the local generals against the left-of-center defenders of constitutional government. President Lyndon Johnson had ordered the invasion to stem what he claimed to be a Communist threat, probably inspired from Cuba, but in this instance it was Castro who derived most political profit. He had nothing to do with the civil war (though the defeated constitutional leaders later established close links with him), yet the spectacle of American troops fighting Dominicans had superb propaganda value for Fidel's "anti-imperialist" strategies at home and abroad.

Both Vietnam and the Dominican Republic were most convenient as revolutionary arguments when Castro convoked the "Tricontinental Conference" in Havana in January 1966 to reach out for Third World leadership. The conference spawned OSPAAAL (Solidarity Organization of the Peoples of Africa, Asia, and Latin America) and OLAS (Latin American Solidarity Organization), both with headquarters in the Cuban capital. Among those attending the conference was an Angolan rebel leader, a poet named Antônio Agostinho Neto, who headed the Popular Movement for the Liberation of Angola (MPLA) in the independence war against Portugal. Neto, a remarkably gifted and intelligent man, and Castro became friends, and soon MPLA guerrillas began to be trained secretly on the Isle of Youth along with a growing number of other African and Latin American revolutionary groups. These were the deep roots of Cuba's subsequent massive military involvement in Angola and, later, Ethiopia. From Brazil the Cubans brought Francisco Julião, the leader of peasant leagues demanding land reform in the drought-parched Brazilian northeast, but he never developed into a serious revolutionary.

A month after the Tricontinental Conference, the Soviet Union, swallowing its distaste for Castro's costly new revolutionary instruments, signed the 1966 trade agreement, bringing two-way commerce that year to $1 billion, a record figure, and granting Cuba $91 million in fresh credit. As usual, his risk calculations were correct, and in the course of 1967, Fidel had the satisfaction of denouncing the Soviet Union and being courted by it at the same time.

In a speech on March 13, he attacked the Venezuelan Communist party, which was loyal to Moscow, for its failure to help the guerrillas fighting the country's elected democratic government; the guerrillas were being betrayed, he charged. In April a session of the reconvened Tricontinental Conference in Havana received a message from Che Guevara, radioed from his Bolivian jungle hideout, with this exhortation: "How closely we could look into a bright future, should two, three, or many Vietnams flourish throughout the world with their share of deaths and their immense tragedies, their everyday tragedies, their everyday heroism and their repeated blows against imperialism." In an implicit message to the Soviets, Che remarked: "The solidarity of all progressive forces of the world with the people of Vietnam is today similar to the bitter agony of the plebeians urging on the gladiators in the Roman arena. It is not a matter of wishing success to the victim of aggression, but of sharing his fate; one must accompany him to his death or to victory."

Undaunted, Soviet Premier Kosygin landed in Havana on June 26, 1967, to spend four days with Castro immediately after his summit meeting with President Johnson in Glassboro, New Jersey, a remarkable gesture of courtesy. In his speech on July 26, Fidel informed the world that in the case of aggression, Cuba would fight alone and never accept a truce, an obvious allusion to the Soviet-American settlement of the 1962 missile crisis. Early in August, Castro presided over a OLAS conference of Latin

Americans, repeating his attacks on the Venezuelan Communist party and charging that the Soviet Union was aiding and supporting reactionary governments in Latin America. He specifically mentioned Venezuela and Colombia (both with elected democratic governments and both fighting Cuban-inspired guerrilla movements), exclaiming indignantly that "if internationalism exists, if solidarity is a word worthy of respect, the least that we can expect of any state of the socialist camp is that it refrain from giving any financial or technical aid to those regimes." A resolution approved by the OLAS conference warned that revolutions and armed insurrection would occur even without Communist parties, and that, in effect, it was not necessary that revolutionary leadership be in Communist hands.

Now Fidel was clearly provoking the Russians, in a game of brinkmanship that he seemed to be playing mainly for the sake of proving his revolutionary theory. He had nothing to gain from confrontation: He had already made his point so many times, the Russians had demonstrated a patience they had never had with anyone else, and therefore there was no need to bait the bear. Yet, it is a Castro trait to see how far he can go; first, with the Americans, then with the Soviets.

On October 9, Castro learned that Che Guevara had been killed and his body mutilated by the Bolivian rangers. He proclaimed official mourning in Cuba, and one should not doubt his grief: He *was* immensely attached to Che, notwithstanding any differences they might have had in trying to hammer together Cuban communism. Whether or not Guevara's death had a subsequent influence on relations between Cuba and the Soviet Union, they hit rock bottom that autumn. Castro, in effect, ignored the fiftieth anniversary of the Russian Revolution by refusing an invitation to go to Moscow, and by sending a delegation headed by the health minister, José R. Machado Ventura (the previous year, Raúl Castro and President Dorticós had gone to Moscow for the anniversary). Fabio Grobart, the Polish-born cofounder of the old Cuban Communist party—not Castro—delivered the main speech at anniversary celebrations in Havana. The Castro brothers, Dorticós, and the rest of the leadership deigned to attend the reception at the Soviet embassy on November 7, but they were greeted by the chargé d'affaires; the ambassador happened to be away in Moscow. Presently, the Soviets decided that the time had come to teach Fidel Castro a lesson.

For Castro and for Cuba, 1968 was the worst year since the revolution: everything seemed to be coming apart, and only through the sheer force of his personality, and thanks to his security mechanisms, was Fidel able to prevent a complete collapse. In their own way, the challenges of 1968 were greater and more complex than the Bay of Pigs and the missile crisis because they dealt with the essence of revolutionary rule. Curiously, however, the Johnson administration was unaware of how extremely vulnerable Castro had become. Perhaps it was too distracted by the Tet offensive in Vietnam and its aftermath to pay much attention to the little island in the Caribbean, not an imminent threat to the United States.

In tribute to Che Guevara, 1968 was designated the "Year of the Heroic *Guerrillero*," which was exactly what Fidel had to be at that juncture to survive. He had the worst of all possible worlds; while the Cuban economy was tottering on the edge of catastrophe because of appalling planning (actually, Castro had no coherent economic plans at all, only grandiose ambitions) and frightening mismanagement, he was antagonizing the Russians, who were Cuba's last hope for saving the revolution. And, to be consistent, Castro crushed the supposedly pro-Soviet faction of the "old" Communists now incorporated in his brand-new Cuban Communist party. In a conversation in August 1967 with K. S. Karol, a French left-wing journalist, Fidel said that "in a year or two" Cuba could be self-sufficient economically because its exports would expand to the point where it would no longer depend on a single market and a single supplier. For this reason, Castro told Karol (whose name and books are banned in Cuba because of his "leftist" criticisms of Fidel), he need not fear the loss of Soviet support as a consequence of his ideological independence and defiance. Reality, however, turned out to be quite different—and quite painful—and he should have known it.

The first reality was the state of the economy. Though as far back as 1960, Castro had created JUCEPLAN, a central agency based on GOSPLAN, the central Soviet planning organization, he was personally responsible for its failure to function adequately. This was so because Fidel's impatience led him into continuous shifts between short-, medium-, and long-term planning as well as into endless improvisations. No concept was given reasonable time to produce results (or to be proven unsatisfactory), overall coordination was therefore lacking, and political or visionary pressures pushed Castro into sudden decisions to undertake grandiose projects the economy could not possibly handle. The notion of producing the largest sugar harvest in history in 1970 was a case in point. At the same time, large capital investments in industrial or agricultural development programs that had not been adequately studied cost Cuba precious resources diverted from normal economic activities. Inevitably, production dropped in all sectors, shortages became even more acute than before, and Cubans were exhorted (or forced) to sacrifice for the revolutionary future.

By 1968, Castro had personally assumed the planning and execution of economic policies, shutting off any alternative ideas and, naturally, brooking no argument. It was an intriguing turning point in his career. Now in his early forties, and no longer the young man of the Sierra, Fidel had become transformed into a total dogmatist ideologically, societally, and econom- ically, in absolute disregard of the experiences of other men and other so- cieties, but also in contemptuous rejection of many Marxist and Soviet views. When the Russians finally imposed their will on him, Fidel knew how to bend with the wind, suspending his public displays of indepen- dence for as long as it appeared to be prudent, but never abandoning his personal dogmatism, ever ready to reassert it. Meanwhile, the economy approached bankruptcy. The ravages of Hurricane Inez in 1967 and the great drought in 1968 aggravated the situation. Cuban production nearly ground to a halt: René Dumont, the French agricultural specialist and the most perceptive foreign observer of the Cuban scene in the late 1960s, remarked later, "There was nothing to buy, for which reason there was no stimulus to work." Elsewhere in the Communist world, notably in Hun- gary and Czechoslovakia, planners were just then beginning to experiment with market economies so that there would be something to buy. In Cuba Castro seemed determined to prove that to go backward in Marxist eco- nomic history represented progress.

On March 31, 1968, the anniversary of the students' attack on the Batista palace, Castro proclaimed a new radical revolution in Cuba, which in a sense was his equivalent of the Chinese Cultural Revolution that was beginning to wind down in China. Although Cuba had no Red Guards and no blood was shed, Fidel moved to nationalize the entire retail trade sector still in private hands—58,012 businesses, ranging from auto mechanics' repair shops to small stores, cafés, and street vendors of sandwiches and ice cream—for reasons of ideology. Like Mao Zedong, Castro must have felt that revolutionary fervor was eroding among his people, and that a power- ful injection of radicalism was urgently needed to make the juices flow again. He called it the "Great Revolutionary Offensive," imposing revolu- tionary purity by eliminating the remnants of the "bourgeoisie" he so de- spised, and by mobilizing Cuban manpower on a gigantic scale (voluntary extra work by everybody) for agricultural production, and especially the record sugar harvest he was planning for 1970.

This latest milestone of the revolution was unveiled by Fidel in one of his most intricate (and long) speeches, which he delivered at Havana Uni- versity before an audience of hundreds of thousands. He unfolded, mostly from memory, massive statistics on milk production and imports, province-by-province rainfall measurements, prices of sugar over the past sixteen years, rising egg production, population growth, fish catch, gross national product comparisons worldwide, per capita incomes, economic de- velopment investments, food distribution from schools to workers' dining rooms, education, and sports. This economic state-of-the-union message, emphasizing the very impressive revolutionary conquests in the social realm, then turned into a bitter indictment of "those who do not work, the loafers, the parasites, the privileged, and a certain kind of exploiter that still remains in our country."

Castro is a master at the use of big and obscure facts and statistics to score political points, heaping scorn and ridicule on the objects of his attacks, and vastly amusing his audience in the process. On this occasion he reported that in Havana "there still are 955 privately owned bars, making money right and left, consuming supplies," which he described as "an incredible thing" nine years after the revolution. He went on for nearly an hour reciting the results of investigations of these bars and other private businesses, offering such conclusions as "the data gathered on hot dog stands . . . showed that a greater number of people who intend to leave the country are engaged in this type of business, which not only yields high profits but permits them to be in constant contact with *lumpen* and other antisocial and counterrevolutionary elements." Pursuing this linkage between hot dogs and counterrevolution, Fidel informed his listeners that "the greatest percentage of those not integrated into the Revolution was among the owners of hot dog stands; of forty-one individuals who answered this item, thirty-nine, or ninety-five point one percent, were counterrevolutionary." As the crowd laughed and applauded, Castro exclaimed, "Are we going to construct socialism, or are we going to construct vending stands? . . . We did not make a Revolution here to establish the right to trade! Such a revolution took place in 1789—that was the era of the bourgeois revolution, it was the revolution of the merchants, of the bourgeois. When will they finally understand that this is a Revolution of socialists, that this is a Revolution of Communists . . . that nobody shed his blood here fighting against the tyranny, against mercenaries, against bandits, in order to establish the right for somebody to make two hundred pesos selling rum, or fifty pesos selling fried eggs or omelets . . ."

The crowd roared its approval when Castro announced that "clearly and definitely we must say that we propose to eliminate all manifestations of private trade!" He had made the point that through the 1968 nationalizations, Cuba had firmly advanced toward true "socialism," and that it would continue hurtling down the ultrarevolutionary road. While the relationship between prices and wages has bedeviled Marxist economics for decades, Castro was naïvely prepared to tell K. S. Karol in 1967 that "it is absolutely necessary to de-mythicize money and not to rehabilitate it. In fact, we plan to abolish it totally."

* * *

At the start of 1968, Fidel Castro's immediate concern was to prevent a serious breakdown in the Cuban economy because of Moscow's refusal to increase petroleum shipments. This was the first result of the Soviet decision to bring Castro politically and ideologically under control; when the Cuban foreign trade minister arrived in Moscow in October 1967 to open talks for a new commercial agreement, he was apprised that fuel deliveries would not be raised to match the 8 percent growth in the annual Cuban demand. Moreover, the Russians rejected the Cubans' proposal for extending the annual trade agreements into a three-year pact—Castro at that stage was keen on a longer Soviet aid commitment—and they became very vague about the date the 1968 agreement would be signed. Fidel reacted with an explosion of private fury at the news of the Soviet petroleum reprisal, and on January 2 he went on television to announce gasoline rationing and to order sugar mills to use alternative fuels. Cuba depended on the Soviet Union for 98 percent of its oil needs, but in his television statement on rationing, Castro said that there was only a "limited possibility" of increased Soviet deliveries, and that, in any event, the "dignity" of the revolution prevented Cuba from begging for more oil.

Ten days later, Fidel showed the Russians that not only would he not be intimidated by them, but that he was determined to remain outspokenly critical about their policies and dogmatically independent. Addressing the closing session of a weeklong International Cultural Congress in Havana, a meeting organized to enhance Cuba's prestige and attended by five hundred intellectuals from seventy countries (including Jean-Paul Sartre, Lord Bertrand Russell, and Julio Cortázar, then the leading Argentine novelist), Fidel Castro was at his oratorial best, urging his guests to help him define how intellectuals can best serve the revolution. It was a most elegant appeal to world public opinion to understand the "real" value of the Cuban experiment, but Castro minced no words in saying what he thought about orthodox Communist parties that "remained completely removed from the struggle against imperialism." He alluded to the fact that at the time of the 1962 missile crisis, European Communist parties failed "to mobilize the masses" in Cuba's support, implying that they preferred to obey Soviet instructions rather than engage in "just combat." Finally, Fidel chastised "these groups" for not raising high "the banner of Che" after his death; they would "never be able to die like him," he said, "nor to be true revolutionaries like him."

On January 25, Castro lowered the boom on the so-called "microfaction" in the new Communist party who were the same old-line leaders whom he had purged for "sectarianism" six years earlier but had later reinstated. Again, going after the "old" Communists was a thinly disguised assault on the Russians because of the deep-rooted associations between the two. Aníbal Escalante, the alleged chief of the "sectarian" movement, was once more the main target of Castro's purge, but this time he and thirty-six others were arrested, tried on charges of "conspiracy" against the revolution, and given prison sentences; Escalante was sentenced to fifteen years. As sensitive as this whole matter of intramural Communist struggle was, the regime never spelled out fully in public what exactly the Escalante conspiracy represented. Nevertheless, word circulated immediately that the "microfaction" had taken it upon itself to convince Moscow to suspend all economic aid to Cuba to force Fidel's ouster and to put in power a "loyal" Communist regime. While it is impossible to ascertain precisely how valid these charges were—interpretation of attitudes is the key to Communist demonologies—Castro is not likely to have invented the entire episode. The Russians are not above staging coups d'état against their best allies if it suits their interests, and Castro was becoming almost intolerable in demanding both ideological independence and massive economic support.

In his Revolutionary Offensive speech on March 13, 1968, Fidel said that the party's Central Committee had resolved not to publish the "microfaction" closed-door court proceedings for the time being (they were never published). However, he noted that although the Escalante group lacked significance as "a political force," its actions "were of a very serious nature . . . as a political intention and as a tendency within the revolutionary movement, a frankly reformist, reactionary, and conservative current." In presenting the case to the Central Committee, Fidel spoke from noon until midnight—his longest speech. Castro said later that in dealing with it, "the revolutionary courts were not as severe as some would have wished but . . . unnecessary severity has never been a characteristic of this Revolution." Unfortunately, this judgment is unlikely to be shared by hundreds of Cubans who were tried by these courts and, twenty or twenty-five years later, remain imprisoned on obscure charges. There are at least three hundred political prisoners—fewer than thirty were released in 1986—and by all accounts they are subjected to extremely harsh treatment. Freed prisoners have reported systematic torture up to the mid-Seventies, and Castro will never be able to erase from his revolution's history the shame of the irrational and capricious reign of terror in his "Gulag South" penal system in the quarter-century after victory.

The only authoritative account of the Escalante affair that I know to exist despite the continuing official silence was given to me in Havana in 1985 by Fabio Grobart, now the chairman of the Historical Institute of the Marxist-Leninist Movement in Cuba. Grobart's account is important because he is close to Castro and because he reopened for me, as Fidel had done earlier, crucial aspects of the Cuban-Soviet disputes in the context of the missile crisis of 1962.

The Escalante group, Grobart told me, "had its own discipline, its own purpose, and its own policy to struggle against the leadership of the party, which is the leadership of Fidel Castro. . . . It was a line that could have even led to the destruction of the Revolution, a division within the Revolution." Specifically, Grobart continued, Escalante's conspiracy "coincided with the problems we had at the time of the introduction in Cuba of nuclear weapons in October 1962.

"At that time," he said, "the Soviet Union committed certain errors, withdrawing these missiles without consulting the Cuban government. And Escalante, as a saboteur, took advantage of [Cuban] disagreements with certain things in the methods of the Soviet Union to make himself appear as a friend of the Soviet Union, and to throw garbage and mud at the Revolution, saying, 'You know, they [Fidel Castro] are anti-Soviet,' and so on. . . . He tried to present himself as the true defender of the Soviet Union, which is what Fidel has never ceased to be. He attempted to gain supporters, to demoralize the party, to weaken the unlimited trust that the party and the people have in Fidel Castro. This was the error, this is the crime that Aníbal Escalante has committed." Escalante served part of his sentence, then died while working as a farm administrator.

Having disposed of Escalante and his group, Castro proceeded to display further annoyance with the Soviet Union by refusing to send a Cuban delegation to the conference of world Communist parties that Brezhnev had organized in Bucharest in February to deal with the internal divisions caused by the split with China. This was a direct slap at the Soviets; in his March speech Castro mentioned the decision not to attend the Bucharest conclave, remarking, however, that "for the moment this is not a fundamental question." The Kremlin retaliated on March 22, 1968, by signing a relatively modest trade agreement with Cuba. Whereas the volume of trade (generally a euphemism for Soviet aid) had risen 23 percent from 1966 to 1967, the increase in 1968 was only 10 percent. Moreover, the accord provided for interest payments by Cuba on a $330 million Soviet credit that financed her adverse trade balance. To make matters worse, this debt would rise very soon, inasmuch as Cuba was committed to deliver 5 million tons of sugar to the Soviet Union during 1969, but its 1968 harvest was only 5.3 million (3 million below the target volume and 1 million less than in 1967), and Castro knew that the 1969 harvest would be below even 5 million tons. He also knew that without considerably increased Soviet assistance, all his ambitious economic development plans would have to be dropped. While the regime's official newspaper *Granma* had proudly asserted on February 2 that "no one can call us a satellite state and that is the reason we are respected in the world," in the end Fidel Castro had little choice but to fall into line, abandoning for at least a time his exhortations for armed revolutionary struggle throughout the Third World.

Relations between Havana and Moscow were very tense. A tough Soviet diplomat, Aleksandr A. Soldatov, was transferred from London to Havana to try to deal with Castro. In Moscow *Pravda* published an editorial denouncing "reactionaries who follow the writings of men who call for revolutionary changes in the entire social system," which seemed to be addressed as much to Fidel as to anybody else. But neither side could risk a complete break, and it was late August before a new rapport began to take shape between them.

CHAPTER
2

Just before midnight on August 20, 1968, Soviet, East German, Polish, Hungarian, and Bulgarian military forces invaded Czechoslovakia to grant it "fraternal" assistance against the reformist Communist regime headed by Alexander Dubček. This was twelve years after Soviet armies drowned in blood an anti-Communist rebellion in Hungary, seven years after an American-organized exiles' brigade invaded Cuba at the Bay of Pigs, and three years after United States Marines and airborne troops landed in the Dominican Republic to intervene in the civil war there against an alleged Communist threat.

Three days later, Fidel Castro—a disputatious ally and client of the Soviet Union and victim of an invasion engineered by the United States— went before television cameras in Havana to explain his and the Cuban revolution's reaction to the occupation of the territory of one socialist nation by the armies of fellow socialist states. He would discuss as well his views on the suddenly ended "Prague Spring," the Czechoslovakian Communist party's experiment with "Marxism with a Human Face"—liberal and nonrepressive Marxism.

I happened to be in Prague on the invasion night, and given my journalistic coverage of the Bay of Pigs and the Dominican intervention—as well as my special interest in the Cuban revolution—I was immensely curious as to where Castro would situate himself in the rising dispute between pro- and anti-Soviet Communist parties around the world over the application of the "Brezhnev Doctrine." This doctrine murkily vested in the Soviet Union the right (actually because Brezhnev said so) to invade neighboring Communist countries if communism were in danger. For Cuba this loomed as a very uncomfortable doctrine if the United States chose to apply it in reverse to protect or assure representative democracy in *its* sphere of influence; theoretically, it could legitimize a new and more successful Bay of Pigs. Thus, Castro's reaction to the Prague invasion was extremely relevant to his own destiny—just as it was relevant to the Soviet Union that most certainly expected solidarity and understanding from its extraordinarily costly Caribbean ally. It was a superb intellectual challenge for Fidel, and he handled it superbly.

Most artfully, he landed on all sides of the issue. Although Czechoslovakia's sovereignty and international law were unquestionably violated, Castro said, "we accept the bitter necessity that required the dispatch of these forces to Czechoslovakia, we do not condemn the socialist countries that took this decision" because it was "imperative to prevent at all costs" that Czechoslovakia move toward "capitalism and into the arms of imperialism." But, he went on, "as revolutionaries . . . we have the right to demand that a consistent position be adopted toward all the questions affecting the revolutionary movement in the world." In other words, Fidel remarked, if Warsaw Pact armies had acted to prevent the destruction of socialism in Czechoslovakia, "will Warsaw Pact divisions be sent to Cuba if Yankee imperialists attack our country, and if our country solicits it when it is threatened by an attack by Yankee imperialists?" In this manner, Castro demonstrated his solidarity with the Soviet Union and the principle of "socialist intervention" while recognizing that international law had been violated by the Russians. Moreover, he made his "acceptance" of the Prague invasion conditional on a new Soviet guarantee of military action to protect Cuba from the United States—on top of the guarantees he believes were made by President Kennedy to Khrushchev as part of the settlement of the missile crisis. The Kennedy-Khrushchev "understanding" on that point was never made explicit, but both governments have respected it for twenty-five years.

In a way, Castro was the only real beneficiary of the destruction of the Prague Spring. By backing the Kremlin when scores of Communist parties condemned the invasion, he established a new claim on Soviet economic aid for Cuba, and used the crisis to end his own dispute with the Russians without losing face. Fidel was also helpful to Brezhnev by announcing that Cuba opposed "all these liberal economic reforms that were occurring in

Czechoslovakia and that are also occurring in other countries of the socialist camp." In this area, however, he was quite sincere because of his dedication to the most rigid forms of Marxist economics and Stalinist planning. He went so far as to express the hope that the Soviet Union would reject the temptation to fall for market economy ideas. He delivered a savage attack on Yugoslavia for sponsoring "bourgeous liberal policies" and being an "imperialist tool" while pretending to be a Communist state. Castro denounced the Yugoslavs for refusing to sell Cuba arms in 1959, and this furious attitude led ultimately to ugly personal clashes between himself and Marshal Tito in the Nonaligned Movement.

Castro's stand on Czechoslovakia notwithstanding, Soviet-Cuban relations did not return to normal overnight. He told me many years later that all the misunderstandings and resentments dating back to the October crisis were not finally resolved until late in 1968. There were still deep mutual suspicions, and the Soviets were still holding back on oil and other deliveries to Cuba, pending the solution of a variety of outstanding problems. In fact, so severe was the Soviet-induced fuel shortage in the autumn of 1968 that Raúl Castro had to array his tanks along beaches in stationary artillery positions because he lacked gasoline for them. Both in his Czechoslovakia speech and other pronouncements, Fidel went on criticizing Communist regimes and parties helping "reactionaries" in Latin America, and continued to talk about armed insurrection in the hemisphere—all of which continued to annoy the Soviets. But his revolutionary sword was now blunted: Che Guevara's death in Bolivia, the collapse of the leftist guerrilla movements that Cuba supported in Venezuela, Colombia, and Guatemala made his rhetoric less credible. And when reform-minded (but not Marxist) generals in Peru overthrew the civilian democratic government in mid-1968, in time both Castro and the Russians formed friendships with the Peruvian military. Moscow sold them tanks and aircraft, and Castro, overcoming in this case his distaste for military regimes, later dispatched his friend Captain Nuñez Jiménez as a superambassador to Lima—clearly with the notion that in the long run the Andes were promising revolutionary territory.

Perhaps the last sticking point between Moscow and Havana was their dispute over Manuel Piñeiro Losada, the red-bearded deputy interior minister in charge of the State Security Department (the political secret police). During the investigations of the Escalante affair early in 1968, Piñeiro accidentally found Escalante himself in a secret meeting with an adviser from the KGB (the Soviet State Security Committee). A corps of KGB advisers was attached to the Cuban Interior Ministry, and its chief took umbrage over Piñeiro's failure to report the incident to him. Raúl Castro, who directed the overall "minifaction" investigation, told this astonishing story to the Central Committee, lifting a corner of the secrecy mantle over the relationship between the two undercover police organizations. Raúl said he had brought up the matter of Escalante's clandestine meeting with the Soviet adviser with Ambassador Soldatov and the chief KGB adviser and discovered that the latter was furious at Piñeiro: The Russians were naturally disturbed that their man had been spotted with Escalante. Raúl said to the chief adviser, "You are almost asking me to arrest Piñeiro because he failed to show you respect, and I don't propose to do it. So the chief replied, 'We are Piñeiro's bosses, not you . . . and how can you think that we would . . . ,'" and I broke in to say, 'We don't believe so, but if you did not think so obtusely, you would interpret this as a warning that for us it would be rather painful to find here a Soviet official, diplomatic or not, involved in a matter of internal character." It is unclear how this dispute was settled, but Piñeiro soon ceased to be deputy interior minister, and became head of the Central Committee's Department of the Americas, one of the most powerful political positions—still closely tied to Cuban intelligence operations.

Even so, relations were still cool in November. Pedro Miret went to Moscow for the Russian Revolution anniversary (Fidel, Raúl, and President Dorticós again would not go). Faustino Pérez, the least ideologically minded of Fidel's old companions, was picked to speak at the celebrations in Havana, and Raúl was the ranking Cuban guest at the Soviet embassy reception; his big brother was still sulking.

Foremost on Fidel Castro's mind toward the end of 1968 was the centennial observation of the start of the first independence war, the present-day economy, and a new drive to tighten domestic political controls. The latter appeared to stem from a concern about possible contagion from the Prague Spring, and late in the year Raúl's Revolutionary Armed Forces Ministry initiated an attack on "bourgeois" intellectuals and "counterrevolutionary" literature. The word was that no "softening" in the revolutionary spirit would be tolerated, and in March 1969, the ministry organized a "National Forum on Internal Order" for its own officers, Interior Ministry specialists, party officials, and the Revolutionary Defense Committee.

All these new efforts were consistent with the 1968 Revolutionary Offensive, but quite obviously Czechoslovakia had a big impact. In a conversation early in 1984, Fidel Castro told me he had thought all along that Dubček and the other "liberal" Communist leaders were acting in deep error and that they should have been "aided" in time. Then we spoke of the recent events in Poland, and Castro said that the Polish Communist party had mismanaged the situation, largely because of corruption, with the resulting emergence of Solidarity, the free trade-union movement. Fidel clearly disliked the Solidarity idea—other Cuban officials hinted that there had been some concern in 1980 and 1981 that it might start catching on in Cuba—and he suggested that General Wojciech Jaruzelski, the Communist premier, had saved Poland from a Soviet invasion by imposing martial rule and outlawing Solidarity. The official viewpoint had quickly spread to the Cuban cinema industry: I heard Cuban directors in private conversation angrily criticize the Polish director Andrzej Wajda for producing *Man of Iron*, the inspirational picture about Solidarity that won a Cannes Film Festival award.

Among many facets of revolutionary Cuba that developed disturbingly in the first Castro decade—and have continued since—included the very same Communist party mismanagement and deadening bureaucracy that Fidel had observed with such alarm in Poland in the pre-Solidarity years. There was also the stolid conformity in thought and word that, for example, made Cuban filmmakers denounce Wajda for his Solidarity movie only because that followed the official line. Even by 1968, it was already commonplace for Castro and *Granma* to deplore periodically and liturgically the excesses of bureaucracy without achieving the slightest progress in streamlining it. In truth, the revolution and the creation of the new Communist party had spawned a new privileged ruling class built around and below Fidel by the party, the security apparatus, and the military—and inevitably this elite reposed on a faithful bureaucracy. The Castro cult of personality was derived from this form of political organization of the society, and it instantly raised the question of whether in a relatively short time Fidel had actually become isolated from the realities of his people.

His initial concept of governance was to commune with the masses through the dialogue mechanism of preaching and teaching, then ask the crowds for approval of proposed policies, and assume that a revolutionary consensus had thereby been achieved. Clearly, however, this was not consultation in any sense of the word, and after a while Castro no longer knew what people thought and what bothered them. Having abolished the "liberal bourgeois freedom of the press" in the early 1960s, he could not learn much from reading his own newspapers and magazines, and palace courtiers do not normally generate or deliver bad news. In the first years of the revolution, a favorite Fidel occupation was to bounce around the countryside in his jeep, stopping here and there, chatting with people, asking with genuine interest what they did and what problems they had. This, too, had its limitations, but it gave Castro some sense of his people. I asked him about this old practice one day in January 1984 when we were driving in his jeep on the outskirts of Havana. He turned around in the front seat where he sat next to the driver and shook his head sadly. "No, I don't get to do that very much anymore," he said. "You know, now I have all the responsibilities of running the state, attending meetings, receiving ambassadors and so on—so there is no time."

As early as 1968, visitors from abroad who spent much time with Castro—such as Frenchmen K. S. Karol and René Dumont—began to wonder whether Fidel insisted on running everything personally because he hungered for total power or whether he was the victim of the system he had devised. Viewing Castro more than a quarter-century after the revolution, the conclusion is that both judgments are correct, and that really he is his own prisoner for life.

This is so because for Fidel Castro the revolution is a permanent struggle, not necessarily a Marxist, socialist or Communist struggle, but a Cuban national struggle. On October 10, 1968, he flew to Oriente province in a downpour to deliver a patriotic speech commemorating the hundredth anniversary of the first Cuban uprising against Spain. The theme throughout was "one hundred years of struggle," with the emphasis on it being only the beginning for Cuba. Not once did Castro mention Marxism-Leninism, socialism, the Communist party or the Soviet Union: His text was dedicated to Céspedes and Martí and all the other heroes of liberation struggles—and to sugar production. Two days earlier, he said proudly as applause soared around him, Cubans had planted exactly 33,230 acres of sugarcane across the island in a single day in tribute to the independence anniversary and the memory of Che Guevara. In Fidel's mind, the record sugar harvest he was planning for 1970 would represent the start of the second "hundred years of struggle," and he knew that the inspiration for the great effort had to be patriotic and not ideological.

To achieve the goal of ten million tons of sugar in 1970, a goal Castro had linked to the "honor of the Revolution," the nation was mobilized as if for war, and he designated 1969 as the "Year of the Decisive Effort." At the same time, Fidel ordered changes in harvest methods so that actual work would start in July 1969 and end in July 1970. Normally, the harvest lasts the three months of winter, but the unprecedented magnitude of the planting and acreage made it necessary to devote the whole year to it. Canecutting thus became the principal and most obsessive economic effort in Cuba during 1969, limiting everything else to an absolute minimum. Industrial and commerce workers, students, old people, and children as well as "brigades" of foreign visitors were assigned to cut cane weekdays and weekends, and holidays were abolished. The army too was drafted for canecutting, and Fidel and all his top companions were televised and photographed cutting cane on certain days as an example to the nation. However, this activity requires skill, endurance, and strength, and the inexperienced "volunteers" often were in the way of the quarter million professional *macheteros*. Nevertheless, "voluntary" work was a political requirement to justify the economic near-paralysis of the country, and so the harvest lasted 334 days. Since drought was a menace to the crop, Castro announced that "we have made a pact with rain." And rain kept its word in 1969, providing just the right pattern.

Fidel was an incurable optimist about the revolution, never discouraged by the initial failures. Not only did he want that record sugar crop in 1970, but speaking on the tenth anniversary of his victory, he pledged that Cuba's overall farm production would increase by no less than 15 percent annually over the next twelve years—the highest growth rate in world history. Given the sharp drop in farm production in the first decade of the revolution, which Castro was the first to acknowledge, this promise seemed totally out of touch with reality. Yet, his word was not questioned. Fidel said that to meet his objectives, Cuba had to import 8,000 tractors annually for ten years, train 80,000 tractor operators, and provide 180,000 farm workers with mechanization skills. When it came to imports, however, Castro again had to rely almost entirely on the Soviet Union and other Communist countries. By 1969 the Russians, having had their way, were ready to resume large-scale aid—and to think again of the strategic importance of Cuba.

The Cuban-Soviet economic relationship was based principally on the concept that deliveries of Cuban sugar would pay at least in part for Soviet shipments of petroleum, machinery, automotive equipment, industrial goods, and just about everything else a modern economy required (weapons were a grant). But even though Moscow calculated the value of the sugar at prices well above the world market (following the example of the United States import quota), Cuba was ten million tons in arrears by 1969 because the crops had been so low in past years. Therefore, Castro opened the year by announcing sugar rationing at home so that more would be available for export in anticipation of the record harvests he projected. He provided the nation with a series of most convincing explanations for the rationing.

Meanwhile, the entire relationship with the Soviets was improving. Deputy Premier Novikov, the Kremlin's full-time expert on Cuba, was on hand for the tenth-anniversary celebrations, and Castro's speech was full of praise for the Soviet Union. In February a very favorable trade agreement was negotiated. In April the love feast continued with a Castro speech on the anniversary of Lenin's birth and praise for the Soviet Union for aiding Cuba to become the "first socialist state" in Latin America. The same day, Fidel presided over the ceremony inaugurating the Soviet-Cuban Friendship Association. In June Carlos Rafael Rodríguez led a delegation of the Cuban Communist party to Moscow for an international conference boycotted by the Chinese, North Korean, Albanian, and North Vietnamese parties (the

Cubans had evidently quietly set aside their professed friendship with Vietnam). Rodríguez, the "old" Communist who became a top *Fidelista* in the new Caribbean communism, pleased his hosts when he declared that "Cuba is convinced of the importance of the unity of action of the Communists" in order to develop "a broader offensive against imperialism and the forces of reaction and war."

Having absorbed Castro back in the ideological and economic fold, the Russians now turned their attention again to the strategic equation in Cuba for the first time since the 1962 missile crisis. An eight-ship Soviet naval squadron arrived in Havana on July 20, and the Castro brothers, along with the entire top leadership, attended a reception given by Rear Admiral Stepan Sokolan, the task force commander. Then, Fidel and his government went on a cruise on the flagship *Groznig;* Fidel and the admiral celebrated the 26th of July anniversary together by cutting cane near Havana along with a contingent of sailors from the cruiser—no doubt, an unforgettable experience. The Soviet defense minister, Marshal Andrei A. Grechko, was the next military chief to wield a machete during a November visit with Fidel and Raúl. (On December 30 the entire staff of the Soviet embassy in Havana spent a day cutting cane in what had become an act of piety for foreigners, comparable to the laying of wreaths at national monuments.) As the decade of the 1960s came to a close, Cuban-Soviet friendship was stronger than ever—with Fidel Castro frequently expressing his gratitude for the economic and military assistance from Moscow. By the end of 1969, the Cubans owed the Russians $4 billion, representing the difference between the value of Soviet deliveries to Cuba and vice versa. Castro himself calculated that during the decade, Cuba had received, free, $1.5 billion in military equipment, including jet aircraft.

Nineteen seventy was called "The Year of the Ten Million Tons"—but it was not. Despite good weather and the incredible effort expended by the whole nation on Fidel Castro's orders, Cuba fell short of the goal, producing 8.5 million tons. Actually, it was a record crop (the previous record was six million tons in 1965), but even this volume was not commensurate with the social and economic costs to Cuban society. It was the result of Castro's astonishing penchant for making commitments and promises without having the slightest idea whether they could be kept—his weakness was the political grand gesture, a prohibitively expensive habit. Apart from the self-inflicted sugar fiasco, he repeated the same error with milk when he promised in December 1968 that production would quadruple within two years; but in 1970 the production was running 25 percent below 1969 levels (probably because the energies of the peasants had been channeled into sugarcane cutting).

However, Castro was painfully honest about the failures. Describing the agricultural shortcomings in a speech on July 26, 1970, he said that "the unquestionable inefficiency of all of us . . . signified that we were incapable of waging what we called the simultaneous battle on all the fronts of production." He admitted that "the heroic effort to increase the sugar production led to imbalances in the economy, in diminished production in other sectors and . . . in an increase in our difficulties." Then he warned somberly that the next five years would be even more difficult, adding that "I want to speak of our inability in the overall work of the Revolution. . . . We must face our responsibility in these problems, and my [responsibility] in particular. . . . Our apprenticeship as leaders of this Revolution has cost too much." Earlier, Fidel had observed that "the battle of the ten million was not lost by the people: It is us, the administrative apparatus, the leaders of the Revolution who lost it. . . . Our ignorance of the problems of the sugar mills prevented us from remedying in time all the difficulties." However, René Dumont, who was in Cuba during that period, remarked that sugar experts who attempted years earlier to call Castro's attention to these problems "were simply sent away."

Still, no reverses could slow him down or diminish his interest in everything, everybody, everywhere. He refused to settle for any kind of status quo. In terms of intellect, a comment Walter Lippmann once made about H. G. Wells might also apply to Fidel: "[He] seemed to win by a constant renewal of effort in which he refused to sink either into placid acceptance of the world, or into self-contained satisfaction with his vision." Thus in the second half of 1970, Castro could turn away from the depressing domestic scene and its economic problems to the international problems and controversies on which he thrived.

Two such situations developed in September 1970, and Fidel was involved in both of them, directly and indirectly. The first was the victory of his friend Salvador Allende Gossens in a three-way race for the presidency of Chile on September 4. Allende, a frequent visitor to Havana, was of Marxist persuasion and headed the Chilean Socialist party, and this was the fourth time he had run for president. In 1970 the CIA and American corporations vainly invested tens of millions of dollars in secret support for Allende's right-wing opponent, Jorge Alessandri, and in anti-Allende propaganda. However, Allende obtained only a 36.3 percent plurality, and a runoff with Alessandri the following month was required under the Chilean constitution.

The Nixon administration's reaction to the idea of a democratically elected Marxist president in Latin America was naturally alarmed revulsion, and the immediate fear in Washington centered on the nightmarish scenario of a Castro-Allende revolutionary axis. Fidel, just as naturally, was delighted for obverse reasons, but he took great care to avoid gloating, which could give Nixon justification to intervene in some way (it was only two years ago that the application of the Brezhnev Doctrine to Czechoslovakia contributed to a certain legitimacy for invasions in one's corner of the world). While the administration drew up overt and covert plans to defeat Allende in the October 24 runoff in the Chilean congress, a new crisis over Cuba suddenly erupted.

About a week after Allende's first victory, a U-2 spy plane brought back photographs showing new barracks, communication towers, and antiaircraft sites being constructed near the naval base of Cienfuegos in southern Cuba. Also appearing in the photographs was a new soccer field, and inasmuch as the Cubans do not play this game, CIA photo interpreters concluded that the field was being prepared for the Russians, who do play it. As it happened, a Soviet naval squadron arrived on a visit to Cienfuegos on May 14, and another squadron appeared on September 9. The second squadron included a submarine tender, a 9,000-ton *Ugra*-class vessel, and two towed barges which the CIA believed were for storing radioactive waste from reactors on nuclear submarines. Finally, a Soviet nuclear submarine was spotted in the general area of Cuba. Instantly, Nixon and National Security Adviser Henry A. Kissinger concluded that the Russians were establishing a nuclear-submarine base in Cienfuegos. Given the memories of the 1962 missile crisis, this was not an unreasonable suspicion, even though the Russians had committed themselves at that time to keep offensive weapons out of the Western Hemisphere. The administration could not be sure that Brezhnev would reject a Khrushchev-type adventure in Cuba.

Nixon and Kissinger kept the U-2 discovery and their conclusions a secret, planning to deal with the Soviets through quiet diplomatic channels. But on September 16, Kissinger summoned Soviet Ambassador Anatoly Dobrynin and confronted him with the aerial photographs. Then, he held a press briefing to discuss the dangers of the Allende victory in Chile and to warn the Soviets against "operating strategic forces out of Cuba, say, Polaris-type submarines." He added that "we are watching these events in Cuba." Now Castro and Allende seemed to form part of an interlocking crisis, but neither of them chose to polemicize with Kissinger. In October the Soviet Union privately assured Kissinger that it was building no bases in Cuba, and TASS subsequently repeated this publicly. Castro has never said a word on the subject in public, and this Cuban minicrisis served chiefly to reassert the 1962 "understanding" between the superpowers over their mutual guarantees on Cuba. Still, the Americans thought that Brezhnev might have been "testing" them by dispatching naval squadrons to Cuba and openly engaging in military construction in Cienfuegos. Fidel certainly would not have minded such a test. On October 24 he learned with vast pleasure that Allende had been confirmed as president by the Chilean congress. He saw this as the beginning of a long and profitable relationship and promised to visit Allende the following year.

In June 1961, when Fidel Castro told a group of writers, artists, and intellectuals that there was full creative freedom in Cuba "within the Revolution," but nothing "against the Revolution," he was establishing a relatively relaxed set of cultural standards. It took the hard-line ideological offensive in 1971, evidently inspired by Castro, to demonstrate how brutally these standards had changed in a decade. For reasons that defy understanding when one considers Castro's own intellectual wealth, shortly after 1961 he was willing to impose (or allow to be imposed) in Cuba a grotesque and repressive travesty of cultural life—always in the name of the revolution.

The onslaught on supposed "counterrevolutionary" intellectuals began around 1965, when the regime started to arrest those it considered as "antisocial elements," especially if they were homosexuals, and to put them in the UMAP forced-labor army battalions. This was done as part of a broader ideological offensive against homosexuals in general during the political power struggles of the 1960s. In 1968, when the political climate turned unusually harsh, specific writers became targets of the party ideologues. The award of a poetry prize to Heberto Padilla and a theater prize to playwright Antón Arrufat that year by the official Union of Cuban Writers and Artists (UNEAC), a Castro creation, led to an internal ideological clash typical of the changing cultural environment. As a compromise, UNEAC decided to publish the prize-winning works along with a note from the union's editorial board, expressing its disagreement with them because "they are ideologically contrary to our Revolution." In murky Marxist ideological jargon, Padilla was accused of "ambiguity" and "antihistorical attitudes," and Arrufat, of disseminating an "imperialist-type" reality. Shortly thereafter, a Writers' and Artists' Congress was held in Cienfuegos, approving a resolution that each writer "contribute to the Revolution through his work and this involves conceiving of literature as a means of struggle, a weapon against weaknesses and problems which, directly or indirectly, could hinder this advance." No Cuban author knew what it meant.

Late in 1969, prize-winning novelist José Lorenzo Fuentes was expelled from UNEAC because of his connections with a Mexican diplomat allegedly employed by the CIA. It seemed as if the party's cultural ideologues were determined to absorb Marxism-Leninism to the point of turning Cuba into a tropical version of an Arthur Koestler novel about Communist police states. In 1970 the most prestigious Cuban novelists and poets suddenly discovered that no publishing house or magazine would publish their work—no explanation given. This mysterious ban would last into the mid-1970s.

The new official line was proclaimed in a "Declaration" issued by the First National Congress of Education and Culture held in Havana in April 1971, and what an ideological monstrosity it was in terms of Cuban cultural life can be gleaned from the published text. "Cultural development" in Cuba, it said, must be aimed at the masses, "contrary to the tendencies of the elites. . . . Socialism creates objective and subjective conditions that render feasible a true creative freedom while rejecting as inadmissible those tendencies that are based on a criterion of libertinage and aimed at concealing the counterrevolutionary poison of works that conspire against revolutionary ideology . . ."

Rambling on page after page, this cultural edict provided that in hiring personnel for universities, mass communications media, and literary and artistic institutions, the candidates' "political and ideological conditions be taken into account." Selectivity in invitations to foreign writers and intellectuals was recommended to avoid the "presence of persons whose work and ideology are at odds with the interests of the Revolution." Moreover, the congress declared that "cultural channels may not serve for the proliferation of false intellectuals who plan to convert snobbism, extravagance, homosexuality, and other social aberrations into expressions of revolutionary art, alienated from the masses and from the spirit of our Revolution." It seemed incomprehensible that Fidel Castro could tolerate such insults against his beloved revolution by his own ideologues.

Yet, Castro evidently approved of the crackdown on Cuban intellectuals because the arrest of the poet Heberto Padilla in March 1971 must have been authorized by him. This arrest led an impressive group of European and Latin American intellectuals, including Sartre and García Márquez, to write Castro demanding Padilla's release. He was freed thirty-seven days later, after reading a statement of self-criticism and urging other writers to do likewise. Even though his friends regarded him as a traitor, Padilla remained in Cuba for a decade, working as a translator of foreign literature. He finally left Cuba in 1981, after García Márquez made another personal appeal to his friend Fidel. Even the obedient UNEAC wrote a letter to Castro protesting the lengthy detention of homosexuals in military forced-labor units, and they were finally sprung. Yet it left an ugly scar on Cuban society. Overall, Castro's shocking cultural policies have dealt a lethal blow to creativity in his country; even in 1986 the island was a wasteland of ideas and a reign of strict self-censorship. It may take generations before Cuba returns to the free cultural age of José Martí.

* * *

In any event, Castro was extremely busy during 1971 with Cuba's foreign relations and probably lacked time to supervise the cultural life personally. Ties with the Soviets were being strengthened daily, but Fidel was still not ready to visit Moscow. Instead, he dispatched President Dorticós to attend the Twenty-fourth Congress of the Soviet Communist party in March, preferring himself to meet in Havana with top Soviet officials. Nikolai Baibakov, the chief Soviet economic planner, came in April, and Soviet Premier Kosygin spent five days in Cuba in October, his second visit to Fidel.

On November 10, Castro flew nonstop to Santiago, Chile, for a ten-day visit that became a three-week stay. It was his first trip abroad in seven years since his last journey to Moscow, and he enjoyed it hugely. It was also the first time he was returning to Latin America in twelve years, and that too pleased him greatly, as he said to his aides aboard the Soviet IL-18 jet airliner equipped with a bed and an office. From the airport, Fidel drove into the city, standing up in a convertible next to Allende and waving to the huge crowds that cheered him in the streets. After several days in Santiago, Fidel embarked on a voyage to the north and south of the long, narrow country, visiting schools and plants, delivering speeches, granting interviews, meeting young and old Chileans, and visibly having a fine time. In the northern city of Antofagasta, he joined a group of folklore musicians for a picture, holding a guitar in one hand and patting the head of the musician next to him with the other.

Nathaniel Davis, then the United States ambassador in Santiago, described Castro's visit as "an extraordinary display of high-level tourism, thinly disguised meddling, and shrewd commentary on the Chilean scene; it was a circus." Fidel stayed longer in Chile than his hosts had expected, almost running out of things to do, but he evidently wanted to have a solid look at Allende's socialist experiment. He went to nine provinces, from the seaport of Valparaíso and the copper mines in central Chile to Río Blanco high up in the Andes and Tierra del Fuego across the Magellan Straits. He talked to workers everywhere, conducted endless dialogues with university students in sports stadiums, discussed theology with "revolutionary priests" in Santiago, donned straw hats and hard hats, kept warm in colorful wool *ruana* blankets, and played a rousing game of basketball in Iquique, wearing a Number Twelve sweatshirt and showing that at the age of forty-five, he was still in great physical shape.

En route home, Castro stopped in Lima to meet Peru's military rulers, with whom Cuba was developing cordial relations, then in Guayaquil, Ecuador, to converse with the aging president, José María Velasco Ibarra, who once upon a time had been a dictator. In all three Pacific Coast countries, Castro offered Cuba's support for their controversial claims to a two-hundred-mile territorial-water limit to protect fisheries and won warm applause in exchange. In Havana Fidel received a hero's welcome after his long absence, but he was already thinking of more foreign travel the following year.

Recalling that South American trip many years later, Castro said he knew the CIA had tried to assassinate him first in Chile, then in Peru and Ecuador. He told an interviewer that weapons for the assassins, who posed as Venezuelan journalists with Venezuelan credentials, had come from the American embassy in La Paz, Bolivia. These arms, he added, ranged from rifles with telescopic sights to machine guns and a television camera with a hidden gun: "It was even in front of me, but they didn't shoot."

On May 3, 1972, Fidel Castro left Cuba aboard his IL-62 jet airliner for a two-month voyage to ten countries on two continents, climaxing the trip with his first visit to the Soviet Union in eight years. In Moscow Castro came face to face with the nascent Soviet-American detente, arriving exactly one month after the departure of his archenemy, Richard Nixon. While Nixon's presence at the Kremlin had resulted in the signing of SALT I, the nuclear arms limitation agreement, Castro's sojourn resulted in Cuba joining the Council for Mutual Economic Assistance (CMEA), the Communist common market, better known as Comecon. By joining CMEA, Cuba became fully integrated into the worldwide Communist economic system, which included the Soviet Union, the six European Communist countries, Outer Mongolia, and Vietnam. The Russians were most skilled at dealing simultaneously with adversaries and friends, even when, like the United States and Cuba, they were each other's mortal enemies. Fidel also received a Soviet marshal's saber as a special accolade, and of course he volunteered no public comments about the Brezhnev-Nixon embrace the previous month. He had become a very sophisticated revolutionary and a fine practitioner of personal diplomacy.

Castro's realignment with Soviet foreign policy included his acceptance of the view (at least publicly) that the time for guerrilla warfare on the Sierra Maestra model had run its course, and that social change in Latin America had to be accomplished by less violent means, which had been argued by the Russians all along. They still thought that Fidel's 1958 victory had been basically an aberration due to his personality and special Cuban conditions, and could not be repeated elsewhere. Chances are that Castro had not given up his personal revolutionary dogmas, yet he, too, had to recognize new realities in the hemisphere. After Che Guevara's death, the aura of romantic revolutions had paled (Camilo Torres, the revolutionary Colombian priest Castro had met in 1948, had been killed in his own guerrilla war, and the Peruvian rebel poet Hugo Blanco had been captured), forcing Fidel to rethink his strategies. Allende's election in Chile and the emergence of the left-wing reformist generals in Peru confirmed the emergence of the new trend. Castro was too intelligent to ignore it, deciding on the route of Cuban statesmanship and the confining of aid to insurgents only in situations where revolutionary conditions already existed.

Consequently, the Seventies saw Castro concentrating on ending the diplomatic and political isolation imposed on him by the United States, and he was doing fine. The visits to Chile, Peru, and Ecuador produced new friendships for Cuba on the west coast of South America. Argentina, Colombia, and Venezuela resumed diplomatic relations with Cuba, leading the Venezuelan guerrilla chief, Douglas Bravo, to accuse Fidel publicly of betrayal (just as Fidel had earlier accused Venezuelan Communists of betraying Bravo). He developed relations with Panama, Jamaica, Barbados, Bahamas, and Trinidad and Tobago in the vital arc of Central America and the Caribbean when he realized that his thirteen-year-old revolution no longer required a cordon of other revolutionary states around Cuba for protection. Soviet protection was more weighty, anyway. Gradually, Castro began reentering Latin American organizations, such as the United Nations Economic Commission for Latin America (ECLA), and the Organization of American States (OAS) prepared to lift diplomatic and trade sanctions against Cuba established in the early 1960s. Finally, the Cubans played an important part in organizing SELA, a new regional economic entity. They were no longer pariahs.

Now Castro set out to expand his friendships in Africa, with visits to Guinea, Sierra Leone, and Algeria. It was his first trip there. Cuba has had considerable involvement in Africa since the first days of the revolution; Che Guevara was the principal emissary to the continent until he vanished from the political scene in 1965. The Cubans had championed the cause of Algerian independence since 1959, sending military and medical supplies to the National Liberation Front (FLN) who fought the French in 1960, then sending a Rebel Army combat battalion to help freshly independent Algeria in its border war with Morocco in 1963. In the Sixties, there were Cuban military missions in Algeria, Ghana, Congo (Brazzaville), and Guinea, and later in Equatorial Guinea, Somalia, and Tanzania. Castro's visit to Sierra Leone was followed by the arrival of a Rebel Army mission to train the national militia.

Along with Che Guevara, he believed that Africa was the future scene of great revolutionary changes with Cuba as guide and mentor. Since much of Africa had either been recently decolonized or was peacefully preparing for independence (or fighting for it as liberation movements were doing in the

Portuguese colonies), Castro saw an extraordinary potential for Cuban revolutionary influence. And African nationalists, mainly of leftist persuasion, welcomed the Cubans: Cuba's cultural origins were partly African, and the Afro-Cuban tradition was powerful. The Cubans offered the perfect alternative to either American or Soviet "imperialism," and finally, Cuba was a sister Third World nation with a triumphant revolution.

By the time Castro set foot in Africa, there was no African left-of-center government or liberation movement that did not have Cuban ties of some sort. Africans were being trained by Cubans in everything from medicine to military organization in their own countries or in schools and camps in Cuba and the Isle of Youth. Fidel was planning for the future, knowing intuitively that Cuba had a perhaps decisive role to play in parts of Africa; this, in turn, gave him and Cuba international importance beyond Latin America. In addition to his Bolívarian dream in his own hemisphere, Castro now aspired for Third World leadership.

In Conakry, Guinea, he found a fellow revolutionary in President Ahmed Sekou Touré, who told a crowd greeting Castro at a sports stadium that "Cuba is the light in Latin America." Their conversations served to coordinate Cuban support for guerrilla movements against Portuguese colonial rule and South Africa, Guinea being a natural transit point from Cuba to the African continent. Active backing for African guerrillas fit Castro's (and the Russians') thinking because anticolonial "national liberation" movements were involved; Khrushchev had been a foremost advocate of such movements. Fidel started his Guinean tour clad in his usual olive-green fatigues, but on the second day he changed into the national dress of white trousers and a white short-sleeved tunic buttoned below his beard at the neck; he kept on his black combat boots, green cap, and military webbing belt for an arresting overall effect. He always delighted in trying on the attire of countries he visited, which, in turn, delighted the local audiences. Sekou Touré awarded Fidel the Order of Fidelity to the People. The *compañeros* whom Castro took along to Africa suggested what his interests in the region were: Juan Almeida, the black vice-president and former Rebel Army Chief of Staff; Manuel Piñeiro Losada, his closest adviser on intelligence and contact with revolutionary movements; Arnaldo T. Ochoa Sánchez, a new-generation officer who would soon command Cuban combat troops in Angola; and Raúl E. Menéndez Tomassevich, once Raúl Castro's deputy in the guerrilla war and now the principal insurgency and militia expert.

In Sierra Leone Castro spent one day with President Siaka Probyn Stevens, a relatively moderate politician, then returned to Conakry for more talks with Sekou Touré. In Algeria Fidel was back in a warm revolutionary environment, visiting for ten days the country Cuba had helped in its independence war. His old friend Ahmed Ben Bella had been overthrown by Houari Boumedienne, but Castro established a good relationship with the new president. As a socialist and revolutionary state, Algeria was Cuba's oldest and best ally as well as a key link both to Africa and the Arab world. Similarly, the Algerians had a military rapport with the Soviet Union, which equipped their armed forces, and Castro and Boumedienne could also share their Third World views on the Russians—not always flatteringly. Algiers was a revolutionary planning and plotting center for Africa, and Fidel devoted much time to this topic, refining his ideas on "anticolonialist struggles" in speech after speech. Finally, Boumedienne was immensely important to Castro as a key leader in the Nonaligned Movement that Fidel hoped to dominate.

From Algiers Castro's Soviet jet took him over the Mediterranean, across Italy, the Adriatic, and Yugoslavia, directly to Sofia, Bulgaria. This was his introduction to non-Soviet European Communist states, called "satellites" by their detractors, and was presumably an instructive experience for Fidel, representing the Cuban variety of Soviet-supported communism. He instantly fell into the mandatory "fraternal" practice of warm embraces with local leaders, but in Bulgaria and throughout the rest of Eastern Europe, Castro evoked genuine interest and excitement among the eager crowds that greeted him in the streets and squares of the old cities. He was different, he was a legend, and he was informal—he appeared to be all the things the traditional Communist leaders were not. For his part Fidel was fascinated to discover old cultures in new but distinctly designed Marxist mantles.

In Sofia his host was Todor Zhivkov, the Bulgarian leader. He had been in power longer than any other Communist in Europe (except for the self-isolated Enver Hoxha of Albania), and never questioned Soviet wisdom. Castro was in gala uniform for official functions and a concert, but seemed even happier dressed in a gray sweatsuit playing basketball with an army team against a civilian team. He dutifully observed military maneuvers and was presented with a Bulgarian-made AK-47 submachine gun and an antique pistol of Bulgarian partisans; his private weapon collection grew as he traveled the world.

From Sofia the Cuban leader flew to Bucharest to meet Rumanian President Nicolae Ceauşescu, also a Communist master of the art of survival but a maverick in foreign policy. It must have intrigued Fidel how Ceauşescu, a next-door neighbor of the Soviet Union, had succeeded for so long in defying the Russians on just about everything: He refused to break relations with Israel after the 1967 Middle East war (unlike the rest of the Soviet bloc), he remained pointedly friendly with the Chinese and Yugoslav Communist heretics, and he roundly condemned the Soviet invasion of Czecho-slovakia, which came five days after his visit to Prague (he also declined to let Rumanian troops participate in the "fraternal" action). Defying Moscow had not been a signal success for Castro, but in common with Ceauşescu he had a dedication to total internal regimentation and an inability to make workers produce well. Fidel may not have learned Ceauşescu's foreign policy secrets; he did learn, however, that Rumania produced excellent wine (a matter of interest to him) when he drank some from a wineskin.

In Hungary there was still another subtle Communist experience for Castro. He came to Budapest a quarter-century after Soviet tanks had destroyed the anti-Communist "freedom fighters" (this expression was born in Budapest in 1956) to find the country with a considerable degree of autonomy from the Russians in internal affairs and a remarkable prosperity stemming from market-economy reforms. These were the reforms Czechoslovakia had sought to imitate and expand before the 1968 invasion, and that Castro so violently denounced in his speech. Whatever impression the Hungarian achievements had left on Castro, he remained just as opposed to market economy ideas in 1986, when he banned the experiment in farmers' markets. Under Janos Kadar, who had ruled Hungary since the rebellion, the country lived in remarkable political and cultural relaxation within the Communist system. Evidently, Castro did not wish to imitate these practices either. From Budapest he took home a saber used in the 1848 independence war and an AK-47 automatic rifle.

Then it was on to Poland—and still another facet of European communism. This was a nation with viscerally anti-Russian (now anti-Soviet) traditions, a powerful nationalism, a formidable Roman Catholic church, a Western European culture, memories of the terrible destruction during its opposition to the Nazis in the war—and with a shockingly corrupt and inept Communist party. The party had so thoroughly antagonized the workers that despite the very real advances in education, public health, and living standards for the impoverished population under the socialist system, there had been violent uprisings. The last one had occurred only two years earlier in Gdańsk, the seaport, where the army and the police had had to fire on rioting workers, killing and maiming scores.

When Castro reached Warsaw, the nation lived in the relative calm between storms. Poland was the only Communist country with an articulate—and tolerated—presence of internal political opposition, and a lively and rich cultural life largely stemming from this phenomenon. Naturally, the political opposition had no access to Castro, and he would probably have disapproved of writers and artists to whom the building of socialism was not the only creative ambition. Despite Fidel's rich mind, the political and cultural barriers between the two nominally Communist states were too great for him to bridge. Thus, the visit was a standard affair, though, as usual, Fidel brightened it with his personality and attitudes. One evening, he took a walk to Warsaw's Old Town Square, stopping to chat with a flower vendor, and dropping in at a basement nightclub called Krokodyl where he promised the young patrons to send a stuffed crocodile from the swamps of the Zapata in Cuba. The youngsters loved it. That evening, a news story was sent out from Warsaw by an American wire service that Castro had suffered a heart attack, and Fidel sputtered in fury that it was another CIA provocation designed to destabilize Cuba—which was probably true.

In the coal region of Silesia, Castro put on the traditional black gala uniform of the miners and received the miners' medal. "If this tour continues at this rate, I'll wind up with as many medals as a hero of labor," he told the miners, winning laughter and applause. He evinced considerably less enthusiasm when he cried, "Long live Red Silesia!" Still in the miners' uniform, Fidel paid a visit the same day to Oświęcim (Auschwitz) and Brzezinka, Nazi concentration-camp sites. At Oświęcim he wrote in the visitors' book: "Capitalist and imperialist ideology was capable of reaching such extremes. . . . What I have seen today, reminds me of what the Yankees are doing now in Vietnam." In Cracow Fidel played basketball with the university team. In Gdańsk he toured the Lenin shipyards (where Solidarity would be born eight years later), then joined General Jaruzelski, the defense minister (who would liquidate Solidarity), at military maneuvers in the coastal area.

On June 13, Fidel Castro arrived in East Berlin, but it is unlikely he knew that it was the nineteenth anniversary of a bloody uprising by East German workers and students against Communist rule. Soviet tanks had put down that first postwar rebellion in Eastern Europe in 1953, a now-forgotten date. Met by East German leader Erich Honecker, Castro was visiting a very prosperous Communist country, contiguous to the Western world even in divided Berlin. He was taken to the Brandenburg Gate, which stands astride the line between the two Berlins, and he could not have missed the Berlin Wall the East Germans had erected in 1961 to prevent their fellow citizens from fleeing to the West. He did not mention the Wall in a speech to East German frontier guards, but he compared the Berlin border zone to the United States naval base at Guantánamo as a foreign enclave. In Merseburg Fidel walked down the street holding hands with schoolchildren, and at Moritzburg Castle he was the guest of the commander of the Soviet Army division stationed near East Berlin. There, his speech chronicled the use of freshly arrived Soviet arms in the Bay of Pigs battle. Castro told me in 1984, that East Germany was the country he admired most in the region: like Cuba, it had to live next to an enemy.

In Czechoslovakia, whose invasion by Warsaw Pact armies four years earlier Castro had rationalized and praised, he was most warmly greeted by the Communist party secretary general, Gustav Husak, the man in charge of "normalizing" the situation to suit the Kremlin. Husak awarded Castro the Order of the White Lion. The venerable Charles University, where the liberalizing spirit of the Prague Spring had been born among its philosophers, then granted him the degree of Doctor *Honoris Causa* in Juridical Sciences. In black robes and cap, and amid medieval splendor, Fidel Castro, the philosopher of revolution, used the opportunity to lecture his audience on the ideological history of his Movement. Confusing the matter once more, he explained that the Moncada Program in 1953 "was not yet a socialist program," although it had been officially described as such in Cuba ever since 1961, when Fidel announced that his revolution *was* socialist in character. However, he said, "the revolutionary process of Cuba is a confirmation of the extraordinary force of the ideas of Marx, Engels, and Lenin." And for the next hour, this man with great gifts of intellect and oratory rewarded one of the world's greatest universities with a tiresome recitation of Marxist-Leninist banalities, praise of the Soviet Union, and denunciations of "imperialism."

Castro landed in Moscow on June 26, his third time there, for a relaxed two-week visit. It marked the official end of a decade of Cuban-Soviet misunderstandings, a fact underlined by the presence of Brezhnev, Kosygin, and Podgorny at Vnukovo-2 Airport, and the award to Fidel the next day of the Order of Lenin. Apart from his meetings with the Kremlin leadership, Castro conferred with Defense Minister Grechko and the Soviet General Staff, again emphasizing the importance of the Cuban-Soviet military ties. He visited four Soviet cities, the space center, an aircraft plant, and agricultural stations, seemingly completely at home now in the Soviet Union. He flew home on July 5, and Cuba's membership in the CMEA, the Communist common market, was announced six days later.

International relations continued to keep Castro busy for the balance of 1972, and much of 1973, while the fragile economy at home seemed to be receiving less of his attention.

In December 1972, Salvador Allende came to Cuba to see Fidel, and they spent pleasant days in the sun chatting and sailing in a cabin cruiser. This was to be their last meeting.

On December 18, Castro was again in Moscow for a week's stay to help celebrate the fiftieth anniversary of the foundation of the Soviet Union. During that week, he was able to put the finishing touches on five major economic and technical assistance agreements with the Russians. Thus, the Cuban debt of some $4 billion was deferred altogether until 1986—virtually a fifteen-year moratorium—and the Cubans would then have twenty-five years over which to repay it. The anticipated Cuban trade deficits in 1973 and 1974 (still not enough sugar was produced) were to be covered by separate credits. The volume of two-way trade between 1973 and 1975 was increased sharply, although no figures were given, which simply meant more Soviet aid to the limping Cuban economy. Additional credits of $390 million were earmarked for economic development. And Moscow agreed to "pay" still more for Cuban sugar and nickel so that the actual debt would rise more slowly. Back in Havana, Castro outlined these accords in a speech on January 3, 1973, commenting that Cuban-Soviet relations were "a model of truly fraternal, truly internationalist, and truly revolutionary relations."

This was not hyperbole. By the mid-1970s, Cuba was receiving about one half of total Soviet economic aid to all the Third World (including Vietnam) as well as probably one half of Soviet military aid to those countries. Because Soviet and Cuban foreign policies were increasingly on the same track in the light of changing world conditions—and not because Moscow was forcing Castro to follow its line as it did in 1968—Fidel could conduct his relations with the Third World as he pleased. He was free to enhance the international standing that he had lost quite perceptibly after the 1962 missile crisis and after Che Guevara's death. Therefore, he literally had the best of all worlds.

In this sense, it should not surprise Americans, as it seems to, that Castro has no intention of ever trading his Soviet relationship for a relationship with the United States. He knows that no American administration (or U.S. Congress) would provide him with the kind of massive economic and military aid he receives from the Soviets—with relatively few questions asked. Moreover, he realizes that the United States would require that he shed his Communist system and his Third World policies to qualify for even the most minimal assistance. Repeatedly, in public statements and in private conversations, Castro insists that apart from political and "moral" considerations, it would make no practical sense for him to sever his friendship with the Russians—even if this friendship has its very trying moments. It would certainly suit him to have easy trade with the United States as well as other arrangements helpful to the island's economy, but not at the price of renouncing his Soviet alliance, such a renunciation having long been an American condition for lifting sanctions against Cuba. He marvels privately that serious American officials and politicians are so naïve as to believe that Washington would be doing him a favor by establishing relations with Havana with basic preconditions. It is simply a perspective that continues to elude most Americans, and this is why all Washington's past attempts at negotiations have led nowhere—except for nonpolitical accords of mutual convenience.

With a free hand in the Third World, Castro began to concentrate on it (aside from his Latin American interests) with great seriousness in the early 1970s, making it a centerpiece of Cuban foreign policy in the ensuing years. After his first visit to Africa in the spring of 1972, in December on his way to Moscow, he stopped in Morocco to meet King Hassan; for his purposes, friendship with socialist Algeria did not preclude a rapport with the Moroccan monarchy. In September 1973, Fidel attended the Fourth Conference of Nonaligned Nations in Algiers, where he became acquainted with most of the Third World leaders, making new friends and a very good impression—even though he was criticized for defending the Soviet Union from accusations that it was as "imperialist" as America.

Then Castro flew to bomb-ravaged Hanoi on his first visit to Vietnam. He felt a special kinship with this country as a fellow victim of the United States. Vietnam peace treaties had been signed in January, and America was now out of the war. This was also his first time in Asia, but Fidel had no desire to go to China, an enemy of his Soviet and Vietnamese friends. He was in Hanoi the day Salvador Allende's regime was ousted by a military coup and the Chilean president was killed in his palace. With Allende's death, Castro lost a personal friend and a vital Latin American ally.

Fidel Castro received the ultimate accolade from the Soviet Union—and further recognition of his international standing—when Leonid Brezhnev arrived in Havana on January 28, 1974. Never before had the highest Soviet leader visited Latin America—and this was a special trip to Cuba, not part of a larger tour.

Brezhnev was Castro's guest for a week, paying him the special tribute of going to Santiago to see the former Moncada barracks (now a school) that Fidel and his rebel band had attacked on July 26, 1953, setting the Cuban revolution in motion. Then he drove with Fidel to the nearby El Siboney farm, where the rebels had assembled for the assault the night before. This was Castro's final vindication: The Russians would never again speak of Moncada as a "putsch" or an "adventure," since now they recognized the Marxist correctness of the *Fidelista* enterprise. In Havana Brezhnev and Castro addressed a million Cubans at a rally on Revolution Square under the statue of José Martí, then signed a grandiloquent Soviet-Cuban "Declaration" of principles. In his speech to Brezhnev, Castro did, in effect, confirm the belief that he was the Kremlin's spokesman in the Nonaligned Movement: "As we have expressed at the conference of nonaligned countries in Algeria, the very existence of the Soviet Union constitutes a brake on the militarist adventures of the aggressive forces of the imperialist world, without which they would have already launched a new effort to divide the planet, and would not have hesitated to invade countries possessing petroleum and other basic raw materials."

A great turning point in the history of the Third World, Africa, and revolutionary movements was the overthrow of the dictatorial regime in Portugal by left-of-center military officers in April 1974. Among the first moves made by the young officers were to announce the end of Portugal's hopeless and ruinous colonial wars and to promise independence as soon as possible for Angola, Mozambique, Guinea-Bissau, and Cape Verde Islands. The most important of these wars was fought in Angola, and in January 1975, the military government and the three rebel factions signed the Alvor agreement (named after the Portuguese town where they met), providing for Angolan independence on November 11.

Because Fidel Castro had had the foresight to support the Marxist MPLA (Popular Movement for the Liberation of Angola) when he first met Agostinho Neto early in the 1960s, Cuba found itself in the extraordinary position of exercising a decisive influence on Angolan events. This, in turn, made the Cubans the most important outside force in African military politics, first in Angola and then in Ethiopia—a state of affairs that continued to prevail in the late 1980s. Fidel Castro was consequently elevated to the status of an unquestioned world leader, calling the shots in Africa and Central America, assuming the responsibility for formulating Third World economic policies toward the industrialized countries, and, of course, becoming more than ever an accursed obsession for the United States. At the same time, Castro stood on the threshold of formally transforming his revolution into a permanent and constitutional Communist institution. He was barely fifty years old.

CHAPTER

3

Contrary to widespread belief, it was Fidel Castro's idea—certainly it was not the Russians'—to engage Cuban combat troops in the civil war in Angola on an absolutely open-ended basis. The year 1986 began the second decade of the large-scale Cuban military presence there. Over 200,000 Cuban troops were rotated through Angola during the first decade, and Castro has pledged to rotate 200,000 more Cubans, if necessary. But contrary to Fidel's own assertions, it was *not* South Africa's armed intervention in the Angolan civil war that forced him to rush his forces to Angola. The truth is that Castro beat everybody to it, entering first the conflict in an impressive display of instinct, imagination, and daring.

After the Alvor agreement on independence, the Angolan civil war broke out among the MPLA and two rival movements. These were the FNLA (National Front for the Liberation of Angola), led by Holden Roberto and backed by the United States, Zaire, China, and South Africa, and UNITA (National Union for the Total Liberation of Angola), headed by Jonas Savimbi and supported by the United States and South Africa. The FNLA and UNITA were obviously anti-Communist. Curiously, the Soviet Union withdrew its support for the MPLA, not trusting it politically and militarily, and Castro was its only faithful friend. He had trained MPLA guerrilla officers in Cuba, and when the internal power struggle erupted, he quietly dispatched 250 Cuban combat advisers in May 1975 to organize MPLA forces. Simultaneously, Flavio Bravo, one of the Castro brothers' oldest Communist friends, and now a top regime official, met in Brazzaville with the MPLA president, Agostinho Neto.

In July the MPLA requested additional aid as both the FNLA and UNITA began to acquire great new strength from United States equipment and advice (the CIA was reported to have spent $31 million in helping anti-Communist factions in 1974 and 1975, until Congress stopped this covert aid). In August South Africa and Portuguese mercenary units crossed from Namibia into southern Angola to protect a border hydroelectric station from the MPLA. Everybody seemed to be involved in Angola against the MPLA, the Cubans being its only, but increasingly effective, defenders. Late in September and early in October, three Cuban freighters delivered military detachments and arms to Angola, and, belatedly, Soviet arms began arriving through Brazzaville. In early October there were about 1,500 Cuban Army personnel in Angola. On October 23, South Africans in force entered Angola in Operation Zulu, and Cuba responded with a troop airlift in "Operation Carlota." A full-fledged Cuban expeditionary force rushed units aboard Cubana Airlines transport planes through Conakry in Guinea (Fidel's friend President Sekou Touré coordinated the operation). At first, the Cuban aircraft tried to refuel in Barbados and Trinidad without receiving permission, then Guyana authorized the refueling when the other countries refused.

There is no question that the Cubans saved MPLA from destruction. They helped it defeat its enemies in the crucial battle for the railroad terminal port of Benguela on November 5, and assured the capture of Luanda—threatened by Holden Roberto's FNLA—in time for Angolan independence on November 11. In February 1976, Castro had fifteen thousand troops in Angola, and the Organization of African Unity (OAU) recognized the MPLA as the government. Castro said that Cuban forces would remain in Angola as long as the MPLA needed them—they were there ten years later as the Angolan regime was unable to survive alone the onslaughts of Jonas Savimbi's UNITA, massively and openly backed by South Africa. In one of the ironies of Angola, Cuban units, because of their discipline, were assigned to guard the oil-producing installations of the American-owned Gulf Oil Corporation in the northern enclave of Cabinda. Then the Soviets quietly purchased the oil from Gulf on the high seas (through a broker in Curaçao), after which the tankers sailed to Cuba. It was much cheaper for the Russians to ship at least some of the oil to Cuba from Cabinda rather than from the Black Sea. The final irony was that when in 1985 South African commandos tried to blow up Gulf Oil's installations, Castro exploded in terrible rage publicly.

The Angolan operation proved that Cuba and the Soviet Union could coordinate overseas military undertakings smoothly in Africa—and presumably elsewhere in the Third World—through rapid deployments of Cuban personnel with small arms and simultaneous (or prepositioned) deliveries of heavy Soviet arms. It is nonsense, however, to suggest that the Cubans are the "Gurkhas of the Soviet empire" (as American administrations have charged), and that their troops were ordered into Angola by the Russians. While in Angola, and later in Ethiopia, Cuban and Soviet interests have obviously coincided, it is not plausible that Moscow even attempted to "order" Cubans there. It would not work practically, and, above all, there is ample evidence that it was all Fidel's idea in the first place.

There was no reason to doubt Castro's veracity when he told Barbara Walters in a television interview in 1977 that "you should not think that the Soviets were capable of asking Cuba to send a single man to Angola. . . . That is totally alien to Soviet relations with Cuba and to Soviet behavior. A decision of that nature could exclusively be taken by our party and our government on our own initiative at the request of the Angolan government. . . . The Soviets absolutely did not ask us. They never said a single word in that sense. It was exclusively a Cuban decision." He added: "We decided to send the first military unit to Angola to fight against South African troops. That is the reason why we made the decision. If we had not made that effort, it is most likely that South Africa would have taken over Angola."

Castro's denial that Moscow had pushed him into Angola is corroborated by Arkady Shevchenko, a senior Soviet diplomat who defected to the United States in 1978. Shevchenko wrote in his memoirs that when he asked Deputy Foreign Minister Vasily Kuznetsov sometime in 1976, "How did we persuade the Cubans to provide their contingent?" Kuznetsov "laughed." Shevchenko added that "after acknowledging that Castro might be playing his own game in sending about 20,000 troops to Angola, Kuznetsov told me that the idea for the large-scale military operation had originated in Havana, not Moscow. It was startling information. As I later discovered, it was also a virtual secret in the Soviet capital."

When I raised the question of Angola in a conversation with Castro in January 1984, he said that "the Angolans asked us for help, and we sent them help, with great effort and great sacrifice. . . . Angola was invaded by South Afrida [which] has the moral condemnation of the entire world. . . . Therefore, we could never have done anything more than just to help Angola against an external invasion from South Africa. . . . In the instances in which we have provided help, outside of Latin America, it was to the countries that have been attacked. It is not the case of actions against governments there, regardless of what the governments are. We helped the people of the Portuguese colonies as everybody was helping them every-where, and when the United Nations helped them. We fight against South Africa when the United Nations condemns South Africa for its aggression in Angola; we helped Ethiopia when it was the target of an external inva-sion aimed at disintegrating the country." Castro also makes the point that, as in Nicaragua, the Cubans provide extensive medical, educational, and technical assistance to Angola along with military help. It is a unique formula, accepted by the Angolans, Ethiopians, Nicaraguans, and others in the Third World because it comes from a Third World country.

Just as Cuba was becoming engaged in Angola, the Ford administration decided that the time might have come to improve relations with Fidel Castro. The idea came from Secretary of State Henry Kissinger, who did not think that the emerging rivalry in Angola would prevent reaching some understanding with Cuba after nearly fifteen years of broken relations and unremitting mutual hostility. Kissinger had succeeded in negotiating an overture to China (and a Nixon trip to Peking) in the midst of the Vietnam War, and he reasoned that it should be possible to establish a dialogue with Castro. The Soviet-American détente was in full bloom, which loomed as another favorable factor in approaching the Cubans. Be-sides, Kissinger had valid reasons to think that Castro would be interested. What he could not predict was the ultimate degree of Cuban involvement in Angola.

Before any action was taken, the State Department undertook a series of confidential conversations with Cuban diplomats at the United Nations to assess Havana's reactions to possible talks, and obtained encouraging rep-lies. Castro was known to be flexible on certain aspects of relations with the United States when it suited him—he allowed the resumption of emigra-tion to the mainland that meant the departure of 250,000 Cubans between 1965 and 1973, and worked out an antihijacking "understanding" with Washington in 1973. Now Kissinger wanted to know how much further the Cuban leader might go. In a speech in March 1975, he pointedly said there was "no virtue in perpetual antagonism" with Cuba. In May he ap-proved a "secret advance probe" policy through which Castro was informed that the United States was considering lifting sanctions against Cuba selec-tively and would suspend RB-71 spy-plane overflights over the island dur-ing preliminary contacts.

On July 29 the United States joined a majority of OAS members at a conference in San José, Costa Rica, to abolish the collective embargo on economic and political ties with the Cubans. This followed fairly detailed secret conversations conducted with Cuban emissaries, who had step-by-step personal instructions from Castro. The American negotiator was Assistant Secretary of State for Inter-American Affairs William D. Rogers, a highly respected international law practitioner, whose cloak-and-dagger meetings with the Cubans ranged from the Pierre Hotel in Manhattan to coffee shops at New York's La Guardia Airport to National Airport in Washington. The most promising aspect of this effort was that both sides agreed in principle that there should be no preconditions to actual negotiations, everything being negotiable. Ramón Sánchez-Parodi, the Cuban emissary, advised Rogers that Castro did not insist that the United States end the economic embargo before negotiations began, a very major change in his stand. Rogers told the Cubans that the fate of the Guantánamo naval base was negotiable as well.

When on August 9 Castro returned to Southern Airways the $2 million ransom the Cuban government had collected three years earlier for a hijacked plane, the administration read this as a positive signal. Ten days later, the State Department announced that American firms based in foreign countries would be allowed to do business with Cuba for the first time in twelve years. This measure had been under consideration for some time, but now it was meant as a firm diplomatic gesture toward Castro. Late in September, Rogers announced that the United States was prepared "to improve our relations with Cuba" and "to enter into a dialogue." But then the whole effort went off the track. In Angola the Cuban military presence assumed large proportions, eliciting public criticism by Kissinger. In Havana Castro chose that particular period to sponsor a "Puerto Rican Solidarity Conference," urging Puerto Rican independence as "a matter of principle," knowing perfectly well that it was an issue calculated to provoke United States anger. At the United Nations, Cuban Ambassador Ricardo Alarcón de Quesada tied it all together in an October speech declaring that "Cuban solidarity with Puerto Rico is nonnegotiable" and "it is Cuba's duty to give effective support to MPLA in Angola."

Late in November, Rogers and Sánchez-Parodi had their last secret meeting. The diplomatic enterprise had collapsed and the traditional hostility returned. The American side claims the Cubans decided to break off the preliminary negotiations without providing a clear explanation; the suspicion is that Castro felt it was impossible to enter into full-fledged negotiations with Washington while the war in Angola was raging. The Cubans say that it was the Americans who severed the contacts out of anger over the Puerto Rican Solidarity episode in Havana, but they do not explain why Castro had so blatantly reopened this extremely controversial issue if indeed he was serious about talks with the United States. Fidel frequently chooses not to explain his actions even to his closest associates, but late in 1985 he told an American visitor—without providing any explanation for the breakdown—that the Kissinger diplomatic effort ten years earlier had come the nearest to a real breakthrough in the whole history of postrevolutionary relations between the United States and Cuba.

The first great milestone in Fidel Castro's revolution were the victory in 1959 and the creation of the new Communist party in 1965 as the ruling political body in Cuba. The next milestone was the "institutionalization of the Revolution," as Castro called it, through the promulgation of a new Cuban constitution on February 24, 1976. Over the past seventeen years, the "Fundamental Law," drafted immediately by the first revolutionary government, and literally thousands of laws and regulations formed the juridical framework of the Cuban state, though no doubt ever existed as to where actual power reposed.

Nevertheless, laws had to be refined, revised, and codified, and, as much as anything else, full legitimacy had to be granted to the socialist character of Cuba and its Communist objectives. Consequently, in October 1974, Blas Roca, secretary general of the old Communist party and a member of the Politburo of the new party, was named chairman of a commission charged with the task of drafting the new constitution. The draft was published six months later, then it was submitted for discussion to millions of Cubans in party and military organizations, labor unions, and youth and women's groups. Clearly, no basic changes resulted from these discussions, which Castro regarded as "direct democracy," but—quite surprising to the regime—people wanted that democracy to be even more direct.

An innovation in the constitutional draft was the creation of a type of local self-government called "Popular Power" (no such governmental structure exists in other Communist countries) capped by a National Assembly with legislative functions described as the "supreme organ of state power." A profound division developed, however, over the method of selecting deputies to the National Assembly. Some advocated direct elections and some favored choice by municipal Popular Power assemblies. In the first instance, voters would at least potentially have a voice in the formulation of main national policies, and the decision-making process would have to be made reasonably visible to the public. In the second instance, membership in the National Assembly could be determined through political manipulation on the local level with candidates nominated from the municipal assemblies or by party and government officials. Inasmuch as the draft also provided for National Assembly deputies "to explain the policy of the state and periodically render account to [the electors]," direct elections could have been disastrous for the central government. So bitter was the internal dispute over this point that the constitution, on which a popular referendum was held on February 15, 1976, failed to spell out the method of election. Only *after* 97.7 percent of the voters had approved the charter, did the Central Preparatory Commission, headed by Fidel Castro, insert the provision that "the National Assembly . . . is composed of deputies elected by the Municipal Assemblies." This was the end of the first and last major attempt to democratize Cuban Marxism.

The constitutional referendum was preceded in December 1975 by the first Congress of the Cuban Communist Party, chaired by Castro and attended by Mikhail Suslov, the chief Soviet Communist ideologue and one of the most powerful members of the Politburo. Castro's report to the Congress was a 248-page document (in book form) chronicling the history of Cuba, the *Fidelista* revolutionary movement, its transformation into socialism, the first ten years of the Communist party, the achievements of the revolution, and the Cuban struggle against "imperialism" and in support of "liberation movements." Although Cuban troops were fighting in Angola even as Castro addressed the Congress, his report made no mention of it; he simply alluded to "the recent constitution of the independent republic of Angola, under the direction of the MPLA, in the midst of strong and heroic struggle against imperialism." The Angolan war was not yet a public event as far as Cuban opinion was concerned.

The constitution itself defined Cuba as a "socialist state of workers and peasants and all other manual and intellectual workers," and the Communist party was "the highest leading force of the society and of the state, which organizes and guides the common effort toward the goals of the construction of socialism and the progress toward a Communist society." First, it hailed José Martí, who "led us to the people's revolutionary victory," then Fidel Castro, under whose leadership the "triumphant revolution" was to be carried forward.

Thus enshrined in the constitutional text, Castro was in effect named Leader for Life as a law; the corollary was that it would be unconstitutional (and not just "counterrevolutionary") to challenge him. Pursuant to constitutional provisions, the National Assembly then elected a thirty-one-member Council of State with Fidel Castro as president and Raúl Castro as first vice-president. As president of the council, Castro became "the Head of State and Head of Government." Total power was therefore legally vested in him as President of Cuba and chairman of the Council of Ministers as well as first secretary of the Communist party and military Commander in Chief.

There was no specific succession procedure, but Raúl was the first vice-president of both the Council of State and the Council of Ministers, the second secretary of the Communist party (no other Communist party in the world has such a post), and defense minister; the rank of General of the Army was also created for him. Succession was thus automatically resolved, and Fidel remarked once in absolute seriousness, "The creation of the institutions has assured the continuity of the Revolution" after his death. He added straightfacedly that he was not really needed anymore, explaining over the years that Raúl was his successor (automatically) because he had the leadership qualities, not because he was his brother. The faithful Dorticós was demoted from the presidency of Cuba to a ministerial post (he later committed suicide).

The 1976 constitution and the Cuban Communist party's First Congress established the permanent character of the Cuban revolutionary state, ruling out any basic structural or ideological changes in the future—barring cataclysms. As time went on, normal societal requirements would be reflected, but the existing structure or philosophy of the state would never be affected. In this sense, the future of Cuba was set in granite. By 1986, after two more quinquennial Communist party congresses, everything remained the same, with Fidel Castro the only and final authority and arbiter of every decision taken in Cuba. The National Assembly held its two annual sessions as prescribed by the constitution, but each session lasted only two or three days.

The decade between 1976 and 1986, the year when Fidel Castro celebrated his sixtieth birthday and the twenty-seventh anniversary of his revolution, was devoted to immense activity in foreign policy and to continually frustrated efforts to energize and organize the Cuban economy and improve the quality of life of the island's ten million citizens after basic needs of health and education had been met equitably. Internationally, Castro scored more successes than defeats, remaining as defiant as ever and gaining a considerable degree of world acceptability and respectability. In fact, his only defeats since the 1970s, when Allende was overthrown in Chile, were the electoral ouster of his friend Prime Minister Michael Manley in Jamaica, and the American invasion of Grenada, where Castro had had great hopes of expanding Cuban influence in the eastern Caribbean.

Grenada was a bitter blow to Castro because his combat platoons and armed workers there hardly resisted the American invasion and were quickly defeated. Castro demoted his military commander and his ambassador in Angola to lowly occupations. At the Havana airport, when the bodies of dead Cubans arrived from Grenada, Fidel stood alone for a long moment of meditation, his shoulders hunched.

With the United States, there were gains and losses. Negotiations with the Carter administration led to the establishment of diplomatic "interests sections" by the Cubans in Washington and by the Americans in Havana in 1977. These "sections," euphemisms for embassies in the absence of actual diplomatic relations, provided instant channels of communication between the two governments. The Carter and Reagan administrations kept the chief of the Cuban interests section at arms' length, but Castro made a point of giving quasi-ambassadorial treatment to the senior American diplomat in Havana, inviting him to palace receptions. When he was informed in 1985 that a new chief of the American interests section had been appointed, he showed great curiosity, asking visiting Americans and diplomats what sort of person the new *Americano* was. Talks with the United States in 1978 led to an agreement that allowed exiles on the mainland to visit their families in Cuba. Tens of thousands took advantage of this agreement (it was suspended in 1985 when Castro became annoyed with the Reagan administration over the establishment of the antiregime Radio Martí operated by the Voice of America). Political negotiations with the Carter administration, personally orchestrated by Castro, fizzled out, and in 1980 the exodus of over 120,000 Cubans to Florida in small boats from the port of Mariel put an end to the diplomatic contacts. Castro encouraged their departure out of anger over an unguarded remark by President Carter that the United States awaited Cuban political refugees with "open arms."

Castro's great moment in the international sun came when he was elected chairman of the Nonaligned Movement for 1979–1982, assuming the formal Third World leadership to which he had so long aspired. He hosted the Nonaligned Movement's summit conference in Havana in September 1979, at which ninety-two heads of state or their representatives were present—from the Communist octogenarian Marshal Tito of Yugoslavia to the deeply religious Islamic president of Pakistan, Mohammad Zia ul-Haq, and Prime Minister Indira Gandhi of India—and he succeeded at all times in keeping the spotlight on himself. Events were going his way everywhere. In March of 1979, a pro-Cuban regime was established on the tiny island of Grenada in the eastern Caribbean by his friend Maurice Bishop, a leftist lawyer and politician of vast personal appeal. In July the *Sandinista* Liberation Front rebels ousted the Somoza dictatorship in Nicaragua; Castro knew Carlos Fonseca, the founder of the *Sandinista* movement who was killed in 1966, well and he provided considerable assistance to the rebels after forcing them to unite in a common front. Now Cuba had revolutionary allies in Central America and on the outer fringes of the Caribbean, just a hop away from strategic Venezuela.

In October Fidel Castro returned to New York for the first time in nineteen years to address the United Nations General Assembly as chairman of the Nonaligned Movement. It was a moment of vindication he savored: In 1960 he had sought refuge in a hotel in Harlem, and now he was spending three days in New York at the twelve-story midtown building belonging to the Cuban mission to the United Nations (complete with living quarters and a school) that Cuba had just purchased for $2.1 million. The Cuban mission employed the third largest number of officials after the United

States and the Soviet Union. Reminiscing about this latest visit to New York and the reception he had held at the Cuban mission, Fidel told me with immense satisfaction, "that first time in 1960, we lived on chicken . . . this time I brought my own lobsters, my own rum . . ." In his two-hour speech at the General Assembly on October 12, Castro—acting as spokesman for the destitute Third World—urged the United States and other "wealthy imperialists" to grant the underdeveloped nations $300 billion over ten years. "If there are no resources for development, there will be no peace," he said, "and the future will be apocalyptic." The fate of the Third World, including its gigantic debt to the industrial nations, became the centerpiece of Castro's foreign policy in the Eighties: He fervently championed it at the 1982 Nonaligned Movement's summit meeting in New Delhi, and in an extraordinary antidebt offensive he mounted in Havana in 1985.

Economic issues were one aspect of Castro's ever-growing "internationalism." Another was military participation in revolutionary confrontations across the globe. From the initial involvement in Angola in 1975, Castro moved on to Ethiopia in 1978, dispatching nearly twenty thousand combat troops to assist the new Marxist regime of Lieutenant Colonel Mengistu Haile Mariam repulse an attack by Somalia on the contested Ogaden region. As it had done in Angola, the Soviet Union provided arms and advisers; this was the second Soviet-Cuban joint military venture. In Third World policies, Castro and the Russians were totally on the same wavelength. When the Soviet Union invaded Afghanistan in December 1979 to save "socialism" in an Asian rerun of applying the Brezhnev Doctrine to Czechoslovakia, Castro stood foursquare behind this action. He rationalized it even though the fierce Afghan resistance to the Russians created great embarrassment for Castro as the Nonaligned Movement's chairman, especially with Moslem nations. But he could also rationalize switching Cuba's early support for the Eritrean secession movement from Ethiopia to the other side once President Mengistu became his ally. Again Castro's activities threatened to create a Soviet-American confrontation. The presence of Soviet-supported Cuban forces in the crucial Horn of Africa nearly led the Carter administration to break off Strategic Arms Limitation Talks with the Soviet Union. Fidel was always in the limelight.

Castro has never made a secret of his support for the *Sandinista* movement before or after its triumph. He has given Nicaragua, his revolutionary junior partner, maximal support in military advisers and civilian technicians. The Cubans had trained the *Sandinistas* in military camps in Pinar del Río and on the Isle of Youth, and it was Fidel who took it upon himself in 1978 to bring together rival factions among Nicaraguan rebels. Unless they were united, he told them at a secret meeting in Havana, Cuba would not supply arms to them in their "final offensive" year. Since the *Sandinista* victory in 1979, Castro continues to stage-manage the Nicaraguan revolution from Havana; Nicaraguan President Daniel Ortega Saavedra is a very frequent guest in Cuba, sometimes publicly, sometimes secretly.

Castro fully supports the leftist guerrillas in El Salvador (he also supports the M-19 guerrillas in Colombia), but he knows that Cuba cannot protect the two Central American countries from a direct United States attack (and he realizes that the Soviet Union would not commit itself to a military defense of Nicaragua as it is committed to do in Cuba). He believes that political settlements are possible both in Nicaragua and El Salvador because endless stalemate is the alternative. But he is also aware that in a total crisis, the United States might try to annihilate him. His concern is heightened by early Reagan administration threats to go "to the source" of Central American upheavals, which it believes is Cuba—and by the invasion of Grenada in 1983.

Early in 1984, in a conversation about possible American military action in Central America and beyond, Castro told me: "We have no means to be able to decide militarily the events there. All our means are defensive. We have no fleet and no air force capable of neutralizing or breaking a blockade by the United States. It's not a question of options, it's a practical question. . . . Besides, from a political viewpoint it would be improper for us to attempt a military participation under such circumstances because it would be justification before American public opinion for a United States agression." Replying to my question whether he was concerned about an American invasion, Castro said: "We have made great efforts to strengthen our defenses. After Grenada, we have made even greater efforts. We are increasing considerably our defense and resistance capability, including the preparation of the people for a prolonged, indefinite war. If the United States deterrent, as the Reagan government has said, is the nuclear force, our deterrent is to make it impossible for this country to be occupied, for an occupation army to be able to maintain itself in our country. First, it would be necessary to fight very hard to occupy our country. But the occupation of our country would not be the end, but the beginning of a much harder and much more difficult war, in which we would be victorious sooner or later at an enormous cost."

Castro still believes that the Salvador civil war can be settled politically by negotiation, and he does not hide the Cuban support for the leftist guerrillas there. Just as he knew most of the Nicaraguan leaders prior to the 1979 victory, he is acquainted with the Salvadoran guerrilla leaders, having tried hard to impose unity on their factions. During a conversation with me early in 1984, he indicated that these leaders had been visiting him in Havana, saying that "I haven't spoken in many months with the principal leaders of the revolutionary movement in El Salvador because all of them are now inside their country, and it's not possible to have any direct contact with them."

In Castro's Third World vision and policies, doctors and teachers are as important as combat troops, and he takes immense pride in explaining that Cuba is the only developing nation willing and ready to help others. He says that the new Cuban generation has an internationalist spirit not found elsewhere. In a conversation about this internationalism, he told me: "Look—when after the triumph of the revolution in Nicaragua we were asked for teachers, there were twenty-nine thousand volunteers. . . . In the beginning, we had no doctors to send to the interior of our country. Today, we have doctors in more than twenty-five countries of the Third World— more than fifteen hundred doctors working in the Third World. And there will be more because we are graduating two thousand doctors annually. It is a new culture, a new morality. . . . It is amazing: You go to our universities, and one hundred percent volunteer for any task. When we needed volunteers to go to Angola, three hundred thousand of them responded. When we needed volunteers for Ethiopia, more than three hundred thousand responded. They were civilian reservists. Now, hundreds of thousands of Cubans have fulfilled internationalist missions. People ask why there are two thousand Cuban teachers in Nicaragua, but who else will do the work that Cubans perform there? How many [people] in Latin America are prepared to go where our teachers go to live with the poorest families, to eat what the poorest families eat themselves, to teach there? You won't find such people. . . . We have more people disposed to go to any place in the world as doctors, as teachers, as technicians, and as workers than the Peace Corps of the United States and all the churches together—and we are a country of only ten million inhabitants."

Since 1985, Castro has devoted an astounding amount of time to the problem of the Third World's debt to the banks and governments of industrialized countries, arguing that the destitute debtors cannot pay what they owe without destroying their economies, and predicting dire consequences if efforts are made to collect the money. In 1986 Latin America alone owed more than $350 billion (much of this to United States banks), and the Third World total debt stood near $750 billion. In speeches, interviews, and at conferences he had sponsored in Havana, Castro turned the debt issue into a powerful political instrument in battling "imperialism," but his efforts have had the positive impact of calling international attention to the gravity of the crisis. Staying excellently informed about all Latin American developments, he was, in effect repeating in 1985 and later exactly what he had done in 1959: warn the industrialized countries of the north that an explosion was in the offing if they did not decisively attack the roots of hemispheric problems. In 1961, President Kennedy took up Castro's challenge by launching the Alliance for Progress; in the late 1980s the "rich" governments would do well to listen again to this Cuban Cassandra.

At home, however, Castro is much less bold and imaginative. He stubbornly refuses to relax the harsh standards of totally centralized planning (even when most of the Communist world has discovered the merits of relative decentralization). Because of his orthodox ideological inflexibility, Castro sees heresy in any attempt to experiment with market forces. Even though the Soviet Union has begun to establish joint industrial production ventures with Western capital, and in 1979 China embarked on a market-economy policy—including allowing private retail stores—Castro will not be budged. The chronic and alarmingly deficient performance of the Cuban economy seems to strengthen his resolve to be faithful to orthodoxy. In June 1986—sounding as he did during the "revolutionary offensive" in 1968—Castro denounced "certain concepts [proposed] by persons, supposedly very Marxist and very versed in Marxism, but really with a capitalist or petit-bourgeois soul." Ideological and political controls over what must be the world's most indoctrinated society tend to tighten rather than relax after the twenty-seven years of revolution.

Even some of Castro's close associates are at a loss to understand the reasons for his new hard-line attitude during the mid-1980s. After the Mariel exodus of 120,000 Cubans in 1980, Castro appeared to have concluded that the nation required a certain relaxation of tensions—the Mariel experience was a trauma because the regime was taken aback by this manifestation of internal resentments—and must be allowed a degree of consumer freedom. Accordingly, many food items were released from rationing and made available at "free stores" for extremely high prices, and uncontrolled farmers' markets were authorized to sell produce in the cities. It was far from a bonanza because foodstuffs remained in short supply as a result of inadequate production, but it seemed a small step toward a liberalization of the economy within Marxism.

As preparations were made during 1985 for the Third Congress of the Communist party, many senior economic planners hoped for still more liberalization and a new policy of decentralization. There was talk about allowing private owners operating through cooperatives to take the government out of the business of running taxis all over Cuba, and about ending clothes and footwear rationing, which had created a state monopoly over shoddy products at an immense cost to the treasury (Cuban women increasingly preferred to have private seamstresses make their dresses). Ideas also circulated about abandoning the ideologically designed youth work brigades whose weekend activities cost more in blankets, boots, mosquito netting, food, and transportation than they brought in farm production. But Castro evidently would have none of it, reemerging as the fierce apostle of the pure revolution while the economy kept deterioriating.

Presumably because Fidel was unable to formulate an economic plan for the next quinquennial party Congress—in part because he was busy presiding over international conferences on Third World debt—the Congress was postponed to February 1986. But even then Castro was still not ready to deal with the key issues, so the most important work of the Congress was postponed till the end of 1986. According to Castro, the most significant achievement of the first part of the Congress was to rejuvenate the leadership. Even this, however, was an illusion. The Politburo contained the same old faces, although several cosmetic changes were made. The new Central Committee was an assemblage of middle-aged men and women drawn from the bureaucracy, party organizations, the armed forces, and Security Services. In a nation where more than half the population was born after the revolution, only 9 percent of the Central Committee membership were under thirty-five; more than 50 percent were over the age of forty-six. Only 18.2 percent were women. And 78.1 percent had a university education, all of which suggests that the party created a new ruling elite, heavy on bureaucrats and administrators (27.5 percent) and full-time party officials (27.1 percent). Twenty percent of the seats went to the armed forces and Interior Ministry Security Services.

In 1986, Fidel Castro imposed an ossification of the regime and society under the guise of keeping the revolutionary fires burning. Almost immediately after the first session of the party Congress ended, Castro ordered the closing of the farmers' markets on the grounds of illicit enrichment and corruption. It was strangely reminiscent of his discovery in 1968 that privately owned hot dog stands were hotbeds of counterrevolution. At the same time, Castro moved even further ahead with the militarization of the Cuban society through the expansion of the Territorial Troop Militias, a highly trained reserve organization exceeding one million people—10 percent of the population. The Militias' 340-page illustrated *Basic Manual* is must reading in Cuba (though it sells for one peso in bookstores), and exercises and maneuvers are conducted continuously in "defense zones" into which the island has been divided. The ever-present threat of invasion by the United States has always justified a high degree of preparedness in Cuba, especially after the Grenada incident, but militarization in the mid-1980s is as much as anything, a political move to strengthen the cohesion of the revolutionary society under Castro's extremely strict leadership.

Castro's revolution was again in trouble with the Soviets in the mid-1980s, chiefly because of Cuba's economic waste and inability to meet its sugar delivery commitments. Soviet emissaries had warned him frankly that this state of affairs could not continue indefinitely, and when the 1985 trade agreement was finally signed in the middle of the year, the communiqué published in the Cuban press said the accord had been reached after "long and difficult negotiations," a most unusual phrase. In fact, Castro had to take the unprecedented step in 1985 and 1986 to purchase for cash sugar from the Dominican Republic to keep going the shipments to the Soviet Union. However, he was able to pay the very low world price in dollars to the Dominicans while being credited by the Soviets for this sugar at a very high, subsidized price.

Politically, Castro shows total deference to the Russians. He dutifully attended the Soviet Communist Party Congress in 1986, holding his first meeting with Gorbachev since the latter became the top Kremlin leader. Then, he found it necessary to visit North Korea to meet Kim il-Sung, the senior Communist dictator in the world who presides over the most repressive Communist society anywhere. His Moscow and Pyongyang speeches sounded like carbon copies of every communist speech delivered that year in the Soviet sphere of influence. Fidel seemed to join the ranks of the great conformists.

Who is Fidel Castro at the age of sixty? In the immediate sense, he is the undisputed and still enormously popular (and even loved) leader of an extremely volatile nation, which he has led for more than a quarter-century to a place in the sun in world affairs and toward what he has called "a life of decency." A grandfather himself, he sees to it that in Cuba all children are healthy and clean and well educated. In 1985 the national goal was a ninth-grade education for every Cuban. None of this was true a generation or so ago. It is a great accomplishment in any society, underdeveloped or industrial. No Third World country approaches Cuban standards in the area of that decent life. None have a higher doctor-population ratio than Cuba or a greater longevity expectation at birth.

But there are other aspects to Castro, and disturbing ones when one takes his intelligence and experience into account. Since around 1980 his behavior suggests that he has few fresh ideas for his aging revolution. Curiously, this man of astonishing daring and imagination and romanticism is allowing—or forcing—his beloved social and human experiment to be locked into obsolete ideological orthodoxy and deadening bureaucratization. His sallies against "bourgeois" tendencies sound quaintly antiquated, if not slightly caricatural, in a world that has changed so much since such expressions were still in vogue. Considering the extent to which national creativeness has been blunted (one hopes not buried) in the name of conformism, Castro faces the danger that his revolution may be decaying.

In the physical sense, much of the early revolutionary construction is already decaying: the paint is peeling off and windows are broken at the great and admirable Camilo Cienfuegos school complex at the foot of the Sierra Maestra, and masses of costly imported equipment are destroyed by the Caribbean weather because the regime remains unable to solve the problem of unloading ships and loading trucks. These are obvious examples: There are others. Mismanagement discourages work and production, and resulting shortages aggravate the problem of low productivity and high absenteeism. It becomes a vicious circle. Bureaucratic corruption and black marketering in the streets are reemerging like a cancer on the body of the revolution that was born so pure, and Fidel Castro inveighs in rage against "vile money." Young people drink too much because there is little else to do in their spare time, and they are not touched by the mystical magic of the revolution as their fathers and mothers were when Fidel Castro was an obsessed young rebel.

He is furious over the immense rate of absenteeism among Cuban workers, low productivity, shoddy quality of industrial goods (in the province of Havana in 1985, one half of the soft drink and beer bottles produced in the first six months were unusable and had to be destroyed), and the shocking waste of materials and resources in industrial plants, farms, and government offices. At the Third Congress of the Communist party in 1986, Fidel lectured and chastised his fellow citizens in a speech lasting five hours and forty minutes. However, he rejects any suggestion that it is the over-centralized system of government and management that is at fault, allows no basic structural changes, and resents outside criticism as "counterrevolutionary." Again, such absolutism in a man who does understand the workings of Cuban society is perplexing, and raises the question of whether he has lost contact with his people on this small island.

Does Fidel Castro have doubts and fears, and can he share them with another human being? Celia Sánchez died of cancer in January 1980, and her passing was not only a personal and emotional tragedy for Fidel, but it deprived him of a safe haven—the opportunity occasionally to be himself as Fidel and not as Commander in Chief. Watching him, almost motionless, hour after hour, listening to hundreds of speeches at the conferences on external debt he had organized in Havana during 1985, it seemed as if Castro was seeking refuge in this environment from pressures elsewhere; perhaps it was an illusion. Still, in a variety of surroundings, Fidel Castro appears a lonely man—frustrated one day, triumphant the next, but lonely, and still searching for something that is impossibly elusive. He might be pondering about the past and the future—and about the verdict of the generations to come. Indeed, he expressed this concern in two sentences at two very crucial moments of his life:

Addressing the judges trying him for the assault on Moncada, the moment of the revolution's birth in 1953, Fidel Castro said: "Condemn me. It does not matter. History will absolve me!"

Speaking about the creative process to Cuban artists and writers in 1961, the year he triumphed at the Bay of Pigs and declared himself a Marxist-Leninist, Fidel Castro said: "Do not fear imaginary judges we have here. . . . Fear other judges who are much more fearsome, fear the judges of posterity, fear the future generations who, in the end, will be responsible for saying the last word!"

Notes

In writing *Fidel, A Critical Portrait,* I have relied on four principal categories of source material, which are discussed here in detail. The four categories are: interviews and conversations in Cuba; interviews and conversations in the United States; Cuban newspapers and periodicals, before and after the revolution; and books and other texts published in Cuba and elsewhere.

I believe that an explanation of the sources and material will be more useful to readers and other researchers than the masses of footnotes that would otherwise have been required. Moreover, some passages and sections are composites of two or more interviews or conversations, occasionally incorporating information from published sources, and it would be neither practical nor helpful to try to identify individual references and quotations. Except for specific identification in the actual text, no data are provided on Castro speeches to which references or allusions are made; given the extraordinary number of speeches he has delivered in his life (plus the fact that some of his texts are not available at all, and some only in the form of fragments), it would have been an onerous and pointless task to catalogue all the references.

Finally, the Chapter Notes are provided as a general guide to the *Portrait* in terms of identifying types and groups of the source material.

I. Interviews and Conversations in Cuba

The bulk of the original material in *Fidel, A Critical Portrait* came from tapes as well as informal interviews and conversations I conducted in Cuba in 1984 and 1985—including long discussions with President Castro in both years. In his case, I have also drawn on my notes from interviews and conversations we had during 1959, and in the course of a tour of the Bay of Pigs battlefield in his company in June 1961.

Formal, taped interview sessions were held during 1985, with seventeen close associates and comrades of Fidel Castro; seven of them were interviewed on two or more occasions. All the interviews were conducted in Spanish, and I translated into English from tape transcripts the portions and quotations appearing in this book. Following is the list of these interviews, with the identification of the subjects:

José R. Fernández, vice-president of the Council of Ministers; minister of education; Alternate Member of the Political Bureau of the Cuban Communist party

Carlos Rafael Rodríguez, vice-president of the Councils of State and Ministers; member of the Political Bureau

Vílma Espín Guilloys, president of the Federation of Cuban Women; member of the Council of State and Political Bureau (wife of Raúl Castro)

Pedro Miret Prieto, vice-president of the Council of Ministers; member of the Council of State and Political Bureau

Armando Hart Dávalos, member of the Council of State; minister of culture; member of the Political Bureau

José R. Machado Ventura, member of the Council of State; member of the Political Bureau and of the party secretariat

Jorge Enrique Mendoza Reboredo, editor of the Communist party organ *Granma;* member of the party's Central Committee

Ramiro Valdés Menéndez, Commander of the Revolution (one of three Rebel Army officers honored with this title); member of the Central Committee; until 1986, vice-president of the Councils of State and Ministers, interior minister, and member of the Political Bureau; demoted without explanation by the Third Party Congress

Guillermo García Frías, Commander of the Revolution; member of the Central Committee; until 1986, vice-president of the Councils of State and Ministers, minister of transport, and member of the Political Bureau; demoted by the Third Party Congress

Fabio Grobart, cofounder of the Cuban Communist party in 1925; member of the Central Committee; chairman of the Institute for the Study of Marxist-Leninist Movement in Cuba; Hero of the Cuban Revolution

Blas Roca Calderío, member of the Central Committee; until 1986, vice-president of the Council of State and member of the Political Bureau; retired because of age by the Third Party Congress; secretary general of the Popular Socialist Party (Communist) until the creation of the new Cuban Communist party under Fidel Castro in 1965

Melba E. Hernández, member of the Central Committee; Heroine of the Cuban Revolution (one of two women participating in the Moncada attack)

Faustino Pérez Hernández, member of the Central Committee; cabinet minister in the first revolutionary government; one of Fidel Castro's two companions after the Alegría de Pío defeat

Universo Sánchez, director of the Environment Protection Office; one of Fidel Castro's two companions (with Faustino Pérez) after the Alegría de Pío defeat

Antonio Nuñez Jiménez, vice-minister of culture; former executive director of INRA; chronicler of Fidel Castro's postwar years

Alfredo Guevara, Cuba's Ambassador to UNESCO; Fidel Castro's university friend; companion at Bogotá uprising in 1948

Conchita Fernández, Fidel Castro's private secretary in the postwar years

Additionally, interviews were taped with Norberto Fuentes, a Cuban journalist and chronicler of the revolution, and with seven citizens in the villages of the Sierra Maestra who played central roles in assuring the survival of Fidel Castro and his companions during the first year of the guerrilla war (these accounts total 210 pages of transcript).

All the above tapes and transcripts were donated to the University of Miami in Coral Gables, Florida.

Taped interviews with President Castro were conducted in and near Havana in January 1984, and the transcripts total 315 pages. Interviews and conversations in February and May 1985 were of an informal nature, not being taped. In every case, I made detailed notes after these meetings. Transcripts of the 1984 Castro interviews also are at the University of Miami. Taped interviews conducted in Cuba added up to 1,964 pages. Apart from the 1985 conversations with President Castro, literally scores of other informal discussions with Cuban personalities would have provided thousands more pages of transcripts; the material was preserved in my notes—some of which must remain confidential.

Among these informal conversations, mention should be made of many meetings with Eugenio Rodríguez Balari, president of the Institute of Consumer Affairs; Pedro Álvarez Tábio and Mario Mencía, historians of the pre-1959 period of the revolution; Ernesto Guevara Lynch, the father of the late Che Guevara; Manuel Moreno Fraginals, a leading Cuban historian who knew Fidel Castro as a student; Pastor Vega, a major Cuban cinema director; Luis Baez and Gabriel Molina, editors of *Granma;* and Lionel Martin, an American author and journalist who has lived in Cuba since 1961. Many conversation partners to whom I am greatly in debt for their time and patience, will not be mentioned here in order to respect their privacy. Among my American friends, Henry Raymont, a colleague and an outstanding expert on Cuba, was a very special adviser.

I have dwelt so much on interviews and conversations in Cuba because they are crucial in any historical or biographical undertaking about the revolution and its figures. This is so because no coherent or comprehensive body of revolutionary literature or history exists in Cuba. As will be seen in the section on books in these Notes, the available works are fragmentary, and historical objectivity is not their best facet. Within the senior ranks of the Cuban bureaucracy, there is strong opposition to a historical reconstruction of the revolution, particularly when it is attempted by foreigners. Ironically, an example of such resistance is the Historical Division of the Council of State, under the Council's secretary, Dr. José M. Miyar Barrueco. For this reason, interviews become absolutely essential, and it must be emphasized that in my case the access resulted from personal instructions by President Castro.

II. Interviews and Conversations in the United States

As in Cuba, interviews in the United States, principally with Cuban exiles, were essential in reconstructing much of the life of Fidel Castro. Most of them were conducted in Miami, touching chiefly on Castro's youth.

José Ignacio Rasco and Juan Rovira provided important insights on young Fidel at the Belén College in Havana. Rasco, Enrique Ovares, and Max Lesnick contributed information on Castro's university days. Ovares was an invaluable source on Castro's involvement in the Cayo Confites expedition and the Bogotá uprising (he was there with Fidel and Alfredo Guevara). Max Lesnick's recollections and interpretations formed a bridge between Castro's activities at Havana University and the start of his revolutionary career. Raúl Chibás, the brother of the late Eddy Chibás, the founder of the *Ortodoxo* party in 1947, spent several days with me in Miami, sharing his memories of Fidel Castro. He knew Castro as a political leader at the university, a young *Ortodoxo* politician, and then as the chief of the revolution. Chibás was twice in the Sierra Maestra with Castro, serving as the treasurer of the 26th of July Movement, and, later, as director of Cuban Railways.

There are 329 pages of transcripts of these taped interviews as well as tapes at the University of Miami.

Other important interviews in Miami and Washington, D.C., were informal in character, and the material was consigned to my notebooks. Most of these conversations were confidential.

III. Cuban Newspapers and Periodicals

The Cuban press before the revolution was extremely important not only in re-creating the political mood of the era, but also for tracking Fidel Castro's public career. The first reference to him in a Havana newspaper appeared in 1944, and news stories about him and articles by him became increasingly frequent over the years, up to the eve of his victory.

This material is identified in the *Portrait* whenever it is relevant; it did not seem useful to include it in the Chapter Notes. The best Cuban press sources were the weekly magazines *Bohemia* and *Carteles,* and the newspapers *Diário de la Marina, El Mundo, El País, La Calle,* and *Alerta.* Nearly complete collections of these publications are at the Library of Congress in Washington, the University of Miami, and other universities in the United States. Prerevolutionary university publications, such as *Saeta* and *Mella,* are very difficult to locate. Newspaper collections from the prerevolutionary period are to be found at the José Martí National Library in Havana, but they are very incomplete, and access is controlled by the authorities. Moreover, no copying equipment is publicly available in Cuba; research in this area is incomparably easier in the United States. Prerevolutionary university publications seem to have been lost, and Cuban researchers seek to obtain copies from the United States.

For the first revolutionary period, the principal Cuban newspaper sources are *Revolución* of the 26th of July Movement and *Hoy,* the Communist organ; they were merged into *Granma* in 1965. All the other Cuban daily newspapers disappeared by 1961. *Revolución, Hoy,* and *Granma* are useful as repositories of speeches by Castro and other revolutionary figures, and for official news, reportage, and editorials. *Bohemia* is the most interesting magazine, publishing fragments of revolutionary history. *Verde Olivo* of the Ministry of Revolutionary Armed Forces and *Moncada* of the Ministry of the Interior provide much of the ideological line. References to items from these publications are made in the *Portrait* when relevant. Fairly complete collections of Cuba's postrevolutionary press exist at the Library of Congress, the University of Miami, etc.; in Havana, access to this material is rather difficult. The few extant collections of *Lunes de Revolución,* the first-rate literary supplement launched in 1959 and stopped in 1961, are in private hands in Cuba. They exist in the United States.

IV. Books

A bibliography at the end of this volume lists the most helpful and interesting Cuban, American, and European books on Fidel Castro and Cuba. Most of them are obtainable at the Library of Congress and university libraries in the United States.

As noted above, however, there is an astonishing paucity of serious and useful and up-to-date books about the Cuban revolution—and especially about Fidel Castro. His own speeches are available in a large number of incomplete or excerpted editions in Cuba and the United States, but none of these are much help to a biographical researcher.

Most of the material about Castro is confined to the previctory period. In my opinion, the most valuable book is *Moncada: Premier Combat de Fidel Castro,* by the French biographer Robert Merle, published in 1965 (and sadly not available in English). Merle interviewed the Castro brothers and most of the survivors of the Moncada attack when their memories were still fresh. Lionel Martin, the American journalist who knows Castro as well as any foreigner, has written a helpful account of his youth and war years in *The Young Fidel,* but historians may find problems with Martin's ideological interpretations, especially on Castro's relations with the Communists; still, it is the only chronicle of this kind. The very long introduction to Castro's *Selected Works, 1947–1958* is a helpful sketch of Fidel against the background of his time.

Fidel Castro himself tells interesting tales about his childhood and youth in *Diary of the Cuban Revolution* by Carlos Franqui, the first editor of *Revolución,* who taped a series of interviews with him in 1959. Franqui's volume, which is essential for the study of the Sierra War, also contains important correspondence of Fidel Che Guevara, Raúl Castro, Celia Sánchez, and others. In 1985, Fidel Castro gave a series of extremely lengthy interviews to a Brazilian Dominican friar, Frei Betto, on the subject of religion. Published in book form in Havana and Rio de Janeiro, these interviews provide fascinating glimpses of Castro's young life as seen by him; references to this text are identified in the *Portrait* and the Chapter Notes.

Mario Mencía, the Cuban revolutionary chronicler, is the author of two very readable books: one, the account of the preparations for the attack on Moncada, and the other, the story of Fidel Castro in prison (the latter is available in English as *Time Was on Our Side).* Inevitably, they reflect a strong ideological bias, but nothing better is available, Mencía's third book, covering the period between Fidel's imprisonment and the landing in Cuba, has not been issued as of 1986, although it is completed. Marta Rojas, a Cuban journalist who covered Castro's trial in 1953, has written three books on Moncada and the judiciary proceedings that offer useful facts and impressions. Antonio Nuñez Jiménez describes selectively Castro's first year in power in *En Marcha con Fidel*, but the book's sycophancy makes it almost unreadable. In terms of Castro's personality, interviews with him by Lee Lockwood in *Castro's Cuba, Cuba's Fidel*, published in 1967, may be the best material of this kind.

Important insights into Castro appear in books by the French journalist K. S. Karol in *Los Guerrilleros en el Poder*, and by the late French agrarian scientist René Dumont in *Cuba est-il Socialiste?* These critiques are from the Left, and even mention of these books is banned in Havana. Comments on Castro by U.S. Ambassador Philip W. Bonsal in *Cuba, Castro and the United States* remain very valid.

So fallow, however, is the field of biographical work on Fidel Castro that the researcher is forced back to original interviewing, with its advantages and its drawbacks.

Chapter Notes

Book One: The Man

CHAPTER 1

Page

19–20 The account of Fidel Castro after the Alegría de Pío battle is taken from interviews with Faustino Pérez and Universo Sánchez.

20–21 The author's conversations with Castro, January 1984; February 1985

21 The visits by Communist emissaries to Castro in Mexico are discussed in detail in Book III.

22 Castro brought up nationalism and patriotism in his speech on October 10, 1978, the hundredth anniversary of the first independence war.

22 The Bolívarian theme was constant at 1985 conferences in Havana on the question of Latin America's external debt.

24 In *Fidel y la Religión,* by Frei Betto, published in Havana in 1985

Page

25   Castro wrote of Robespierre in a wartime note to Celia Sánchez.

CHAPTER 2

29–35   Faustino Pérez and Universo Sánchez interviews
35   This was a taped *MacNeil-Lehrer News Hour* interview by Robert MacNeil in February 1985.

CHAPTER 3

37   The Soviet aid figures are calculated at this level by the U.S. government, and not questioned by Cuban officials.
37–38   Castro's discussion of "internationalism" in taped interview with the author in January 1984
38   Castro discussed concealing his Marxism-Leninism in the course of his report to the First Congress of the Cuban Communist party in 1975.
39   The statistics in Castro's speeches on public health appeared in articles in *Granma* in July 1985.
40–41   Castro discussed speechmaking in interviews with Carlos Franqui in *Diary of the Cuban Revolution*.
41   Castro's secret speeches were mentioned in private conversations with the author by the president's close associates in Havana.
41–42   The article that made Castro angry was by Juan Luis Cebrián, editor-in-chief of *El País* of Madrid, published in January 1985.
42–43   On Castro's furies, sources included Ramiro Valdés, and on foul language, Carlos Rafael Rodríguez.
43   Wayne Smith, former chief of the U.S. Interests Section in Havana, in an interview on the reasons for Mariel
44   The author's private interviews on Mexican trip

CHAPTER 4

46   Castro's concerns with his men were discussed in interviews with Melba Hernández and Pedro Miret. Also in Robert Merle's book on Moncada
46–47   Castro discussed the chaplain with Frei Betto.
47   The author accompanied Castro to the hospital.
47   Armando Hart's account of the meeting was in an interview with the author in May 1985.
48–49   The strike fiasco was discussed with the author by Castro in an

Page

informal meeting and by Faustino Pérez in a taped interview in 1985.

49   Billington discussed Castro in *Fire in the Minds of Men—Origins of the Revolutionary Faith* (New York: Basic Books, Inc., 1980), pp. 8–9.

49–50   For Castro's views on revolutionary armies, sources range from Régis Debray's *Revolution in the Revolution?* to Fred C. Judson's *Cuba and the Revolutionary Myth* and Fidel's speeches at the Debt Conference of Personalities, Havana, August 1985.

51   For March 1959 NSC meeting, see Pamela S. Falk's *Cuban Foreign Policy,* published in 1986.

52   Faustino Pérez interview with the author in 1985

54   The best estimates on Cuban political prisoners come from Amnesty International and the Americas Watch Committee. Castro's comments were in a conversation with the author in 1985.

55–56   The reconstruction of the Sorí-Marín episode is based on the author's interviews in Miami and Havana.

56   The Cubela story was reconstructed from the author's interviews in Havana and Washington.

57–58   The author discussed Che Guevara with his father, Ernesto Guevara Lynch, in Havana in 1985.

63   Vice-President Fernández told of his first encounters with Castro in interviews with the author in Havana in 1985.

64   Castro discussed "institutions" in interviews with the author in 1984.

65   Rodríguez made the comments on Castro in an interview with the author in 1985.

CHAPTER 5

69   Correspondence between Fidel and Che appears selectively in Carlos Franqui's *Diary of the Cuban Revolution;* it is kept in its entirety in the Historical Division of the Council of State, in wartime archives established by Celia Sánchez.

69–70   Castro talked about loneliness in a 1977 television interview with Barbara Walters; he repeated some of it to the author in 1984.

72   Piñeiro told the author about his New York days in a conversation at the presidential palace in 1985.

Page

75 Castro brought up the question of personality cult in conversations with the author in 1984 and 1985.

76 The author and his wife were the recipients of Castro's food gifts and cuisine advice at their Havana house in May 1985.

77–78 Castro explained his information systems in a conversation with the author in 1985.

79 Castro explained his absence from the Moscow funeral in a 1985 interview with *Playboy* magazine.

80–81 The vignettes on Castro's activities came from the author's interviews with Conchita Fernández in 1985.

Book Two: The Young Years

CHAPTER 1

98 Castro spoke of Céspedes in a speech on October 10, 1968, the centennial anniversary of the first independence war.

102 Castro discussed his mother's religious devotion in interviews with Frei Betto in 1985.

CHAPTER 2

105 Castro discussed his age in a conversation with the author in 1985.

107 Castro told the story about the "Jew" bird to the author in 1985.

113 Castro told stories about his appendicitis in interviews with Carlos Franqui in 1959, and in a conversation with the author in 1985.

115 Raúl Castro is quoted by Robert Merle.

116 Castro's letters from prison appear in Luis Conte Agüerro's *Cartas del Presidio*, in Robert Merle's book on Moncada, and in Mario Mencía's *Time Was on Our Side*.

CHAPTER 3

118 The Avellaneda Literary Academy story comes from an interview with José Ignácio Rasco.

121–122 Castro discussed his family origins in his letters from prison.

124 Castro discussed the Bible in interviews with Frei Betto.

CHAPTER 4

139–142 The relationship between Castro and Alfredo Guevara was described by the latter in an interview with the author in 1985.

Page

CHAPTER 5

147 The birth of the *Ortodoxo* party, with Castro in attendance, is chronicled by Luis Conte Agüerro in *Eduardo Chibás.*

149–150 Castro discussed his ideological evolution with Arturo Alape in Havana in 1981.

151 Boardinghouse evenings were described to the author in Miami by George Volsky, who attended some of these soirées.

159–160 Alfredo Guevara told the Demajagua bell story in an interview with the author in 1985.

CHAPTER 6

165 Moreno Frajinals discussed Fidel's revolutionary impluses in conversations with the author in 1985.

169 Castro's discussion of Perón was included in an interview with the Colombian journalist Arturo Alape in 1981.

173–181 Castro's initial involvement and participation in the *Bogotázo* is described in this book in a narrative section constructed from four main elements. The first two are an account by Fidel included in Arturo Alape's *El Bogotázo: Memorias del Olvidio,* and a separate, lengthy interview he granted Alape. The other two are interviews conducted by the author with Alfredo Guevara in Havana in 1985, and with Enrique Ovares in Miami in 1984. Castro, Guevara, and Ovares were the principal Cuban delegates to the Bogotá student congress.

CHAPTER 7

185–188 The account of Castro's early married life is based on conversations in Havana with personal friends and acquaintances who asked not to be identified.

190–192 The material on Fidel's denunciation of the gangsters and his escape from Havana is based on the author's interviews in Miami with Max Lesnick in 1984 and 1985, and on conversations in Havana in 1985 with Castro's personal friends, who have requested anonymity.

194 Castro's versions of his ideological progress are imprecise and not always consistent. The three versions in this section are from a 1977 interview with Barbara Walters, a 1981 interview with Arturo Alape, and 1985 interviews with Frei Betto. It is impossible to say whether Fidel did not recall exactly what he had said to different persons at different times or whether he

Page

tended to tailor versions to his conversational partners, assuming that there would be no comparison of texts. Presumably, Castro does not wish his ideological history to be crystal clear. Alfredo Guevara's comments were made to the author in 1985.

195–197 The material on Castro's law practice is based on interviews conducted by Mario Mencía with Jorge Aspiazo, and on the author's conversations with Fidel's personal friends in Havana.

203–208 The section on Castro as candidate is based on interviews in Miami with Raúl Chibás and Max Lesnick, and in Havana with Conchita Fernández, Lionel Martin, and Moreno Frajinals.

Book Three: The War

CHAPTER 1

215–217 The material on the incipient period of the organization of Castro's revolutionary Movement is based on the author's interviews in Havana with Melba Hernández, Pedro Miret, and Ramiro Valdés.

221–231 The section covering the first phase of the revolutionary Movement is based on lengthy interviews conducted by the author in Havana with Melba Hernández and Pedro Miret; Mario Mencía's interviews with Jorge Aspiazo; Mencía's *El Grito de Moncada*, Robert Merle's *Moncada*, and Lionel Martin's *The Young Fidel*.

CHAPTER 2

235–238 Pedro Miret and Melba Hernández were the principal sources for the section on the development of the military wing of the Movement. They discussed it in interviews with the author in Havana in 1985. Additional material came from Mencía, Merle, and Martin books.

241–247 The best material on the preparations for Moncada is found in Merle and Mencía books, in the transcripts of Castro's trial, and in Marta Rojas's accounts. Pedro Miret and Melba Hernández supplied eyewitness details in interviews with the author.

CHAPTER 3

248–256 The story of the days and hours preceding the attack on Moncada is best documented by Merle and Mencía. In interviews

Page

with the author, Melba Hernández provided additional recollections—such as Castro's determination to organize weddings.

256–257 The Movement's ideology at the time of Moncada was explained by Castro to Frei Betto in their 1985 interviews. He makes a clear ideological separation between himself and his comrades at that stage.

258 Haydée Santamaría's recollections are contained in a documentary filmed by the Cuban Cinema Institute. A tape of the soundtrack is at the University of Miami along with other tapes related to the production of my book.

CHAPTER 4

269–281 The best account of the Moncada and Bayamo attacks and their aftermath is contained in Robert Merle's book. Useful material is also found in Marta Rojas's *La Generación del Centenário en el Juicio del Moncada,* in Castro's trial testimony, and his newspaper and radio interviews after being captured. Melba Hernández, Pedro Miret, and Ramiro Valdés offered me valuable eyewitness accounts in interviews in 1985.

CHAPTER 5

282–300 Material on Castro's imprisonment, trial, and incarceration is best culled from Merle's *Moncada,* Marta Rojas's *La Generación del Centenário* and *La Cueva del Muerto,* Mencía's *Time Was on Our Side,* Fidel Castro's own *History Will Absolve Me,* and Conte Agüerro's *Cartas del Presidio.* I found much fresh descriptive material in my interviews with Pedro Miret and Melba Hernández, and in a conversation with Raúl Castro—all in Havana in 1985. Fidel contributed anecdotes about his prison life when I accompanied him to the Isle of Youth (Pines) in May 1985.

296–298 The text of "History Will Absolve Me," as it is now widely known, was reconstructed by Fidel from memory during his imprisonment. Marta Rojas, who attended the trial and wrote about it for *Bohemia,* took some notes, but they are very fragmentary. Inasmuch as Castro quotes himself in the written version, it is really irrelevant whether it is a verbatim rendition of his actual words before the judges. Though immensely articulate and equipped with an amazing memory, Castro is a stylist and a perfectionist, and one must assume that he had polished considerably the written and now official version. No substance changes are known to exist.

Page

CHAPTER 6

305–314   The incarceration period is best covered by Mencía and in Fidel's prison correspondence appearing in Conte Agüerro's book and in the appendix in Merel's *Moncada*. Again, Pedro Miret and Melba Hernández provided new material in their 1985 interviews with the author. Ramiro Valdés added details.

315–317   Castro's friends were the principal sources for the account of his divorce proceedings and the custody fight over Fidelito.

322–324   Fidel's reorganization of the Movement is chronicled in a series of articles by Mario Mencía, appearing in *Bohemia* in Havana during 1985; and in original news stories and articles in *Bohemia* and *La Calle* in May–July 1955. Additional information was provided in the author's interviews with Pedro Miret, Melba Hernández, Ramiro Valdés and Armando Hart in Havana in 1985, and with Max Lesnick in Miami.

CHAPTER 7

325–340   Fidel Castro's stay in Mexico is chronicled in Mario Mencía's *Bohemia* articles in 1985; in *De Tuxpán a La Plata,* an account published by the Historical Section of the Central Political Directorate of the Revolutionary Armed Forces in 1979; in General Bayo's memoirs; and in the author's interviews in Havana in 1985 with Pedro Miret, Melba Hernández, Ramiro Valdés, Faustino Pérez, and Universo Sánchez. Also, Max Lesnick was interviewed in Miami, as was Ben S. Stephansky in Washington, D.C.

333–338   Most of the material on the early relations between Fidel and Che Guevara is based on *Che Guevara: Años Decisivos* by Hilda Gadea, Che's first wife; Ernesto "Che" Guevara's own recollections in Volume I of his *Escritos y Discursos,* published in Havana in 1977; Mencía's *Bohemia* articles; the *"Tuxpán"* book; and the author's interviews with Pedro Miret, Melba Hernández, and Universo Sánchez.

340–343   Castro's tour in the United States is described in Mencía's
342–343   *Bohemia* articles in 1985, and in Havana newspapers in 1955. The question of Fidelito was discussed with the author by Fidel Castro's friends in Havana in 1985.

CHAPTER 8

347–371   Material on the preparations for the invasion appears in Mencía's articles; the *Tuxpán* book; General Bayo's memoirs; Teresa Ca-

Page

suso's *Cuba and Castro* (including Fidel's romantic interests); and the author's interviews with Pedro Miret, Melba Hernández, Universo Sánchez, Faustino Pérez, and Ben S. Stephansky.

369–370   Castro's dealings in Mexico with Communist emissaries from Cuba are reported by Lionel Martin in connection with the Osvaldo Sánchez Cabrera visit; the visit by Flavio Bravo, much more important, is contained in an unpublished interview by Mario Mencía.

CHAPTER 9

372–376   The voyage of the *Granma* is described in the *Tuxpán* book and in Volume I of Che Guevara's *Escritos y Discursos*. New details came in the author's interviews with Universo Sánchez and Ramiro Valdés. To gain a visual impression of the landing area, I visited the "shipwreck" spot and crossed the mangrove (now a narrow causeway makes crossing it infinitely easier) to the shore during a tour of Oriente with my wife in May 1985. Pedro Álvarez Tábio, the historian of the Sierra war at the Council of State, who accompanied us, described the landing and the mangrove in a taped interview there.

377–395   The section on the Rebel Army's first six weeks in Cuba is based largely on the author's interviews with Faustino Pérez and Universo Sánchez, who were with Fidel after the Alegría de Pío debacle, and on interviews with Pedro Álvarez Tábio, the historian. Invaluable material came in an interview in Havana with Guillermo García. In the Sierra Maestra, I interviewed Ángel Pérez Rosabal, Mario Sariol, Argelio Rosabal, and Argeo González who were among the first peasants to meet and help Fidel's band. My wife and I toured the Alegría de Pío battlefield, and followed part of Fidel's route into the Sierra Maestra by car and jeep and on foot; it is impossible to understand the Castro war in the mountains without becoming acquainted with the terrain, even slightly. The battle of Alegría de Pío is described in some detail in Che Guevara's and Raúl Castro's war diaries.

CHAPTER 10

396–426   The military history of the first full year of the Sierra war is fairly well documented; the political and ideological history less well. Probably the most complete and objective chronological account of the crucial period between *Granma*'s landing on December 2, 1956, and February 20, 1957, when Castro's guerrilla army became consolidated, was published in four consecutive special issues of the newspaper *Granma:* January 3, January 17, February 23, and February 27, 1979. The accounts were written by Pedro Álvarez Tábio and Otto Hernández, incorporating material from Raúl Castro's and Che Guevara's war diaries, and other sources. Additional material on this period appeared in a special anniversary issue of *Bohemia* on December 3, 1976. Álvarez Tábio and Hernández described the battle of Uvero in *El Combate de Uvero,* published in Havana in 1980. Much important material came from the author's interviews with Faustino Pérez, Universo Sánchez, Guillermo García, José R. Machado Ventura, and Ramiro Valdés. Carlos Franqui's *Diary of the Cuban Revolution* contains significant Sierra correspondence (including operational orders and reports) of Fidel and Raúl Castro, Che Guevara, Celia Sánchez, Frank País, and others. The full body of this correspondence is in the archives of the Council of State.

The political and ideological aspects of the revolutionary struggle during 1957 are inadequately explained in published sources—such as Lionel Martin's *The Young Fidel* or the Bonachea-Valdés Introduction to the *Selected Works of Fidel Castro, 1947–1958.* Important correspondence throwing some light on many acute political problems affecting the Movement are included in the Franqui *Diary.* A useful discussion on this topic is found in *The Unsuspected Revolution* by Mario Llerena. Herbert L. Matthews touches lightly on it in *The Cuban Story* in the context of his visit to Castro in the Sierra in February 1957; his book incorporates the text of his dispatches to *The New York Times,* which were the first direct reports on the *Fidelista* guerrillas. Che Guevara comments on many political aspects of the war in his *Escritos y Discursos* volumes.

Fresh material on the politics of the war resulted from the author's interviews with Faustino Pérez in Havana in 1985, and Raúl Chibás in Miami in 1984. Communist party attitudes toward the Sierra war were discussed by Blas Roca and Fabio Grobart in the author's interviews with them in Havana in 1985.

427–430   The story of the secret CIA involvement in the Sierra war was

Page

reconstructed from my own knowledge as a *New York Times* reporter in Cuba in 1959; this information was considered privileged by me at the time. It was subsequently confirmed in Washington by senior CIA and State Department officials on a confidential basis.

CHAPTER 11

433–459   The military history of the victory year—1958—is amply documented in sources ranging from Franqui's *Diary* (and its entries) to war diaries by Raúl Castro, Che Guevara, and Camilo Cienfuegos. The author obtained additional, detailed information in interviews with Faustino Pérez, Guillermo García, Vílma Espín, Universo Sánchez, and José R. Machado Ventura. To gain a visual impression of the war, my wife and I climbed to the Castro wartime command post at La Plata with Pedro Álvarez Tábio, the historian, and Colonel Arturo Aguillera, Fidel's wartime adjutant, who provided a running historical commentary as we went up and down the Sierra Maestra.

436–445   As in the case of the previous years, very little is available from public sources on the politics of the war—notably on the dissensions within the 26th of July Movement, the still painful controversy over the failed general strike in April 1958, and Castro's relations with the Communists. All these remain sensitive subjects in Cuba; Ramiro Valdés, the former interior minister, remarked in a speech in 1977 that it would be harmful to the revolution to bring the full strike story into the open while many of the *compañeros* linked with it were still alive. A bitter denunciation of the strike was written by Che Guevara; it appears in *Escritos y Discursos*. These questions are discussed by Mario Llerena and by the late Manuel Urrutia Lleó, revolutionary Cuba's first president, in *Fidel Castro and Company, Inc.*

Fidel Castro was still angry about the failure of the 1958 strike when the subject came up in a conversation I had with him in February 1985. I was able to gain insights into many of the political problems of the last year of the war in interviews in Havana with Faustino Peréz, Armando Hart, Jorge Enrique Mendoza, and Ramiro Valdés, and in Miami with Raúl Chibá I discussed the Communist party's attitude toward Castro

Page

ing 1958, and in conversations with Blas Roca and Fabio Grobart in Havana in 1985.

445–459   Aside from published sources, material on the Batista final offensive and the Castro counteroffensive came from interviews by the author with Colonel Arturo Aguilera and Pedro Álvarez Tábio in the Sierra Maestra. Raúl Chibás told me of attending the meeting between Castro and General Eulogio Cantillo, the Batista commander in Oriente, on December 28.

Book Four: The Revolution

CHAPTER 1

463   The quotation by Carlos Rafael Rodríguez is from *Letra con Filo,* published in 1983.

The decision to assassinate Castro is reported in an internal CIA memorandum submitted to Director of Central Intelligence Allen W. Dulles in August 1960.

464   Nuñez Jiménez made the comments on Castro and Lenin in a conversation with the author in 1985.

467   Enrique Oltuski reported his conversations with Che Guevara in an article in *Lunes de Revolución* in June 1959.

471–475   Secret negotiations with the "old" Communists by Castro and his top associates were described for the author in taped interviews in Havana in 1985, by Blas Roca, Fabio Grobart, and Alfredo Guevara.

476–478   The creation, existence, and activities of the "hidden government" in 1959, are discussed in taped interviews with the author in 1985, by Alfredo Guevara, Antonio Nuñez Jiménez, Jorge Enrique Mendoza, and Conchita Fernández.

CHAPTER 2

469–482   Material on the initial phase in the relations between Castro and the United States is drawn from Ambassador Bonsal's *Cuba, Castro and the United States* as well as from my own reporting at the time in Havana and Washington for *The New York Times.* In 1959 and 1960, I had numerous "background" conversations with Bonsal, and I have retained my notes.

490   The episode on Castro's meeting with the CIA official was first reported by Finance Minister López-Fresquet; I heard further details from my CIA sources in Washington.

Page

494–495　Nuñez Jiménez is the source for quotations from Castro's "secret speeches."

497　Nuñez Jiménez and Vice-President J. R. Fernández are the sources for the account on the modernization of the Rebel Army and the militia.

500–501　Castro's activities and movements were described in an interview by Conchita Fernández in 1985; much of the material is firsthand, from the author's own reporting in Havana at the time.

501–502　Nuñez Jiménez is the source for Castro's visits to the swamps.

503–506　Most of the material on Urrutia's demise comes from the author's own reporting in Havana in 1959.

507–508　On the first Soviet contacts with revolutionary Cuba, material comes from the author's own notes at the time. Nuñez Jiménez is the source for the story on Castro's vodka-and-caviar meeting with Alexeiev.

CHAPTER 3

514　Castro talked about Hemingway in an interview with the author in 1984.

CHAPTER 4

533　Castro's comments about John F. Kennedy, expressed at great length, were part of an interview with the author on January 28, 1984.

535　The author was Castro's American companion on the visit to the Pioneers' camp.

540–541　Castro discussed the Alliance for Progress in interviews with the author on January 28 and 29, 1984.

543　Ramiro Valdés discussed the preinvasion security precautions in an interview with the author in 1985.

549–554　Vice-President J. R. Fernández, then the military field commander, described the unfolding of the battle and his own and Castro's movements in interviews with the author in Havana in 1985.

CHAPTER 5

564–567　Castro's dealings with Cuban intellectuals were reconstructed from numerous conversations in Havana in 1984 and 1985, and from the text of his "Words to the Intellectuals" on Jun 30, 1961.

Page

575–576   Valuable material on the Escambray fighting came from interviews with Norberto Fuentes, a noted Cuban journalist, who
has studied this period in depth.

578–589   Castro's account of the October crisis was the central part of an
interview with the author on January 28 and 29, 1984. I believe it to be the most comprehensive version ever supplied by
Castro to a foreign writer.

Book Five: The Maturity

CHAPTER 1

597–598   Castro discussed the "errors" of the revolution in an interview
with the author on January 29, 1984.

CHAPTER 2

616–617   Raúl Castro's report on the investigation appears in *Bohemia* in
November 1968.

626   Nathaniel Davis's comments on the Castro visit to Chile are
contained in *The Last Two Years of Salvador Allende*.

CHAPTER 3

637–640   Apart from Castro's own public and private statements to that
effect, reliable information from United States, Portuguese,
French, and Eastern European diplomatic sources corroborates
the claim that the intervention in Angola was a Cuban idea.

640–642   Partial material on U.S.-Cuban secret diplomacy comes from
the author's interview with William D. Rogers, at the time
Assistant Secretary of State for Inter-American Affairs. Additional information is from Cuban diplomatic sources.

647–648   Castro discussed Central America and "internationalism" in an
interview with the author on January 29, 1984.

BIBLIOGRAPHY

I. Books

Aguirre, Sergio, *Raíces y Significación de la Protesta de Baragua*. La Habana: Editorial Política, 1978.

Alape, Arturo. *El Bogotázo: Memorias del Olvidio*. La Habana: Casa de las Américas, 1983.

Alba, Victor. *Los Sudamericanos*. Mexico City: Costa-Amic, 1964.

Álvarez Tábio, Pedro and Otto Hernández. *El Combate de Uvero*. La Habana: Editorial Gente Nueva, 1980.

Argenter, José Miró. *Crónicas de la Guerra*. La Habana: Instituto del Libro, 1970.

Artime, Manuel F. *Traición!* Mexico City: Editorial Jus, 1960.

Baez, Luis. *Camino de la Victoria*. La Habana: Casa de las Américas, 1975.

Baez, Luis. *A Dos Manos*. La Habana: Unión de Escritores y Artistos de Cuba, 1982.

Baez, Luis. *Guerra Secreta*. La Habana: Editorial Letras Cubanas, 1978.

Baliño, Carlos. *Documentos y Artículos*. La Habana: Instituto de Historia del Movimiento Comunista y de la Revolución Socialista de Cuba, 1976.

Barnet, Miguel. *Gallego*. La Habana: Editorial Letras Cubanas, 1983.

Batista, Alberto Reyes, ed. *Cuentos Sobre Bandidos y Combatientes.* La Habana: Editorial Letras Cubanas, 1983.

Batista, Alberto Reyes. *Los Nuevos Conquistadores.* La Habana: Instituto Cubano del Libro, 1976.

Bayo, Alberto. *Mi Aporte a la Revolución Cubana.* La Habana: Ejército Rebelde, 1960.

Benjamin, Jules Robert. *The United States and Cuba.* Pittsburgh: University of Pittsburgh Press, 1974.

Bethel, Paul D. *Cuba y los Estados Unidos.* Barcelona: Editorial Juventud, S.A., 1962.

Betto, Frei. *Fidel y la Religión.* La Habana: Publicaciones de Consejo de Estado, 1985.

Bonsal, Philip W. *Cuba, Castro and the United States.* Pittsburgh: University of Pittsburgh Press, 1971.

Brennan, Ray. *Castro, Cuba and Justice.* Garden City, N.Y.: Doubleday and Company, 1959.

Brzezinski, Zbigniew. *Power and Principle.* New York: Farrar, Straus and Giroux, 1983.

Buckley, Tom. *Violent Neighbors.* New York: Times Books, 1984.

Cantor, Jay. *The Death of Che Guevara.* New York: Alfred A. Knopf, 1983.

Carrillo, Justo. *Cuba 1933.* Miami, Fla.: Institute of Interamerican Studies, University of Miami, 1985.

Castro Ruz, Fidel. *El Pensamiento de Fidel Castro,* Vol. I, Books 1 and 2. La Habana: Editorial Política, 1983.

Castro Ruz, Fidel. *History Will Absolve Me.* New York: Lyle Stuart, 1961.

Castro Ruz, Fidel. *Informe de Comité Central del Partido Comunista Cubano al Primer Congreso.* La Habana: Comité Central del Partido Comunista de Cuba, 1975.

Castro Ruz, Fidel. *La Revolución de Octubre y la Revolución Cubana.* La Habana: Departamento de Orientación Revolucionária de Comité Central del Partido Comunista de Cuba, 1977.

Castro Ruz, Fidel. *Political, Economic and Social Thought of Fidel Castro.* La Habana: Editorial Lex, 1959.

Castro Ruz, Fidel. *Revolutionary Struggle 1947–1958: Selected Works of Fidel Castro,* eds. Roland E. Bonachea and Nelson P. Valdés. Cambridge, Mass.: The M.I.T. Press, 1972.

Castro Ruz, Fidel. *The World Economic and Social Crisis.* Havana: Publishing Office of the Council of State, 1983.

Casuso, Teresa. *Cuba and Castro.* New York: Random House, 1961.

Christian, Shirley. *Nicaragua.* New York: Random House, 1985.

Collins, John M. *American and Soviet Military Trends.* Washington, D.C.: Georgetown University, 1978.

Conte Agüerro, Luis. *Cartas del Presidio*. La Habana: Editorial Lex, 1959.

Conte Agüerro, Luis. *Eduardo Chibás*. Mexico City: Editorial Jus, 1955.

Crankshaw, Edward, and Jerrold Schecter. *Khrushchev Remembers*. Boston: Little, Brown and Company, 1974.

Crassweller, Robert D. *Trujillo: The Life and Times of a Caribbean Dictator*. New York: The Macmillan Company, 1966.

Cross, James Eliot. *Conflict in the Shadows*. Garden City, N.Y.: Doubleday and Company, 1963.

Cuba. *Constitution of the Republic of Cuba*. Havana: Editorial Política, 1981.

Cuba. Fuerzas Armadas Revolucionarias. *Manual Básico del Miliciano de Tropas Territoriales*. La Habana: Editorial Orbe, 1981.

Cuba. Fuerzas Armadas Revolucionarias. *Moncada 26 de Julio*. La Habana: Ediciones Yara, Julio, 1971.

Cuba. Fuerzas Armadas Revolucionarias. Sección de História de la Dirección Política Central. *De Tuxpán a La Plata*. La Habana: Editorial Orbe, 1979.

Daniel, James and John G. Hubbell. *Strike in the West*. New York: Holt, Rinehart and Winston, 1963.

Davis, Nathaniel. *The Last Two Years of Salvador Allende*. Ithaca, N.Y.: Cornell University Press, 1985.

Debray, Régis. *The Chilean Revolution*. New York: Pantheon Books, 1971.

Debray, Régis. *Revolution in the Revolution?* New York: Monthly Review Press, 1967.

Djilas, Milovan. *Conversations with Stalin*. New York: Harcourt, Brace and World, Inc., 1962.

Donovan, John. *Red Machete*. Indianapolis: Bobbs-Merrill, 1962.

Draper, Theodore. *Castroism: Theory and Practice*. Frederick A. Praeger, 1965.

Draper, Theodore. *Castro's Revolution: Myths and Realities*. New York: Frederick A. Praeger, 1962.

Dreier, John C. *The Organization of American States and the Hemisphere Crisis*. New York: Harper and Row, 1962.

Dubois, Jules. *Danger over Panama*. Indianapolis: Bobbs-Merrill, 1964.

Dubois, Jules. *Fidel Castro: Rebel Liberator or Dictator?* Indianapolis: Bobbs-Merrill, 1959.

Dumont, René. *Cuba est-il Socialiste?* Paris: Éditions du Seuil, 1970.

Duncan, Raymond W. *The Soviet Union and Cuba*. New York: Praeger, 1985.

Einaudi, Luigi R., ed. *Latin America in the 1970s*. Santa Monica, Calif.: Rand, 1972.

Ely, Roland T. *Cuando Reinaba Su Majestad el Azúcar*. Buenos Aires: Editorial Sudamericana, 1963.

Erisman, H. Michael. *Cuba's International Relations*. Boulder, Colo.: Westview Press, 1985.

Falk, Pamela S. *Cuban Foreign Policy*. Lexington, Mass.: Lexington Books, 1986.

Fernández, Manuel. *Religión y Revolución en Cuba*. Miami, Fla.: Saeta Ediciones, 1984.

Fraginals, Manuel Moreno. *El Ingenio* (3 Vols.). La Habana: Editorial de Ciencias Sociales, 1978.

Franco, Victor. *The Morning After*. New York: Frederick A. Praeger, 1963.

Franqui, Carlos. *Diary of the Cuban Revolution*. New York: Viking Press, 1980.

Franqui, Carlos. *Family Portrait with Fidel*. New York: Random House, 1984.

Franqui, Carlos. *The Twelve*. New York: Lyle Stuart, Inc., 1968.

Fuentes, Norberto. *Hemingway in Cuba*. Secaucus, N.J.: Lyle Stuart, Inc., 1984.

Gadea, Hilda. *Che Guevara: Años Decisivos*. Mexico City: Aguilar, 1972.

García, Manuel Rodríguez. *Sierra Maestra en la Clandestinidad*. Santiago de Cuba: Editorial Oriente, 1981.

Gerassi, John. *Fidel Castro*. Garden City, N.Y.: Doubleday and Company, 1973.

Gerassi, John. *The Great Fear*. New York: The Macmillan Company, 1963.

Gómez, Máximo Baez. *Invasión y Campaña de las Villas*. La Habana: Editorial Militar, 1984.

Gonzalez, Edward, and David Ronfeldt. *Post Revolutionary Cuba in a Changing World*. Santa Monica, Calif.: Rand, 1975.

Goodsell, James Nelson. *Fidel Castro's Personal Revolution in Cuba: 1959–1973*. New York: Alfred A. Knopf, 1975.

———. *Grenada: The World Against the Crime*. La Habana: Editorial de Ciencias Sociales, 1983.

Gray, Richard Butler. *José Martí, Cuban Patriot*. Gainesville: University of Florida Press, 1962.

Guevara, Ernesto "Che." *Che: Selected Works of Ernesto Guevara*, eds. Roland E. Bonachea and Nelson P. Valdés. Cambridge, Mass.: The M.I.T. Press, 1969.

———. *El Diario del Che en Bolivia*. La Habana: Instituto del Libro, 1968.

———. *Escritos y Discursos*, Vols. 1–9. La Habana: Editorial de Ciencias Sociales, 1977.

Hageman, Alice L., and Philip E. Wheaton, eds. *Religion in Cuba Today*. New York: Association Press, 1971.

Haig, Alexander M., Jr. *Caveat*. New York: The Macmillan Company, 1984.

Harris, Richard. *Death of a Revolutionary*. New York: W. W. Norton and Company, Inc., 1970.

Hart, Armando Dávalos. *Cambiar las Reglas del Juego*. La Habana: Editorial Letras Cubanas, 1983.

Haverstock, Nathan A., and Richard C. Schroeder. *Dateline Latin America*. Washington, D.C.: The Latin American Service, 1971.

Hemingway, Ernest. *Selected Letters 1917–1961*. New York: Charles Scribner's Sons, 1981.

Huberman, Leo, and Paul M. Sweezy. *Cuba: Anatomy of a Revolution*. New York: Monthly Review Press, 1960.

Instituto de História del Movimiento Comunista y de la Revolución Socialista de Cuba, Anexo al Comité Central del PCC. *Cuba y la Defensa de la Republica Española (1936–1939)*. La Habana: Editorial Política, 1981.

James, Daniel. *The First Soviet Satellite in the Americas*. New York: Avon, 1961.

Johnson, U. Alexis. *The Right Hand of Power*. Englewood Cliffs, N.J.: Prentice-Hall, Inc., 1984.

Judson, Fred C. *Cuba and the Revolutionary Myth*. Boulder, Colo.: Westview Press, 1984.

Karol, K. S. *Los Guerrilleros en el Poder*. Barcelona: Seix Barral, S.A., 1972.

Kemp, Geoffrey. *Some Relationships Between U.S. Military Training in Latin America and Weapons Acquisition Patterns*. Cambridge, Mass.: The M.I.T. Press, 1970.

Kenner, Martin, and James Petras. *Fidel Castro Speaks*. New York: Grove Press, 1969.

Kern, Montague, et al. *The Kennedy Crises*. Chapel Hill: University of North Carolina Press, 1983.

Lewis, Oscar, Ruth Lewis, and Susan M. Rigdon. *Four Men*. Urbana: University of Illinois Press, 1977.

Lewis, Oscar, Ruth Lewis, and Susan M. Rigdon. *Four Women*. Urbana: University of Illinois Press, 1977.

Light, Robert E., and Carl Marzani. *Cuba vs. the C.I.A.* New York: Marzani and Munsell, Inc., 1961.

Llerena, Mario. *The Unsuspected Revolution*. Ithaca, N.Y.: Cornell University Press, 1978.

Lockwood, Lee. *Castro's Cuba, Cuba's Fidel*. New York: The Macmillan Company, 1967.

López-Fresquet, Rufo. *My Fourteen Months with Castro*. Cleveland: The World Publishing Company, 1966.

Mallin, Jay, ed. *"Che" Guevara on Revolution*. New York: Dell, 1970.

Martí, José. *Obras Completas*. La Habana: Editorial de Ciencias Sociales, 1975.

Martin, Lionel. *El Joven Fidel*. Barcelona: Ediciones Grijalbo, S.A., 1982.

Massó, José Luis. *Cuba: 17 de Abril*. Mexico City: Editorial Diana, S.A., 1962.

Matthews, Herbert L. *The Cuban Story*. New York: George Braziller, 1961.

Mazlish, Bruce. *The Meaning of Karl Marx*. New York: Oxford University Press, 1984.

Medvedev, Roy. *Khrushchev*. Garden City, N.Y.: Anchor Press/Doubleday and Company, 1983.

Mella, J. A. *Documentos y Artículos*. La Habana: Instituto Cubano del Libro, 1975.

Mencía, Mario. *El Grito de Moncada*. La Habana: Editorial Política, 1983.

Mencía, Mario. *Time Was on Our Side*. La Habana: Editorial Política, 1982.

Méndez, M. Isidro. *Martí*. La Habana: Fernández y Cia., 1941.

Merle, Robert. *Moncada: Premier Combat de Fidel Castro*. Paris: Robert Laffont, 1965.

Mesa-Lago, Carmelo. *Cuba in the 1970s*. Albuquerque: University of New Mexico Press, 1974.

Mesa-Lago, Carmelo. *The Economy of Socialist Cuba*. Albuquerque: University of New Mexico Press, 1981.

Mesa-Lago, Carmelo, ed. *Revolutionary Change in Cuba*. Pittsburgh: University of Pittsburgh Press, 1971.

Miller, Warren. *90 Miles from Home*. Boston: Little, Brown and Company, 1961.

Mills, C. Wright. *Listen, Yankee*. New York: McGraw-Hill Book Company, Inc., 1960.

Ministerio de Fuerzas Armadas (Cuba), Dirección Política. *Moncada 26 de Julio*. La Habana: 1971.

Molina, Gabriel. *Diaria de Girón*. La Habana: Editorial Política, 1984.

Monahan, James, and Kenneth O. Gilmore. *The Great Deception*. New York: Farrar, Straus and Company, 1963.

Montaner, Carlos Alberto. *Fidel Castro y la Revolución Cubana*. Barcelona: Plaza and Janes, S.A., 1984.

Moreno, José A. *Che Guevara on Guerrilla Warfare: Doctrine, Practice and Evaluation.* Pittsburgh: University of Pittsburgh Press, 1970.

Morray, J. P. *The Second Revolution in Cuba.* New York: Monthly Review Press, 1962.

Nolan, David. *The Ideology of the Sandinistas and The Nicaraguan Revolution.* Coral Gables, Fla: University of Miami Press, 1984.

Nuñez Jiménez, Antonio. *Cuba, Cultura, Estado y Revolución.* Mexico: Presencia Latinoamericana, 1984.

Nuñez Jiménez, Antonio. *En Marcha con Fidel.* La Habana: Editorial Letras Cubanas, 1982.

Nuñez Jiménez, Antonio. *Geografía de Cuba.* La Habana: Editorial Lex, 1959.

Oswald, J. Gregory, and Anthony J. Strover, eds. *The Soviet Union and Latin America.* New York: Frederick A. Praeger, 1970.

Padilla, Heberto. *Fuera de Juego.* Rio Piedras, P.R.: 1971.

Partido Comunista de Cuba. *El Movimiento Obrero Cubano: Documentos y Artículos.* Vol. I: 1865–1925. La Habana: Editorial de Ciencias Sociales, 1975.

Peñabaz, Manuel. *Girón 1961.* Miami, Fla.: Daytona Printing, 1962.

Pérez, Louis A., Jr. *Cuba Between Empires.* Pittsburgh: University of Pittsburgh Press, 1983.

Petras, James F., and Robert La Porte, Jr. *Cultivating Revolution.* New York: Random House, 1971.

Pflaum, Irving P. *Arena of Decision.* Englewood Cliffs, N.J.: Prentice-Hall, Inc., 1964.

Phillips, R. Hart. *Cuba: Island of Paradox.* New York: McDowell, Obolensky, 1960.

Phillips, R. Hart. *The Cuban Dilemma.* New York: Ivan Obolensky, Inc., 1962.

Plank, John, ed. *Cuba and the United States.* Washington, D.C.: The Brookings Institution, 1967.

Pritchett, V. S. *The Myth Makers.* New York: Random House, 1979.

Ranelagh, John. *The Rise and Fall of the C.I.A.* New York: Simon & Schuster, 1986.

Reckord, Barry. *Does Fidel Eat More Than Your Father?* New York: Frederick A. Praeger, 1971.

Riding, Alan. *Distant Neighbors.* New York: Alfred A. Knopf, 1985.

Ripoll, Carlos. *Harnessing the Intellectuals: Censoring Writers and Artists in Today's Cuba*. New York: Freedom House, 1985.

Rivero, Nicolas. *Castro's Cuba*. Washington, D.C.: Luce, 1962.

Roa, Raúl. *Aventuras, Venturas y Desventuras de un Mambi*. La Habana: Instituto del Libro, 1970.

Rodríguez, Carlos Rafael. *Letra con Filo*. La Habana: Editorial de Ciencias Sociales, 1983.

Rodríguez, Carlos Rafael. *Palabras en los Setenta*. La Habana: Editorial de Ciencias Sociales, 1984.

Rodríguez, Gerardo Morejón. *Fidel Castro*. La Habana: P. Fernandez y Cia., 1959.

Rojas, Marta. *La Cueva del Muerto*. La Habana: Unión de Escritores y Artistas de Cuba, 1983.

Rojas, Marta. *El Que Debe Vivir*. La Habana: Casa de las Américas, 1978.

Rojas, Marta. *La Generación del Centenário en el Juicio del Moncada*. La Habana: Editorial de Ciencias Sociales, 1979.

Rojas, Ursinio. *Las Luchas Obreras en el Central Tacajó*. La Habana: Editorial Política, 1979.

Ruíz, Hugo. *Angola*. La Habana: Editorial de Ciencias Sociales, 1982.

Sánchez Arango, Aureliano. *Reforma Agraria*. La Habana: Frente Nacional Democrático, 1960.

Sandford, Gregory, and Richard Vigilante. *Grenada: The Untold Story*. Lanham, Md.: Madison Books, 1984.

Sarabia, Nydia. *Voisin: Viajero de la Ciencia*. La Habana: Editorial Científico-Técnica, 1983.

Sartre, Jean-Paul. *Sarter on Cuba*. New York: Ballantine Books, 1961.

Sauvage, Léo. *Autopsies du Castrisme*. Paris: Flammarion, 1962.

Sauvage, Léo. *Che Guevara*. Englewood Cliffs, N.J.: Prentice-Hall, Inc., 1973.

Schlesinger, Arthur M., Jr. *Robert Kennedy and His Times*. Boston: Houghton-Mifflin Company, 1978.

Scott, Peter Dale, Paul L. Hoch, and Russell Stetler, eds. *The Assassinations*. New York: Random House, 1976.

Shevchenko, Arkady. *Breaking with Moscow*. New York: Alfred A. Knopf, 1985.

Sowell, Thomas. *Marxism: Philosophy and Economics*. New York: William Morrow and Company, 1985.

Suárez, Andrés. *Cuba: Castroism and Communism, 1959–1966*. Cambridge, Mass.: The M.I.T. Press, 1967.

Suchlicki, Jaime. *The Cuban Revolution*. Coral Gables, Fla.: University of Miami, 1968.

Suchlicki, Jaime. *University Students and Revolution in Cuba*. Coral Gables, Fla.: University of Miami, 1969.

Suchlicki, Jaime. *Cuba from Columbus to Castro*. New York: Charles Scribner's Sons, 1974.

Taber, Robert. *M-26, Biography of a Revolution*. New York, Lyle Stuart, 1961.

Thayer, Charles W. *Guerrilla*. New York: Harper and Row, 1963.

Thomas, Hugh. *História Contemporánea de Cuba*. Barcelona: Ediciones Grijalbo, S.A., 1982.

Thomas, Hugh S., Georges A. Fauriol, and Juan Carlos Weiss. *The Cuban Revolution 25 Years Later*. Boulder, Colo.: Westview Press, 1984.

Tomasek, Robert D., ed. *Latin American Politics*. Garden City, N.Y.: Doubleday and Company, 1966.

Torras, Jacinto. *Obras Escogidas*. Vol I. La Habana: Editorial Política, 1984.

Ungar, Sanford J. *Estrangement: America and the World*. New York: Oxford University Press, 1985.

U.S. Commission on C.I.A. Activities Within the United States. *Report to the President*. Washington, D.C.: U.S. Government Printing Office, 1975.

Urrutia Lleó, Manuel. *Fidel Castro and Company, Inc*. New York: Frederick A. Praeger, 1964.

Valdés, Nelson P., and Edwin Lieuwen. *The Cuban Revolution*. Albuquerque: University of New Mexico Press, 1971.

Valenta, Jiri, and Herbert J. Ellison, eds. *Grenada and Soviet/Cuban Policy*. Boulder, Colo.: Westview Press, 1986.

Volman, Sacha. *¿Quién Impondrá la Democracia?* Mexico City: Centro de Estudios y Documentación Sociales, 1965.

Wald, Karen. *Children of Che*. Palo Alto, Calif.: Ramparts Press, 1978.

Weyl, Nathaniel. *Red Star over Cuba*. New York: Devin-Adair Company, 1960.

Wilkerson, Loree. *Fidel Castro's Political Programs from Reformism to "Marxism-Leninism."* Gainesville: University of Florida Press, 1965.

Wyden, Peter. *Bay of Pigs*. New York: Simon & Schuster, 1979.

Yevtushenko, Yevgeny. *A Precocious Autobiography*. New York: E. P. Dutton and Company, 1963.

Yglesias, José. *Down There*. New York: The World Publishing Company, 1970.

Zeitlin, Maurice, and Robert Scheer. *Cuba: Tragedy in Our Hemisphere*. New York: Grove Press, Inc., 1963.

II. Hearings, Reports, Pamphlets

Centro de Estudios Sobre América. *Cuadernos de Nuestra America: Enero–Julio de 1984*. La Habana: Ediciones Cubanas, 1984.

Communist Party of Cuba. *Cuba-Chile*. La Habana: Ediciones Políticas, 1972.

Communist Party of Cuba. *The Invasion of Granada: Statements of the Party and the Revolutionary Government of Cuba Concerning the Events*. Havana: 1983.

Douglas, Maria Eulalia. *Guía Temática del Cine Cubano*. (Producción I.C.A.I.C.) 1959–1980. La Habana: Ministerio de Cultura, 1983.

Greer, Germaine. *Women and Power in Cuba*. Granta, 1985.

Grobart, Fabio. *El Cincuentenario de la Fundación del Primer Partido Comunista de Cuba*. La Habana: Comité Central del Partido Comunista de Cuba, Julio 1975.

Institute of Interamerican Studies. *The Cuban Studies Project: Problems of Succession in Cuba*. Coral Gables, Fla.: University of Miami, 1985.

Johns Hopkins University School of Advanced International Studies. *Report on Cuba: Findings of the Study Group on United States–Cuban Relations*. Boulder, Colo.: Westview Press, 1984.

National Bipartisan Commission on Central America. *Report to the President*. Washington, D.C.: 2201 C St. N.W., 1984.

Publicaciones de Consejo de Estado. *Celia, Heroína de la Revolución Cubana*. La Habana: Editorial Política, 1985.

Organization of American States. *The Situation on Human Rights in Cuba. Seventh Report*. Washington, D.C.: Secretariat General of the Organization of American States, 1983.

Smith, Wayne S. *Castro's Cuba: Soviet Partner or Nonaligned?* Washington, D.C.: The Wilson Center, 1984.

Smith, Wayne S. *Selected Essays on Cuba*. Washington, D.C.: Johns Hopkins School of Advanced International Studies, 1986.

Unión de Escritores y Artistas de Cuba. *Ponencias: Forum de la Narrativa–Novella y Cuento*. La Habana: Unión de Escritores y Artistas de Cuba, 1984.

U.S. Central Intelligence Agency. *Cuban Chronology*. Springfield, Va.: National Technical Information Service, April 1979.

U.S. Central Intelligence Agency. *Directory of Cuban Officials*. Springfield, Va.: National Technical Information Service, January 1979.

University of Miami. *The Miami Report: Recommendations on United States Policy Toward Latin America and the Caribbean.* Coral Gables, Fla.: University of Miami, 1984.

III. Documents and Speeches

Communist Party of Cuba. *2nd Congress of the Communist Party of Cuba.* Havana: Political Publishers, 1981.

Foreign Broadcast Information Service. *3rd Congress of the Communist Party of Cuba.* Washington, D.C.: U.S. Department of Commerce, February 7 and 10, 1986.

Partido Comunista de Cuba. *Informe del Comité Central del PCC al Primer Congreso.* La Habana: 1975.

IV. Congressional Hearings

U.S. House of Representatives. Committee on Internal Security. *The Theory and Practice of Communism. Part 5: Marxism Imposed on Chile–Allende Regime.* Washington, D.C.: U.S. Government Printing Office, 1974.

U.S. Senate. Foreign Relations Committee. *Executive Sessions, Vol XIII, Parts 1 and 2.* Washington, D.C.: U.S. Government Printing Office, 1961. (Made public December 1984.)

U.S. Senate Select Committee on Intelligence Activities. *Alleged Assassination Plots Involving Foreign Leaders.* Washington, D.C.: U.S. Government Printing Office, 1975.

U.S. Senate. Select Committee on Intelligence Activities. *Covert Action.* Washington, D.C.: U.S. Government Printing Office, 1976.

U.S. Senate. Select Committee on Intelligence Activities. *The Investigation of the Assassination of John F. Kennedy: Performance of the Intelligence Agencies. Book V: Final Report.* Washington, D.C.: U.S. Government Printing Office, 1976.

U.S. Senate. Select Committee on Intelligence Activities. *Supplementary Detailed Staff Reports on Foreign Aid and Military Intelligence.* Washington, D.C.: U.S. Government Printing Office, 1976.

INDEX